Advances in Volcanology

An Official Book Series of the International Association of Volcanology and Chemistry of the Earth's Interior

More information about this series at https://link.springer.com/bookseries/11157

Atsushi Toramaru

Vesiculation and Crystallization of Magma

Fundamentals of the Volcanic Eruption Process

Springer

Atsushi Toramaru
Department of Earth
and Planetary Sciences
Kyushu University
Fukuoka, Japan

The Japanese version of this book was supported by JSPS KAKENHI Grant Number JP18HP5236 (Grant-in-Aid for Publication of Scientific Research Results) in FY2018. The English version was supported by Kyushu University Practical Volcano Special Education Base Project of the Ministry of Education, Culture, Sports, Science and Technology, Japan in FY2019.

ISSN 2364-3277 ISSN 2364-3285 (electronic)
Advances in Volcanology
ISBN 978-981-16-4208-1 ISBN 978-981-16-4209-8 (eBook)
https://doi.org/10.1007/978-981-16-4209-8

This Springer imprint is published by the registered company Springer Nature Singapore Pte Ltd.
The registered company address is: 152 Beach Road, #21-01/04 Gateway East, Singapore 189721, Singapore

To Y. T.

Preface for English Version

This book is the English version of Vesiculation and Crystallization of Magma—Fundamentals of Volcanic Eruption Process—which was published in Japanese in 2018. This book aims at advancing the comprehensive understanding of fundamental processes, vesiculation and crystallization of magmas, for researchers and graduate students. As I mention in the preface in the Japanese version, I feel the need for the detailed illustration of those fundamental processes in magmatism, because the related works have been produced at a high rate, covering a wider area. In such circumstances, it is difficult for the researchers of the field (including the graduate students) to understand the base of the studies by getting back to fundamental physics and chemistry behind those. Many errors and typos in the Japanese version have been corrected in the English version through careful reading by Mizuki Nishiwaki (Kyushu University), and members of Volcanology seminar of Earthquake Research Institute, University of Tokyo. Unfortunately, because of limited number of cited articles, I could not cite a number of important articles in the reference.

Fukuoka, Japan — Atsushi Toramaru

Preface

Studying volcanic phenomena and elucidating their mechanisms and laws require various types of approaches, including geophysical observation, geological surveys, geochemical analyses, and modeling. Note that analysis based on materials science is one such approach, and research in this area has been actively ongoing, particularly over the last 20 years. For example, correspondence between eruption styles, changes in eruption styles, and eruptive materials have been investigated; moreover, eruption parameters, such as the ascent velocity of magma, have been quantitatively estimated. For their chemical composition and texture, volcanic products record the vesiculation, crystallization, and movement magma experiences in all their processes from generation to accumulation to ascent to eruption. To quantitatively decipher the record of magma movement, an understanding of its key process, i.e., the vesiculation and crystallization of magma, is necessary. However, there is no book comprehensively explaining the scientific procedure and background for achieving such an understanding; therefore, I have felt the need for a book such as this.

Using conventional observational research on volcanos, such as geophysical monitoring, we can obtain data about the occurrence of volcanic earthquakes and volcano deformation. However, to interpret these data, knowledge about the state of magma under the ground is necessary. Moreover, for modeling volcanic conduit flow using computational simulation and calculating convection and fractional crystallization in magma chambers, a basic knowledge of materials science is necessary. Furthermore, before conducting individual observations and developing simulation strategies, cooling of the magma and solubility of volatiles in the magma are extremely fundamental aspects that we should have learned in advance.

This book focuses on the vesiculation and crystallization processes of magma. Vesiculation and crystallization of magma are basic processes that appear at every scene of volcanic eruption. In magma chambers deep underground where cooling crystallization occurs, water as well as other volatiles that have not been incorporated into crystals get concentrated into a liquid and then vesiculate at a certain time point, leading to overpressurization in magma chambers and supply of buoyancy for magma ascent. In magma that is ascending in a volcanic conduit, decompression-induced vesiculation occurs and expansion of the generated gas governs the ascent and exit velocities from the vent. Moreover, as a result of the decrease in water content in liquid caused by decompression-induced vesiculation, the melting point of minerals increases and decompression-induced crystallization occurs. The formation of bubbles and crystals changes the viscosity and magma structure, thus providing feedback to magma movement. In eruptive materials on the surface, temperature and pressure paths the magma has experienced as well as the rate of changes are recorded as the texture of bubbles and crystals. Therefore, the appropriate use of a methodology by which such information is read out from the texture of eruptive materials allows the estimation of magma movement along with eruption.

Developing this methodology requires both encyclopedic knowledge of materials such as rocks and minerals and fundamental approaches for interpreting data obtained by analyzing such materials. For example, Newtonian mechanics, which we learned in high school, is a fundamental approach for interpreting data on ball trajectories. Thus, the fundamental approaches for interpreting data on volcanic materials can be roughly divided into two concepts. One concept is referred to as “equilibrium theory,” which is based on thermodynamics and is integrated into equilibrium phase diagrams. This concept is based on assuming the concept of an equilibrium state and discusses the state without motion. “Disequilibrium theory” is the other concept that discusses motion from one state to another state. It is called “kinetics.” Focusing on these two concepts, this book explains the foundations of fundamental approaches by which we interpret natural materials.

In the fields of applied physics and materials science, a number of textbooks have been published on equilibrium phase diagrams and kinetics. However, it is somewhat difficult to apply the context of such books to volcanic materials because they only explain generalities or deal with materials and situations different from those of volcanic materials. Moreover, I am unaware of any book that comprehensively explains kinetic theories that have been independently developed for magma. This seems to be a major barrier to advancing research and education to the next step.

On the other hand, textbooks published in Japan, and not only those on volcanology, are generally compact and often do not contain detailed explanations of the derivation of equations and the underlying fundamental physics. In particular, in recent years, while easy-to-understand books are valued and popular, difficult-to-understand books tend to be avoided. I have been worrying that this tendency obscures the detailed processes of inquiry and prevents the understanding of underlying assumptions and fundamental principles. Aimed at overcoming these demerits, I have attempted to write this book in as much detail as possible and in an easy-to-understand manner as possible. Although Torahiko Terada (physicist in Japan 1878–1935) commented on Rayleigh's Theory of Sound, which could be casually glanced through while ensuring his hands in his pockets, I gained no confidence that this work has turned out to be such a book.

Based on this thinking, this book has the following structure. Part I introduces instances the author has experienced, to share volcanic phenomena for understanding with readers. Moreover, it contains motivation and inspiration that we gained from natural phenomena to learn about vesiculation and crystallization of magma. Part II explains the vesiculation of magma, whereas Part III describes the crystallization of magma. In both Parts II and III, the equilibrium theory is first explained, followed by the explanation of the disequilibrium theory. Note that the equilibrium theory of crystallization is not explained because it overlaps with discussion on vesiculation, and there are already several textbooks available on the subject, and space is limited. The chapters on the vesiculation process of magma deal with the conditions of magma vesiculation, mechanism of bubble formation, growth of bubbles, temporal development of bubble size distribution, and various processes related to bubbles (secondary growth, coalescence, deformation, detachment, ascent, as well as shrinkage, collapse, and oscillation of bubbles). The chapters on the crystallization process explain the cooling crystallization of magma, crystallization triggered by vesiculation (decompression-induced crystallization), and crystal size distribution (CSD). Part IV, the last chapter on application, introduces attempts to understand the behavior of magma by applying the discussion in earlier chapters to eruptive materials and rocks obtained from actual volcanos. In Appendix, several points—derivation and calculation of equations, useful concepts and summaries, an outline of physical properties, and a conceptual basis for textural analysis—required to understand the contents of the primary text are explained. Although a number of symbols appear in this book, I have attempted to not use the same symbol with different meanings and have provided a list of symbols to help avoid confusion.

Unfortunately, this book cannot cover multiple points that are indispensable to understanding volcanic phenomena from a materials science viewpoint, and that (one feels) should be comprehensively explained elsewhere. These points include a comprehensive introduction to and examination of the chemical composition of observed crystals, the origins of various crystal forms, a detailed explanation of and data on the relevant physical properties (although they are partly discussed), and the coupling of phase change and fluid mechanics.

I would like to express my gratitude to Mie Ichihara (University of Tokyo), Masatoshi Ohashi (University of Tokyo), Kazuya Yamakawa (University of Tokyo), Takafumi Maruishi (Kyoto University), and Mizuki Nishiwaki (Kyushu University) who verified the physics and derivation of equations, thus reporting fundamental mistakes and inappropriate expressions and providing countless suggestions and pieces of advice in two review camps that covered complete 8 days in total. Without their support, this book would certainly have been a far less accurate one; however, it is impossible to deny that many mistakes remain regardless of my best, and the author takes full responsibility for this book. Moreover, I would like to thank Dr. Ichihara for useful discussions on the energetics of nucleation, Dr. Ohashi for useful discussions on the coalescence and deformation of bubbles, Mr. Nishiwaki for the calculation of viscosity in Appendix, and a technical associate,

Kazuhiko Shimada, for obtaining the electron micrographs. Moreover, I would like to express my appreciation to Mika Komatsu of University of Tokyo Press, who provided an opportunity to publish this book, offered words of encouragement, and edited this complicated book.

Fukuoka, Japan Atsushi Toramaru

Contents

Notations[1]

$A(R)$	Function of R of the Laplace transform
$A(R)$, $A(l)$	Coefficient of the Fokker-Planck equation (m s^{-1})
A_+	Condensation rate
A_-	Evaporation rate
A_A	One-dimensional measurement ratio (in Appendix 11.9) (DS)
$\mathcal{A}$	Area, cross-sectional area (m^2)
$\tilde{A}$	Constant defining the increasing rate of the nucleation rate (Eq. 9.22) (DS)
Av	Avrami number (Eq. 7.118) (DS)
a	Kinetic factor of the crystal number density ($m^{-3}K^{-3/2}s^{3/2}$)
$a(X)$	Activity (DS)
$a(T, P, X)$	Activity (DS)
a_0	Intermolecular distance (m)
a_k	Ratio of solubility coefficient
a_M	Effective radius of a molecule (m)
$a_{Melt}^{H_2O}$	Activity of H_2O in liquid (DS)
$a_{Melt}^{OH^-}$	Activity of OH^- in liquid (DS)
$a_{Melt}^{O^{2-}}$	Activity of O^{2-} in liquid (DS)
a_p	Minimum open width of a path in permeability (m)
$B(R)$, $B(l)$	Coefficient of the Fokker-Planck equation (m^2 s^{-1})
$B_0 = B(x, y)$	Beta function
B_C	$B(R)$ at the critical size (m^2s^{-1})
$\tilde{B}$	Constant defining the increasing rate of the crystal growth rate (Eq. 9.122) (DS)
Bo	Bond number (DS)
b	Scale of bubble pressure by the interfacial tension $2\gamma/R(0)P_G(0)$ (in the main text) (DS)
b_p	Shape factor in permeability (DS)
b_v	Damping coefficient of bubble oscillation (Eq. 6.116)
$\tilde{b}(R)$, $\tilde{b}(R, R')$	Function of R, R' in the Laplace transform

[1]MOL = molecule; NOM = number of molecule; DS = dimensionless, No = number.

C	Concentration (NOM m^{-3})
C_0	Reference concentration (NOM m^{-3})
C_{eq}	Equilibrium concentration (NOM m^{-3})
C_{exs}	Amount of exsolving volatiles (NOM m^{-3})
C_{int}	Concentration at the gas/liquid or solid/liquid interface (NOM m^{-3})
C^*_{int}	Balanced concentration (NOM m^{-3})
C_P	Specific heat at constant pressure (J kg^{-1})
C_R	Interfacial equilibrium concentration at the radius (NOM m^{-3})
C_S	Concentration in the solid phase (NOM m^{-3})
C_∞	Concentration at infinite distance from bubbles (NOM m^{-3})
ΔC_e	Effective degree of supersaturation giving the maximum water exsolution rate (NOM m^{-3})
$\tilde{C}$	DS concentration (e.g., wt%) (DS)
$\tilde{C}_W$	Water concentration (wt%)
$\tilde{C}_{W0}$	Reference water concentration (wt%)
$\tilde{C}'_W$	Thermodynamic factor expressing pressure dependence of water solubility wt% (Pa^{-1})
$\lvert\dot{\tilde{C}}_W\rvert$	Water exsolution rate wt% (s^{-1})
Ca	Capilary number (DS)
c, c_1, c_2, c_3	Arbitrary coefficients or constants
c_a	Activity coefficient (Eq. 2.10)
c_{CO_2}	Solubility constant of CO_2
c_G	Coefficient of the growth rate in CSD model 1 (Eq. 9.170)
c_M	Coefficient in the relational equation between eruption column height and decompression rate
c_{H_2O}	Solubility constant of H_2O
c_h	Henry constant (Pa^{-1})
c_J	Coefficient of the nucleation rate in CSD model 4 (Eq. 9.175)
c_M	Coefficient in the relational equation between eruption column height and discharge rate
c_{Mo}	Coefficient of bubble ascent velocity as a function of bubble volume
c_U	Coefficient in the relational equation between eruption column height and exit velocity
c_x, c_y	Rate constant in fluxing
D	Diffusivity (m^2 s^{-1})
D_v	Diameter of a spherical particle in 3D space (in Appendix 11.9) (m)
DF	Degree of deformation of bubbles, $(L_1 - L_3)/(L_1 + L_3)$ (DS)

d_p	Distance between Projections (m)
$F(R)$, $F(R, t)$	Size distribution function (No m^{-4})
F_0	Nucleation Density $F(0)$, $F(0, t)$ (No m^{-4})
$\mathbf{F}_B$	Buoyancy (N)
F_E	Equilibrium size distribution function (No m^{-4})
F_{ext}	Size distribution function in extended volume in Avrami model (No m^{-4})
F_S	Steady-state size distribution function (No m^{-4})
F_{ric}	Friction force N
$\mathcal{F}$	Helmholtz free energy (J)
$\mathcal{F}_V$	Helmholtz free energy per unit volume
$\hat{F}$	Laplace transform of $F(R, t)$
f, f_1, f_2	Arbitrary function
G	Growth rate (m s^{-1})
G_0	Constant characterizing crystal growth rate (m s^{-1})
$\bar{G}_0$	Constant characterizing crystal growth rate (m $s^{-1-\bar{n}}$)
G_1	Constant growth rate in CSD model 1 (m s^{-1})
G_{MAX}	Maximum of crystal growth rate (m s^{-1})
G_{PEAK}	Extremum of master curve of crystal growth rate (m s^{-1})
G_R	Migration velocity including the effect of fluctuation in size space (m s^{-1})
$\mathcal{G}$	Gibbs free energy (J)
$\mathcal{G}_V$	Gibbs free energy per unit volume
g	Gravitational acceleration (m s^{-2})
$g(x)$	Arbitrary function of x
Δg^*	Difference in Gibbs free energy of crystallization component between solid and liquid (J MOL^{-1})
$\dot{H}$	Extraction rate of heat (J s^{-1})
$\mathcal{H}$	Enthalpy (J)
h	Enthalpy per molecule (J)
h_{CO_2}	Henry constant of CO_2
h_{H_2O}	Henry constant of H_2
$\hbar$	Planck constant $\approx 6.626 \times 10^{-34}$ (J s)
I	Flux
I_{in}	Injection rate of crystals (No $m^{-4}s^{-1}$)
I_{out}	Extraction rate of crystals (No $m^{-4}s^{-1}$)
i	Integer
J	Nucleation rate (No $m^{-3}s^{-1}$)
J_0	Pre-exponential factor of nucleation rate (No $m^{-3}s^{-1}$)
$\bar{J}_0$	Constant characterizing crystal growth rate (m $s^{-1-\bar{m}}$)
J_{MAX}	Maximum nucleation rate (No $m^{-3}s^{-1}$)
J_{PEAK}	Extremum of master curve of nucleation rate (No $m^{-3}s^{-1}$)
J_S	Steady-state nucleation rate (No $m^{-3}s^{-1}$)
K	Reaction constant
K_D	Solid-liquid distribution coefficient

K_{ij}	Coalescence frequency of size i and size j m^3 (No $^{-1}s^{-1}$)
$K(v', v - v')$	Coalescence kernel of size v' and size $v - v'$ (No $^{-1}s^{-1}$)
K_W	Solubility coefficient of water: $P = K_W(\tilde{C}_W/\tilde{C}_{W0})^2$
$\mathcal{K}$	Transition frequency (s^{-1})
k	Integer
k_B	Boltzmann constant (J K^{-1})
k_{CO_2}	$c_{CO_2}^{-1}$
k_E	Elastic constant in viscoelastic model (Pa)
k_T, k_m, k_s	Thermal conductivity (W m K^{-1})
k_{H_2O}	$c_{H_2O}^{-1}$
k_p	Permeability (m^2)
L	Effective radius of liquid film disc between bubbles (m)
L	One-dimensional measured value (in Appendix 11.9) (m)
L_1, L_2, L_3	Length of major, intermediate, and minor axes of deformed bubble (m)
L_D	Width of fissure (m)
L_E	Length of fissure in the longitudinal direction (m)
L_L	Ratio of 1D measured value (in Appendix 11.8) (DS)
L_p	Length of Projection (m)
$\mathcal{L}(f)$	Laplace transform of function f
$\mathcal{L}^{-1}(g)$	Inverse Laplace transform of function g
l	Size (m or No)
l_{ij}	Coordinate transformation matrix of vector
$\hat{l}_{ij}$	Inverse matrix of l_{ij}
M_i, $i = 0, 1, 2, 3$	Moment of size distribution function (m^{-i+3})
M^*	Complex modulus of elasticity (Pa)
M_I	Imaginary part of M^*(Pa)
M_R	Real part of M^*(Pa)
Mo	Morton number (DS)
exponent m	Solubility exponent or permeability exponent (kg)
m	Particle mass (kg)
m_{CO_2}	Molecular weight of CO_2 (kg)
m_{H_2O}	Molecular weight of m_{H_2O} (kg)
$\bar{m}$	Exponent in Avrami model
N	Number density of bubbles or crystals (No m^{-3})
$N(l, t)$	Number density distribution function (No m^{-3})
N_0	Molecular number density or number density of monodisperse bubbles (No m^{-3})
N_1	Monomer number density $N(1, t)$ (No m^{-3})
N_{Av}	Avogadro's number
N_{bub}	Bubble number density (No m^{-3})
N_C^*	Number density of critical nucleus size $\approx N_E^*$ (No m^{-3})
$N_E(l)$	Equilibrium number density distribution function (No m^{-3})
N_E^*	Value of $N_E(l)$ at critical size (No m^{-3})

N_L	1D measured number (in Appendix 11.9) (No m^{-1})
N_{mic}	Microlite number density (No m^{-3})
N_{PE}	Number density of preexisting bubbles (No m^{-3})
N_V	Particle number per unit volume (No m^{-3})
n	Dummy integer
n	Population density (No m^{-3})
n_G	Number of molecules in bubbles
n_k	Population density of size k (No m^{-3})
n_v	Population density of size v (No m^{-3})
$\bar{n}$, $\bar{n}_a$, $\bar{n}_r$	Exponent in Avrami model
P	Pressure (Pa)
P_{CO_2}	Partial pressure of CO_2 (Pa)
P_f	Final pressure of decompression (Pa)
P_G	Gas pressure (Pa)
P_G^*	Gas pressure in saturated state (Pa)
P_{H_2O}	Partial pressure of water (Pa)
P_{max}	Liquid pressure at r_{max}(Pa)
P_n	Pressure giving maximum nucleation rate (Pa)
ΔP_n	Pressure range giving maximum nucleation (Pa)
P_∞	Liquid pressure at infinite distance from bubbles (Pa)
$\mathcal{P}_x$	Probability of finding crystal nucleus
$\tilde{P}$	Scaled pressure (DS)
Pe	Péclet number (DS)
p	Growth rate exponent: $R \propto t^p$ (DS)
$p(x)$	Function in ordinary differential equation
Q	Activation energy of diffusion (J mol^{-1})
$\mathcal{Q}$	Amount of heat (J)
q	Activation energy of diffusion (J MOL^{-1})
$q(x)$	Function in ordinary differential equation
$\tilde{q}$	q scaled by $k_B T_0$ (DS)
R	Bubble (crystal) radius (m)
R_0	Scale characterizing crystal size or bubble size (m)
R_1	Characteristic size in CSD model 1 (m)
R_2	Characteristic size in CSD model 2 (m)
R_3	Characteristic size in CSD model 3 (m)
R_4	Characteristic size in CSD model 4 (m)
R_A	Radius of spherical particle in two-dimensional measurement (in Appendix 11.9)
R_B	Bubble radius in mechanical equilibrium with overpressure in bubbles ΔP_G (m)
R_C	Critical bubble (crystal) radius (equilibrium bubble (crystal) radius) (m)
R_G	Size characterizing exponential growth rate against R (m)
R_m	Mean particle size (m)
R_P^*	Scale of equilibrium bubbles in supersaturated liquid Eqs. (4.10), (4.14) (m)

R_{PE}	Bubble radius of preexisting bubbles (m)
R_S	Size defining slope of CSD (m)
R_T^*	Scale of equilibrium crystals in undercooled solution (Eq. 7.28) (m)
R_X^*	Scale of equilibrium crystals in supersaturated liquid (Eq. 7.27) (m)
$\mathcal{R}$	Gas constant (J K^{-1})
$\dot{R}$	Bubble radius changing rate *dR/dt* (m s^{-1})
$\ddot{R}$	Bubble radius changing acceleration d^2R/dt^2 (m s^{-2})
$\tilde{R}$	Scaled bubble radius (DS)
$\tilde{R}_{CT}$	R_C scaled by R_T^* (DS)
$\tilde{R}_{CX}$	R_C scaled by R_X^* (DS)
Re	Reynolds number (DS)
r	Distance from the center of bubble or crystal in the radial direction (m)
r_{max}	Distance r giving the maximum value of liquid pressure r (m)
S	Entropy (J K^{-1})
St	Stefan number: Eq. (7.123) (DS)
s	Entropy per molecule (other than Laplace transform) (J $K^{-1}MOL^{-1}$)
s	Dummy variable in Laplace transform (only in Laplace transform)
$s_f(\theta)$	Shape factor of heterogeneous nucleus DS (DS)
T	Temperature (K)
T_b	Temperature at the boundary of intrusion (K)
T_{GPEAK}	Temperature giving G_{PEAK} (K)
T_{JPEAK}	Temperature giving J_{PEAK} (K)
T_L	Liquidus temperature (K)
$\Delta T_L(C, T)$	Degree of supercooling (K)
T_L'	Thermodynamic factor expressing water-content dependence of liquidus (K $wt\%^{-1}$)
T_{Lpl}'	Thermodynamic factor expressing water-content dependence of plagioclase liquidus (K $wt\%^{-1}$)
T_{Lpx}'	Thermodynamic factor expressing water-content dependence of pyroxene liquidus (K $wt\%^{-1}$)
T_{m0}	Initial temperature of intruding magma (K)
T_n	Temperature giving maximum nucleation rate (K)
ΔT_n	Degree of supercooling giving maximum nucleation rate (K)
T_{s0}	Initial temperature of country rock (K)
t	Time (s)
t_0	Arbitrary time constant (s)
$t_A(R)$, $t_A(R, R')$	Function of R, R' in Laplace transform (s)
t_{Av}	Timescale of Avrami model (s)
t_{ap}	Approaching time of bubble (s)

t_c	Characteristic time of bubble collapse (s)
t_{col}	Characteristic time of Ralyeigh collapse (s)
t_{dec}	Characteristic time of decompression (s)
t_{dif}	Characteristic time of diffusion (s)
t_{difG}	Characteristic time of diffusion growth (s)
$t_{degasDif}$	Characteristic time of water exsolution by diffusion growth (s)
$t_{degasVis}$	Characteristic time of water exsolution by viscosity growth (s)
t_{ex}	Positive: time characterizing injection rate of crystals (s)
t_{ex}	Negative: time characterizing extraction rate of crystals (s)
t_{fd}	Time for drainage of liquid film between bubbles (s)
t_g	Characteristic time of crystal growth (s)
t_{in}	Characteristic time of inertial expansion (s)
t_J	Time characterizing exponential nucleation rate (s)
t_{ki}	Kinetic timescale (s)
t_n	Time giving maximum nucleation rate (s)
t_{re}	Characteristic time of viscoelasticity (s)
$t_{residence}$	Residence time of crystals (s)
t_{sr}	Characteristic time of bubble shape relaxation (s)
t_{th}	(Thermal) Characteristic time of cooling (s)
t_{vis}	Characteristic time of viscosity-limited expansion (s)
t_x	Characteristic time of crystallization (s)
t_z	Time characterizing zonal structure Eqs. (7.180), (7.182) (s)
t^*	Time constant of bubble coalescence (s)
t_n^*	Termination time of nucleation (s)
$\tilde{t}$	Scaled time (DS)
$\tilde{t}^*$	t^* scaled by bubble growth/expansion time (DS)
$\tilde{t}_z$	t_z scaled by t_g (DS)
δt_n	Duration of maximum nucleation rate (s)
U_{st}	Stokes velocity (m s^{-1})
U_t	Terminal velocity (m s^{-1})
U_{VE}	Exit velocity at vent (m s^{-1})
u_r	Velocity around bubbles in the radial direction (m s^{-1})
V	Volume (m^3)
V_{ext}	Extended volume in Avrami model (m^3)
v	Volume of bubbles (m^3)
v_0	Initial bubble size of monodisperse bubbles (m^3)
v_G	Molecular volume in gas phase (m^3 MOL^{-1})
v_m	Partial molecular volume of gas component in liquid phase (m^3)
v_S	Molecular volume of crystallization component (m^3)
v_x, v_y, v_z	Particle velocity in x, y, z direction (m s^{-1})
W	Changing rate of transition number (No $m^{-3}s^{-1}$)

$W_\circ$	Transition probability (s^{-1})
$\mathcal{W}$	Work (J)
w	Number of possible states (DS)
X	Molar fraction (DS)
$\tilde{X}$	Scaled molar fraction (DS)
x	Distance used in Appendix 11.8 (m)
x_D	Effective diffusion distance (m)
y	C_{eq}/C (DS)
y	Constant used in Appendix 11.8 (Eq. 11.105)
$y(R)$	Function of R (DS)
Z	Zeldovich factor (m^{-1})
$\mathcal{Z}(\zeta)$	Function determining the limiting conditions of second nucleation
α_1	Magnitude of nucleation barrier (DS)
α_2	Magnitude of the initial equilibrium (saturated) state (DS)
α_3	Magnitude of decompression time against diffusion growth time (DS)
α_4	Magnitude of decompression time against viscosity relaxation time (DS)
α_s	Shape factor (DS)
β_1, β_2	Cooling-rate exponent of growth rate (Eq. 9.8) (DS)
β_3, β_4	Cooling-rate exponent of nucleation rate (Eq. 9.9) (DS)
β_{ij}	Correction coefficient of bubble coalescence efficiency (DS)
Γ	Crystal nucleation barrier (DS)
$\Gamma(x)$	Gamma function of x (DS)
Γ_q	$\Gamma/\tilde{q}$ (DS)
γ	Interfacial tension or interfacial energy (N m^{-1}, J m^{-2})
γ_0	Macroscopic interfacial energy (N m^{-1}, J m^{-2})
γ_a	Activity coefficient
γ_h	Ratio of specific heat (DS)
γ_{GL}	Gas-liquid interfacial energy (N m^{-1}, J m^{-2})
γ_{SG}	Solid-gas interfacial energy (N m^{-1}, J m^{-2})
γ_{SL}	Solid-liquid interfacial energy (N m^{-1}, J m^{-2})
Δ	Degree of supersaturation (DS)
$\delta(x)$	Delta function of x
δ_T	Value for Tolman correction (m)
$\mathcal{E}$	Internal energy (J)
ε_P	Poynting correction for nucleation supersaturation pressure (Eq. 5.89) (DS)
ϵ	Strain (DS)
$\dot{\epsilon}$	Strain rate (s^{-1})
$\dot{\epsilon}_{rr}$	Strain rate in the radial direction (s^{-1})
ζ	Parameter of the limiting conditions of second nucleation (DS)
η	Viscosity (Pa s)

η^*	Complex viscosity (Pa s)
η_{I}	Imaginary part of η^* (Pa s)
η_{R}	Real part of η^* (Pa s)
Θ_{D}	$t_{\mathrm{dif}}/t_{\mathrm{dec}} = \alpha_3^{-1}$ (DS)
Θ_{V}	$t_{\mathrm{vis}}/t_{\mathrm{dec}} = \alpha_4^{-1}$ (DS)
κ	Thermal diffusivity (m s^{-2})
Λ	Dimensionless number defined by Eq. (6.113) (DS)
λ	Ratio between exponential increasing rates of nucleation rate and that of growth rate (DS)
λ	Parameter in Stefan problem (in Appendix 11.8) (DS)
μ	Chemical potential (J MOL^{-1})
$\mu_{\mathrm{A}}^{\mathrm{B}}$	Chemical potential of B in A (J MOL^{-1})
ν	Kinematic viscosity (m^2s^{-1})
ξ	Cooling rate exponent of crystal number density (Eq. 9.10) (DS)
Π	Dimensionless number defined by Eq. (6.114) (DS)
ρ	Density (kg m^{-3})
ρ_{g}, ρ_{G}	Density of gas (kg m^{-3})
ρ_{m}, ρ_{L}	Density of liquid (kg m^{-3})
σ	Stress (Pa)
σ_{ij}	Stress with respect to coordinate x_i, x_j (Pa)
σ_{rr}, $\sigma_{\theta\theta}$, $\sigma_{\phi\phi}$	Stress in spherical coordinate (Pa)
τ	Time constant (s)
τ_0	Time constant determining second nucleation conditions (DS)
τ_2	Time constant determining second nucleation conditions (DS)
ϕ	Volume fraction
ϕ_{c}	Critical volume fraction (DS)
ϕ_{e}	Effective vesicularity giving maximum water exsolution rate (DS)
ϕ_{G}	Volume fraction converted into dissolved gas $v_{\mathrm{G}}C$ (DS)
ϕ_{n}^*	Volume fraction at nucleation termination (DS)
ϕ_{obs}	Observed volume fraction (DS)
ϕ_{s}	Volume concentration of crystallization component $v_{\mathrm{S}}C$ (DS)
ϕ'_{T}	Thermodynamic factor of crystal volume fraction to temperature (K^{-1})
φ	Phase angle
φ_{M}, φ_{η}	Phase lag angle in viscoelastic body
$\varphi(x)$	Dummy function
χ_G	Cooling rate exponent of growth rate in CSD model 1 (DS)
χ_J	Cooling rate exponent of nucleation rate in CSD model 4 (DS)
χ_{OS}	Growth exponent of Ostwald ripening (DS)

χ_p	Cooling rate exponent of Projection spacing (DS)
χ_t	Time exponent in power law distribution (DS)
χ_U	Volume exponent of bubble ascent velocity (DS)
χ_v	Volume exponent in power law distribution (DS)
Ψ_C	Coalescence frequency coefficient
Ψ_F	Fragmentation frequency coefficient
$\psi(x)$, $\psi(x,t)$	Arbitrary function
ω	Angular frequency (s^{-1})
ω_G	Angular eigen frequency of bubble oscillation in viscous fluid (s^{-1})
ω_L	Angular eigen frequency of bubble oscillation in viscoelastic fluid (s^{-1})
ω_{re}	Characteristic natural angular frequency of viscoelastic relaxation, t_{re}^{-1} (s^{-1})
$\langle\rangle$	Average quantity
$\sim$	Scaled (DS)
$\hat{}$	Variable after Laplace transform
$\propto$	Proportion (y is in proportion to x; $y \propto x$)

Part I
Introduction

Inspired by Nature

1

Abstract
Volcanic eruptions have the diversities in the time scales within 4.6 GA of the Earth's age, the magnitude of eruptive volume, the style (how magmas erupt). Together with these, the tectonic setting where a volcano locates and the chemical compositions of magmas characterize the eruption. The complexity resulting from these diversities make it difficult to share our recognitions and interests on volcanic eruptions. Before learning the fundamental processes such as vesiculation and crystallization of magmas in volcanic eruptions, we need to have our common recognitions and interests to share for scientific approaches. In this chapter, I will show specific examples of eruptions to share the scientific interests for latter chapters.

1.1 1986 Izu-Oshima Eruption

1.1.1 Changes in Eruption Styles

In 1986, a great eruption occurred at Izu-Oshima after it had remained dormant for 30 years. The eruption led to the unprecedented evacuation of all inhabitants from the island. First, Mt. Mihara, a central crater (vent A) in the caldera, produced glowing lava (fire) fountains, thus exhibiting a Hawaiian and Strombolian like eruption (Fig. 1.1, left image).

The lava fountains that started on November 15 were one of the largest lava fountain eruptions and reached a height of 1 km. The fountain height and discharge rate had a fluctuating rhythm of a cycle of several seconds (Fig. 1.1, right chart). However, no remarkable changes were observed in volcanic earthquakes, and there seemed to be no danger for people in visiting there unless they came close to the crater. Tourists visited Gojinkajaya, a souvenir store on the rim of the caldera, and took commemorative photographs against the background of lava-fountaining Mt. Mihara.

The force of the lava fountains gradually decreased over several days and the eruption eventually shifted to a Strombolian-like eruption, in which large bubbles exploded in the magma in the crater with explosive sounds (similar to a death cry) (Fig. 1.2, upper image). The intermittent explosion of magma balloons (at which many large bubbles concentrate), occurred at intervals less than 10 s and continued for 10–20 min, composing isolated episodes (Fig. 1.2, lower chart). The episodes repeatedly occurred at intervals of 2–3 h. The interval between the explosions from the central crater and that between

A. Toramaru, *Vesiculation and Crystallization of Magma*, Advances in Volcanology,
https://doi.org/10.1007/978-981-16-4209-8_1

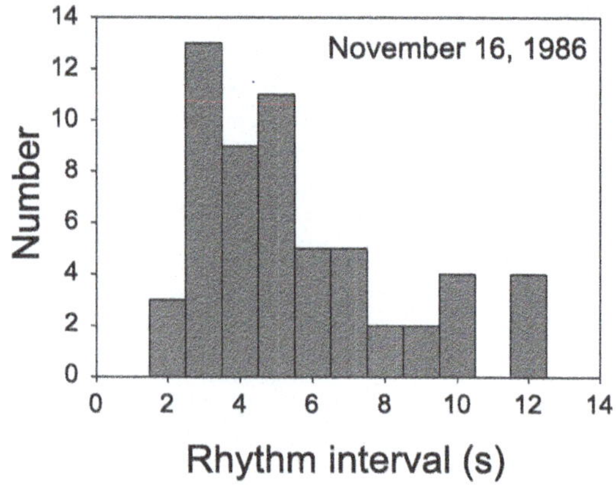

Fig. 1.1 (Left image) Lava fountain eruption from the vent A (the central crater) of 1986 Izu-Oshima eruption. (Taken by Motomaro Shirao on November 17, 1986. Used with permission by Motomaro Shirao.) (Right chart) Cycles found in fluctuation (rhythm) of the discharge rate of the lava fountain eruption around 17:15, November 16, 1986. Analyzed on the basis of a video image taken by a survey group of Nihon University (courtesy of Tatsuro Chiba). Total number of events: 58, average interval: 5.56 s, and variance: 2.7 s

the episodes gradually increased. On the next day (November 21), the number of the explosive sound drastically decreased.

Such changes in eruption styles are related to the properties of gas-charged magma and the amount of magma supply. The magma of Izu-Oshima was basaltic, relatively less viscous, and with low SiO_2 content and a high temperature. The appearance of magma (similar to a liquid), mixed with gases and spouting into air, is nothing other than a "gfountain." If the amount of ascending gas-charged magma decreased, gas bubbles remained in the magma for a longer time, coalesced into larger ones, and cyclically burst out of the crater. The temperature of magma that filled a volcanic conduit falls with time and the viscosity of the magma increases. Then, when large bubbles rise and swell in the conduit, magma balloons cannot sufficiently expand against the internal pressure and become unstable because of the accumulated pressure in bubbles, resulting in failure and explosion at some point in time.

On the evening of the next day (November 21), eruption occurred, at first, at the foot of Mt. Mihara within the caldera, and the vent extended to Kengamine, the rim of the mountain (Fig. 1.3).[1] This eruption was a sort of Plinian eruption having a more than 10 km-high eruption column; moreover, it was a fissure eruption from fissures made by magma itself breaking the crust and a flank eruption from the side of the mountain. The highest height the eruption column reached was reported to be either 12 or 13 km (Japan Meteorological Agency 1987).

Subsequently, the vent extended to the caldera rim and fissure vents (aligned craters C) appeared at the outside of the caldera and threatened to advance toward Motomachi, a built-up area at the foot. This situation forced all the inhabitants to evacuate from the island during the night. In the end, the eruptive activity decreased by the next morning (I can talk about it now); however, the inhabitants were compelled to evacuate for a further month. Analyses of earthquakes, gravity change, and crustal deformation subsequently revealed that magma erupted from the aligned craters B and the aligned craters C was present as a dike at a depth before the eruption, which was a tabular intrusion of magma under the ground that ascended to the surface.

[1]Currently, there is a tourist trail around the vent, and anyone can see the huge vent close up.

Fig. 1.2 (Upper image) Strombolian eruption from the vent A (the central crater) of 1986 Izu-Oshima eruption. Taken by an observation team member, Motomaro Shirao, on November 20, 1986. Used with permission by Motomaro Shirao. (Lower chart) Explosion cycles of magma balloons from 11:22 to 11:45 on November 20, 1986. Analyzed on the basis of a video image taken by an observation team member, Susumu Umino. Total number of explosions: 101, average interval: 6.16 s, and variance: 2.5 s

1.1.2 Bubbles and Crystals

Watching the Izu-Oshima eruption directly or through a video image, we can understand that volcanic eruptions are nothing but the explosive expansion of magma. A direct cause of explosive expansion of magma is the magma vesiculation process. The vesiculation of magma is a phenomenon in which volatiles (composed principally of water and carbon dioxide but predominantly comprising water in the case of volcanoes belonging to island arcs[2] such as Izu-Oshima) dissolving in liquid magma (silicate melt) deep underground under high pressure vaporize at low pressure near the surface and form gases (gas phases). This phenomenon is similar to the phenomenon observed when you open a bottle of beer,

[2]A chain of islands occurring along a trench, where a plate is subducting, on the continent side.

Fig. 1.3 (Left) Subplinian eruption from the aligned craters B of 1986 Izu-Oshima eruption. Upper photo was taken by Katsuyuki Abe. Lower photo was taken by Yasuji Kurokawa. (Right) The condition of the aligned craters B, as it appeared on the Kengamine (the rim of the mountain), immediately after the eruption ended. The day before the eruption, the authors observed explosion of magma balloons from the vent A here. (A picture postcard made by the Volcanological Society of Japan). Photo taken by Motomaro Shirao. Used with permission by Motomaro Shirao

champagne, or carbonated drink. In the case of magma, about 3–4 wt% of water can dissolve in magma at a depth of 4 km and a pressure of 1000 atm

We are aware of how bubbles form because we usually see (and relish) the bubbling process of beer and carbonated drinks in our daily life. I, however, felt frustrated because I could not observe with my own eyes the role of gases in volcanic eruptions and magma movement in volcanic conduits.

When I went up close to the central crater on November 20, there was a feeling under my foot and sounds of scoriae.[3] Various sized magma fragments that were scattered indicated that the explosion caused fragmentation. I picked up some glassy scoriae and took them to my lab. Scoriae that were several centimeters across were embedded in the resin and cut to make specimens for cross-sectional observations. Figure 1.4 shows a backscattered electron (BSE) image of the cross-sections of the scoria, obtained by an electron microscope. The black part is voids and the white part is solid material that was liquid magma (and then solidified as glass containing a small amount of crystals) or crystals (phenocryst). This image shows that (i) bubbles vary in size,[4] (ii) they are relatively spherical in shape, (iii) they coalesced in some places, and (iv) the vesicularity (the volume fraction of bubbles) is extremely high. These facts, which observed in the eruptive materials produced from lava fountains, will acts as a reference for observing eruptive materials produced from volcanoes of other eruption styles.

[3]For highly vesicular volcanic rock, relatively dark colored (nearly black) one is called scoria and relatively light colored one is called pumice. The color is reasonably associated with the chemical composition: pumice is rich in SiO_2 and scoria is poor in SiO_2. The relationship between the color and the chemical composition is unclear. It seems that glass compositions and crystallinity of groundmass are related. Scoria is relatively rich in fine crystals called microlites, whereas pumice hardly contains them.

[4]It often follows a power law distribution.

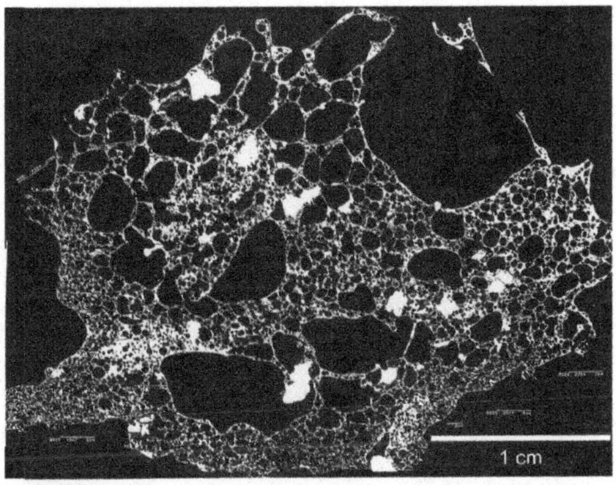

Fig. 1.4 Backscattered electron (BSE) image of the scoria of 1986A. A mosaic image consisting of about 50 electron microscope images. About 3 cm in width

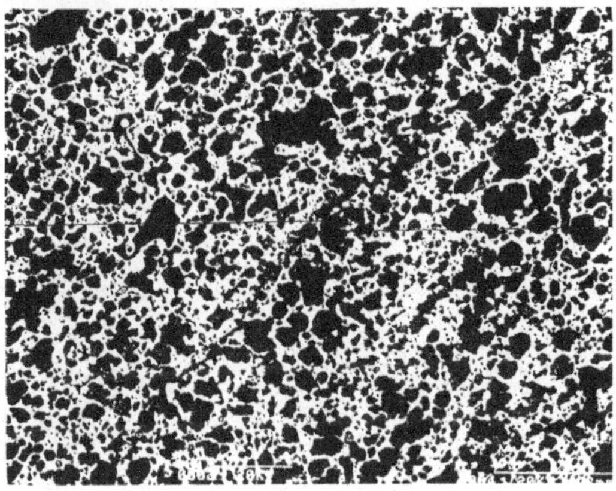

Fig. 1.5 1986B scoria. About 1 cm in width

Next, what is the texture[5] of scoriae emitted from the aligned craters B, a Subplinian eruption completely different from the eruption style of the vent A. Scoria fragments picked up our by research group member on the peripheral road were observed using an electron microscope. Figure 1.5 shows that a vesicular texture of eruptive materials produced from the Subplinian eruption was completely different from that of the scoria from the vent A, i.e., the lava fountain eruption. It features (i) bubbles that look roughly uniform in size, (ii) a number of bubbles are not spherical in shape, (iii) many of them have coalesced, and (iv) their vesicularity is high (but not extremely high). The detailed observation revealed that, while the scoria from the vent A contained less number of fine crystals (microlite; see Sect. 1.3.6) in their groundmass glass, the scoria from the aligned craters B contained more number of microlites (Fig. 1.6).

The difference in texture and morphology of vesiculated volcanic rocks is related to the difference in eruption styles; furthermore, it must be connected with the magma vesiculation process, which is a driving force for volcanic eruptions and the movement of magma in volcanic conduits, which is invisible. Thus, the texture of highly vesicular volcanic rock contains information on the magma movement at eruption. Furthermore, I expect that the detailed examination of texture of vesiculated volcanic rocks will quantitatively clarify invisible magma movement in conduits or will reveal a factor controlling the magma movement.

[5]In material science, microstructure crystals and bubbles form are called texture.

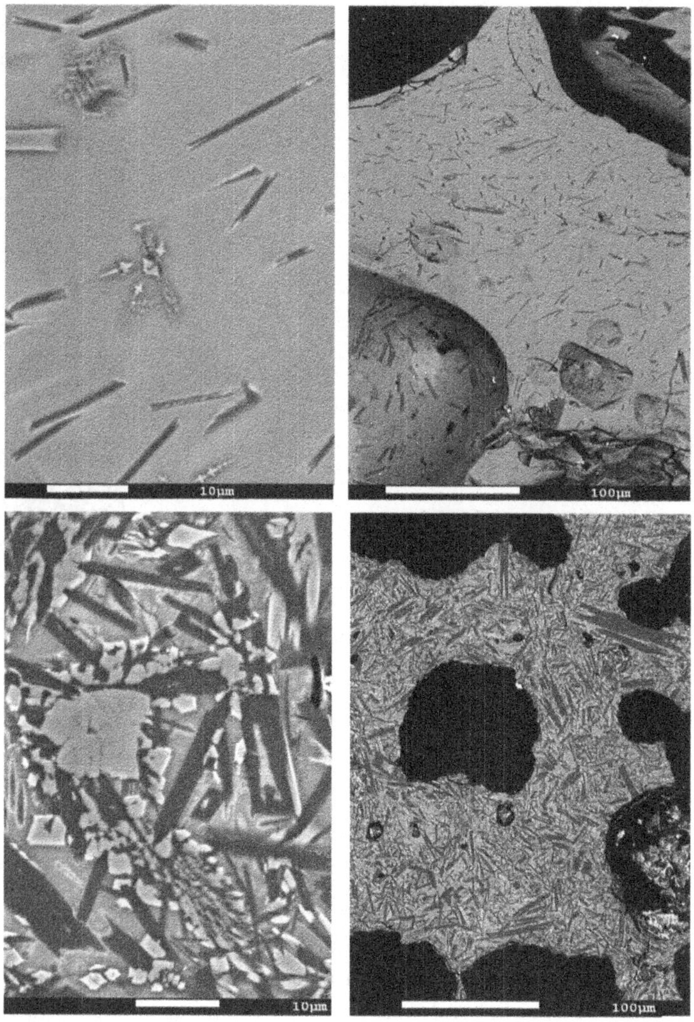

Fig. 1.6 (Upper images) Groundmass glass in 1986A scoria. Very small microlites are found sparsely. (Lower images) Groundmass glass in 1986B scoria. Many microlites are contained. Scale bar = 10 μm (left), 100 μm (right)

1.2 Pumice and Plinian Eruption

1.2.1 Eruption Emitting Pumice or Scoria

Highly vesicular volcanic rocks (pumice and scoria) are not always emitted from volcanic eruptions. Eruptions emitting them are limited to certain styles of eruptions, i.e., Plinian eruptions or eruptions called large-scale pyroclastic flows (ignimbrite). In Plinian eruptions, a volcano expels volcanic plumes of pyroclasts such as highly vesicular volcanic rock and volcanic ash or tephra, into the atmosphere to a height of >10 km. A typical example of Plinian eruption is the Towada volcano.

1.2.2 Towada Volcano

For the post-caldera Plinian eruptions of the Towada volcano, Yukio Hayakawa (then at Gunma University) investigated the geology of the volcano in detail, compiled the eruption chronology, and successfully quantified the volcanic intensity of the past eruptions using a method that had just been developed at that time (Hayakawa 1985). To quantify the volcanic intensity of both real-time eruptions, which can be surveyed and practically observed, and a large number of past eruptions, from deposits and eruptive materials is one of the effective approach to understand the phenomenon of volcanic eruptions.

Although how to define the magnitude and intensity of eruptions is a different issue, the height of eruption column can be a measure of intensity. Walker (1980) proposed an idea that eruptions could be quantified using the spread of eruptive materials and the fraction of relatively fine particles (<1 mm in diameter) to pyroclastic particles (including pumice and ash).[6] Wilson et al. (1980) proposed that for characterizing Plinian eruptions, it is easily understood that the height of eruption column could be used as a measure of the volcanic intensity by focusing attention on the mechanical behavior of volcanic plumes. Moreover, Sparks et al. suggested that the height of eruption column of past volcanic eruptions could be estimated through geological investigations of eruptive materials (Carey and Sparks 1986). At that time the Towada volcano was the only volcano whose volcanic intensity was systematically and quantitatively understood in detail for a single volcano.

The height of eruption column estimated from volcanic particles of the Towada volcano ranges from 12 to 30 km, which covers the range of Subplinian eruptions and typical Plinian eruptions. The analysis of chemical compositions of magma showed that the magma is felsic magma containing 61–74 wt% (dacitic to rhyolitic) of SiO_2. (For the Izu-Oshima Subplinian eruption (1986B) described above, the eruption column height was 12–13 km, and the magma was basaltic andesite magma containing 53–54 wt% of SiO_2.)

Let us look at the pumice of typical felsic magma from a Plinian eruption (the eruption column height = 30 km; SiO_2 = 71 wt%) as an example (Fig. 1.7). A comparison with the scoria of the Izu-Oshima eruption revealed that the bubbles of the pumice are obviously smaller in size. While the scoria of the Izu-Oshima Subplinian eruption are about 100 μm, the majority of pumice of the Towada volcano are about 10 μm. Furthermore, the bubbles retain a spherical shape, suggesting that the magma expands until the bubble walls become extremely thin. Note that the bubble walls partially broke and some bubbles coalesced. Cusp, which remains after bubble coalescence, of bubble walls does not look relaxed because of surface tension. The predominant small-sized bubbles look relatively uniform in size,[7] while large bubbles are unevenly distributed.

Pumice from Plinian eruptions in the post-caldera stage of Towada volcano was investigated in the same manner, and images were obtained in the order of the eruption column height (Fig. 1.8). These images show that bubble size gradually increased as the eruption column height decreased from about 30 to 25 km and from 19 to 12 km. This fact suggests that there is a correlation between the factor determining the bubble size deep underground and the volcanic intensity at the Earth's surface.

Note that the texture of pumice seems to be related to both the volcanic intensity and to the chemical compositions of magma. Chemical compositions of magma considerably affect the properties of magma. In particular, properties such as viscosity (how easily magma flows) and diffusion coefficient (how easily molecules migrate) may strongly influence bubble formation; therefore, the effect has to be carefully examined.

To quantitatively compare correlation, graphs with the bubble number density (BND) on the vertical axis and the eruption column height and SiO_2 content on the horizontal axes are shown in Fig. 1.9. BND equals the inverse number of the cube of bubble size if vesicularity[8] is nearly the same, and BND is a

---

[6] Eruptive materials, i.e., pyroclastic particles have a size distribution and produce deposits on the ground surface with a certain thickness and spreading area. When you mention the spread of eruptive materials and the ratio of a certain particle, you have to define the thickness and spatial location of interest in advance.

[7] It often has an exponential distribution.

[8] Vesicularity is the volume fraction of void (bubbles or vesicles) in vesicular volcanic rocks. See Sect. 10.2.4.

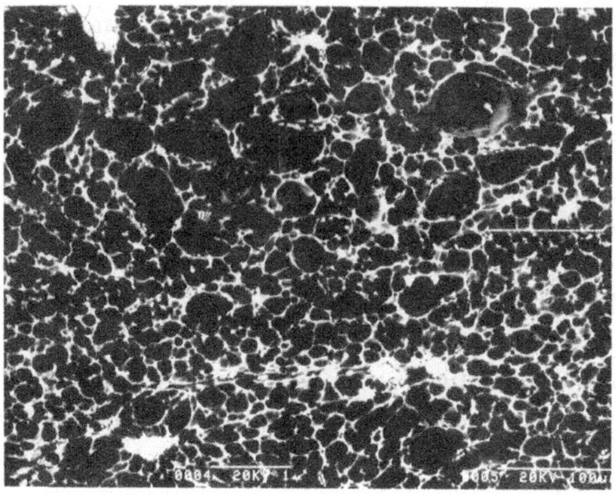

Fig. 1.7 A2 pumice of Towada volcano. Eruption column height = 30 km, SiO_2 = 71 wt%. Upper left part of A2 of Fig. 1.8

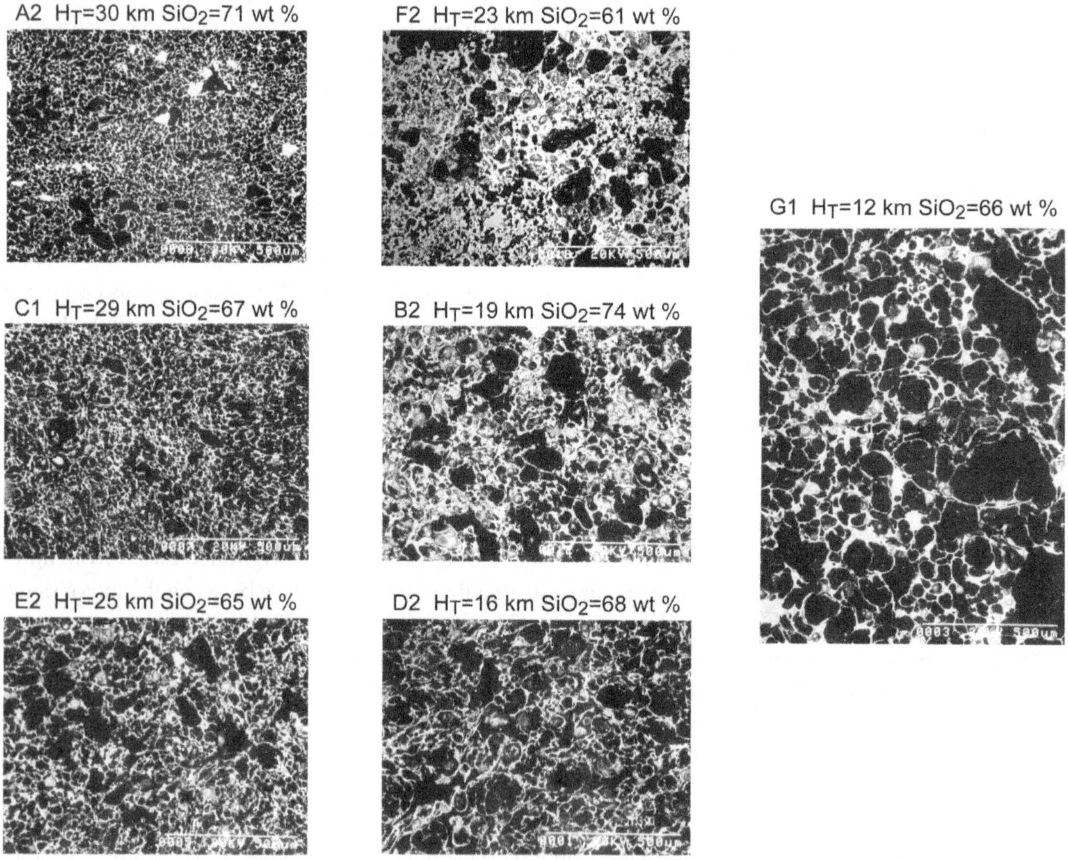

Fig. 1.8 Pumice of Towada volcano. Scales are made uniform for comparison (Width of images = 1 mm) (Toramaru 1990)

more substantial quantity in considering bubble formation mechanism. Based on the figure, the texture of highly vesicular volcanic rock, i.e., BND, seems to be related to both the eruption column height and the chemical composition. The chemical composition of magma considerably affects the viscosity of melt and diffusion coefficient of H_2O, a component of bubbles, in melt and, in turn, considerably affects bubble

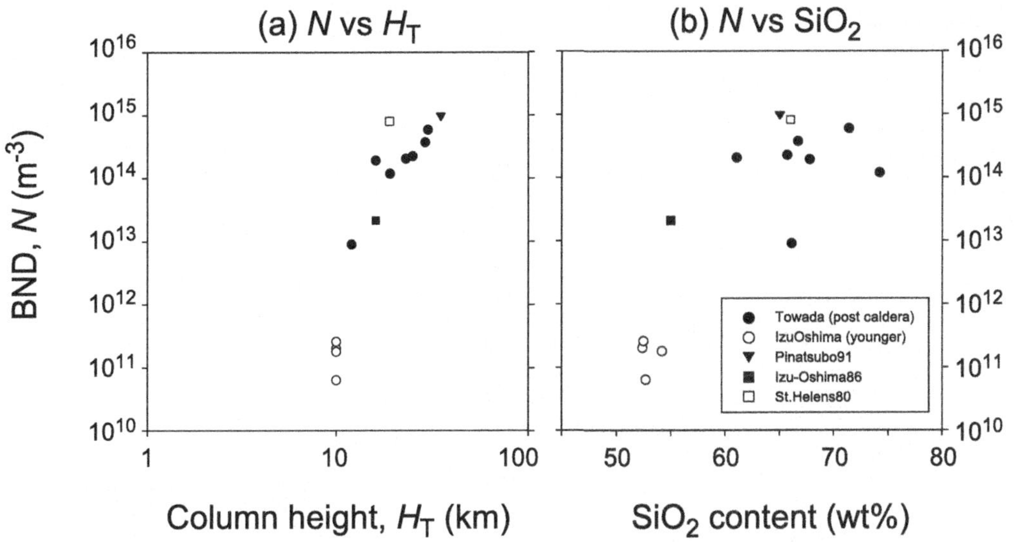

Fig. 1.9 Relationships between bubble number density (BND) and eruption column height (H_T) and between BND and chemical composition (SiO_2 wt%). Reprinted from (Toramaru 2006), Copyright 2006, with permission from Elsevier

formation. This relationship between BND (bubble size) and the eruption column height may indirectly express the influence of physical properties of magma. To understand the movement of invisible magma in conduits, we have to consider the influence of the physical properties of magma. For this purpose, in the following chapters, we will understand bubble formation in magma, i.e., vesiculation process of magma with the role of physical properties of magma in mind as well as examine the influence. Then, we will consider the relationship once again.

The intensity of Plinian eruptions and scaling law for volcanic eruptions

It is an extremely important problem to determine the intensity of huge explosive eruptions, notably Plinian eruptions. There should be a special relationship between the eruption intensity and the eruption mass (magnitude), which is closely related to a natural mechanism determining the eruption intensity, i.e., a law hidden in nature. The relationship holding between them is a sort of scaling laws. The discovery of scaling laws has a special importance for scientific development. A famous example of the relationship holding in the field of earth science is a relationship between the magnitude of an earthquake and the characteristics of the hypocenter or fault parameters. This is a ground for considering earthquakes as a mechanical problem of faults; however, such a relationship has not yet been determined for volcanic eruptions.

Several years after the suggestion of a method estimating the eruption column height from the dispersal patterns of eruptive materials from past eruptions by Carey and Sigurdsson (1989) compiled geological data of Plinian eruptions that have occurred all over the world and examined what determined the eruption column height. Consequently, the eruption column height has been found to be positively correlated with the total erupted mass, as shown in Fig. 1.10.

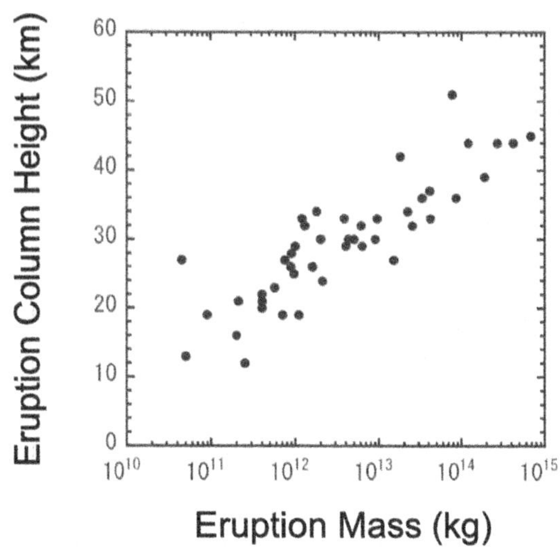

Fig. 1.10 Relationship between eruption column height and total erupted mass (from Carey and Sigurdsson 1989) Adapted by permission from Springer Nature

The intensity of Plinian eruptions and scaling law for volcanic eruptions:2

This figure beaten me because a course of my research I vaguely had in my mind at that time, that is, a discovery of a scaling law for volcanic eruptions, had been already made. The eruption column height had been found to be proportional to the discharge rate to the power of 1/4 via a fluid-dynamic mechanism; however, the presence of correlation of the total erupted mass or the integral of the discharge rate over eruption duration with the discharge rate itself was not self-evident. This correlation informed me something that what was not self-evident.

They argued that the increase in eruption duration caused the erosion of volcanic conduits by magma flow, the enlargement of cross-sections of conduits, substantial increase in discharge rate (because the discharge rate was proportional to cross-sections of conduits), and increase in the eruption column height in the end. In fact, eruptive materials from large eruptions comprised not only essential materials, in other words, magma involved in the eruption, but also lithic fragments, which might form the wall of conduits. This conclusion was based on the intuition of volcanologists who had examined eruptive materials by the fieldwork.

However, believe that there is another fundamental factor determining the total erupted mass. This factor must be related to the discharge rate for a special reason. To date, there have been no reports that can quantitatively explain the empirical scaling law holding between these data in nature.

1.2.3 Widespread Volcanic Ash

Plinian eruptions emit both highly vesicular volcanic rock and volcanic ash[9] Sometimes, fine particles such as volcanic ash can be carried to very long distances areas and get deposited there. Such volcanic ash is called widespread volcanic ash (or tephra) and is an important geological indicator used in tephrochronology to determine the age of eruption (Machida and Arai 2003). The major constituent of the widespread volcanic ash is bubble-wall glass shards, which are several dozen micrometers in size. Fig. 1.11 shows an electron micrograph of bubble-wall volcanic ash with various shapes. This image

[9]Particles measuring <1 mm in diameter are called volcanic ash.

Fig. 1.11 Electron micrograph of bubble-wall AT widespread volcanic ash (collected at the Aso somma). Micropumice, a small pumice with small bubbles, was also found

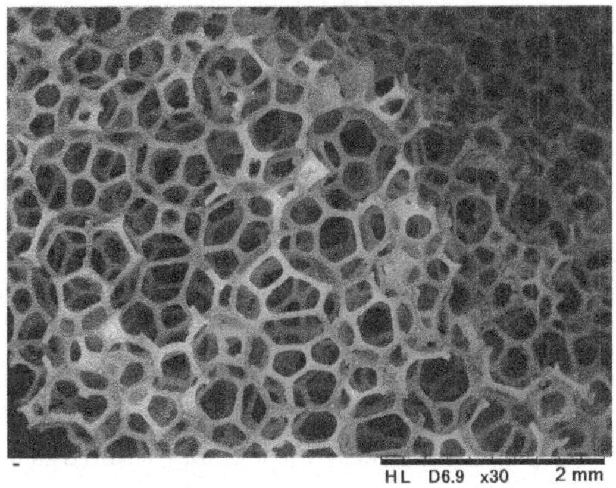

Fig. 1.12 Scanning electron micrograph of reticulite in Hawaii

clearly tells us that liquid foam occupied with gas bubbles is fragmented into volcanic ash by a volcanic eruption. Interestingly, the size of bubbles estimated from these bubble-wall glass shards is larger than that of bubbles found in highly vesicular volcanic rock. Such volcanic ash is found in eruptive materials from Plinian eruptions, particularly, eruptions that form calderas.

1.2.4 Reticulite: The Ultimate Pumice

Reticulite is a characteristic of ejecta from lava fountain eruptions of volcanos in Hawaii (Hawaiian eruptions), which eject even less viscous basaltic lava than that of the 1986 Izu-Oshima lava fountain. Measuring several millimeters to several centimeters of particle size, reticulite looks like a sponge with a vesicularity up to 98% and hardly deforms; however, it is very fragile. Nevertheless, when a 10-cm or more thick reticulite layer is deposited on the ground, the layer will be preserved relatively well.[10] Figure 1.12 shows an electron micrograph of reticulite collected in Hawaii. The size of bubbles is very uniform, and almost all of bubble walls have broken; only the liquid at the Plateau border has solidified and remains in the glass frames. The abovementioned bubble-wall volcanic ash can be considered as fragmented sil-

[10]On a wall at the side of Hawaiian Volcano Observatory (HVO) of the United States Geological Survey before the 2018 Kilauea activity, one could found a reticulite layer of a large-scale eruption from the Kilauea vent at around 1500 (estimated by the carbon 14 dating method) (Swanson et al. 2012).

icate melt having a reticulite-like structure. Magma producing the bubble-wall volcanic ash is rhyolitic magma of the highest viscosity, unlike basaltic ones. Because rhyolitic magma does not result in reticulite suggests that the fragmentation process of highly viscous magma in explosive eruptions is violent. Note that non-explosive eruptions with highly viscous magma sometimes produce a reticulite-like structure in a lava flow of obsidian.

Extremely less viscous magma can produce some types of tephra, known as Pele's hair and Pele's tear, named after Pele, the Hawaiian fire goddess. Pele's hair is a fibrous glass of a certain thickness, whereas Pele's tear is a tiny solidified drop of glass. Both of them are formed when liquid viscosity and surface tension of glass within a certain range is blown off by gas and stretched out. Shimozuru (1994) noticed a similarity to an ink jet process and thought of a dimensionless Pele number. Similar to the Capillary number, the Pele number is the ratio of the Weber number to the Reynolds number. The Reynolds number is the ratio of the inertial force to the viscous force, whereas the Weber number is the ratio of the inertial force to the surface tension. The inertial force is a force that stretches the liquid, the viscous force is a force that prevents progress of extremely ill-balanced deformation of liquid, and the surface tension is a force that maintains the outer shape of the droplets (spherical). Shimozuru (1994) determined that Pele's hair would be produced for a larger Pele number and Pele's tear would be produced for a smaller Pele number. In natural volcanic eruptions, liquid is extended by flow of the surrounding volcanic gases and air and results in the formation of Pele's hair and tear. Various Pele's hairs and tears are produced because of the complex forces produced by complex gas flow and rotation of droplets themselves.[11]

1.2.5 Volcanic Eruptions as Global Phenomena

The discovery of the scaling law between discharge rate and total erupted mass by Carey and Sigurdsson (1989) is based on a method estimating the height of an eruption column from dispersal patterns of eruptive materials. In the method, a fluid dynamic model for motion in an eruption column was developed and the motion of particles following the model is tracked. To be more specific, how far the particles are blown off, where they fall out is determined by a height, and where the particles have reached in the eruption column (neutral buoyancy level, height of umbrella-cloud) (Fig. 1.13). The dispersal patterns of particles found in practical geological surveys were able to be explained to some extent by calculation based on the idea, and the eruption column height was estimated, i.e., this idea is qualitatively the same with ballistic dynamics: a particle emitted from a higher height horizontally will fly farther away Carey and Sigurdsson (1989).

During World War II, Takeshi Minakami estimated, based on this idea, the height to which the explosion of Mt. Asama spouted magma into the air from eruptive materials produced by the eruption (Minakami 1942). A story has been told that Minakami, who is known to have been all over the world to classify and name volcanic earthquakes, conducted such a study because he had only balances and tape measures at that time.[12]

---

[11] Glass fiber is a material used in our daily life without special recognition. It is used to transmit information, e.g., optical communication lines; although Pele's hair is similar in size, a glass fiber can be extremely long. Glass fiber is manufactured by extending liquid glass (silicate melt) mechanically while holding both ends, thus resulting in a fine fiber with a uniform diameter. When melt is extended, a part that has already become thin by extending will harden and become less capable of deforming because of a cooling effect, while the other part that remains hot and thick will be still capable of deforming and become thin. In this manner, extending high-temperature liquid naturally yields fibers of uniform diameter.

[12] I was shocked again at this story. The idea of quantitatively estimating the intensity of an explosion from the dispersal patterns of eruptive materials produced from the explosion, in other words, the idea of quantification of eruption on the basis of eruptive materials, had already been born in Japan during the war. It is a matter of regret that this idea had not developed in Japan since then.

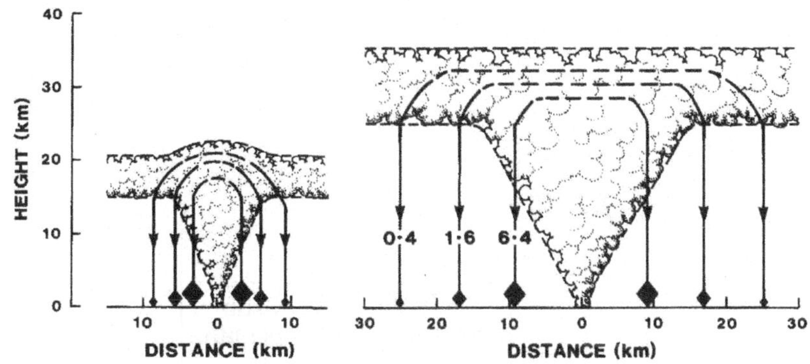

Fig. 1.13 Difference in particles' transportation and deposition process according to eruption column height and particle size. (adapted from Carey and Sparks 1986) Reprinted by permission from Springer Nature

Just at that time when I read the draft of Carey and Sigurdsson (1989), the Kagoshima International Conference on Volcanoes was held in Kagoshima, Japan. After the Conference, fortunately for me, Shigeo Aramaki invited me to a field trip to southern Kyushu and Aso areas with the Conference's participants, including Sigurdsson, Sparks, Schmincke, and the young Pyle. We observed local deposits, including the Osumi pumice fall deposit, which erupted from a typical Plinian eruption, as well as the subsequent Ito pyroclastic flow deposit. The team members seemed to easily take in what they observed even if these deposits were local deposits. They had conducted field work of volcanos all over the world on their own feet and had established their own standards or scales in their minds for observing eruptive deposits. I have imagined that they have a lot of drawers in their minds, which are arranged in order of magnitude or characteristics of eruptions. When they encountered a new eruptive deposit, they did not randomly place the data in a drawer but verified the data according to information linked to the drawers to place the new data in a suitable drawer. After examining many eruptions all over the world, they would think about the universal characteristics of natural phenomena. Consequently, they compiled numerous eruptions into that paper by pursuing universality in eruptive phenomena. Similarly, in earth science, where a global point of view is important, one's attitude to natural phenomena will considerably affect the quality of research.

1.2.6 Variety and Unified Classification of Explosive Eruptions

Unified classification is important to systematically organize various phenomena using a few indicators. If this is possible, the indicators used can be said to play considerably essential roles in understanding nature because not all indicators can be used to systematically organize various natural phenomena.

The previous section on the 1986 Izu-Oshima basaltic eruption stated that lava fountain eruptions and Subplinian eruptions that spout eruption columns of > 10 km produce different eruptive materials. Moreover, the previous section on the Towada volcano introduced a typical felsic Plinian eruptions in terms of texture and column height.

This section summarizes the relationship between eruption styles and eruptive materials. Volcanic eruptions are broadly classified into two types: eruptions that produce fragmented eruptive materials or pyroclasts (such as highly vesicular volcanic rock (pumice and scoria) and volcanic ash) and eruptions that discharge unfragmented lava. The former eruptions are known as explosive eruptions and the latter are known as nonexplosive eruptions. It was Walker who attempted to classify explosive eruptions by deposit characteristics of eruptive materials. He noticed that he could easily organize eruption styles by plotting data obtained from geological surveys and analysis of eruptive materials on a diagram with a dispersion index on the horizontal axis and fragmentation index on the vertical axis, as shown in Fig. 1.14.

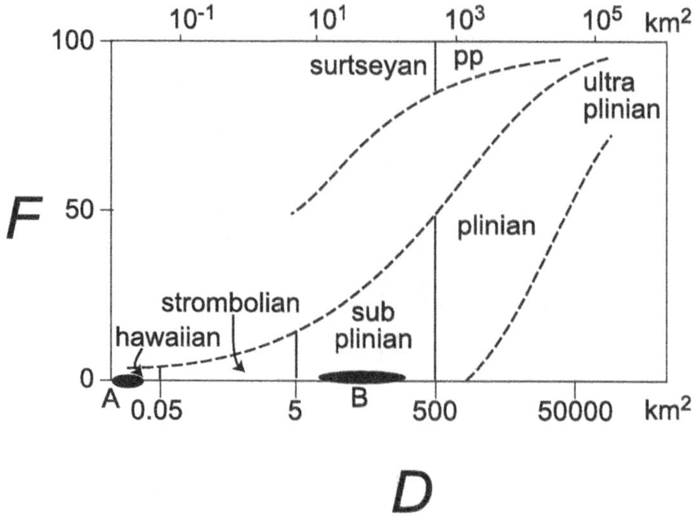

Fig. 1.14 Walker diagram with dispersion index on the horizontal axis and fragmentation index on the vertical axis (Walker 1980). Dispersion index D is the area enclosed by the isopach contour representing 1/100 of the maximum thickness. Fragmentation index F is mass percentage (%) of particles that are smaller than 1 mm collected at the point where the main axis of dispersion intercepts the isopach representing 1/10 of the maximum thickness. Approximate values, based on data of Mannen (1999), of the B eruption of 1986 Izu-Oshima were plotted. Values of the A eruption are estimates; the abbreviation pp means phreatoplinian, i.e., Plinian eruption involving phreatomagmatic eruption

This "well-organized" diagram suggests that the diagram comprehends an underlying physical mechanism in nature. Plotting the 1986 Izu-Oshima eruption on this diagram indicated that the lava fountain from the central crater is located in a Hawaiian-type eruption and the fissure eruption is located in Subplinian eruption (or Plinian eruption according to an estimation of the degree of dispersion), which largely agrees with our observation. This Walker diagram has the advantage that eruption styles of unobservable past eruptions can be inferred from deposits. In this manner, the history of past eruptions has been deciphered somewhat. Being conscious of such correspondence between eruptive materials and eruption styles, this book aims at providing a basis to consider the relationship between bubble texture in eruptive materials and eruption styles.

1.3 Lava Dome and Pyroclastic Flow

1.3.1 Eruptions that Do Not Produce Pumice or Scoria: Non-explosive Eruptions

Note that not all volcanic eruptions produce pumice and scoria, i.e., lava flows and lava domes. Let us look at a typical example: Heisei eruption of Fugendake, Unzen volcano, Nagasaki.

1.3.2 Heisei Eruption of Unzen Volcano

In 1990, soon after the transition from the Showa era to the Heisei era in Japan, Fugendake of Unzen volcano erupted. A series of eruptions started with increase in volcanic gas activity and phreatic eruptions. After several months of an inactive period, Vulcanian eruptions with ejecting volcanic bombs and

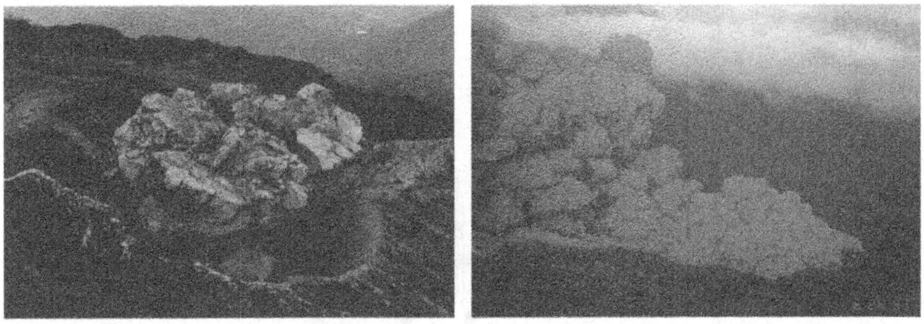

Fig. 1.15 Heisei eruption of Unzen volcano. (Left) A lava dome developed on the vent. (Right) Pyroclastic flow caused by collapse of the dome. (A picture postcard made by the Volcanological Society of Japan) Photo taken by Setsuya Nakada. Used with permission by Setsuya Nakada

phreatomagmatic eruptions occurred. Although these eruptions were not so large and occurred only several times, they were followed by the emergence of a lava dome and pyroclastic flows caused by collapse of the dome. The pyroclastic flows occurred as much as several thousand times by the end of the eruption in 1995; we have never forgotten this volcanic disaster that produced tragic consequences.

There are various eruption styles of volcanos: Plinian eruptions that produce eruption columns, umbrella-shaped volcanic plumes (umbrella cloud) and pyroclastic fall deposit; Hawaiian and Strombolian eruptions that discharge active lava fountains; and lava flow eruptions that effuse relatively less viscous magma. Moreover, lava domes that are masses of viscous lava extruded from vents are well known. A major feature of the Heisei eruption of Unzen volcano is that, although the eruption was a lava dome-style, it produced pyroclastic flows caused by collapse of the dome.

The word "glava dome" gives an impression that magma does not spout out or flow out but that sticky, highly viscous lava forms a dome, similar to a rice cake, on the vent. However, an image of the actual lava dome, shown in the left image of Fig. 1.15, shows cracked rock blocks rather than a rice cake shape. Furthermore, the rock blocks partially collapsed and ran down on the slope as volcanic plumes (surge) comprise gases and fine pyroclastic particles as shown in the right image of Fig. 1.15.

The shape of the cracked lava dome informs that this magma has very complex properties. Before appearing at the vent, the lava should flow up and ascend in the underground conduit, which can be seen in the shape of uncracked part of the lava dome. Thus, this magma behaves as a fluid until the moment when it appears at the Earth's surface; then, it partly behaves as a solid that can be broken. Furthermore, the magma has the property of being smashed into fragments such that it produces volcanic plumes that run down the slope. These complex properties are not exclusive to the magma of the Heisei eruption of Unzen volcano but common to magma in general. The Heisei eruption of Unzen volcano was an easy-to-understand case because the properties are connected to phenomena occurring on the surface, i.e., the shape of dome and generation of pyroclastic flows.

1.3.3 Temporal Change of Discharge Rate

Magma flow accompanied by the growth of a lava dome has a beginning and the end, of course, and does not progress at a constant rate. Figure 1.16 shows the discharge rate calculated from the change in shape of the lava dome (Nakada et al. 1999). During the eruption period from 1991 to 1995, two peaks of eruptions were identified. Such fluctuating changes in discharge rate were obviously observed at the Mount St. Helens Volcano, US, and the Soufrière Hills Volcano on the Caribbean island of Montserrat. Fluctuating phenomena such as the ascent rate of magma have been attracting interest from volcanologists in recent years because of the problem of a mechanism in volcanic conduits and magma. Melnik and Sparks (1999)

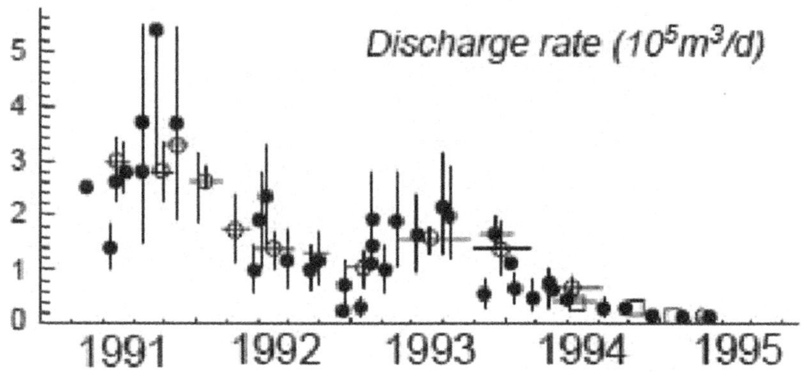

Fig. 1.16 Change in the discharge rate of Heisei eruption of Fugendake, Unzen volcano. (Nakada et al. 1999) Reprinted by permission from Springer Nature

Fig. 1.17 Pyroclastic surge deposit (the granular layer at the bottom) and the upper block-and-ash flow (BAF) deposit produced by the Heisei eruption of Unzen volcano on June 23, 1993. A cross-section of a tree trunk caught up and burned in a pyroclastic flow is found at the left side of the picture. Scale: approx. 1 m

reported that the fluctuating phenomena are derived from unstable magma flow caused by change in viscosity of magma because of the formation of crystals or bubbles, and Ida (1996) and Maeda (2000) separately believed that the phenomena are caused by oscillatory deformation of the conduit wall in response to change in pressure at the lower part of the conduit. These thoughts will be examined in subsequent chapter.

1.3.4 Block-and-Ash Flow

As a lava dome grows, the dome will become unstable and liable to collapse gravitationally because of the steepening at the vicinity of the summit. Low temperature, water-poor dacitic magma behaves in an elastic matter rather than as a fluid. The magma changes its shape with slow, gradual changes as if it is fluid; at the vicinity of the summit, it forms a "grosette" structure with repeating brittle fracture and folding. With decrease in the temperature, the magma becomes elastic and is cracked in places that have highly deformed. Such a lava dome with multiple cracks frequently collapses again and again under unstable conditions at the vicinity of the summit margin. A pyroclastic flow caused by collapse of lava domes is called a block-and-ash flow because it comprises rock blocks and volcanic ash. Deposits formed by block-and-ash flows have a unique structure (Fig. 1.17).

1.3.5 Interior of Pyroclastic Particles

Then, what texture does the magma of Heisei eruption of Unzen volcano, in which its complex characteristics seem to control phenomena on the surface, have? What texture do the pyroclastic particles produced from the magma have? Fig. 1.18 shows optical micrographs of a pyroclastic particle collected on June 24, 1993. These micrographs indicate that the dome-collapse pyroclastic particles contain only a very few bubbles. They instead contain phenocrysts, and many crystals are contained in the groundmass. The crystals in groundmass vary in size from large ones (50–100 μm) to very small ones (several micrometers).

The magma of the Heisei eruption of Unzen volcano is dacitic magma containing 68 wt% of SiO_2. Highly vesicular volcanic rock produced from Towada volcano's Plinian eruption has a similar chemical composition but contains many bubbles, which informs us that vesiculation occurred in the texture. The texture of the Unzen volcano's pyroclastic particle and the texture of Towada's vesicular rock are appreciably different. This suggests correlation between texture of eruptive materials and eruption styles. Furthermore, fine crystals, called microlite, may help us to understand the mechanism of the eruption style of lava domes and the factors controlling difference between eruption styles.

1.3.6 Microlite

Microlites are fine crystals that are < 100 μm and present in volcanic rocks. Recall that microlites are contained in scoriae of the 1986 Izu-Oshima Subplinian eruption. Detailed investigation reveals that the number density of microlites of the Unzen eruption is one to several orders of magnitude lower than that of the microlites of the Izu-Oshima Subplinian eruption; conversely, the size is larger. The number density is the number of crystals present in unit volume and is deeply related to the crystal formation process. This is very interesting when considering the meaning of the bubble number density (BND) of highly vesicular volcanic rock, which correlates with the eruption column height. Moreover, note that microlites are almost not contained in basaltic lava fountains (SiO_2-poorest volcanic rock) from the summit vent of 1986 Izu-Oshima eruption. Furthermore, microlites are almost not contained in felsic magma (SiO_2-rich volcanic rock) of a typical Plinian eruption of the Towada volcano.

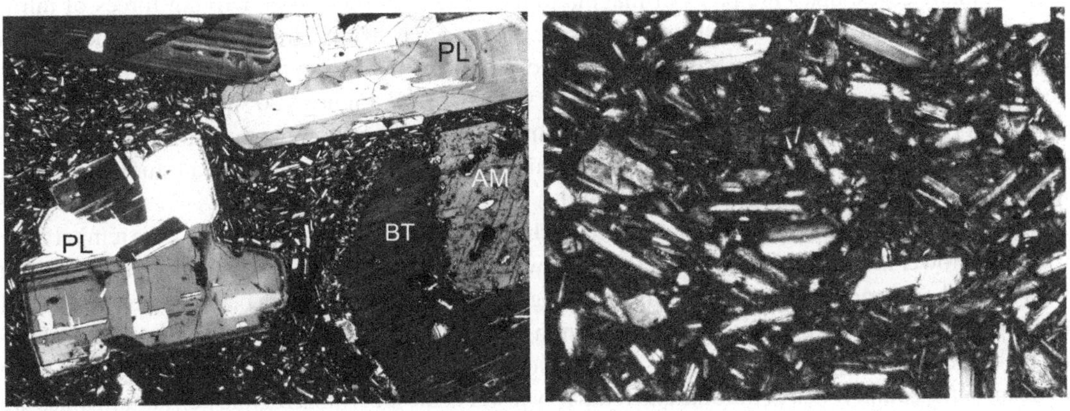

Fig. 1.18 Optical micrographs (transmitted light, crossed nicols) of the internal texture of a particle (approx. 1 cm) in a pyroclastic flow deposit produced by Heisei eruption of Unzen volcano on June 24, 1993. (Left) Width: 5 mm. PL: plagioclase, BT: biotite, AM: amphibole. (Right) Width: approx. 0.6 mm. Microlites are densely found in groundmass filling space between phenocrysts

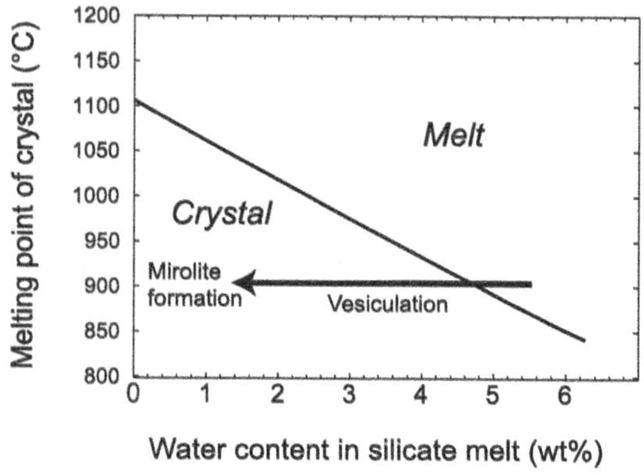

Fig. 1.19 Influence of water content in melt on melting points of crystals (plagioclase) of Heisei eruption of Unzen volcano

Generally, crystals are formed when liquid cools down and solidifies; however, microlites in volcanic products, pumice, and scoria, and particles contained in pyroclastic flows are not formed by cooling crystallization (Cashman and Blundy 2000). Such microlites are formed because of the increase in the melting point, which occurs because the H_2O present in magma[13] or liquid silicate (hereafter referred to as melt) comes out as a gas and decreases because of the vesiculation of magma. Figure 1.19 shows the influence of water content in melt on the melting point of crystals. As the vesiculation of magma progresses, water content decreases, melt becomes supersaturated, and crystals are finally formed. This suggests that the correlation of microlite texture with eruption styles is related to the vesiculation process of magma, and the possibility of quantitatively estimating relationships between the vesiculation process of magma and eruption styles or volcanic intensity and invisible phenomena occurring in volcanic conduits and magma chambers from eruptive materials.

1.4 Crystallization and Vesiculation of Magma that Cooled down and Solidified

1.4.1 Dike as a Place for Cooling Crystallization Experiment in Nature

A dike is a product formed by magma intruding into a brittle-fractured mantle or crust and solidifying there. In many cases, dikes are paths of magma to the surface (feeder dike). Driving forces of intrusion and ascent of magma include the overpressure of a magma chamber and buoyancy of the magma itself. Magma-forming dikes may contain bubbles or not many bubbles before intrusion. Bubbles that are present in magma before intrusion change their shape by flow. Figure 1.20 shows a dike of basaltic andesite at Hirado island, Nagasaki. Similarly, magma solidified at shallow depths can be exposed at the ground surface as intrusive rock after erosion of the upper strata. Consequently, we can study the history of magma that cooled down and solidified at depths.

Magma in a dike crystallizes by being cooled by the surrounding rock (country rock), the temperature of which is lower than that of the magma. In a narrow dike similar to the one shown in Fig. 1.20, magma is cooled by heat conduction. Although convection generally occurs in magma being cooled because of temperature differences, the effect of convection in such a narrow dike is almost negligible considering the viscosity of magma and the timescale of cooling. Therefore, the cooling rate is expected to be higher near the country rock of the dike and relatively lower at the innermost region, i.e., natural dikes are regarded as products of crystallization experiments in which the cooling rate varied spatially in a systematic manner.

[13] Magma is molten rock and generally contains not only silicate liquid but also crystals and bubbles.

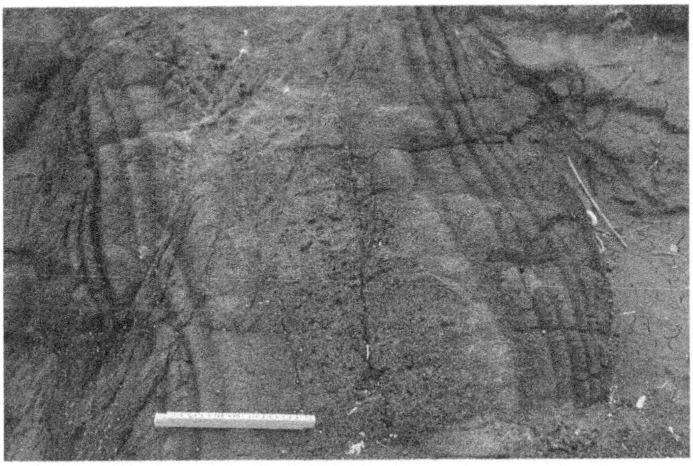

Fig. 1.20 Dike containing bubbles formed by second boiling, found in Hirado island, Nagasaki. Scale: 20 cm

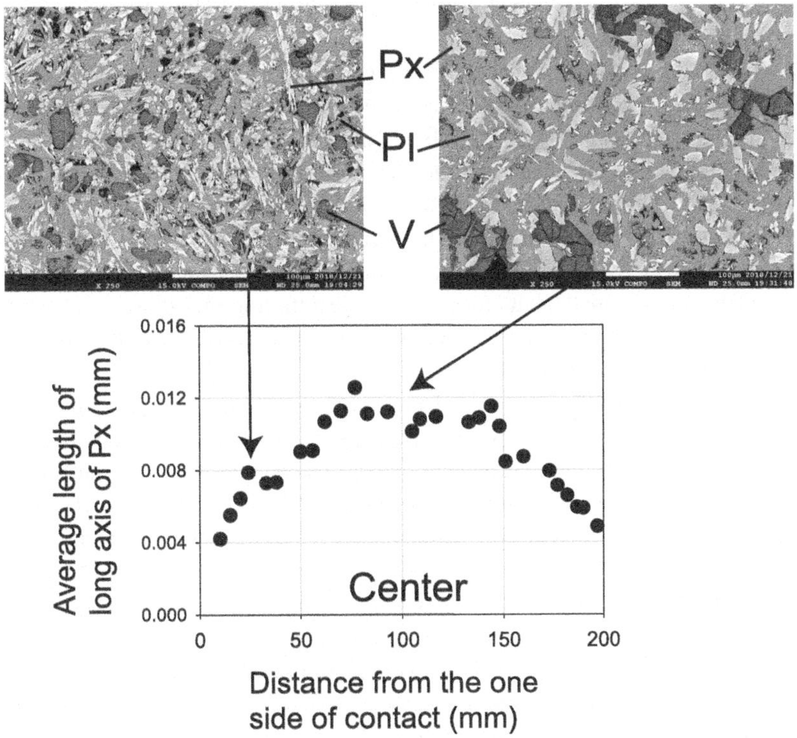

Fig. 1.21 Change in texture of groundmass in the dike found in Hirado island, Nagasaki (Yoshida 2008 MS). (Upper images) BSE (Backscatter Electron) images, taken by an electron microscope, of the surroundings and the center of the dike. Image width is 0.5 mm. The brightest crystals are pyroxene (Px). Gray needle-like crystals between pyroxene are plagioclase (Pl). Dark gray parts were bubbles or vesicle (but now are amygdaloidals filled with secondary minerals such as zeolite) (V). (Lower graph) Change in size of groundmass pyroxene. Note that sizes are average values

How has the crystal texture of the dike at Hirado changed with distance? In the lower graph in Fig. 1.21, the pyroxene size found in the groundmass is plotted as a function of the distance from boundary of the dike (Yoshida 2008). This figure shows that the size linearly increases with distance from the boundary with the country rock toward the center of the dike and then becomes nearly constant at the center. Because groundmass pyroxene is considered to not be present at the time when magma intruded, this distance dependence is a result of a natural experiment that shows how crystallization of pyroxene, i.e., nucleation and growth process is affected by cooling rate. The aim of this book, which explains the crystallization process of magma, is to explain or understand such a natural experiment result.

1.4.2 Vesiculation in Dikes

Vesiculation of magma occurs not only when magma ascends and decompresses during eruptions but surprisingly also when magma cools down, crystallizes, and solidifies, which is easily understood by imagining that when you place water in the refrigerator to make some blocks of ice, you will obtain ice blocks in which air dissolved in water is trapped and frozen in the form of white bubbles. The dike at Hirado has characteristic stripes parallel to the boundary with the country rock. The detailed studies revealed that the difference in the abundance of bubbles produced this striped pattern, and the cause of these cyclic bubbles is the vesiculation of magma associated with both cooling and crystallization.

As melt crystallizes, anhydrous crystals such as plagioclase and pyroxene form. Consequently, in magma in which crystallization progresses, volatiles such as water are increasingly concentrated in the residual melt. Then, when crystallization progresses to some extent, the residual melt is saturated with water, and then supersaturated by further crystallization, resulting in vesiculation and bubble formation, as shown in Fig. 1.20. Such vesiculation triggered by cooling crystallization of melt is known as second boiling.

In a magma chamber that is located deep underground, the temperature inevitably decreases and water gets gradually concentrated in the melt because of crystallization. Then, bubbles are formed somewhere in the magma chamber. Because bubble formation causes volume expansion, the magma chamber is about to expand and exerts a force on the surrounding; however, the magma chamber cannot easily expand because the surrounding is rock. Both cooling and crystallization unavoidably progresses and magma vesiculation and bubble formation inevitably proceeds. In this manner, the pressure of the magma chamber is increasingly increased, and at a point (the critical state) where the surrounding rock cannot withstand the pressure with all its strength, the surrounding rock cracks and the magma flows out through the crack. The magma that allows to somewhat increase its volume is decompressed (i.e., solubility decreases and a degree of supersaturation increases) and increasingly vesiculates, leading to the increase in the fraction of a gas phase. The vesiculated magma decreases its density, expands the crack, and ascends towards the surface by buoyancy at an accelerated pace, which is the onset of an eruption.

Thus, second boiling has been considered as an overpressure that triggers eruptions (Bowen 1956). An evidence that second boiling occurred in a dike is seen in the rock and can be observed. Then, does an evidence that second boiling occurred in a magma chamber that triggers volcanic eruptions remain in eruptive materials? A previous section described vesicular texture in pumice and scoria formed by explosive eruptions. As shown there, vesicular texture is generally complex and to discover evidences for a specific elementary process in vesicular texture may require working hypothesis, theoretical and experimental support and some luck. However, it is important to observe vesicular texture by thinking that there is something. In the last chapter for applications, we will try to understand a process that triggers eruptions by observing vesicular texture of pumice from the viewpoint of pursuing an evidence of second boiling.

1.4.3 Magma Vesiculation as a Driving Force for Volcanic Explosions

Vesiculation process of magma is often said to be a driving force for volcanic explosions. Let us look at that qualitatively and intuitively here. The solubility of H_2O in melt increases with increase in pressure (Fig. 1.22, left). Therefore, with ascent and decompression of magma, H_2O dissolved in melt under high pressure becomes soluble, or the limit where H_2O can dissolve to the maximum reaches supersaturation

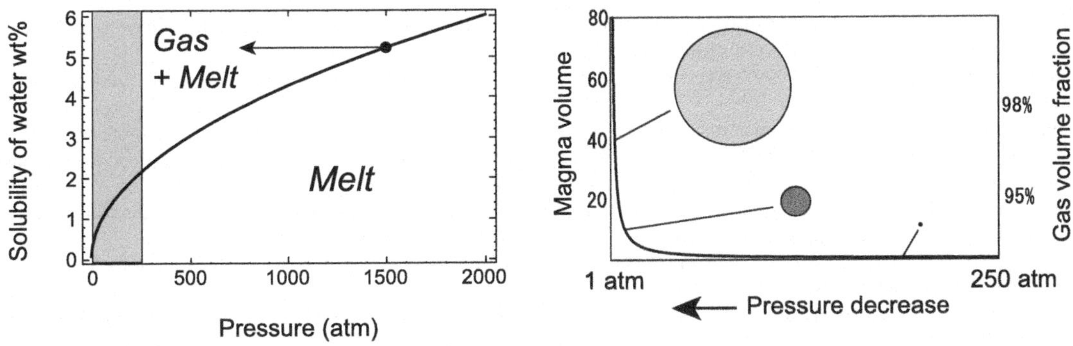

Fig. 1.22 The solubility of H_2O in melt and change in the volume of magma. The range of pressure in the right graph is of the range shown in the shaded square in the left graph of solubility. Drawn with enlarging from 40 atm

and separates from the magma as bubbles at a point. If the vesiculated magma continues ascending and being decompressed, the gas phase then expands. With the expansion of the gas phase, the volume of the magma containing bubbles increasingly grows as the magma approaches the surface. In particular, when the magma is close to the surface and the pressure further lowers, the volume violently enlarges as large as several dozen times what was at depths (Fig. 1.22, right). This is a source of explosive power. Let us recall that pumice and scoria produced from lava fountains and Plinian eruptions has expanded almost tenfold.

Moreover, bubbles coalesce with each other and are about to escape to the surface or to a lower pressure part by buoyancy by leaving melt underground. Coalescence and enlargement of bubbles is recorded in cross-sections of scoriae produced by Izu-Oshima lava fountain eruptions, as described earlier. Furthermore, the explosions of magma balloons on November 20 may be the ultimate state. Although bubbles retain their shape by surface tension as long as they are small, bubbles enlarged by growth, expansion, or coalescence can easily change their shape under the influence of magma flow and interaction with surrounding bubbles or crystals. This is qualitatively recognized from the picture of a cross-section of pumice or scoria, which indicates the relationship between the size and shape of bubbles. A variety of eruption styles seen at the surface is attributed to the involvement of such complex factors.

References

Bowen NL (1956) The evolution of the igneous rocks. Dover Publications

Carey S, Sigurdsson H (1989) The intensity of plinian eruptions. Bull Volcanol 51:28–40

Carey S, Sparks RSJ (1986) Quantitative models of the fallout and dispersal of tephra from volcanic eruption columns. Bull Volcanol 48:109–125

Cashman K, Blundy J (2000) Degassing and crystallization of ascending andesite and dacite. Phil Trans Roy Soc London A: Math Phys Eng Sci 358(1770):1487–1513

Hayakawa Y (1985) Pyroclastic geology of Towada volcano. Bull Earthq Res Inst Univ Tokyo 60:507–592

Ida Y (1996) Cyclic fluid effusion accompanied by pressure change: implication for volcanic eruptions and tremor. Geophys Res Lett 23(12):1457–1460

Japan Meteorological Agency (1987) Report on the research about volcano phenomena in natural disasters—the 1986 Izu-Oshima volcano

Machida H, Arai F (2003) Atlas of tephra in and around Japan. University of Tokyo Press, p 360. (in Japanese)

Maeda I (2000) Nonlinear visco-elastic volcanic model and its application to the recent eruption of Mt. Unzen J Volcanol Geotherm Res 95(1–4):35–47

Mannen K (1999) Total Grain Size Distribution and Total Mass of TB Tephra of the Izu-Oshima 1986 Eruption. Japan Kazan 44(2):55–70 (in Japanese with English abstract)

Melnik O, Sparks RSJ (1999) Nonlinear dynamics of lava dome extrusion. Nature 402:37–41

Minakami T (1942) On the distribution of volcanic ejecta (Part 1.). The distribution of volcanic bombs ejected by the recent explosions of Asama. Bull Earthq Res Inst 20:62–92
Nakada S, Shimizu H, Ohta K (1999) Overview of the 1990–1995 eruption at Unzen Volcano. J Volcanol Geotherm Res 89:1–22
Shimozuru D (1994) Physical parameters governing the formation of Pele's hair and tears. Bull Volcanol 56:217–219
Swanson DA, Rose TR, Fiske RS, McGeehin JP (2012) Keanakakoi Tephra produced by 300 years of explosive eruptions following collapse of Klauea's caldera in about 1500 CE. J Volcanol Geotherm Res 215:8–25
Toramaru A (1990) Measurement of bubble size distributions in vesiculated rocks with implications for quantitative estimation of eruption process. J Volcanol Geotherm Res 43:71–90
Toramaru A (2006) BND (bubble number density) decompression rate meter for explosive volcanic eruptions. J Volcanol Geotherm Res 154:303–316
Walker GPL (1980) The Taupo pmice: product of the most powerful known (ultraplinian) eruption? J Volcanol Geotherm Res 8:69–94
Wilson L, Sparks RSJ, Walker GPL (1980) Explosive volcanic eruptions—IV The control of magma properties and conduit geometory on eruption column behavior. Geophys J Roy Ast. Soc 63:117–148
Yoshida M (2008) Vesicle layering structures of dikes in Hirado Island, Nagasaki, Japan. Analysis of rock textures and proposal of the formation mechanism—Master thesis, Department of Earth and Planetary Science, Kyushu University, p 82. ((in Japanese with English abstract)

Part II
Vesiculation of Magma

In Chap. 1 of Part I, we learned from the textural observation of highly vesiculated volcanic rocks (pumice and scoria) that bubble formation in magma, i.e., vesiculation of magma, has an extremely close relationship with the appearances of eruptions. Moreover, we learned that the expansion of gas generated by magma vesiculation is a driving force for volcanic explosions and that magma vesiculation progresses by cooling, and solidification of magma plays a role in triggering eruptions. Conditions for magma vesiculation are dependent on the solubility of water (the major component of the gas phase that vesiculates in magma) in silicate melt. When magma satisfies the vesiculation conditions, bubbles form. The forming bubbles then grow and expand and after experiencing complex processes such as deformation and coalescence, they reach the surface or cool down and solidify under the ground. In Part II, we discuss the details of a series of the magma vesiculation process to understand physical processes underlying eruption phenomena and construct tools for interpreting natural eruptive materials.

Chapter 2 starts with the description of thermodynamics and equilibrium phase diagrams to understand conditions for magma vesiculation. Theoretically reproducing experimentally estimated equilibrium phase diagrams helps you to understand the thermodynamic significance of equilibrium phase diagrams. Furthermore, a system containing carbon dioxide is explained because you can understand phenomena such as degassing and fluxing from the point of view of gas species and pressure change in the system.

Chapter 3 explains the nucleation of bubbles, one of the elementary processes when vesiculation actually occurs. A phenomenon called nucleation is the transformation of the molecular states of volatiles from a microscopic world of molecules to a macroscopic world, as well as the unfamiliar ways of thinking, such as size space, dynamics of molecular clusters, and master equations, which will be used to deal with nucleation. It is the critical nucleus that is the important concept, which is based on the macroscopic equilibrium thermodynamics explained in Chap. 2 and is related to thermodynamics of fluctuation that handles the fluctuation of microscopic molecules. In that sense, the critical nucleus is a symbolic concept, showing that the nucleation process is related to both the microscopic and macroscopic worlds. While this chapter aims to derive the nucleation rates of bubbles, it is impossible to deny that somewhat complex calculation will be required to fundamentally understand the matter. Chapter 4 describes the growth of bubbles, which is involved with various constraints and elementary processes, such as viscous flow of silicate melts around bubbles, molecular diffusion in liquid, and equilibrium concentration at a gas-liquid interface controlled by gas pressure in bubbles. Behind this, there are thermodynamics and the problems of dynamics handling temporal change and mass transfer. This chapter aims to systematically understand the processes of growth and expansion of bubbles using theories for the processes using time constants and their ratio or controlling parameters and comparing the theoretical results with experimental results. Chapter 5 attempts to provide a complete picture of the vesiculation process of magma. For this purpose, bubble size distribution is explained by two description methods: the Eulerian view and the Lagrangian view. Although they are somewhat technical, these methods of calculation can provide physical meanings to vesicular texture finally observed at the surface. Chapter 6 focuses on various issues such as secondary growth and deformation of bubbles, coalescence of bubbles, exsolution, and detachment and ascent of bubbles by buoyancy from substrates, as well as collapse and oscillation of bubbles. The purpose of this chapter is, via such issues, to imagine the unobservable behavior of magma behind complex vesicular texture observed in nature and to obtain suggestions for many unsolved research topics.

Conditions for Magma Vesiculation 2

Abstract

In the previous chapter, we learned by observing highly vesiculated volcanic rocks (pumice and scoria) that bubble formation in magma, i.e., the vesiculation of magma, has an extremely close relationship with appearances of eruptions. Moreover, we learned that the expansion of gas generated by magma vesiculation is a driving force for volcanic explosions and that magma vesiculation progresses by cooling and solidification of magma as well as plays a role in triggering eruptions. Thus, the conditions (to be precise, the necessary conditions) for magma vesiculation, which play an important role in volcanic eruptions, are dependent on the solubility of water (H_2O) (the major component of the gas phase that vesiculates in magma) in silicate melt. Then, how has the solubility of water in melt been understood? Moreover, what types of conditions (such as temperature and pressure, etc.) influence the solubility? In this chapter, we will increase our understanding of the solubility of water in magma.

2.1 Significance of Equilibrium Theory

Equilibrium is the fundamental underlying consideration of disequilibrium processes, i.e., the kinetics of nucleation and growth of bubbles and crystals. Note that equilibrium is summarized into equilibrium relationships between phases as a function of temperature, pressure, and composition and is expressed via equilibrium phase diagrams or simply phase diagrams. The problem of disequilibrium arises in the process in which when temperature, pressure, or composition changes, a state shifts to another stability field after crossing one of the phase boundaries in an equilibrium phase diagram. One of the reasons why equilibrium theory is necessary is that stable phases can be identified and change of the combination of phases can be graphically understood based on equilibrium phase diagrams. Moreover, for considering kinetics, the evaluation of the force that drives a state to another new equilibrium state is necessary (the degree of supercooling or of supersaturation). However, it can be qualitatively understood by the location on the equilibrium phase diagram. Note that quantitative evaluation of the driving force is to estimate difference in free energy; moreover, the estimation should be based on thermodynamic considerations required for drawing equilibrium phase diagrams, which is another reason why equilibrium theory is required.

A. Toramaru, *Vesiculation and Crystallization of Magma*, Advances in Volcanology,
https://doi.org/10.1007/978-981-16-4209-8_2

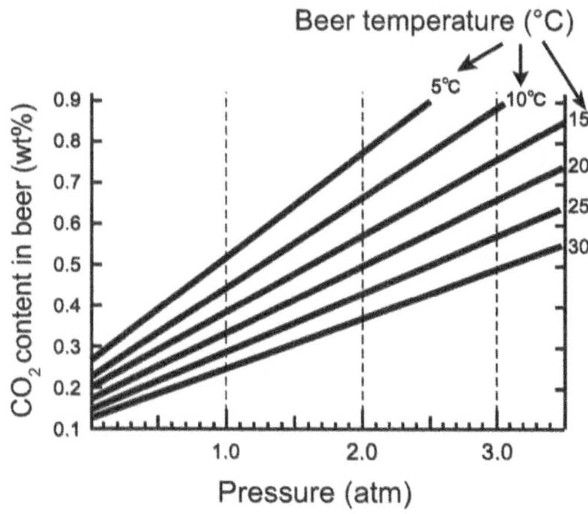

Fig. 2.1 The solubility of CO_2 in beer (Biiru no umasa wo saguru, edited by Kirin Brewery Company, Limited) Adapted with permission by Kirin Holdings Company, Limited

This chapter explains equilibrium theory related to the formation of bubbles of volatiles (water and carbon dioxide) in silicate melt because of changes in temperature and pressure. The stability field of a gas phase, defined by equilibrium theory, is defined by an equilibrium phase diagram or a solubility curve. The purpose of this chapter is to derive the solubility curve as a function of temperature, pressure, and concentration of volatiles.

2.2 Solubility of Gas Components in Liquid and Henry's Law

The fact that solubility, the maximum amount of gas components that can dissolve in liquid, increases with the pressure is expressed by the empirical Henry's law. This is well represented in Fig. 2.1, showing the solubility of carbon dioxide in beer because the pressure of gas P, i.e., motion of molecules in gas, is balanced with motion of gas molecules in liquid (in a sense of the average of molecular groups). Because the activity of a molecular group of a gas component in liquid is proportional to the concentration of the gas, C, the balance can be written as follows:

$$C = c_{\mathrm{h}} P \tag{2.1}$$

where c_{h} is a constant of proportionality.[1] This is the so-called Henry's law.[2] From this chart, in the case of beer and carbon dioxide, the constant of proportionality c_{h} is found to increase with decrease in temperature. Henry's law is called phase equilibria because it represents the equilibrium (balance) relationship of chemical components between a liquid phase and a gas phase. Although the absolute values of c_{h} cannot be determined only by thermodynamic consideration, they are related to an activity coefficient.

Then, how about the solubility of gas in silicate melt? For carbon dioxide, the relationship between the pressure and the solubility is linear as written in Eq. (2.1). However, in the case where water dissolves in silicate melt, the relationship is not linear as shown below. The relationship between the pressure and the solubility is as follows:

[1] Values of constants depend on units of concentration C and pressure P. Because units will vary according to, for example, ease of intuitive understanding in the problem setting and ease of handling in physical calculation, the explanation here will be continued without determining the unit.

[2] The inverse of c_{h}, i.e., a coefficient of an equation rewritten as the form of $P =$, is called the Henry coefficient or the Henry constant (sometimes c_{h} in $C = c_{\mathrm{h}} P$ is called the Henry constant).

$$C \propto P^m \tag{2.2}$$

In this equation, the exponent m is 1/2 at low pressure under about 100 MPa and approaches 1 with increase in pressure. This is because of a reaction occurring when H_2O dissolves in silicate melt as has been observed in the next section. As water is the predominant gas component contained in magma under island arcs erupting explosively,[3] let us examine this in more detail.

2.3 Dissolution Reaction of Water in Silicate Melt

In silicate melt, the solubility of water was systematically investigated by Burnham et al. (Burnham 1975). Fig. 2.2 shows the result in which the solubilities against pressure are represented by curves that do not follow the normal Henry's law, in which they should be straight lines. The reasons are explained below.

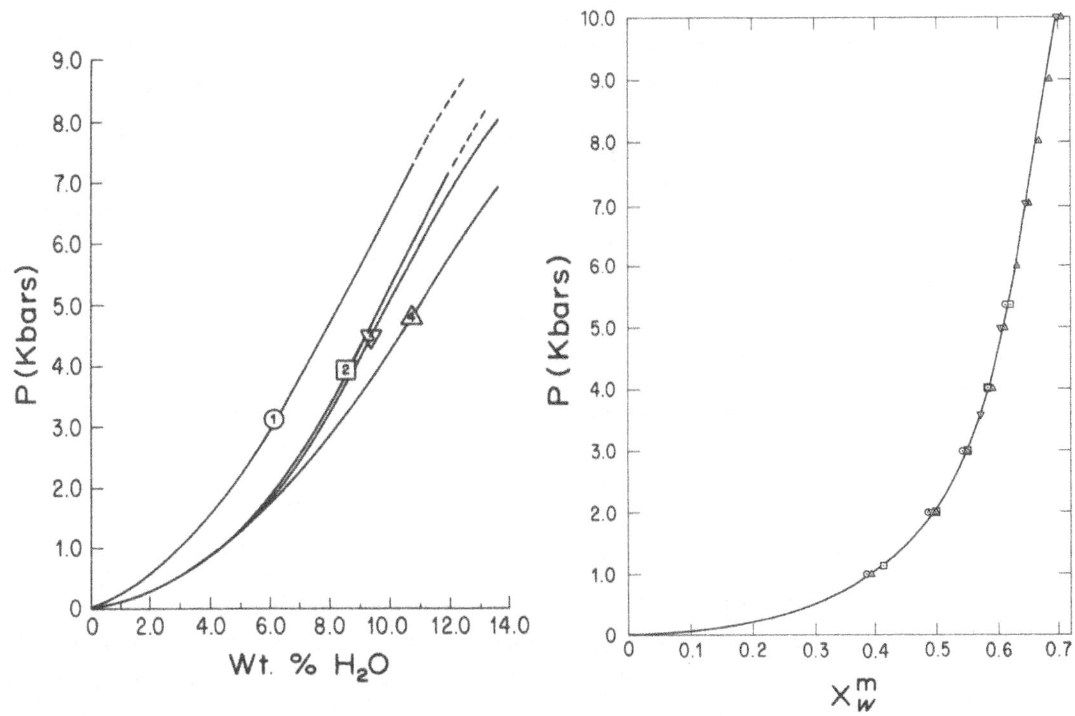

Fig. 2.2 The solubility of water in magma. The vertical axis represents pressure (kbars) (1 kbar = 100 MPa). The experimentally determined solubility of liquids having various compositions (Burnham 1979). The curve with the circle 1 is of basaltic melt (Columbia River basalt), the curve with the square 2 is of andesitic melt (Mt. Hood andesite), the curve with the inverted triangle 3 is of albite melt, and the curve with the triangle 4 is of granitic (rhyolitic) melt (Harding pegmatite). (Left chart) The horizontal axis represents water content in magma (wt%). (Right chart) Plot normalized by the reduced molarity by albite. The reduced molarity by albite ($NaAlSi_3O_8$) is the ratio of the water molarity to the Al molarity in actual melt. The method of calculation slightly varies according to whether water molar fraction is more than or less than 0.5. From EVOLUTION OF THE IGNEOUS ROCKS: Fiftieth Anniversary Perspectives by Hatten Schuyler Yoder. Copyright © 1979 by Princeton University Press. Reprinted by permission

[3]In magma from oceanic ridges, carbon dioxide is present more compared to water (Schmincke 2000).

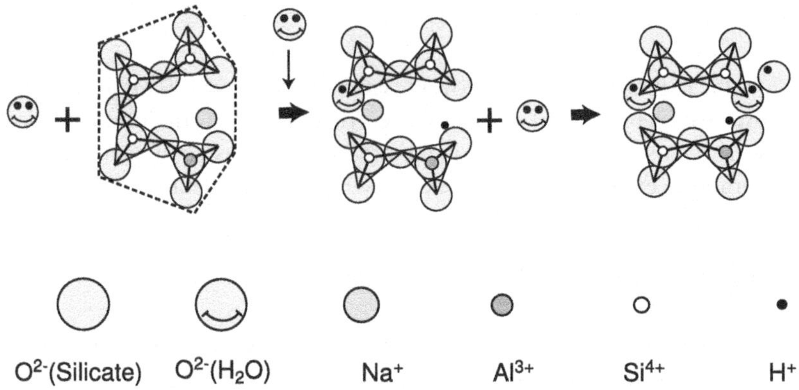

Fig. 2.3 Dissolution process of water molecules in magma (Burnham 1979). Note that a bridge between oxygen atoms at the left end is broken after the dissolution reaction. From EVOLUTION OF THE IGNEOUS ROCKS: Fiftieth Anniversary Perspectives by Hatten Schuyler Yoder. Copyright © 1979 by Princeton University Press. Reprinted by permission

The dissolution reaction of water in silicate melt comprises a process in which H_2O molecules in gas components dissolve in melt as they are

$$H_2O_{Gas} \rightarrow H_2O_{Melt} \tag{2.3}$$

a process in which water molecules dissociate into hydrogen ions and hydroxyl groups in melt, and a process in which the hydrogen ions react with oxygen ions of bridging oxygen.

$$H_2O_{Melt} \rightarrow H^+ + OH^- \tag{2.4}$$

$$H^+ + O^{2-} \rightarrow OH^- \tag{2.5}$$

Bringing the last two processes together yields the following reaction formula,

$$H_2O_{Melt} + O^{2-} \underbrace{\rightarrow}_{K} 2OH^- \tag{2.6}$$

which is called the dissociation reaction. Here, let us represent the equilibrium constant by K. The latter process is a process in which a water molecule is separated into two species, H^+ and OH^-; cuts networks of SiO_2; and makes bonds to oxygen belonging to bridging oxygen (Fig. 2.3). To provide direct evidence for this, it is necessary to detect molecular vibration, which confirms the formation of molecular bonding. However, to measure the viscosity of melt or the dependence of electric conductivity on water content, which are sensitive to the network structure of SiO_2, can be an alternative to prove it, although in an indirect manner. The first process occurs following the normal linear Henry's law, and the solubility linearly increases with increase in pressure. This can be known from the solubility of CO_2 in melt in which the second process is not involved (Fig. 2.14). The process that is predominant between these two processes depends on an equilibrium constant in the second process. As seen later, after all, under the condition where the second process predominately occurs (K is sufficiently large) at the relatively high temperature and relatively low pressure, this process controls the solubility.[4]

[4]Under high pressure conditions where water content is high, bridging oxygen that is supposed to be available in the dissociation reaction has been used by less content of water than the solubility, therefore, the second process will not occur.

Assuming that the dissociation reaction completely occurs, all water molecules change into two species, H^+ and OH^-, and dissolves in melt. That is, the second reaction controls the solubility, and the chemical equilibrium can be thermodynamically expressed by

$$\mu_{\mathrm{Melt}}^{\mathrm{H_2O}} + \mu_{\mathrm{Melt}}^{\mathrm{O^{2-}}} = 2\mu_{\mathrm{Melt}}^{\mathrm{OH^-}} \quad (2.7)$$

$$\ln a_{\mathrm{Melt}}^{\mathrm{H_2O}} + \ln a_{\mathrm{Melt}}^{\mathrm{O^{2-}}} - 2\ln a_{\mathrm{Melt}}^{\mathrm{OH^-}} = \mathrm{const.} \quad (2.8)$$

The latter equation can be rewritten as follows:

$$\frac{a_{\mathrm{Melt}}^{\mathrm{H_2O}} a_{\mathrm{Melt}}^{\mathrm{O^{2-}}}}{\left(a_{\mathrm{Melt}}^{\mathrm{OH^-}}\right)^2} = \mathrm{const.} \quad (2.9)$$

Here, assuming that solution is an ideal solution, because the activity of OH^- is proportional to the molar fraction of OH^- in liquid (see the Box of Sect. 2.5.1 on chemical potential), $a_{\mathrm{Melt}}^{\mathrm{OH^-}} \propto X_{\mathrm{Melt}}^{\mathrm{OH^-}}$ holds. The molar fraction of OH^- is equal to the molar fraction of water considering the dissociation of water (Eq. (2.4)), i.e., as $a_{\mathrm{Melt}}^{\mathrm{OH^-}} \propto X_{\mathrm{Melt}}^{\mathrm{H_2O}}$ holds and oxygen is sufficiently present, the molar fraction and the activity can be set at 1 (Stolper 1982). Therefore, Eq. (2.9) is rewritten as follows:

$$a_{\mathrm{Melt}}^{\mathrm{H_2O}} = c_{\mathrm{a}} \left(X_{\mathrm{Melt}}^{\mathrm{H_2O}}\right)^2 \quad (2.10)$$

where c_{a} is a coefficient.[5] This is the activity model by Burnham.

The Burnham model assumes that the solution has albite composition ($NaAlSi_3O_8$). In this case, dissociation reaction stoichiometrically occurs in proper quantities; therefore, the system can be assumed to be a binary system of water and albite. The occurrence of dissociation reaction was subsequently confirmed by studies on spectra of molecular bond vibration measured by NMR and FT-IR. Using this activity model, the solubility of water in silicate melt will be proportional to the square root of the pressure. The solubility in basalt (3 wt% at 1 Kb) is slightly lower than that in rhyolite (4 wt%).

Stolper (1982) proposed on the basis of FTIR data of water in silicate glasses that H_2O is dissolved in melt not in the form of only hydroxyl groups (OH^-), as in Burnham's model, but in the form of hydroxyl groups and molecular water (H_2O) because the equilibrium constant K of Eq. (2.6) is small. Because of the dissociation reaction and the solubility being 4 wt%, water will be present in the forms of hydroxyl groups and molecular water in an almost equal ratio (left chart of Fig. 2.4).

Subsequently, Nowak and Behrens (1995) and Shen and Keppler (1995) revealed using in situ observation of not-quenched melt at high temperature that this partitioning of reaction between hydroxyl groups and molecular water results from the quenching of melt into glass. Their research demonstrated that a reaction constant in the reaction formula (2.6), which determines the ratio of molecular water to hydroxyl groups, depends on the temperature and that the dependence considerably changes when crossing the boundary at a glass transition point (right chart of Fig. 2.4). It is considered that, at low

[5]Noted that this coefficient c_{a} is different from the activity coefficient γ_a appearing in general textbooks on thermodynamics. In this book, the coefficient c_{a} is handled as a constant independent of temperature and pressure. As seen below, the dependence of temperature and pressure is handled by being included in the chemical potential, which is a term other than the activity, at the standard state (Eq. (2.39)). However, the activity coefficient γ_a appearing in textbooks on thermodynamics is defined as being dependent on temperature and pressure, as described in the Box in Sect. 2.5.1. In any case, when calculating the chemical potential as a function of temperature and pressure, there is no problem as long as both temperature and pressure are related to partial molar quantity, which is a partial derivative of the chemical potential. The same result will be obtained in the end.

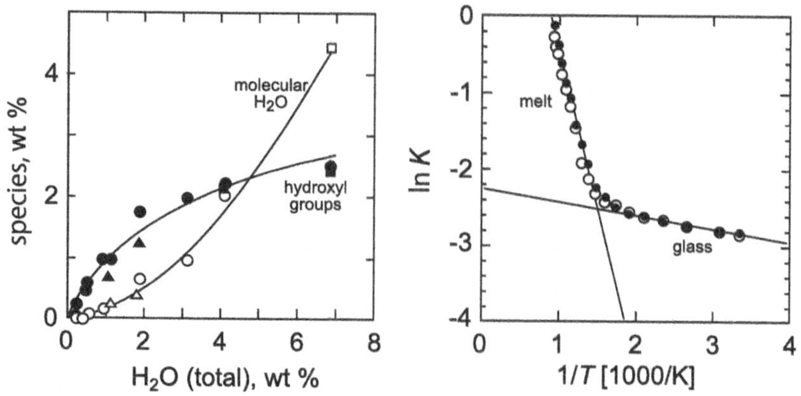

Fig. 2.4 (Left chart) The pressure dependence of the fractions of chemical species (molecular water and hydroxyl groups) when water dissolves in silicate melt (Stolper 1982). The horizontal axis shows the total water content (wt%) and the vertical axis shows the amount of each chemical species (wt%). Black circles and white circles denote rhyolitic glass, black triangles and white triangles denote basaltic glass, and black squares and white squares denote albite glass. Reprinted from (Stolper (1982)), Copyright 1982, with permission from Elsevier. (Right chart) The temperature dependence of the dissociation reaction constant $K = X_{OH}^2/X_{H_2O}X_O$ (Nowak and Behrens 1995). White circles (○) denote the heating experiment and black circles (•) denote the cooling experiment. Pressure, 150 MPa; water content, 4.14wt%. Reprinted from (Nowak and Behrens 1995), Copyright 1995, with permission from Elsevier

pressure (5 wt% of water content or lower) in which bridging oxygen is sufficiently present, most of the water dissolves in melt as hydroxyl groups. At higher pressure, albite species that react with hydroxyl groups have already been consumed such that water will dissolve as molecular water. Consequently, the dependence of the solubility on the pressure will be linear.

The temperature dependence of an equilibrium constant is related to change in Gibbs free energy $\Delta\mathcal{G}$, standard enthalpy $\Delta\mathcal{H}$, and entropy ΔS of the reaction via the following equation.[6]

$$\ln K = -\frac{\Delta\mathcal{G}}{\mathcal{R}T} = -\frac{\Delta\mathcal{H}}{\mathcal{R}T} + \frac{\Delta S}{\mathcal{R}}. \tag{2.12}$$

A slope shown in the right chart of Fig. 2.4 can be $\Delta\mathcal{H}/\mathcal{R}$ and an intercept $\Delta S/\mathcal{R}$. From experimental results, the fact that $\Delta\mathcal{H} = 33.6$ kJ/mol and $\Delta S = 29.8$ J/mol at a melt state (Nowak and Behrens (2001) revised these values as 35.0 and 27.2, respectively) and $\Delta\mathcal{H} = 1.52$ kJ/mol and $\Delta S = -18.6$ J/mol K at a low-temperature glass state was obtained (Nowak and Behrens 1995). The values for K are approximately 1 at 1000 K and 1.85 at 1500 K (Nowak and Behrens (2001) revised these values as 0.416 and 1.69, respectively).

[6] A differential form of this equation

$$\frac{d\ln K}{dT} = \frac{\Delta\mathcal{H}}{\mathcal{R}T^2} \tag{2.11}$$

is called the van't Hoff equation.

The beginning of thermodynamics and the three laws of thermodynamics

Thermodynamics was developed in the 17th century along with the industrial revolution. It started as a science to elucidate what is heat, or the actual state of heat, and to answer the practical question of how efficiently we can extract mechanical work from heat via a heat engine.[a] As demonstration between experiment and theory, thermodynamics was built up as a science to respond to both social demands with industrial revolution and to consistently understand natural thermal phenomena. An end point of thermodynamics as a basic science is that all natural thermal phenomena can be explained if a small number of variables and three laws are assumed. The small number of variables include intensive variables (quantities that do not depend on the system size and show the intensity of physical contribution) such as temperature T, pressure P, and chemical potential μ, as well as extensive variables (quantities that depend on the system size) such as entropy S, volume V, and the number of molecules N. The precise descriptions of the three laws is left to other textbooks on thermodynamics; the symbolic and brief descriptions as a basis of thermodynamics are as follows:

The first law: The internal energy $\mathcal{E}$ of a system[b] changes by the heat $\mathcal{Q}$ flowing into the system and the work $\mathcal{W}$ done on the system.

$$\Delta \mathcal{E} = \mathcal{Q} + \mathcal{W} \tag{2.13}$$

The second law: When a system changes, the entropy S does not change or increases.

$$\Delta S \geq 0 \tag{2.14}$$

The third law: The entroepy of a perfect crystal is zero at zero absolute temperature.

[a]This story is written in detail in *Butsurigaku towa nandarouka* in Japanese (*What will be physics* in English) by Shin-Itiro Tomonaga, and *Personaggi e scoperte nella fisica classica: dalla caduta dei gravi alle onde elettromagnetiche* by Emilio Segrè.

[b]A system is a substance or a collection of substances being focused on.

2.4 Change of Solubility with Pressure: Burnham's Model

Figure 2.5a schematically shows the change of the solubility as a function of the pressure (same pressures in both gas and liquid phase). In a system containing water at the water content corresponding to point A, water completely dissolves in melt, i.e., water is in saturation at the pressure corresponding to state (point) A. At a pressure higher than the pressure corresponding to point A, more water can dissolve in melt. The saturation concentration decreases with decrease in pressure P (the curve AB). When the pressure is decreased, the system changes to the direction represented by the vector AX. At a pressure lesser than the pressure corresponding to point A, the system shifts to the supersaturated state in which, thermodynamically speaking, melt and gas (non-zero amount of gas phase must be present) coexist in an equilibrium. This is decompression-induced vesiculation, and volatiles dissolving in magma separate out and vesiculate as magma ascends.

Because water dissolving in melt is considered to almost dissociate into hydroxyl groups at high temperature and relatively low pressure, let us derive the solubility as a function of pressure using Burnham's activity model (Eq. (2.10)) based on the thermodynamics of phase equilibrium between the gas phase and the liquid phase. This corresponds to the determination of AB and the phase boundary

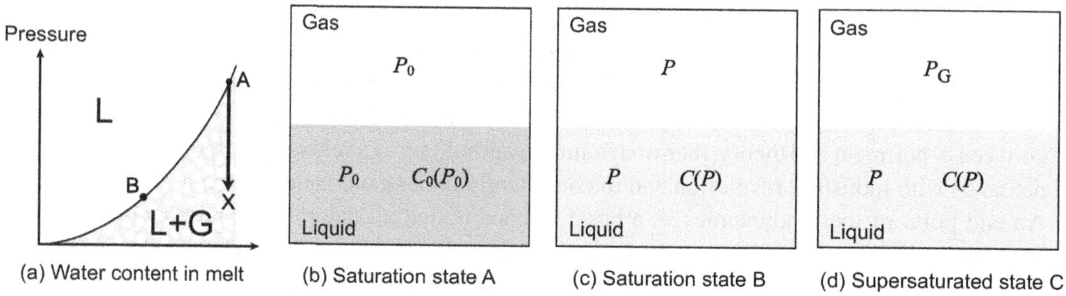

Fig. 2.5 **a** Phase equilibrium relationship as a function of pressure—water content. Corresponding to the solubility of H_2O in melt shown in Fig. 2.2. **b** Saturation state at the condition A. **c** Saturation state at the condition B. **d** Equilibrium state C when the gas phase and the liquid phase have different pressures

curve of the gas-liquid equilibrium. First, assuming that water completely dissolves in melt as hydroxyl groups at low pressure (given a constant temperature (T_0)), gas-liquid equilibrium conditions (see the Box in this section) at saturation state A (start point, Fig. 2.5b) and the saturation state B (arbitrary point, Fig. 2.5c) are expressed as follows:

$$\mu_{\mathrm{Melt}}^{\mathrm{H_2O}}(P_0, X_0) = \mu_{\mathrm{Gas}}^{\mathrm{H_2O}}(P_0) \tag{2.15}$$

$$\mu_{\mathrm{Melt}}^{\mathrm{H_2O}}(P, X_{\mathrm{eq}}(P)) = \mu_{\mathrm{Gas}}^{\mathrm{H_2O}}(P) \tag{2.16}$$

where μ is the chemical potential. X is the molar fraction of water in liquid and has a relationship of $X = C_0/N_0$ (where N_0 is the reduced number of molecules of albite per unit volume of melt (Fig. 2.2)) with the water content C_0 (the number of molecules per unit volume of melt. $\tilde{C}$ denotes wt%.). Subscripts eq and 0 denote values in equilibrium and initial values, respectively. Considering $X_0 = X_{\mathrm{eq}}(P_0)$, $X_{\mathrm{eq}}/X_0 = (C_{\mathrm{eq}}/N_0)/(C_0/N_0) = C_{\mathrm{eq}}/C_0$ holds. Subtracting Eq. (2.16) from Eq. (2.15) yields

$$\mu_{\mathrm{Melt}}^{\mathrm{H_2O}}(P_0, X_0) - \mu_{\mathrm{Melt}}^{\mathrm{H_2O}}(P, X_{\mathrm{eq}}(P)) = \mu_{\mathrm{Gas}}^{\mathrm{H_2O}}(P_0) - \mu_{\mathrm{Gas}}^{\mathrm{H_2O}}(P) \tag{2.17}$$

As the left-hand side (L.H.S) passes through the intermediate state (P, X_0), $0 = -\mu_{\mathrm{Melt}}^{\mathrm{H_2O}}(P, X_0) + \mu_{\mathrm{Melt}}^{\mathrm{H_2O}}(P, X_0)$ is inserted in the left-hand side.

$$\begin{aligned} \text{L.H.S.} =\ & \mu_{\mathrm{Melt}}^{\mathrm{H_2O}}(P_0, X_0) - \mu_{\mathrm{Melt}}^{\mathrm{H_2O}}(P, X_0) \\ & + \mu_{\mathrm{Melt}}^{\mathrm{H_2O}}(P, X_0) - \mu_{\mathrm{Melt}}^{\mathrm{H_2O}}(P, X_{\mathrm{eq}}(P)) && (2.18) \\ =\ & \int_{\mu(P, X_0)}^{\mu(P_0, X_0)} d\mu + \int_{\mu(P, X_{\mathrm{eq}})}^{\mu(P, X_0)} d\mu && (2.19) \\ =\ & \int_P^{P_0} \left(\frac{\partial \mu}{\partial P}\right)_{X_0} dP + \int_{X_{\mathrm{eq}}}^{X_0} \left(\frac{\partial \mu}{\partial X}\right)_P dX && (2.20) \\ =\ & \int_P^{P_0} v_{\mathrm{m}}(X_0) dP + k_{\mathrm{B}} T_0 \left[\ln a(X_0) - \ln a(X_{\mathrm{eq}})\right] && (2.21) \\ =\ & v_{\mathrm{m}}(X_0)(P_0 - P) + k_{\mathrm{B}} T_0 \ln \left(\frac{X_0}{X_{\mathrm{eq}}}\right)^2 && (2.22) \end{aligned}$$

in which Burnham's activity model (Eq. (2.10)) is used. Moreover the right hand side (R.H.S.) becomes below:

$$\text{R.H.S} = \int_{P}^{P_0} v_{\text{G}} dP = k_{\text{B}} T_0 \ln \frac{P_0}{P} \tag{2.23}$$

where v_{G} is the volume of one molecule of a gas component in the gas phase, and the gas phase is assumed to be an ideal gas satisfying the following equation

$$P v_{\text{G}} = k_{\text{B}} T \tag{2.24}$$

Furthermore, the partial molar volume of H_2O in melt vm is assumed to be constant.[7] As the left-hand side equals the right-hand side,

$$\frac{C_{\text{eq}}}{C_0} = \left(\frac{P}{P_0}\right)^{\frac{1}{2}} \exp\left[\frac{v_{\text{m}} P_0}{2k_{\text{B}} T_0}\left(1 - \frac{P}{P_0}\right)\right] \tag{2.25}$$

is yielded, which represents the relationship between the solubility and the pressure. The term of exp is, as seen from the derivation procedure, the effect of compressibility of water molecules in the melt (the so-called Poynting correction; see Sect. 3.6.2). Neglecting this term yields a well-known relationship in which the solubility is proportional to the square root of the pressure (the simplified Burnham model).

$$C_{\text{eq}}(P) = C_0 \left(\frac{P}{P_0}\right)^{\frac{1}{2}} \tag{2.26}$$

Phase equilibrium conditions and conservation laws under disequilibrium state

The conditions under which phase I and a phase II are in equilibrium are determined by the fact that thermodynamic potential (Appendix (Chap. 11) 11.1) for selected independent variables is at the minimum. This is equivalent to the conditions in which the entropy becomes the maximum (Callen 1960), e.g., in a system in which the temperature and the pressure are constant, the conditions are determined by the fact that change in the sum total of Gibbs free energy of the phase I and the phase II will be the local minimum ($\delta\mathcal{G} = \delta\mathcal{G}_{\text{I}} + \delta\mathcal{G}_{\text{II}} = 0$). Consequently, it is concluded that the equilibrium conditions are such that intensive variables in the two phases are equal to each other.

$$T_{\text{I}} = T_{\text{II}} \quad \text{thermal equilibrium} \tag{2.27}$$

$$P_{\text{I}} = P_{\text{II}} \quad \text{mechanical equilibrium} \tag{2.28}$$

$$\mu_{\text{I}}^i = \mu_{\text{II}}^i \quad \text{chemical equilibrium} \tag{2.29}$$

These equilibrium conditions mean thermal equilibrium, mechanical equilibrium, and chemical equilibrium; they are plainly expressed by intensive variables such as temperature, pressure, and chemical potential of each component i. These three equilibrium conditions are, as seen below, related to three conservation laws, i.e., the conservation of energy, conservation of momentum, and conservation of mass.

---

[7] Although the partial molar volume v_{m} is the volume when X_0 of water dissolves, it is not written as $v_{\text{m}}(X_0)$ but as simply v_{m}. v_{m} is a function of pressure in general, and the subsequent calculation will be more precise if the functions given in Burnham and Davis (1974) and Burnham (1975) are used. A more precise solubility can be calculated by giving these molar volumes as a function of temperature and pressure and then numerically integrating them.

In the disequilibrium state, differences in intensive variables cause flows of energy, momentum, and mass. The simplest setting for the magnitude of these flows is that the flux I (flow per unit area per unit time) is proportional to difference in an intensive variable as a driving force, Δ.

$$I_T \propto \Delta T \quad \text{energy flow} \tag{2.30}$$

$$I_P \propto \Delta P \quad \text{momentum flow} \tag{2.31}$$

$$I_\mu \propto \Delta \mu \quad \text{mass flow} \tag{2.32}$$

The relationships between these driving forces and the flows are, in the case of continuum, known as Fourier's law for heat ($I_T = -k_T dT/dx$), the stress-strain rate relation (constitutive equation) for momentum, and Fick's law for mass. Moreover, the same holds true for the relationship between voltage and electric current. These quantities are conserved everywhere within a continuum even if they flow. For example, the heat capacity $\rho C_P T$ has to satisfy the following general conservation law in a 1D space.

$$\rho C_P \frac{\partial T}{\partial t} = -\frac{\partial I_T}{\partial x} \tag{2.33}$$

Substituting Fourier's law into I_T in the equation above gives the heat conduction equation (thermal diffusion equation). We can understand from these that the thermodynamic understanding of the equilibrium state is the fundamental basis of general conservation laws.

2.5 Solubility When Bubbles and Liquid Have Different Pressure

2.5.1 General Case Where Bubbles and Liquid Are Not in Mechanical Equilibrium

For subsequent discussion, this section derives chemical equilibrium relationship, i.e., saturation concentration in the general case where pressure of a gas phase and that of a liquid phase differ (state C in Fig. 2.5d). The fact that pressure of gas and that of liquid differ indicates that, if the gas-liquid interface is flat, they are not in a mechanical equilibrium state (see the Box in the previous section). When the interface is curved, such as bubbles, even if gas and liquid are in mechanical equilibrium, the pressure of the gas and that of the liquid differ because of the interfacial tension, and the gas pressure has a special relationship with the liquid pressure via a bubble radius. This case will be handled later. Moreover, in other words, in the case where they are not in mechanical equilibrium via a flat interface, i.e., where the interface is in motion, the pressure of bubbles differs from that of liquid. The pressure within bubbles at the time is determined by an equation of motion for bubble expansion, i.e., time evolution of the bubble radius, which can be described by the Rayleigh-Plesset equation, mentioned later, and an equation of state of the gas.

Now letting the gas pressure be $P_{\rm G}$, Eq. (2.23) should be integrated from $P_{\rm G}$ to P_0, as seen from the discussion above, while Eqs. (2.20)–(2.22) will be integrated from P to P_0, with no change. Consequently, in Eq. (2.25), pressure in the term of exp is the pressure in liquid P and pressure in the term before exp will be the pressure of gas $P_{\rm G}$. Therefore, Eq. (2.25) can be rewritten as follows:

$$\frac{C_{\rm eq}(P_{\rm G})}{C_0} = \left(\frac{P_{\rm G}}{P_0}\right)^{\frac{1}{2}} \exp\left[\frac{v_{\rm m} P_0}{2k_{\rm B} T_0}\left(1 - \frac{P}{P_0}\right)\right] \tag{2.34}$$

To clarify the relationship of this with saturation concentration $C_{\mathrm{eq}}(P)$ of bubbles that is in dynamic equilibrium with an arbitrary liquid pressure P via a flat interface, given by Eqs. (2.25), (2.34) is divided by Eq. (2.25). Consequently, the ratio of equilibrium concentration $C_{\mathrm{eq}}(P_G)/C_{\mathrm{eq}}(P)$ in the case where the gas pressure P_{G} and the liquid pressure P differ is given as follows:

$$C_{\mathrm{eq}}(P_{\mathrm{G}}) = \left(\frac{P_{\mathrm{G}}}{P}\right)^{\frac{1}{2}} C_{\mathrm{eq}}(P) \tag{2.35}$$

When examining this equation on the basis of $C_{\mathrm{eq}}(P)$, the solubility in gas—liquid mechanical equilibrium at a pressure P, and considering the influence of the gas pressure $P_{\mathrm{G}}(\neq P)$), the solubility is found to be dependent on the power of 1/2 of the pressure ratio P_{G}/P.

Chemical potential and Gibbs free energy

Change in chemical potential of a component i is attributed to the infinitesimal changes in temperature, pressure, and molar fraction.

$$d\mu_i = -s_i dT + v_i dP + \left(\frac{\partial \mu_i}{\partial X_i}\right) dX_i \tag{2.36}$$

s_i and v_i in this equation are defined by the following, which are equal to partial molar entropy and partial molar volume, respectively.

$$s_i = -\left(\frac{\partial \mu_i}{\partial T}\right)_{P,X_i} \quad v_i = \left(\frac{\partial \mu_i}{\partial P}\right)_{T,X_i} = \left(\frac{\partial V}{\partial N_i}\right)_{P,T,N_{j\neq i}} \tag{2.37}$$

The last equation can be obtained from an equation on Gibbs free energy in Maxwell's relations. $(\partial V/\partial N_i)_{P,T,N_{j\neq i}}$ defines the partial molar volume of the component i. Moreover, when temperature, pressure, and molar fraction are given, chemical potential can be given by

$$\mu_i = \mu_0 + k_{\mathrm{B}} T \ln a(T, P, X_i) \tag{2.38}$$

where μ_0 is chemical potential of a pure substance under standard conditions. Other general textbooks describe that in an ideal solution the activity coefficient γ_a in the activity $a(T, P, X_i) = \gamma_a X_i$ will be a function of temperature and pressure.[a] This book adopts the following expression to separate dependence on composition and that on temperature and pressure.

$$\mu_i = \mu_{0,i}(T, P) + k_{\mathrm{B}} T \ln a(X_i) \tag{2.39}$$

The activity $a(X_i)$ in this equation is a function only of the composition in melt (the molar fraction of each component, X_i). $\mu_{0,i}(T, P)$, however, is the chemical potential of a pure substance ($X_i = 1$), is a function of temperature and pressure, and is independent of the composition.

[a]Partial molar volume $v_i = (\partial \mu_i/\partial P) = (\partial \mu_0/\partial P) + k_{\mathrm{B}} T(\partial \ln \gamma_a/\partial P)$ holds. The first term gives the molar volume of a pure substance and the second term, the derivative of the activity coefficient, gives difference between the molar volume and the partial molar volume. Using data of the partial molar volume, this effect will be considered in this expression, even if the activity coefficient is handled as a constant.

Chemical potential is Gibbs free energy per unit molecule and because Gibbs free energy is $\mathcal{G}$ and the number of molecules of each composition is N_i,

$$\mathcal{G} = \sum_{i=1}^{n} N_i \mu_i \tag{2.40}$$

is obtained. Therefore, chemical potential can be defined as partial molar Gibbs energy as follows:

$$\mu_i = \left(\frac{\partial \mathcal{G}}{\partial N_i} \right)_{T,P,N_{j \neq i}} \tag{2.41}$$

2.5.2 Case Where Bubbles and Liquid Are in Dynamic Equilibrium

Because pressure of gas and that of liquid is different indicates they are not in mechanical equilibrium if the gas—liquid interface is flat. However, if the gas exists as bubbles with curved interfaces, they can be in mechanical equilibrium via the interfacial tension. In the mechanical equilibrium, the pressure of gas is uniquely related to the liquid pressure and the bubble radius via the interfacial tension by the following Laplace equation (Sect. 4.6 on the Rayleigh-Plesset equation).

$$P_{\mathrm{G}} - P = \frac{2\gamma}{R} \tag{2.42}$$

This equation indicates that the pressure of the interior of bubbles is raised by the contribution of the interfacial tension, $2\gamma/R$, as compared with the surrounding liquid pressure. This can be really realized by remembering the situation in which a balloon bursts.

Thus, because the pressure P_{G} within the bubbles is a function of the radius R (i.e., $P_{\mathrm{G}}(R)$), the equilibrium concentration $C_{\mathrm{eq}}(P_{\mathrm{G}}(R)) = C_R(R)$,[8] can be obtained by replacing $C_{\mathrm{eq}}(P_{\mathrm{G}}(R))$ by $C_R(R)$ in Eq. (2.34). Here, P_{G} in the light hand side of that equation is given by Eq. (2.42). Moreover, using the equilibrium concentration and the Laplace Equation (2.42), for the equilibrium concentration $C_{\mathrm{eq}}(P)$ (so-called solubility) when mechanical equilibrium holds through the flat interface under the pressure P, the equilibrium concentration $C_{\mathrm{eq}}(P)$ at the interface of a bubble with the radius R

$$C_R(R) = \left(1 + \frac{1}{P} \frac{2\gamma}{R} \right)^{\frac{1}{2}} C_{\mathrm{eq}}(P) \tag{2.43}$$

holds. This is the Gibbs-Thomson equation, which represents the interface with smaller radius of curvature, thus yielding a higher equilibrium concentration (Fig. 2.6).

Note that while in the diffusion-controlled region of bubble growth, mechanical equilibrium almost holds, mechanical equilibrium comes to be lost in the viscosity-controlled region. If not in mechanical equilibrium, Eq. (2.35), in which the interfacial equilibrium concentration is considered as a function of the gas pressure rather than a function of the bubble radius, it should be used for interfacial equilibrium concentration.

[8]Write $C_R(R)$ when we clearly express C as a function of R.

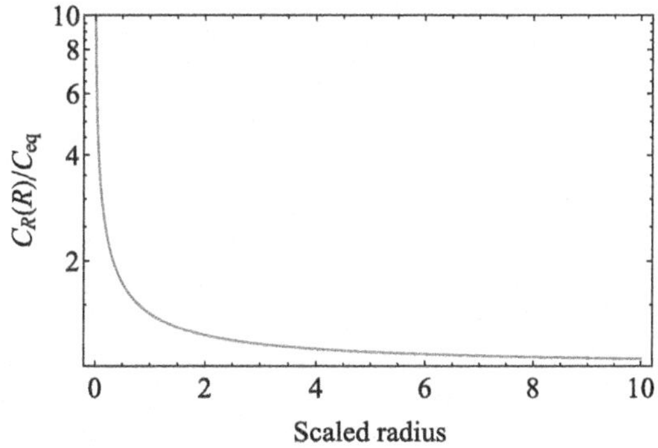

Fig. 2.6 Equilibrium concentration (Eq. (2.43)) at a bubble interface in dynamic equilibrium in supersaturated solution. The bubble radius is scaled by $2\gamma/P$

2.6 Pressure Dependence of Solubility of Water in the Case of Incomplete Dissociation

Up to this point, we have handled the case in which water molecules are completely dissociated following Burnham's model. However, water molecules are, in fact, dissociated into molecular water and hydroxyl groups following the equilibrium constant of dissociation reaction. Let us verify the influence by the following procedure. When considering dissolution reaction as molecules (Eq. (2.3)) and dissociation reaction (Eq. (2.6)) because we cannot know the details of the dissociation reaction in liquid, we will evaluate it via the equilibrium constant. As a course of action, the pressure dependence of dissolution as molecular water will be thermodynamically calculated and a quantity ratio of molecular water and hydroxyl groups involved in dissociation reaction is connected by the equilibrium constant.

Consider first molecular water. In the integral directly determining the pressure dependence of the solubility (Eq. (2.21)), assuming the activity of an ideal solution for molecular water $a(X) = X$, the solubility for molecular water (corresponding to Eq. (2.25)) is given by the following:

$$\frac{X_{\mathrm{m}}}{X_{\mathrm{m0}}} = \left(\frac{P}{P_0}\right) \exp\left[\frac{v_{\mathrm{m}} P_0}{k_{\mathrm{B}} T_0}\left(1 - \frac{P}{P_0}\right)\right] \tag{2.44}$$

In this equation, the dissociation of water molecules is not yet considered. Therefore, the term of (P/P_0) and the term of exp are not raised to the power of 1/2. Equilibrium molar fraction X_{m} is the molar fraction of H_2O, which dissolves as molecular water and X_{m0} is its reference value.

In the case where dissociation reaction occurs but dissociation is incomplete, water dissolves as hydroxyl groups, the molar fraction of which is denoted as X_{OH}. Because two hydroxyl groups form one water molecule (and one oxygen atom), the number of molecules of water is half of that of hydroxyl groups. The molar fraction of the total water content X_{W} is the sum of the molar fraction of molecular water and that of hydroxyl groups, i.e.,

$$X_{\mathrm{W}} = X_{\mathrm{m}} + 0.5 X_{\mathrm{OH}} \tag{2.45}$$

Assume that equilibrium constant K corresponding to dissociation reaction (Eq. (2.6)) is

$$K = \frac{X_{\mathrm{OH}}^2}{X_{\mathrm{m}} X_{\mathrm{Oxy}}} \tag{2.46}$$

X_{Oxy} shows the molar fraction of oxygen contributing to the reaction in liquid and can be approximated as $X_{\mathrm{Oxy}} = 1$ (Stolper 1982). By this equation, X_{OH} is related to X_{m} and substituting it into Eq. (2.45)

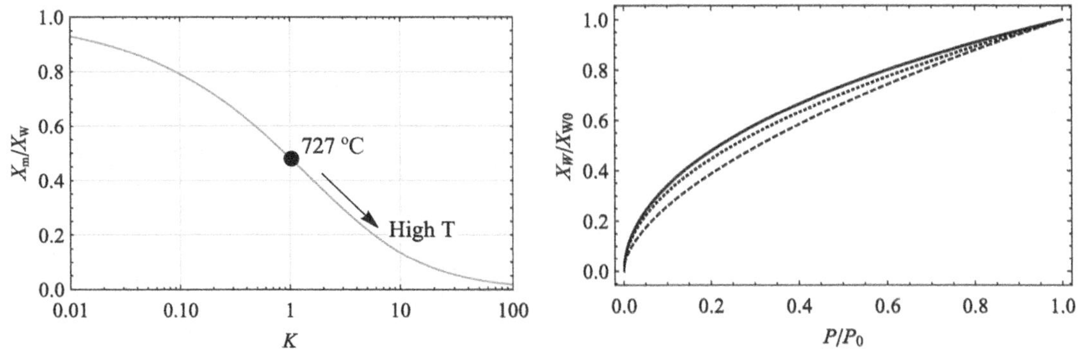

Fig. 2.7 (Left chart) Relationship between the equilibrium constant K and the fraction of molecular water in melt. Assuming that the total water content $X_W = 0.45 \approx 4.14$ wt%. According to Nowak and Behrens (1995), because at 150 MPa K is about 1 at 1000 K (revised value of K by Nowak and Behrens (2001) shifts to 1163 K for $K = 1$. So, related discussion will be modified.) and soars at higher temperature, it is considered that in magma under conditions related to eruptions, water is almost completely decomposed into hydroxyl groups as the first approximation. (Right chart) Pressure-dependence of the solubility of water in magma. Values of water content and pressure are normalized by values at the initial saturation pressure (150 MPa, $X_W = 0.45 \approx 4.14$ wt%). The solid line represents Burnham's model (Eq. (2.25)), the dotted line the simplified Burnham's model (Eq. (2.26)), and the broken line the reaction model in the case of $K = 2$

gives the relationship between the total molar fraction of dissolving water X_W and the molar fraction of molecular water X_m as follows.

$$X_W = X_m + 0.5K^{\frac{1}{2}} \cdot X_m^{\frac{1}{2}} \tag{2.47}$$

The quantity ratio of molecular water X_m to the total water content X_W becomes zero when K approaches infinity, as shown in the left chart of Fig. 2.7. In other words, all of the water dissolve in melt in the form of hydroxyl groups and that corresponds to Burnham's model. When $K \to 0$, all of the water dissolves in the form of molecular water and that corresponds to the linear Henry's law. K has an obvious temperature dependence as seen earlier. The dependence on composition or pressure is an issue to be addressed in future.

Because the total molar fraction of water and the molar fraction of molecular water at the reference state $P = P_0$ are X_{W0} and X_{m0}, respectively,

$$X_{W0} = X_{m0} + 0.5\, K^{\frac{1}{2}} \cdot X_{m0}^{\frac{1}{2}} \tag{2.48}$$

holds. Normalizing X_W and X_m by X_{W0} gives $\tilde{X}_W = X_W/X_{W0}$ and $\tilde{X}_m = X_m/X_{W0}$, and their relation can be written as follows:

$$\tilde{X}_W = \tilde{X}_m + 0.5\, K_0^{\frac{1}{2}} \cdot \tilde{X}_m^{\frac{1}{2}} \tag{2.49}$$

where K_0 is

$$K_0 = \frac{K}{X_{W0}} \tag{2.50}$$

Equation (2.44) can then be rewritten as follows:

$$\tilde{X}_m = \left(\frac{X_{m0}}{X_{W0}}\right) \cdot \left(\frac{P}{P_0}\right) \exp\left[\frac{v_m P_0}{k_B T_0}\left(1 - \frac{P}{P_0}\right)\right] \tag{2.51}$$

$$\frac{X_{m0}}{X_{W0}} = \frac{1}{4}\left[-0.5\, K_0^{1/2} + (0.25\, K_0 + 4)^{1/2}\right]^2 \tag{2.52}$$

From Eq. (2.49), in the case where dissociation reaction completely occurs, $\tilde{X}_m^{1/2} \ll 0.5K_0^{1/2}$, i.e., when $X_m \ll 0.25K$, we will determine that it can be approximated by the simplified Burnham's model. That is, $\tilde{X}_W \approx 0.5K_0^{1/2} \cdot \tilde{X}_m^{1/2}$, $\tilde{X}_m = (P/P_0) \cdot \exp((1 - P/P_0)v_m P_0/(k_B T_0))$ is obtained, which is equal to Eq. (2.25).

The pressure-dependent change of the solubility in a reaction model can be obtained from Eqs. (2.49)–(2.52) by giving a value of X_W at K and P_0 (X_{W0}). Based on the high-temperature state, $K > 1$ is assumed, which corresponds to 0.4 or less of a fraction of molecular water (left chart of Fig. 2.7). In the right chart of Fig. 2.7, some solubility models explained so far are shown and compared to each other. They show similar trends but there are differences in solubility by about 20%.

2.7 Change in Solubility by Temperature

Let us examine the influence of temperature on equilibrium relationship between silicate melt and water. Figure 2.8 shows an experimentally estimated equilibrium phase diagram. Figure 2.9a shows a schematic equilibrium phase diagram, made with reference to experimental results, with temperature on the vertical axis and water content in the system on the horizontal axis. Moreover, this is one of the eutectic systems, and liquidus of water as one of the end members, has a negative slope . Note that pressure of the system is constant, i.e., we consider an isobaric process ($P = P_0 = P_G$) with increasing temperature (the vector AY in Fig. 2.9a). This corresponds to a state in which low-temperature felsic (Al- or Si-rich) magma is heated by high-temperature mafic (Fe- or Mg-rich) magma when they are mixed with each other at the magma mixing event. Moreover, in this case, the state shifts from a saturation state to a supersaturation state and vesiculation occurs. Here, the temperature dependence of the solubility of water is important. The following a gas-liquid phase boundary F_H-E_H in Fig. 2.9a (i.e., curves in the left chart of Fig. 2.8 and L/L + V boundary in the right chart of Fig. 2.8) is thermodynamically determined.

Describing the equilibrium relationship in the state A (Fig. 2.9b) using temperature and composition (the molar fraction of water) as independent variables yields

$$\mu_{\text{Melt}}^{H_2O}(T_0, X_0) = \mu_{\text{Gas}}^{H_2O}(T_0) \tag{2.53}$$

Similarly, the equilibrium relationship in the state D (Fig. 2.9c) will be

$$\mu_{\text{Melt}}^{H_2O}(T, X_{eq}(T)) = \mu_{\text{Gas}}^{H_2O}(T) \tag{2.54}$$

Subtracting each other of two equations above yields

$$\mu_{\text{Melt}}^{H_2O}(T_0, X_0) - \mu_{\text{Melt}}^{H_2O}(T, X_{eq}(T)) = \mu_{\text{Gas}}^{H_2O}(T_0) - \mu_{\text{Gas}}^{H_2O}(T) \tag{2.55}$$

As both the two independent variables change when shifting from the start point (T_0, X_0) to (T, X_{eq}), we assume an intermediate point (T, X_0) and rewrite the left-hand side (L.H.S.) as follows:

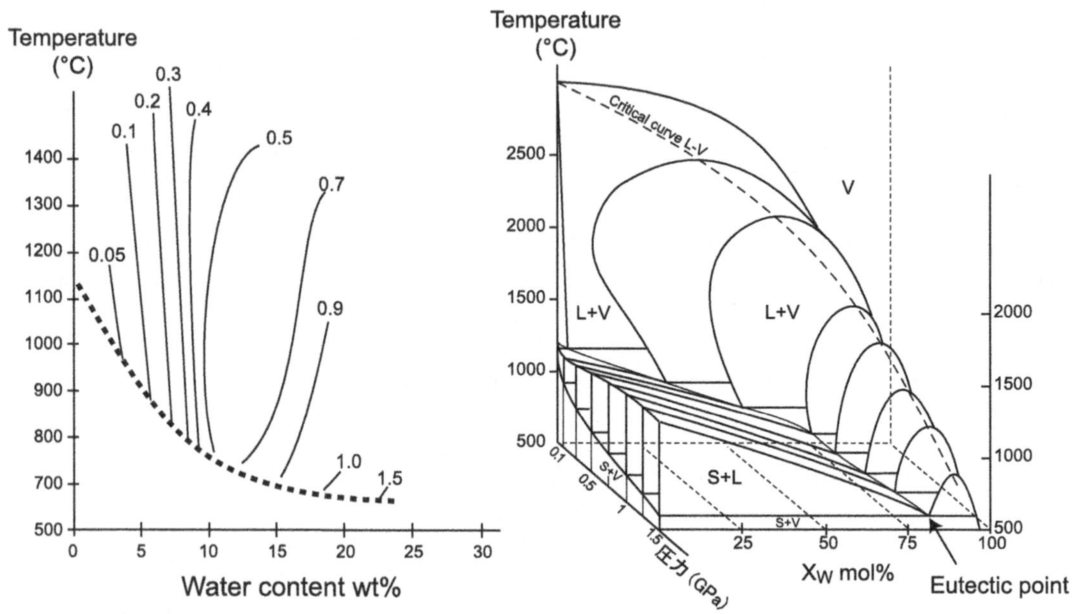

Fig. 2.8 Experimentally estimated equilibrium phase diagram of the albite-H_2O binary eutectic system (Paillat et al. 1992). (Left chart) Pressure-dependent change of liquid (L) /liquid + vapor (L+V) boundaries (solid lines) in the low water content region. Numbers above the lines denote pressure (GPa). The dotted line represents pressure-dependent change of eutectic points. It shows that eutectic points move to higher temperature and lower H_2O content with decrease in pressure. (Right chart) Temperature-pressure-molar fraction of water equilibrium phase diagram in a 3D space. The vapor phase and the liquid phase are not distinguished in higher pressures making a single phase of mixture, and critical points are generated. A line connecting the critical points is represented by a broken line as a critical curve L-V. Adapted by permission from Springer Nature (Paillat et al. 1992), COPYRIGHT (1992)

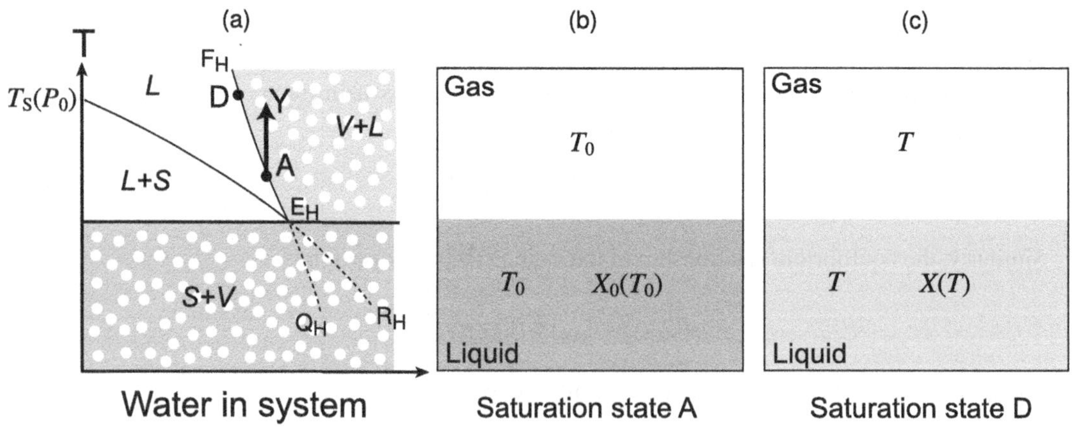

Fig. 2.9 **a** Phase equilibrium relation seen as a function of temperature-water content. **b** Saturation state at the point A in (**a**). **c** Saturation state at the point D in (**a**). Assuming isobaric processes

$$\text{L.H.S.} = \mu_{\text{Melt}}^{H_2O}(T_0, X_0) - \mu_{\text{Melt}}^{H_2O}(T, X_0) + \mu_{\text{Melt}}^{H_2O}(T, X_0) - \mu_{\text{Melt}}^{H_2O}(T, X_{\text{eq}}(T)) \tag{2.56}$$

$$= \int_T^{T_0} \left(\frac{\partial \mu_{\text{Melt}}^{H_2O}}{\partial T} \right)_{X_0} dT + \int_{X_{\text{eq}}}^{X_0} \left(\frac{\partial \mu_{\text{Melt}}^{H_2O}}{\partial X} \right)_T dX \tag{2.57}$$

$$= -s_{\text{Melt}}^{H_2O}(X_0)(T_0 - T) + k_B T \left[\ln a(X_0) - \ln a(X_{\text{eq}}) \right] \tag{2.58}$$

where $s_{\mathrm{Melt}}^{\mathrm{H_2O}}(X_0)$ is entropy at X_0, i.e., at the reference state.[9] Moreover, the right-hand side (R.H.S.) will be

$$\text{R.H.S.} = \int_T^{T_0} \left(\frac{\partial \mu_{\mathrm{Gas}}^{\mathrm{H_2O}}}{\partial T} \right)_{X_{\mathrm{eq}}} dT = -s_{\mathrm{Gas}}^{\mathrm{H_2O}}(T_0 - T) \tag{2.62}$$

Here, using Burnham's activity model (Eq. (2.10)) in Eq. (2.58) gives

$$T - T_0 = -\frac{k_\mathrm{B} T}{\Delta s} \ln \left(\frac{X_{\mathrm{eq}}}{X_0} \right)^2 \tag{2.63}$$

where Δs is $s_{\mathrm{Gas}}^{\mathrm{H_2O}} - s_{\mathrm{Melt}}^{\mathrm{H_2O}}(X_0)$. Given the latent heat $\Delta \mathcal{H} = N_{\mathrm{Avogadro}} \cdot \Delta s \cdot T = 20$ kJ/mol, plotting Eq. (2.63) yields Fig. 2.10, in which experimental results by Yamashita (1999) are plotted and can be explained.[10]

Finally, in Burnham's model in which dissociation reaction completely proceeds, the solubility of water as a function of the temperature is given by

$$\frac{C_{\mathrm{eq}}}{C_0} = \frac{X_{\mathrm{eq}}}{X_0} = \exp \left[-\frac{\Delta h_0}{2 k_\mathrm{B} T_0} \left(1 - \frac{T_0}{T} \right) \right] \tag{2.66}$$

Giving consideration to incomplete dissociation reaction, we can use here the procedure used in the pressure dependence. To be more specific, the equilibrium of reaction in which water dissolves in melt in the form of molecular water can be described instead of Eq. (2.63) by

$$T - T_0 = -\frac{k_\mathrm{B} T}{\Delta s_1} \ln \left(\frac{X_\mathrm{m}}{X_0} \right) \tag{2.67}$$

---

[9] If we assume an intermediate point (T_0, X_{eq}),

$$\begin{aligned} \text{L.H.S.} =& \mu_{\mathrm{Melt}}^{\mathrm{H_2O}}(T_0, X_0) - \mu_{\mathrm{Melt}}^{\mathrm{H_2O}}(T_0, X_{\mathrm{eq}}(T)) \\ &+ \mu_{\mathrm{Melt}}^{\mathrm{H_2O}}(T_0, X_{\mathrm{eq}}(T)) - \mu_{\mathrm{Melt}}^{\mathrm{H_2O}}(T, X_{\mathrm{eq}}(T)) && (2.59) \\ =& \int_{X_{\mathrm{eq}}}^{X_0} \left(\frac{\partial \mu_{\mathrm{Melt}}^{\mathrm{H_2O}}}{\partial X} \right)_{T_0} dX - \int_{T_0}^{T} \left(\frac{\partial \mu_{\mathrm{Melt}}^{\mathrm{H_2O}}}{\partial T} \right)_{X_{\mathrm{eq}}} dT && (2.60) \\ =& k_B T_0 \left[\ln a(X_0) - \ln a(X_{\mathrm{eq}}) \right] + s_{\mathrm{Melt}}^{\mathrm{H_2O}}(X_{\mathrm{eq}})(T - T_0) && (2.61) \end{aligned}$$

is obtained and it will be difficult to calculate because X_{eq} appears in the term $s_{\mathrm{Melt}}^{\mathrm{H_2O}}(X_{\mathrm{eq}})$, other than the term of the activity.

[10] Here, Δs is change in entropy per molecule associated with water exsolution and corresponds to the latent heat associated with vesiculation Δh_0.

$$\Delta h_0 = T_0 \Delta s(X_0) = T_0 \left(s_{\mathrm{Gas}}^{\mathrm{H_2O}} - s_{\mathrm{Melt}}^{\mathrm{H_2O}}(X_0) \right) \tag{2.64}$$

$$\frac{dT}{dX_{\mathrm{eq}}} = -2 \frac{k_\mathrm{B} T^2}{\Delta h_0 X_{\mathrm{eq}}} \tag{2.65}$$

As seen from the equations above, in an endothermic reaction $\Delta h_0 > 0$, i.e., $dT/dX_{\mathrm{eq}} < 0$ holds. Therefore, the slope of the boundary of L/L+V corresponding to a liquidus of a eutectic system phase diagram in a temperature—composition (albite—water) space becomes negative.

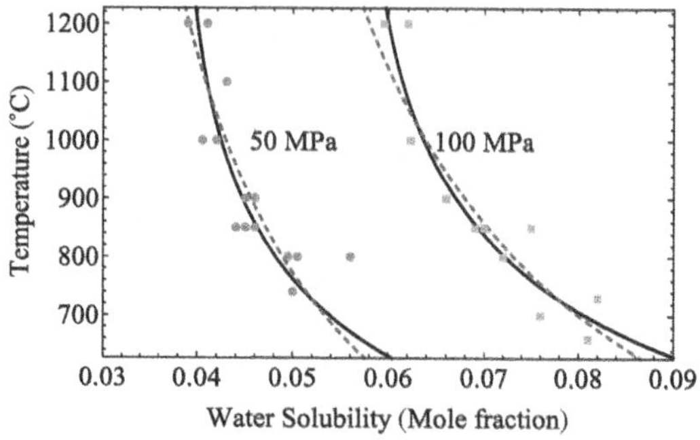

Fig. 2.10 Comparison between the experimental data by Yamashita (1999) and Eq. (2.63). Assuming that $X_0 = 0.078$ at $T_0 = 1000$ K at 100 MPa and $X_0 = 0.052$ at $T_0 = 1000$ K at 50 MPa. The molar fraction here is defined by the ratio to one atom of oxygen. The solid line represents Burnham's model (Eq. (2.63)). The broken line represents a calculation result obtained by the reaction model ($\Delta\mathcal{H}_1 = 13$ kJ/mol, $K = 1$)

where Δs_1 is entropy of reaction in which water dissolves in the form of molecular water (2.3)) and molar enthalpy at T_0 in the dissolution reaction as molecular water is $\Delta\mathcal{H}_1 = \Delta s_1 T_0 \times N_{\mathrm{Avogadro}}$. Using them, we obtain $\tilde{X}_\mathrm{m}$ corresponding to Eq. (2.51) as

$$\tilde{X}_\mathrm{m} = \left(\frac{X_{\mathrm{m}0}}{X_{\mathrm{W}0}}\right) \exp\left[-\frac{\Delta\mathcal{H}_1}{\mathcal{R}T_0}\left(1 - \frac{T_0}{T}\right)\right] \tag{2.68}$$

Substituting Eqs. (2.68) and (2.52) into Eq. (2.49) yields the solubility. Note that the equilibrium constant K is assumed to be constant regardless of temperature and the enthalpy of the dissociation reaction is assumed to be constant.[11] (Figure 2.10). In the dissociation model, the latent heat of dissolution of molecular water $\Delta\mathcal{H}_1$ is estimated at 13 kJ/mol and the heat of the dissociation reaction of water molecules is estimated at 33 kJ/mol. In the case where $K = 1$, shown in Fig. 2.10, because about 20% of water remains as molecular water, the total effective latent heat is $-13 + 0.8 \times 33 = 13.4$ kJ/mol. In the simplified Burnham's model, however, the latent heat of dissolution in melt $\Delta\mathcal{H}$ is assumed to be 20 kJ/mol. This value is somewhat smaller than the latent heat of dissolution of molecular water in the dissociation reaction model.

From the discussion above (Eqs. (2.25) and (2.66))), we can more generally describe the influence of pressure and temperature on the solubility of water using Burnham's model as the following equation of the solubility of water as a function of temperature and pressure:

$$\frac{C_\mathrm{eq}}{C_0} = \left(\frac{P}{P_0}\right)^{\frac{1}{2}} \exp\left[\frac{v_\mathrm{m} P_0}{2k_\mathrm{B} T_0}\left(1 - \frac{P}{P_0}\right) + \frac{\Delta h_0}{2k_\mathrm{B} T_0}\left(\frac{T_0}{T} - 1\right)\right] \tag{2.69}$$

2.8 The Influence of Water on Melting Points of Crystals and Decompression-Vesiculation Induced Crystallization

In general, the freezing point of a pure solid $T_\mathrm{S}(P_0)$ (corresponding to an end point of the liquidus of a binary system) (Fig. 2.9) decreases when another component is added (freezing point depression). The freezing point of crystals in magma decreases when water is added. This is represented by the solid—

[11] If considering temperature dependence of the dissociation reaction constant, the results do not change much. The reason is that at a temperature above 1000 K, the dissociation proceeds 60% or more.

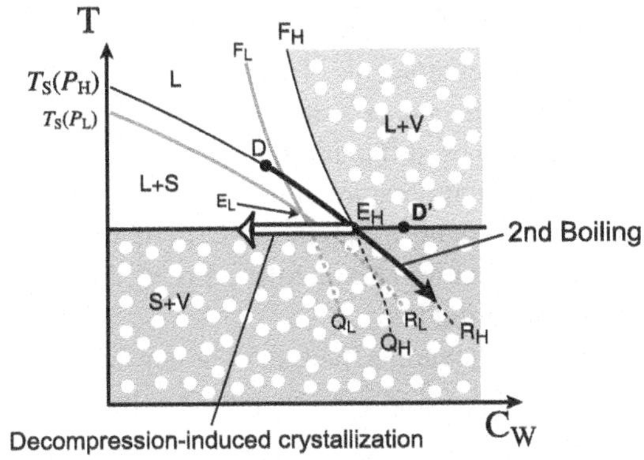

Fig. 2.11 Phase equilibrium reaction seen as a function of temperature—water content. Representing the case in which $P_0 = P_H$ and $P_0 = P_L$ in Fig. 2.9. The relative path of decompression-vesiculation induced crystals (decompression crystals) to the phase boundary is indicated by a gray arrow. The path in which vesiculation occurs by cooling crystallization (second boiling) is indicated by a black arrow

liquid phase boundary curve $T_S(P_H)$- E_H in Fig. 2.11. Moreover, freezing point depression is dependent on pressure. Taking the albite—H_2O system as an example, let us examine it thermodynamically (Burnham 1979).

When we deal with the crystallization process of albite in the albite-water system, we can think the equilibrium condition in the albite—H_2O system comprising a single-component (albite) solid phase and a two-component liquid phase. This is the same procedure as the equilibrium of the above-examined system comprising a single-component gas phase and a two-component liquid phase can be used. First, we handle chemical equilibrium under constant-pressure conditions to obtain the relationship between freezing points (liquidus T_L) and water content. Subsequently, it will be extended to the case where water content is determined by chemical equilibrium involving decompression-induced vesiculation.

When the albite—H_2O system is considered as a binary eutectic system, chemical equilibrium conditions of albite—liquid are given by

$$\mu_{\text{Solid}}(\text{pure Ab}, T, P) = \mu_{\text{Melt}}(\text{pure Ab}, T, P) + k_B T \ln a_{\text{Melt}}^{\text{Ab}} \tag{2.70}$$

Rewriting it yields

$$k_B T \ln a_{\text{Melt}}^{\text{Ab}} = \mu_{\text{Solid}}(\text{pure Ab}, T, P) - \mu_{\text{Melt}}(\text{pure Ab}, T, P) \tag{2.71}$$

Generally, based on a reference state (T_0, P_0), the chemical potential μ at a certain temperature and pressure (T, P) can be given as $\mu(T, P) = \mu(T_0, P_0) - s(T - T_0) + v(P - P_0)$. Assuming that the heat of fusion (per unit molecule) of albite at the reference state is Δh_{Ab},

$$\begin{aligned}\Delta h_{\text{Ab}} = \ & \mu_{\text{Melt}}(\text{pure Ab}, T_0, P_0) - \mu_{\text{Solid}}(\text{pure Ab}, T_0, P_0) \\ & + (s_{\text{Solid}}^{\text{Ab}} - s_{\text{Melt}}^{\text{Ab}})T_0\end{aligned} \tag{2.72}$$

holds. Therefore, the conditions of chemical equilibrium are as follows:

$$k_B T \ln a_{\text{Melt}}^{\text{Ab}} = -(\Delta h_{\text{Ab}} - T\Delta s_{\text{Ab}} + P\Delta v_{\text{Ab}}) \tag{2.73}$$

where Δs_{Ab} is fusion entropy per molecule of albite at the reference state ($s^{\text{Ab}}_{\text{Solid}} - s^{\text{Ab}}_{\text{Melt}}$) and Δv_{Ab} is volume change associated with fusion per molecule of albite at the reference state. Because the reference state is the melting point T_{Ab} of dry albite at a pressure of 1 atm, $\Delta h_{\text{Ab}} = \Delta s_{\text{Ab}} T_{\text{Ab}}(X^{\text{W}}_{\text{Melt}} = 0)$ holds. In Eq. (2.73), pressure P is assumed to be sufficiently larger than a pressure of 1 atm.

In Eq. (2.73), temperature T is the temperature at the chemical equilibrium (albite liquidus: T^{Ab}_{L}) and shows the relationship between pressure P and the activity of albite a^{Ab}_{M}. Moreover, it is related to the molar fraction of water, as shown below, and therefore Eq. (2.73) shows the relationship between the molar fraction of water and albite liquidus.

Replacing $a^{\text{Ab}}_{\text{Melt}}$ in Eq. (2.73) with the molar fraction of water $X^{\text{W}}_{\text{Melt}}$ (the same as X_{W} so far) gives T^{Ab}_{L} as a function of albite liquidus and $X^{\text{W}}_{\text{Melt}}$. For this purpose, the following Gibbs-Duhem relation is used.

$$s dT - v dP + \Sigma n_i d\mu_i = 0 \tag{2.74}$$

Given that T and P are constant, applying it to a liquid phase yields

$$X^{\text{W}}_{\text{Melt}} d \ln a^{\text{W}}_{\text{Melt}} + X^{\text{Ab}}_{\text{Melt}} d \ln a^{\text{Ab}}_{\text{Melt}} = 0 \tag{2.75}$$

To this equation, using the simplified Burnham's model and a self-evident conservation law

$$a^{\text{W}}_{\text{Melt}} = c_{\text{a}} \left(X^{\text{W}}_{\text{Melt}} \right)^2, \tag{2.76}$$

$$X^{\text{W}}_{\text{Melt}} + X^{\text{Ab}}_{\text{Melt}} = 1 \tag{2.77}$$

gives

$$a^{\text{Ab}}_{\text{Melt}} = \left(X^{\text{Ab}}_{\text{Melt}} \right)^2 = \left(1 - X^{\text{W}}_{\text{Melt}} \right)^2 \tag{2.78}$$

Here, because the activity of albite (X^{Ab}_{Melt}) at dry state equals 1, the coefficient is 1. Using this equation for Eq. (2.73)) and arranging it with respect to T yields

$$T^{\text{Ab}}_{\text{L}}(X^{\text{W}}_{\text{Melt}}, P) = \frac{\Delta h_{\text{Ab}} + P \Delta v_{\text{Ab}}}{\Delta s_{\text{Ab}} - k_{\text{B}} \ln(1 - X^{\text{W}}_{\text{Melt}})^2} \tag{2.79}$$

which shows freezing point depression caused by adding water. Figure 2.12 shows this equation graphically.

Freezing point depression, described so far, conversely indicates that the melting point of crystals increases with decrease in water content. Therefore, when water exsolves from melt as magma vesiculates, melt crystallizes, which is known as decompression-induced crystallization. Substituting the solubility of water as a function of pressure into the molar fraction of water $X^{\text{W}}_{\text{Melt}}$ in Eq. (2.79) yields the melting point as a function of partial pressure of water (water saturation pressure) in vesiculated magma. Moreover, substituting the relationship between the solubility of water and pressure into the term of pressure yields the melting point as a function only of water content under water saturation conditions. For example, using the solubility of water obtained by assuming that T is constant and neglecting the term of exp

$$P = P_0 \left(\frac{\tilde{C}}{\tilde{C}_0} \right)^2 \tag{2.80}$$

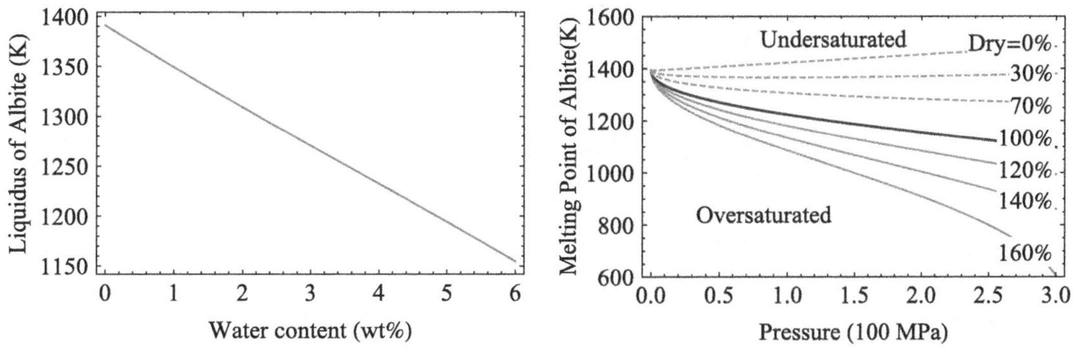

Fig. 2.12 The influence of the water content in melt on the melting point of albite. (Left chart) The melting point of albite against the water content in the case where the water content changes according to the solubility relation (equilibrium dehydration). (Right chart) Change of the melting point of albite against the pressure in the case where the water content is undersaturated, saturated and oversaturated in the solubility. Dry means anhydrous melt. The melting point of albite at a pressure of 1 atm is 1391 K, the latent heat is 62 kJ/mol, and Δv is 1.37×10^{-5} m^3/mol (calculated with 2200 kg/m^3 as the density of albite melt, 2500 kg/m^3 as the solid phase of albite, and 0.252 kg as the molecular weight). The reference point of the solubility of water in melt is assumed to be dacite melt, i.e., 6 wt% ($= \tilde{C}_0$) and 0.5 ($= X_0$) as the molar fraction at 200 MPa($= P_0$) (Fig. 2.2)

yields the following approximate expression.

$$T_{\mathrm{L}}^{\mathrm{Ab}}(\tilde{C}) = \frac{\Delta h_{\mathrm{Ab}} + P_0 \Delta v_{\mathrm{Ab}}(\tilde{C}/\tilde{C}_0)^2}{\Delta s_{\mathrm{Ab}} - k_{\mathrm{B}} \ln(1 - X_0 \tilde{C}/\tilde{C}_0)^2} \tag{2.81}$$

Moreover, substituting Eq. (2.69), the solubility of water as a function of temperature and pressure, into Eq. (2.79), we can obtain an equation that describes how pressure changes with decrease in temperature in a system where three phases, i.e., crystals, gas, and melt, coexist in equilibrium. This is useful for estimating increase in pressure in magma chambers where cooling crystallization occurs at depth.

In an equilibrium phase diagram, the decompression-induced crystallization passes via the eutectic point of the gas phase and albite from the gas—liquid coexisting region to the gas—solid coexisting region. The eutectic point is defined by the intersection curve of solubility surface (Eq. (2.69)) and the liquidus surface (Eq. (2.79)). As temperature does not change in isothermal decompression-induced vesiculation, the liquidus and solubility curves themselves relatively move with decompression-induced vesiculation in each temperature—concentration plane in the equilibrium phase diagram (Relative relationship is shown in Fig. 2.13). In other words, the gas—liquid phase equilibrium curve (F_{H}-Q_{H}) moves toward lower water content, and the solid—liquid phase equilibrium curve (T_{S}-R_{H}) moves toward lower temperature. Because the degree of the latter is small, point A that maintains the gas—liquid equilibrium at pressure $P5$ in the beginning goes via the eutectic point $E_{\mathrm{H}}(P4)$ of gas, liquid, and solid at $P4$, which is lower than $P5$ in pressure, and enters the region supersaturated with the solid phase, where crystallization occurs (decompression-induced crystallization). It undergoes nucleation at a certain degree of supercooling (a certain degree of supersaturation) and then approaches a eutectic point determined by pressure ($E_{\mathrm{H}}(P2)$ in this figure) via bubble growth and crystal growth. The increase in temperature results from the latent heat of crystallization, and the degree of the increase in temperature depends on combination of chemical equilibrium and conservation of energy in the crystallization. Thus, in nature, both vesiculation and crystallization occur as magma ascends, and the mechanical properties of magma change and thereby affect eruption styles. Change of the eutectic point $E_{\mathrm{H}}(P)$ as a function of water

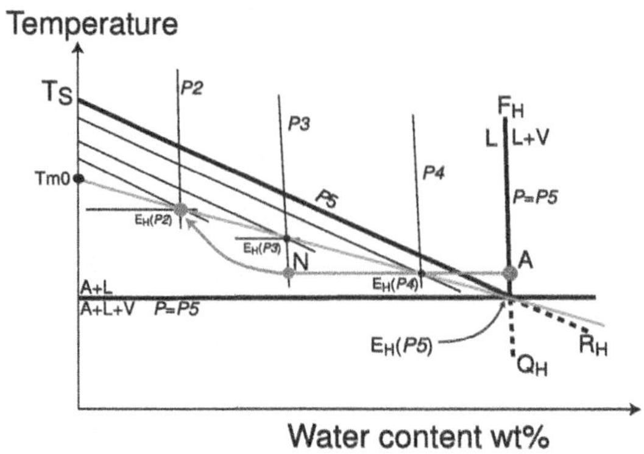

Fig. 2.13 Change in the equilibrium phase diagram associated with decompression, equilibrium, and vesiculation and vesiculation/crystallization paths. T_{m0} is the melting point of a pure solid at a pressure of 1 atm. N is a point where crystals nucleate. $E_H(P5) \sim E_H(P2)$ corresponds to the dotted line in the left chart of Fig. 2.8

content is shown in Fig. 2.13. This is a projection of the locus of the eutectic point in the temperature—pressure—water content phase diagram (Fig. 11.1 in the Appendix (Chap. 11)) onto the temperature—water content plane.

2.9 Concentration of Volatiles and Vesiculation Caused by Cooling Crystallization

When water does not dissolve in the solid phase (crystals), H_2O concentrates in the remaining liquid as cooling crystallization proceeds. Let us look this in the equilibrium phase diagram show in Fig. 2.11. As liquid starts cooling crystallization from point D, liquid changes along the liquidus, crosses the vapor-liquid equilibrium curve at the albite—water eutectic point E_H, and enters a stable condition of the liquid phase + vapor phase. Such a phenomenon of vesiculation with cooling is known as second boiling. This phenomenon exactly resembles the situation in which when you make ice cubes in a freezer, air that cannot dissolve in ice appears as small bubbles and makes the ice cubes smoky.

In cooling the crystallization of magma, the volume decreases if only crystallization proceeds; therefore, the magma chamber is decompressed. As crystallization further proceeds, the temperature of magma decreases further below the magma-volatile eutectic point, magma gets supersaturated with volatiles, the magma then vesiculates, and it then gets overpressurized.

This overpressurization process is controlled by the mechanical properties of rocks surrounding the magma chamber and the rate of crystallization. When the surrounding rocks can freely deform or magma can freely move or expand such overpressurization does not occur. However, in a situation where magma cannot expand at all because the surrounding rocks are rigid, pressure of magma increases with crystallization. The temperature, pressure, and water content in the residual liquid (i.e., the total amount of crystals) at that time change toward lower temperature and higher pressure along the eutectic point. As far as an equilibrium state is assumed, when crystallization initiatively proceeds by temperature change (i.e., cooling), vesiculation dependently proceeds and the system is at the eutectic point where the vapor phase, liquid phase, and solid phase coexist. In fact, a path on the equilibrium phase diagram deviates from the eutectic point because the kinetics of crystal growth and bubble growth has an influence because of time scale. A summary of the equilibrium phase diagram of the water-albite system at low water content is shown in Fig. 11.1 in the Appendix (Chap. 11).

2.10 A System Containing Carbon Dioxide

2.10.1 Solubility of Carbon Dioxide

The dissolution of carbon dioxide into silicate melt only comprises the following dissolution reaction.

$$CO_{2Gas} \rightarrow CO_{2Melt} \tag{2.82}$$

Therefore, the solubility of carbon dioxide in silicate melt increases linearly with the gas pressure in accordance with Henry's law. Moreover, at low pressure, the solubility of carbon dioxide is known to be much lower than the solubility of water (Fig. 2.14), i.e.,

$$\tilde{C}_{CO_2}(P_G) = c_{CO_2} P_G \tag{2.83}$$

holds. For intuitive understanding, we use the unit wt% for the unit of the content of CO_2 and write it as $\tilde{C}_{CO_2}$ (in this book C is used when the unit is the number of molecules per unit volume of melt). The unit of pressure P_G is Pa. In this case, a constant of proportionality c_{CO_2} (the inverse of Henry's constant) is 5×10^{-10} (wt%/Pa) for rhyolite. Note that the constant of proportionality is known to considerably vary according to melt composition, whereas that of ultramafic melt varies twofold to threefold (Papale 1997; Newman and Lowenstern 2002 and references therein).

Nevertheless, a simplified solubility of water can be expressed similarly as follows:

$$\tilde{C}_{H_2O}(P_G) = c_{H_2O} P_G^{\frac{1}{2}} \tag{2.84}$$

The constant of proportionality c_{H_2O} is 4×10^{-4} (wt%/Pa1/2).

2.10.2 Solubility in a System Containing Water and Carbon Dioxide

In the case where a system containing both water and carbon dioxide, the solubility of each depends on partial pressure. Therefore, the following equations corresponding to Eqs. (2.84) and (2.83) are as follows.

$$P_{H_2O} = k_{H_2O} \tilde{C}^2_{H_2O} \tag{2.85}$$

$$P_{CO_2} = k_{CO_2} \tilde{C}_{CO_2} \tag{2.86}$$

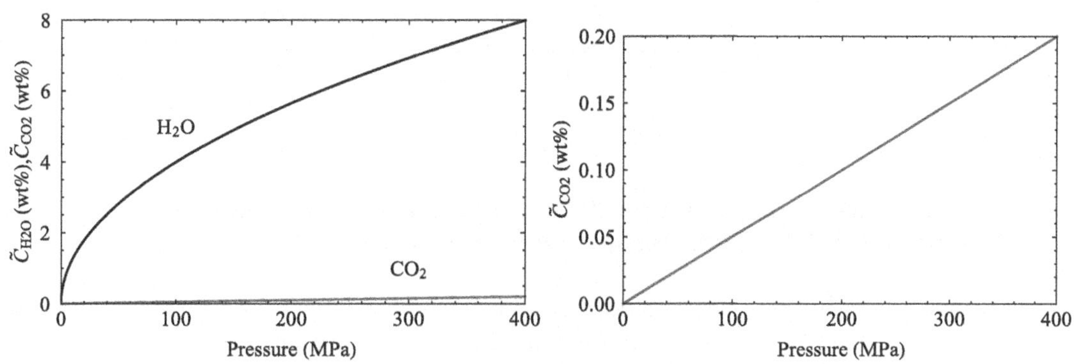

Fig. 2.14 Difference in the solubility in silicate melt between water and carbon dioxide

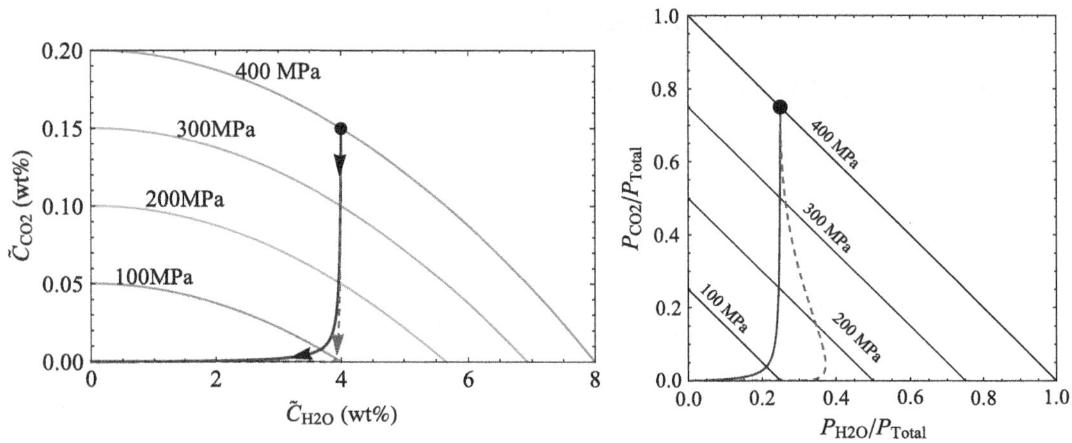

Fig. 2.15 (Left chart) Constraint relationship between the solubility of water and that of carbon dioxide in the system where water and carbon dioxide coexist. The solid line represents the composition path of closed system degassing and the broken line represents the composition path of fractional (Rayleigh) degassing. (Right chart) Partial pressure of each, in gas in equilibrium with liquid. The coordinates of the intersection of the curves with the isobars correspond to partial pressure

where solubility coefficients $k_{H_2O} = c_{H_2O}^{-2} = 6.25 \times 10^6$ (Pa/wt% ^{2}) and $k_{CO_2} = h_{CO_2}^{-1} = 2 \times 10^9$ (Pa/wt%). Furthermore, both P_{CO_2} and P_{H_2O} are the partial pressure of carbon dioxide and water, respectively, and have the following relationship with total pressure P

$$P_{H_2O} + P_{CO_2} = P \tag{2.87}$$

Substituting Eqs. (2.85) and (2.86) into this equation yields the relationship between the solubility of water and that of carbon dioxide.

$$\tilde{C}_{CO_2} = \frac{P}{k_{CO_2}} - \left(\frac{k_{H_2O}}{k_{CO_2}}\right) \tilde{C}_{H_2O}^2 \tag{2.88}$$

Figure 2.15 shows this equation graphically. Using the values above for silicate melt, the value of the ratio of solubility coefficients k_{H_2O}/k_{CO_2} is calculated at 0.003125.

This indicates that the presence of carbon dioxide considerably affects the relationship between the solubility of water and the total pressure of gas. For example, if 4 wt% of water alone is contained in a system, the system at a total pressure of 100 MPa, i.e., at a depth of about 4 km, is almost under a saturation condition and not supersaturation. However, if just a little amount (e.g., 0.03 wt%) of carbon dioxide is present in a system, the system is sufficiently supersaturated under the total pressure and vesiculation occurs.

2.10.3 Gas Composition Change with Progress of Vesiculation Relationship Among Total Pressure, Partial Pressure, and Solubility

As vesiculation of magma proceeds with decrease in pressure, how does the composition of the gas contained in melt change? Let us consider the case where vesiculation proceeds while maintaining a chemical equilibrium. First, we introduce molarity (mol/kg) x_0 and y_0 corresponding to the initial mass content of water and carbon dioxide in melt ($\tilde{C}_{H_2O}(P_0)$ and $\tilde{C}_{CO_2}(P_0)$) (wt%) under the initial pressure P_0. The reason is that because molar mass of water and that of carbon dioxide differ, the amount of gas exsolving is directly related to the pressure of the gas phase via the equation of state of an ideal gas (described with molarity). Assuming that the molarity of water and that of carbon dioxide at any pressure P ($< P_0$) are x and y_0, when molar mass of water and that of carbon dioxide are written as $m_{H_2O} = 0.018$ (kg) and $m_{CO_2} = 0.044$ (kg), the relationship between the molarity of the mass content can be expressed by $x = 0.01 \times \tilde{C}_{H_2O}(P)/m_{H_2O}$ and $y = 0.01 \times \tilde{C}_{CO_2}(P)/m_{CO_2}$. Rewriting the relationship of the solubility (Eqs. (2.85) and (2.86)) and the relationship between the total pressure and the partial pressure (Eq. (2.87)) with x and y gives

$$P_x = P_{H_2O} = 10^4\, m_{H_2O}^2 k_{H_2O} x^2 = k_x x^2 \tag{2.89}$$

$$P_y = P_{CO_2} = 10^2 m_{CO_2} k_{CO_2} y = k_y y \tag{2.90}$$

$$P = P_{H_2O} + P_{CO_2} = k_x x^2 + k_y y \tag{2.91}$$

where $k_x = 10^4 m_{H_2O}^2 k_{H_2O} \approx 2 \times 10^7$ (Pa/(mol/kg)^2) and $k_y = 10^2 m_{CO_2} k_{CO_2} \approx 8.8 \times 10^9$ (Pa/(mol/kg)). Moreover, $\tilde{C}_{H_2O} = 4$ (wt%) is equivalent to 2.2 (mol/kg). The initial conditions are as follows:

$$P_{x0} = k_x x_0^2 \tag{2.92}$$

$$P_{y0} = k_y y_0 \tag{2.93}$$

$$P_0 = k_x x_0^2 + k_y y_0 \tag{2.94}$$

2.10.3.1 Gas Composition and Partial Pressure in Closed System Degassing

Consider the case in which a gas phase is absent at an initial condition. The partial pressure of each gas in a gas phase generated in the decompression-induced vesiculation process is nothing but the ratio of molar amounts of gases exsolving, i.e., the ratio of $x_0 - x$ to $y_0 - y$.

$$\frac{P_x}{P_y} = \frac{x_0 - x}{y_0 - y} \tag{2.95}$$

From the relationship of the solubility (Eqs. (2.89) and (2.90)), the following equation is obtained.

$$\frac{P_x}{P_y} = \frac{k_x}{k_y}\frac{x^2}{y} \tag{2.96}$$

Assuming that the right-hand sides of these two equations are equal to each other, solving the obtained equation with respect to y gives the following equation representing change in the content of carbon dioxide y with change in the water content x attributed to the progress of vesiculation:

$$\frac{y}{y_0} = \frac{a_k x_0 \left(\frac{x}{x_0}\right)^2}{1 - \left(\frac{x}{x_0}\right) + a_k x_0 \left(\frac{x}{x_0}\right)^2} \tag{2.97}$$

where a_k is the ratio of solubility coefficients, defined by the following equation, and controls pressure change at fluxing, as seen later.

$$a_k = \frac{k_x}{k_y} = \frac{100 m_{H_2O}^2 k_{H_2O}}{m_{CO_2} k_{CO_2}} \tag{2.98}$$

For the silicate melt, ak will be 0.0023 (mol/kg) using the values above. Change of gas composition with gas exsolution at this time is plotted in Fig. 2.15. The figure shows that the content of carbon dioxide starts to decrease first. In the current case, the gas exsolving is maintained in the system and influences the following gas—liquid equilibrium state. This is called closed system degassing (Anderson et al. 1989).

The relationship between the total pressure of gas in equilibrium with melt and the water content in the melt is given by Eq. (2.94) at the initial state and by Eq. (2.91) during vesiculation. Furthermore, as Eqs. (2.89) and (2.92) and (2.90) and (2.93) give $P_x/P_{x0} = (x/x_0)^2$ and $P_y/P_{y0} = y/y_0$, respectively,

$$\frac{P_x}{P_0} = \frac{P_x}{P_{x0}}\frac{P_{x0}}{P_0} = \frac{P_{x0}}{P_0}\left(\frac{x}{x_0}\right)^2 \tag{2.99}$$

$$\frac{P_y}{P_0} = \frac{P_y}{P_{y0}}\frac{P_{y0}}{P_0} = \frac{P_{y0}}{P_0}\left(\frac{y}{y_0}\right) \tag{2.100}$$

are obtained. From Eqs. (2.92), (2.93), and (2.94), the coefficients of the equations above, P_{x0}/P_0 and P_{y0}/P_0, are given by

$$\frac{P_{x0}}{P_0} = \frac{a_k x_0\left(\frac{x_0}{y_0}\right)}{a_k x_0\left(\frac{x_0}{y_0}\right) + 1} \tag{2.101}$$

$$\frac{P_{y0}}{P_0} = \frac{1}{a_k x_0\left(\frac{x_0}{y_0}\right) + 1} \tag{2.102}$$

Therefore, each partial pressure in gas can be calculated with a parameter $a_k x_0$ for the initial mole ratio x_0/y_0. In the right chart of Fig. 2.15, the partial pressure of each component in gas in equilibrium with the melt are shown. This shows that carbon dioxide is relatively rare in the melt and is relatively abundant in gas.

2.10.3.2 Gas Composition and Partial Pressure in Rayleigh Degassing

In the case where gas is removed at the instant of vesiculating, as the partial pressure ratio of gases is determined by the amounts of gases exsolving during an infinitesimal time, dx and dy

$$\frac{P_x}{P_y} = \frac{dx}{dy} \tag{2.103}$$

are obtained. Combining this equation with Eq. (2.96) yields

$$\frac{dy}{dx} = \frac{y}{a_k x^2} \tag{2.104}$$

Regarding this equation as a differential equation with respect to x and y and integrating it with respect to x from x_0 to x and with respect to y from y_0 to y gives

$$y = y_0 e^{\frac{1}{a_k}\left(\frac{1}{x_0} - \frac{1}{x}\right)} \tag{2.105}$$

This can be called fractional degassing or Rayleigh degassing; it shows that carbon dioxide drastically decreases (left chart of Fig. 2.15). The partial pressure of each in the gas phase at that time is calculated by

$$P_x = \frac{a_k x^2}{a_k x^2 + y} P_0 \tag{2.106}$$

$$P_y = \frac{y}{a_k x^2 + y} P_0 \tag{2.107}$$

and is shown in the right chart of 2.15. In the case of Rayleigh degassing, because CO_2 selectively exsolves, CO_2 content in the melt drastically decreases and thereby the partial pressure of CO_2 in the gas phase in equilibrium decreases and that of H_2 increases in relative terms.

The above discussion about the system containing both water and carbon dioxide is a very simplified one, i.e., an ideal solution is assumed for dissolution in melt and an ideal gas is assumed for a gas phase. Although it nearly matches with experimental results at lower pressure, chemical interaction between water and carbon dioxide in melt and non-ideality of the gas phase have to be discussed at higher pressure.

2.10.4 Gas Composition and Change of Pressure with Addition of Carbon Dioxide-Rich Fluid

In recent years, the evidence that CO_2-rich fluid is added to melt at depths near the mantle (fluxing) have been disclosed from studies on melt inclusions in minerals (e.g., Yoshimura and Nakamura 2013). Now, let us consider a situation under an initial condition in which molarity x_0 of H_2O, corresponding to saturation concentration, is dissolved alone as a volatile in 1 kg of silicate melt under a pressure P_0 (Fig. 2.16). Assuming that molarity Δy_0 of gas (high-temperature fluid) comprising CO_2 alone at the pressure P_0 is added to it, because the partial pressure of H_2O in this gas is 0, H_2O in the melt is about to exsolve in the gas phase, while CO_2 in the gas is about to dissolve in the melt. Consequently, contents of gas components in the silicate melt change to reach equilibrium with the gas phase. Assuming that total pressure in that process is P and partial pressures of H_2O and CO_2 in the gas phase are P_x and P_y, respectively, under the conditions above, let us understand the fundamental characteristics of

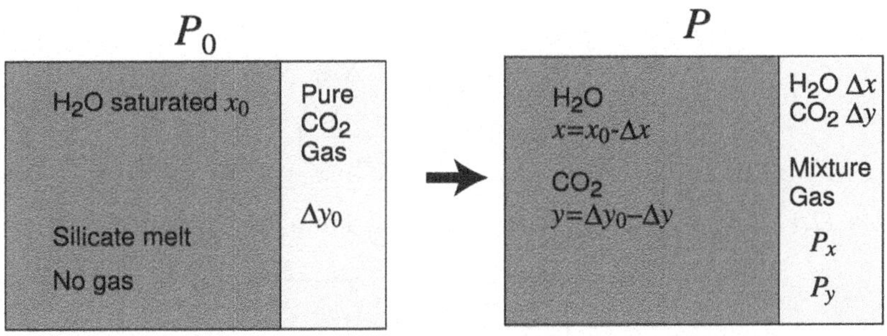

Fig. 2.16 Schematic figure of CO_2 fluxing. (Left) The initial conditions. x_0, saturation concentration at pressure P_0, of H_2O dissolves in silicate melt as the only volatile and gas (high-temperature fluid) consisting of CO_2 alone is added to the melt. (Right) Composition changes as silicate melt and gas are about to reach equilibrium. The total pressure P_x at that time is the sum of P_x and P_y, the partial pressure of each in the gas, and is proportional to the molar ratio of the gas components

concentration of each gas component in the melt, the pressure of the system, and the respective partial pressures after the system has reached the equilibrium state. Note that temporal change has led to the equilibrium, and the volume of the system and the temperature are assumed to be constant.

2.10.4.1 Analysis About Equilibrium State

The relationship of solubility described so far gives the following initial condition

$$P_0 = k_x x_0^2 \tag{2.108}$$

Because we now assume that there is no gas that is in equilibrium with melt, this is a hypothetical condition. Saturation contents $x_s(P_x)$ and $y_s(P_y)$ at partial pressures P_x and P_y yields the following:

$$P_x = k_x x_s^2 \tag{2.109}$$

$$P_y = k_y y_s \tag{2.110}$$

They change with time in the process leading to the equilibrium state. Although molarity of the gas is Δy_0 at the initial condition, when the contents of gas components in the melt are x and y, $x_0 - x$ $(= \Delta x)$ of H_2O and $\Delta y_0 - y$ $(= \Delta y)$ of CO_2 are present in the gas and the total gas molarity changes to $x_0 - x + \Delta y_0 - y$. Given an ideal gas, because the ratio of molarity of gases will be the ratio of pressure,

$$P = \frac{x_0 - x + \Delta y_0 - y}{\Delta y_0} P_0 \tag{2.111}$$

is obtained, and because partial pressures of H_2O and CO_2 are the ratio of the molarity,

$$P_x = \frac{x_0 - x}{\Delta y_0} P_0 \tag{2.112}$$

$$P_y = \frac{\Delta y_0 - y}{\Delta y_0} P_0 \tag{2.113}$$

are obtained.

In the equilibrium state, saturation concentration x_s (Eq. (2.109)) and y_s (Eq. (2.110)) determined by partial pressure are equal to the concentrations in liquid $x = x_e$ and $y = y_e$, respectively, i.e.,

$$x_e = \left(\frac{P_x}{k_x}\right)^{1/2} = x_0 \left(\frac{x_0 - x_e}{\Delta y_0}\right)^{1/2} \tag{2.114}$$

$$y_e = \frac{P_y}{k_y} = a_k x_0^2 \left(\frac{\Delta y_0 - y_e}{\Delta y_0}\right) \tag{2.115}$$

holds. Here, Eqs. (2.112), (2.113), and (2.108) were used. From these equations, we obtain

$$\frac{x_e}{x_0} = \frac{1}{2}\left(\frac{x_0}{\Delta y_0}\right)\left[-1 + \sqrt{1 + 4\frac{\Delta y_0}{x_0}}\right] \tag{2.116}$$

$$\frac{y_e}{x_0} = \frac{a_k x_0}{1 + a_k x_0 \frac{x_0}{\Delta y_0}} \tag{2.117}$$

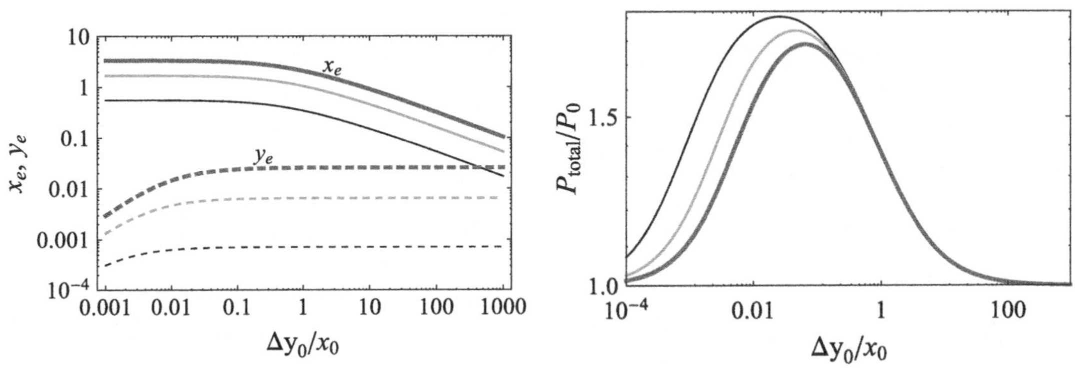

Fig. 2.17 Equilibrium gas composition and total pressure after CO_2 fluxing. (Left chart) Equilibrium gas composition (H_2O: x_e, CO_2: y_e) is plotted as a function of the amount of fluxing $\Delta y_0/x_0$. As this value increases, the amount of fluxing increases. The thickness of the lines represents difference of the parameter $a_k x_0$. It substantially represents difference in x_0, the initial water content in silicate melt, because a_k is a constant value, 0.0023. Fine line: 1 (wt%), gray fine line: 3 (wt%), thick line: 6 (wt%). (Right chart) Corresponding total pressure P/P_0

As seen from these equations, the equilibrium concentration is determined only by the ratio of the initial water content to the amount of fluxing $\Delta y_0/x_0$ and relative gas distribution $a_k x_0$ (left chart of Fig. 2.17).

Although the water content dissolving in silicate melt changes minimally when fluxing is less, the water content decreases with increase in fluxing and the content of carbon dioxide increases in relative terms. $x_0/\Delta y_0$ is the ratio of the amount of water dissolving in the melt to the amount of carbon dioxide added to the melt, and ak represents the ratio of solubility coefficients as given by Eq. (2.98). Because the value of a_k is large, the CO_2 added is relatively easy to dissolve. The value for silicate melt is about 0.0023, which is smaller than 1; therefore, as compared with water, CO_2 is extremely hard to dissolve in silicate melt. Thus, the total pressure of the gas increases after reaching equilibrium (right chart of Fig. 2.17). As seen from the figure, the total pressure increases by fluxing and reaches a maximum at a certain value of fluxing. A value corresponding to the maximum increase in pressure rises monotonously in the range $0.01 << \Delta y_0/x_0 < 0.08$ when the initial water content ranges from 0.1 (wt%) to 10 (wt%) (the position of the horizontal axis that gives the maximum moves the right). Furthermore, the magnitude of increase in pressure P_{max}/P_0 is larger as the initial water content is lesser (the maximum moves downward). This indicates that fluxing at low pressure produces relatively large increase in pressure. However, as expected, the absolute value of increase in pressure under high pressure is larger than that under low pressure.

2.10.4.2 In the Case of Gas Following Linear Solubility

To investigate factors in increasing pressure, we generally examine two gas components (x, y), the solubility of which is linearly dependent on pressure, i.e., $P_x = k_x x$ and $P_y = k_y y$ hold. By the same analysis as used so far, pressure P in equilibrium state against pressure at the initial condition P_0 is given as follows:

$$\frac{P}{P_0} = \frac{\frac{x_0}{\Delta y_0}}{1 + \frac{x_0}{\Delta y_0}} + \frac{1}{1 + a_k \frac{x_0}{\Delta y_0}} \tag{2.118}$$

which shows that parameters $x_0/\Delta y_0$ and a_k control the behavior of increase in pressure (Fig. 2.18). Note that Fig. 2.18 shows that in the case where fluxing fluid is easier to dissolve, $a_k > 1$ leads to decrease in pressure and in the case where fluxing fluid is more difficult to dissolve, $a_k < 1$ leads to increase in pressure.

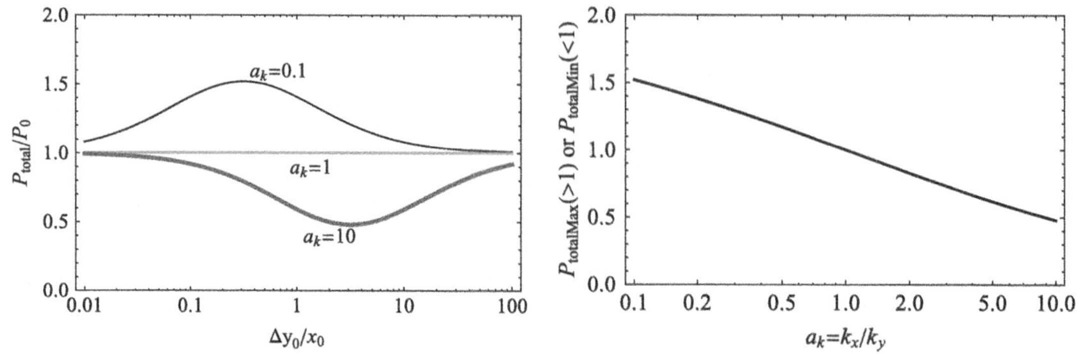

Fig. 2.18 (Left chart) Change in pressure by fluxing in the equilibrium state in the case of linear solubility. The amount of fluxing is on the horizontal axis and the ratio of solubility coefficient a_k is a parameter. Fine line: $a_k = 0.1$, medium-fine line: $a_k = 1$, thick line: $a_k = 10$. (Right chart) Relationship between the maximum pressure change and the ratio of solubility coefficient. When a_k is exactly equal to 1, the same amount dissolves against the same partial pressure. In this case, pressure does not change. In the case where fluxing fluid is easier to dissolve, $a_k > 1$ leads to decrease in pressure and in the case where fluxing fluid is more difficult to dissolve, $a_k < 1$ leads to increase in pressure

2.10.4.3 Analysis on Temporal Change

Let us consider the process leading to an equilibrium state. Assuming that the process proceeds to the equilibrium state through diffusion of gas components in liquid and regarding it as a simple first-order reaction:

$$\frac{dx}{dt} = c_x(x_s(P_x) - x) \tag{2.119}$$

$$\frac{dy}{dt} = c_y(y_s(P_y) - y) \tag{2.120}$$

Here, a local equilibrium at the interface is assumed and rate constants c_x and c_y depend on a local equilibrium concentration and a diffusion coefficient at the interface. Moreover, x_s and y_s are saturation concentrations at local time and xe and ye are equilibrium solutions of dynamic Eqs. (2.119) and (2.120). Temporal change until the system reaches the equilibrium state is described by Eqs. (2.119) and (2.120), and using the following normalized time $\tilde{t}$ and concentrations $\tilde{x}$ and $\tilde{y}$

$$\tilde{t} = c_x t \tag{2.121}$$

$$\tilde{x} = \frac{x}{x_0} \tag{2.122}$$

$$\tilde{y} = \frac{y}{x_0} \tag{2.123}$$

yields

$$\frac{d\tilde{x}}{d\tilde{t}} = \left(\frac{x_0}{\Delta y_0}\right)^{1/2} (1 - \tilde{x})^{1/2} - \tilde{x} \tag{2.124}$$

$$\frac{d\tilde{y}}{d\tilde{t}} = \frac{c_y}{c_x} a_k x_0 \left[1 - \left(\frac{x_0}{\Delta y_0} + \frac{1}{a_k x_0}\right) \tilde{y}\right] \tag{2.125}$$

As seen from these equations, in addition to the parameters $x_0/\Delta y_0$ and $a_k x_0$ that characterize the equilibrium state described above, the ratio of rate constants c_y/c_x controls the behavior of the solutions. Figure 2.19 shows an example of the calculation.

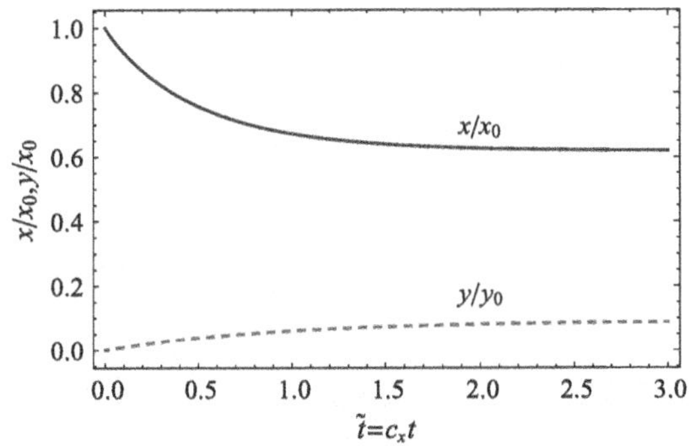

Fig. 2.19 Temporal change until the system reaches the equilibrium gas composition after CO_2 fluxing. $\Delta y_0/x_0 = 1$, $a_k x_0 = 0.1$, $c_y/c_x = 1$. The solid line represents change of x/x_0 and the broken line represents change of y/x_0

References

Anderson AT, Newman JS, Williams SN, Druitt TH, Skirius C, Stolper E (1989) H_2O, CO_2, CI, and gas in Plinian and ash-flow Bishop rhyolite. Geology 17:221–225

Burnham CW (1979) The importance of volatile constituents. In: Yoder HS (ed) The evolution of the igneous rocks. Princeton University Press, pp 439–482

Burnham CW (1975) Water and magmas: a mixing model. Geochim Cosmochim Acta 39:1077–1084

Burnham CW, Davis NF (1974) The role of H_2O in silicate melts: II. Thermodynamic and phase relations in the system NaAl-Si_3O_8-H_2O to 10 kilobars, 700°C to 1100°C. Am J Sci 274:902–940

Callen HB (1960) Thermodynamics. Wiley, New York

Newman S, Lowenstern JB (2002) VolatileCalc: a silicate melt-H_2O-CO_2 solution model written in Visual Basic for excel. Comp Geosci 28(5):597–604

Nowak M, Behrens H (1995) The speciation of water in hapologranitic glasses and melts determined by in situ near-infrared spectroscopy. Geochim Cosmochi Acta 59:3445–3450

Nowak M, Behrens H (2001) Water in rhyolitic magmas: getting a grip on a slippery problem. Earth Planet Sci Lett 184(2):515–522

Paillat O, Elphick SC, Brown WL (1992) The solubility of water in $NaAlSi_3O_8$ melts: a re-examination of Ab-H_2O phase relationships and critical behaviour at high pressures. Contrib Mineral Petrol 112:490–500

Papale P (1997) Modeling of the solubility of a one-component H_2O or CO_2 fluid in silicate liquids. Contrib Mineral Petrol 126:237–251

Schmincke HU (2000) Vulkanismus, Wissenschaftliche Buchgesellschaft, Darmstadt, p 264

Shen A, Keppler H (1995) Infrared spectroscopy of hydrous silicate melts to 1000°C and 10 kbar: Direct observation of H_2O speciation in a diamond-anvil cell. Am Mineral 80:1335–1338

Stolper E (1982) The speciation of water in silicate melts. Geochim Cosmochim Acta 46:2609–2620

Yamashita S (1999) Experimental study of the effect of temperature on water solubility in natural rhyolite melt to 100 MPa. J Petrol 40:1497–1507

Yoshimura S, Nakamura M (2013) Flux of volcanic CO_2 emission estimated from melt inclusions and fluid transport modelling, Earth Planet. Sci Lett 361:497–503

3 Mechanism of Bubble Formation

Abstract

In the previous chapter, we learned that, for bubble generation, a supersaturated state is required. So, how do bubbles actually form in a supersaturated liquid? How do these bubbles grow to bubbles that we can see? To answer these questions and understand bubble formation, it is absolutely necessary for us to consider the meaning of the size and number of bubbles present in highly vesiculated volcanic rocks such as pumice and scoria, described in Chap. 1. These concepts are necessary for us to understand the dominant factors controlling eruption styles and intensity from the bubble texture in highly vesiculated volcanic rocks. In this chapter, we will understand the mechanisms of such bubble formation and behavior of bubbles in liquids based on both theories and experiments.

3.1 Energetics of Bubble Nucleation

In liquids that are supersaturated with a component that can vesiculate (a volatile component), nucleation is the process by which bubbles form embryos. In supersaturated liquids, nucleation does not always occur even if they are slightly supersaturated. Moreover, we already experience it at the table. When you open a bottle or a can of carbonated drinks that are gently carried, you will hear a soft pop and will not see active bubble formation. However, if you open a bottle or a can of the same drink that was subject to impact before opening, you know well that the results will be disastrous when you open it. As shown in Chap. 2, an opened bottle of beer is in supersaturation; however, it seems to require some special energy to vesiculate. In other words, an opened bottle of beer containing carbon dioxide can be stable because it is in supersaturation without work corresponding to the special energy.[1] If we assume that disturbance such as shock does work and gives the special energy required for vesiculation, we will realize what has happened.

[1] The word stable used in this sentence actually means metastable. Water can be stable as liquid water even if it is cooled down to 0 ℃, i.e., the freezing point. This state is a metastable state and such water is sometimes called supercooled water. Let us consider a metastable state from a point of view of energy under gravity as an analogy. When you place a cuboid on a horizontal plane, you can do it in two ways: where the longest sides are perpendicular to the horizontal plane and where the longest sides are parallel to the horizontal plane. Although the latter has lower potential energy and is stable, the former is stable unless pressed by an external force that is larger than a certain degree of force.

A. Toramaru, *Vesiculation and Crystallization of Magma*, Advances in Volcanology,
https://doi.org/10.1007/978-981-16-4209-8_3

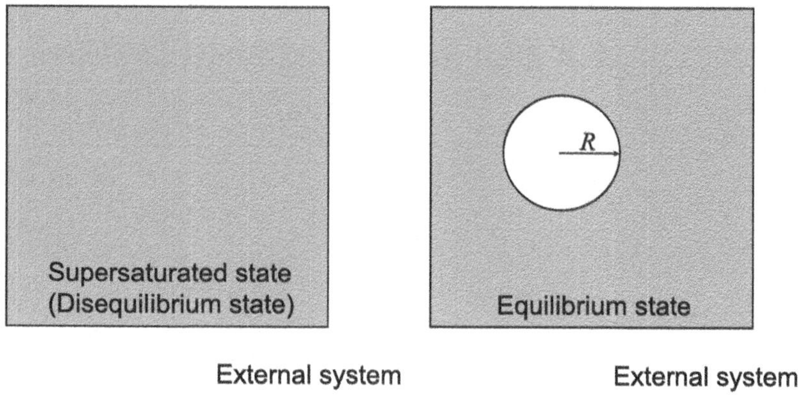

Fig. 3.1 (Left) State A. A system of melt in a disequilibrium, supersaturated (metastable) state. (Right) State B. A melt + gas system in which one bubble is formed and a gas—liquid equilibrium holds at the interface of the bubble. Changes occur by interaction with the external system

Another important thing about bubble formation that we learned from our daily life is that bubble formation is probabilistic. Bubbles are intermittently and discontinuously generated in terms of time. For such a phenomenon, it is important to determine the probability of a phenomenon (bubble generation in this case) happening, i.e., the number of bubble generation per unit time. Here, let us thermodynamically consider the probability of bubble generation by focusing on special energy and work. For thermodynamic consideration, a system should be set first. A system is a substance or a collection of substances to be considered. By comparing a supersaturated state before vesiculation (left of Fig. 3.1) with a state in which only one bubble forms after vesiculation (right of Fig. 3.1),[2] the special energy required to form a bubble can be studied.

3.1.1 Thermodynamics of Fluctuations

In Fig. 3.1, probability associated with the change of state can be related to a change in the system's entropy in thermodynamic terms (Landau and Lifshitz 1958). Boltzmann (1844–1906) provided a definition of a thermodynamic variable entropy S from a microscopic point of view:

$$S = k_B \ln w \tag{3.1}$$

where w is the maximum value of the number of possible states.[3] Note that entropy S used here is of a composite system combining a system of interest and an external system.

w/w_0, the quotient of the number of possible states w divided by the number of possible reference states w_0, gives the realization probability of the state. This probability can be applied to disequilibrium systems, thus yielding the probability of transition between two states of the system. From Eq. (3.1),

$$\text{Probability} = \frac{w}{w_0} = \exp\left(\frac{\Delta S}{k_B}\right) \tag{3.2}$$

[2]This bubble is a very small one comprising 100–1000 water molecules.

[3]The number of possible states indicates the number of possible combinations, under a given total energy of a designated system, of the arrangement of different energy states that microscopic parts constituting the system contain. The maximum value of it corresponds to the arrangement that has the most ways of combination, i.e., the case that has the maximum arrangement of microscopic energy states is an equilibrium state, and entropy at the state reaches a maximum, given by Eq. (3.1).

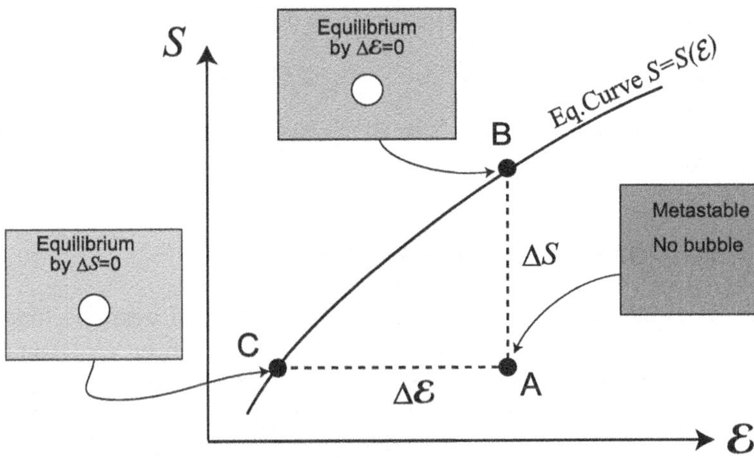

Fig. 3.2 Relationship between change in entropy and internal energy for considering fluctuations

an be obtained, where ΔS is the difference of entropy between any system and the reference system, $S - S_0$. Although calculating entropy difference thermodynamically is not something that easily comes to mind, we will determine that this entropy difference can be related to the external work required to cause the change or can be related to change in free energy.

If you consider a change in entropy corresponding to a change (A→B) in a characteristic of interest in a system, the characteristic here is the presence of an equilibrium bubble. B is not unique because it depends on a process that leads to B. Assuming that the volume V is constant, we represent the relationship between entropy S and internal energy $\mathcal{E}$ by $S(\mathcal{E})$. This relationship is shown in Fig. 3.2: point A is in a metastable state, whereas points B and C are in equilibrium states with a bubble. Note that it is sufficient that change in entropy from the metastable state to the equilibrium state B can be estimated.

By handling infinitesimal change of curve $S(\mathcal{E})$, change in entropy is determined as the product of a slope of $S(\mathcal{E})$ at point C and $\Delta\mathcal{E}$, i.e.,

$$\Delta S_{\mathrm{A}\to\mathrm{B}} = \left(\frac{\partial S}{\partial \mathcal{E}}\right)_V |\Delta\mathcal{E}| = \frac{\Delta\mathcal{E}_{\mathrm{C}\to\mathrm{A}}}{T} = -\frac{\Delta\mathcal{E}_{\mathrm{A}\to\mathrm{C}}}{T} \tag{3.3}$$

From the above, the change probability of the system can be expressed with internal energy as follows:

$$\text{Probability} \propto \exp\left(-\frac{\Delta\mathcal{E}}{k_{\mathrm{B}}T}\right) \tag{3.4}$$

Because $\Delta\mathcal{E}$is change under constant entropy, $\Delta\mathcal{E} = \mathcal{W}$ holds as per the first law of thermodynamics. In this equation, $\mathcal{W}$ is called reversible work; if the system is in thermal equilibrium, T_0, in the equation $\mathcal{E} = \mathcal{F} - TS$, it can be replaced by T because temperature does not change. Now, $\Delta\mathcal{E}$ changes under constant S, i.e., $\Delta\mathcal{E} = \Delta\mathcal{F}$ holds. Therefore,

$$\Delta\mathcal{E} = \Delta(\mathcal{E} - T_0 S) = \Delta\mathcal{F} \tag{3.5}$$

holds and change in internal energy, i.e., work, is equal to change in Helmholtz free energy $\mathcal{F}$.[4]

[4]Conventionally, the probability of nucleation has been evaluated by translating internal energy into work (reversible work) under constant entropy and using it based on Landau and Lifshitz (1958). This book uses free energy to determine direct connection to calculations that are handled later.

Moreover, if both temperature and pressure do not change, $\mathcal{E} - T_0 S + P_0 V = \mathcal{G}$ holds in the same manner, and

$$\Delta\mathcal{E} = \Delta\mathcal{G} \tag{3.6}$$

holds, which indicates that work is equal to change in Gibbs free energy $\mathcal{G}$ under the condition where both P and T are constant.

3.1.2 Energy of Bubble Generation

Because an isothermal process can be assumed, we obtain change in Helmholtz free energy. As precondition, both thermal equilibrium and chemical equilibrium between gas and liquid phases are assumed. Because the pressure of the liquid phase is P and the pressure of the gas phase remains constant at saturation pressure P_G^*,[5] this indicates that a bubble is reversibly formed while ensuring chemical equilibrium. Figure 3.3 shows a process making a gas phase with a constant pressure P_G. If we assume a spherical bubble and consider mechanical equilibrium, P_G^* will change with change in the radius R via Laplace's equation (2.42). Therefore, an equilibrium path to obtain change in energy virtually requires (or assumes) $R =$ to be constant. However, it is unclear what thermodynamics paths are followed when nuclei are formed (Tanaka et al. 2015).

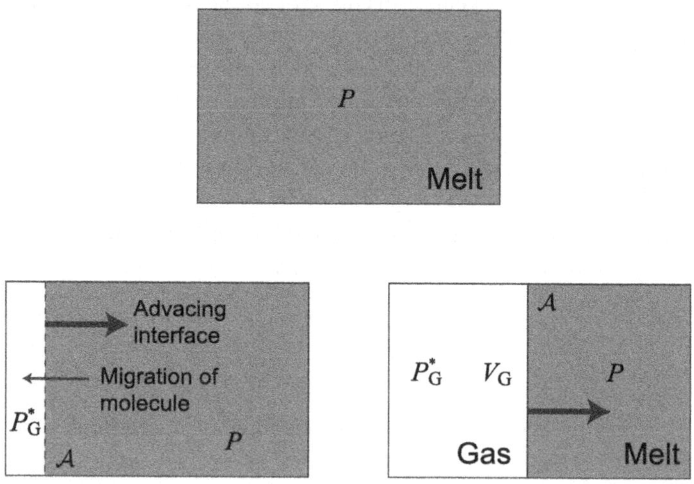

Fig. 3.3 (Above) Melt in a supersaturated (metastable) state. (Below) The interface (area A) between melt and gas moves. Pressure of the gas phase P_G^* is larger than pressure of the melt phase P

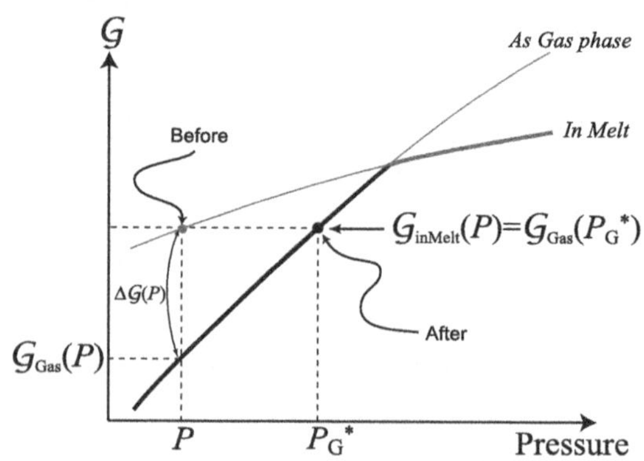

Fig. 3.4 Change in Gibbs free energy with pressure in the cases where the gas phase is formed (As Gas phase) and where the gas is dissolved in melt (In Melt). The state represented by the thick lines is more stable and has a lower free energy than the state represented by the fine lines

[5] The mark $*$ is used as a mark representing saturation states.

Writing $\mathcal{F}$ at a metastable state before bubble formation as $\mathcal{F}_1$ and at a state after bubble formation as $\mathcal{F}_2$ and assuming that the entire volume is constant (V = = constant), $\mathcal{F}_1$ and $\mathcal{F}_2$ can be expressed as follows:

$$\mathcal{F}_1 = \int_0^V (-P)\,dV' = -PV \tag{3.7}$$

$$\mathcal{F}_2 = \int_0^{V_G} (-P_G^*)\,dV' + \int_0^{V-V_G} (-P)\,dV' + \gamma\mathcal{A} \tag{3.8}$$

$$= -P_G^* V_G - P(V - V_G) + \gamma\mathcal{A} \tag{3.9}$$

Therefore,

$$\Delta\mathcal{F} = \mathcal{F}_2 - \mathcal{F}_1 = -(P_G^* - P)V_G + \gamma A \tag{3.10}$$

is obtained, where $P - P_G^*$ ($= \Delta\mathcal{F}_V^*$) is the difference between the Helmholtz free energy of a gas component per unit volume in gas and that in liquid; γ is gas—liquid interface energy (J/m^2); and $\mathcal{A}$ is an area of the interface. $\gamma\mathcal{A}$ is an increment of energy generated by forming new gas—liquid interface.

If we apply Gibbs free energy to this case (Fig. 3.4), then that of a liquid phase component does not change because pressure of the liquid phase component does not change and remains P. Consider a gas phase component and assume chemical equilibrium and thermal equilibrium. As the pressure of the gas phase component changes from P in liquid to P_G^* in the gas phase, the resultant change in $\mathcal{G}$ is $\mathcal{G}_G(P_G^*) - \mathcal{G}_{inMelt}(P)$. Because Gibbs free energy is effective as a thermodynamic potential under a constant pressure, we need to subtract the contribution by an increment of pressure of gas $(P_G^* - P)V_G$ from the change in $\mathcal{G}$. In this manner, we obtain

$$\mathcal{G}_G(P_G^*) - \mathcal{G}_{inMelt}(P) - (P_G^* - P)V_G + \gamma\mathcal{A} \tag{3.11}$$

Here, we assume that the gas phase component occupies the same volume V_G in both liquid and gaseous states. $\mathcal{G}_G(P_G^*)$ is $\mathcal{G}$ of the gas phase with saturation pressure P_G^* and $\mathcal{G}_{inMelt}(P)$ is $\mathcal{G}$ when the gas component is present in the liquid phase with pressure P. Because the liquid and gas phases are in chemical equilibrium with respect to the gas component, $\mathcal{G}_G(P_G^*) - \mathcal{G}_{inMelt}(P) = 0$ holds, and

$$\Delta\mathcal{G} = -(P_G^* - P)V_G + \gamma\mathcal{A} \tag{3.12}$$

is obtained, which agrees with Eq. (3.10). Moreover, expanding $\mathcal{G}_G(P_G^*)$ as

$$\mathcal{G}_G(P_G^*) = \mathcal{G}_G(P) + \frac{\partial\mathcal{G}_G}{\partial P}(P_G^* - P) = \mathcal{G}_G(P) + V_G(P_G^* - P)$$

and substituting this equation into Eq. (3.11) can yield an approximation

$$\Delta\mathcal{G} = \mathcal{G}_G(P) - \mathcal{G}_{inMelt}(P) + \gamma\mathcal{A} \tag{3.13}$$

Using Eq. (3.6), it yields $\Delta\mathcal{G} = \mathcal{G}_G(P_G^*) - \mathcal{G}_{inMelt}(P) + \gamma\mathcal{A} = \gamma\mathcal{A}$, which is incorrect.

We now state the interpretation of Eqs. (3.10) and (3.13). The first terms of the right-hand side, $-(P_G^* - P)V_G$ and $\mathcal{G}_G(P) - \mathcal{G}_{inMelt}(P)$, are the difference in Gibbs free energy $\Delta\mathcal{G}_V^* V_G$ obtained from the difference for states where the gas component exists in a supersaturated state (pressure in both the gas phase and the liquid phase is P) and are negative values because the gas phase state ($\mathcal{G}_G(P)$) is lesser than the state where dissolved in liquid as molecules ($\mathcal{G}_{inMelt}(P)$) (Fig. 3.4). The absolute value become larger as the degree of supersaturation increases. If this volumetric character is the only state change associated with bubble formation, it should not require special energy. A problem is the presence of an interface between a bubble and the surrounding liquid, and the energy required for the formation of the interface. It is the second term $\gamma\mathcal{A}$, which corresponds to the energy. The energy of the interface per unit area γ has a positive value because the energy state increases.

The primary points in consideration of the bubble nucleation process

1. Probability that a bubble nucleus with a critical radius is generated by thermal fluctuation is expressed as follows:

$$\text{Probability} \propto \exp\left[-\frac{\Delta\mathcal{F}^*}{k_B T}\right] \tag{3.14}$$

$\Delta\mathcal{F}^*$ in this equation is the Helmholtz free energy equivalent to the reversible work when a bubble with a critical radius R_C is formed and is estimated from thermodynamic consideration as follows:

$$\Delta\mathcal{F}^* = \frac{4\pi}{3} R_C^2 \gamma \tag{3.15}$$

2. A critical radius R_C satisfies gas—liquid mechanical equilibrium conditions and chemical equilibrium conditions. The mechanical equilibrium conditions are expressed using Laplace's equation

$$P_G^* - P = \frac{2\gamma}{R_C}, \tag{3.16}$$

which indicates the pressure within a bubble P_G^* and the pressure of melt P are in balance via the gas—liquid interfacial tension. The chemical equilibrium conditions are the relation (Eq. 3.21) of the solubility of a gas component having pressure within a bubble P_G^* into melt having a pressure P and given by

$$P_G^* = P_0 \left(\frac{C}{C_0}\right)^2 \exp\left[-\frac{v_m P_0}{k_B T_0}\left(1 - \frac{P}{P_0}\right)\right] \tag{3.17}$$

3. Nucleation rate can be evaluated as the product of the number of critical bubbles N^* (1/m^3) and transition frequency $\mathcal{K}^*$ (1/s), $J = \mathcal{K}^* \cdot N^*$, or the product of flux in size space, i.e., cluster migration velocity G_R^* (m/s) and the value of cluster size distribution function of critical size (1/m^4), $J = G_R^* \cdot F_E^*$.
4. Starting from a master equation, we describe the time evolution of size distribution of molecular clusters (bubbles) using Fokker-Planck equations in size space. We then calculate equilibrium size distribution and use it to evaluate nucleation rate as a solution of steady flux in Fokker-Planck equations.

3.2 Homogeneous Nucleation

Summarizing the previous section, we can write energy budget $\Delta\mathcal{F}$ as a process in which the liquid changes from supersaturated liquid to a saturated state because of the formation of a spherical bubble from Eq. (3.10) as follows:

$$\Delta\mathcal{F}(R) = -\frac{4\pi}{3} R^3 (P_G^* - P) + 4\pi R^2 \gamma \tag{3.18}$$

where R is the radius of the bubble, $4\pi R^3/3$ is the volume of the bubble V_G, and $4\pi R^2$ is the surface area of the bubble $\mathcal{A}$. The first term on the right-hand side indicates a gain (minus) of energy by the gas component coming out of the supersaturated liquid and forming the gas phase, while the second term on the right-hand side indicates a loss (plus) of energy by the interface between the gas and the liquid forming. Note that the formation of a graph of $\Delta\mathcal{F}$ as a function of the bubble size leads to Fig. 3.5. In this figure, the first and second terms on the right-hand side of the equation are plotted such that their contribution can be seen. This energy budget is a function of the radius of the bubble, and the first term contains the square of R and the second term contains the cubic of R. Therefore, when R is small, the second term is dominant and the equation changes along the square of R; however, when R increases, the first term is dominant and the equation decreases along the cubic of R (because of the negative coefficient). Note that a radius corresponding to the maximum value $\Delta\mathcal{F}^*$ is set as R_C.

This change in Helmholtz free energy has two important roles: change in entropy, which determines the probability of bubble formation from a disequilibrium (the metastable supersaturated state), and thermodynamic potential (Appendix 11.1), which determines the course the formed bubble takes afterward in an equilibrium (the bubble will take the course having a lower free energy).

Assuming that after the probability of fluctuations, a bubble happens to form in supersaturated melt and its radius is smaller than R_C (point A). This bubble that is accidentally formed moves toward a lower energy state following the curve of $\Delta\mathcal{F}$, i.e., it moves from point A to the direction indicated by an arrow in Fig. 3.5. This is the direction of the bubble radius decreasing (leftward on the horizontal axis), and the bubble will steadily shrink and disappear in the end.

Then, let us assume that following the probability of fluctuations, a bubble with a radius larger than R_C forms (point B). In this case, the direction of the system energy decreasing is, as indicated by the arrow from point B, the direction of the bubble radius increasing (rightward on the horizontal axis), i.e., the bubble will become increasingly large and the energy budget will be able to become negative. In this case, bubble formation is guaranteed in terms of energy and a bubble with a radius sufficient to be visible to the naked eye will appear if the suitable observation conditions are prepared.

As described above, because the fate of the bubble is considerably affected by whether the radius of a bubble forming accidentally is smaller or larger than R_C, we call R_C the critical radius.[6] Critical means being a point at which certain phenomenon suffers a definite change. Nucleation is nothing but the

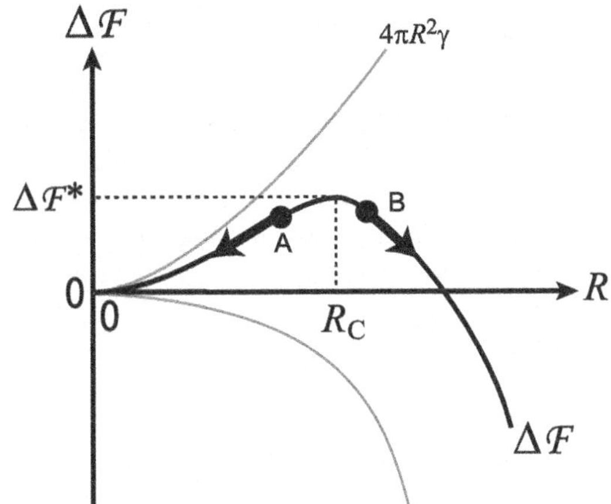

Fig. 3.5 Change in energy associated with bubble formation as a function of a radius of a bubble R

[6]A critical nucleus is a bubble nucleus or a crystal nucleus having a critical radius, which is called the critical nucleus radius or simply the critical radius, i.e., the radius of a critical nucleus.

formation of a bubble with a radius larger than the critical radius by fluctuations or by a chance power (thermal fluctuation). As in the case here, homogeneous nucleation is the process by which a bubble forms and stably grows in a homogeneous liquid. The maximum value of energy, $\Delta\mathcal{F}^*$, corresponding to the critical radius, is none other than the special energy for bubble formation.

Let us understand the critical radius in more detail. Because the critical radius corresponds to a maximum value of a third-order function with respect to R in Eq. (3.18), it corresponds to the case where a derivative of Eq. (3.18) equals 0.

$$\frac{d\Delta\mathcal{F}(R)}{dR} = -4\pi R^2(P_G^* - P) + 8\pi R\gamma = 0 \tag{3.19}$$

i.e.,

$$R_C = \frac{2\gamma}{P_G^* - P} = \frac{2\gamma}{\Delta\mathcal{F}_V^*} \tag{3.20}$$

or $R_C = 2\gamma/\Delta\mathcal{G}_V^*$. This is Laplace's equation describing mechanical equilibrium between a bubble and liquid. The gas phase pressure is P_G^*, pressure of a gas phase being in chemical equilibrium, the liquid pressure of which is P and the water content of which is C. Generally, from chemical equilibrium conditions (Eq. 2.34) where the pressure of the gas phase and that of the liquid phase differ, P_G^* can be defined as follows:

$$P_G^* = P_0 \left(\frac{C}{C_0}\right)^2 \exp\left[-\frac{v_m P_0}{k_B T_0}\left(1 - \frac{P}{P_0}\right)\right] \tag{3.21}$$

Difference in pressure $P_G^* - P$ corresponds to the supersaturation pressure. Assuming C is the reference saturated content C_0, neglecting the exp term of change in liquid pressure (Poynting correction), and using the simplified Burnham model, $P_G^* = P_0$ is obtained, and supersaturation pressure will be $P_0 - P$, which is equal to the amount of decompression from the saturated state.

Substituting this into Eq. (3.18) yields the special energy required for bubble nucleation $\Delta\mathcal{F}^*$ as follows:

$$\Delta\mathcal{F}^* = \frac{4\pi}{3}(R_C)^2\gamma \tag{3.22}$$

This equation reveals that the magnitude of the special energy is 1/3 of the magnitude of surface energy when a bubble with the critical radius forms. As seen above, for nucleation to occur, the size of a bubble formed by thermal molecular fluctuations needs to exceed the critical radius. Therefore, the smaller the critical radius and the lower the special energy, the easier nucleation should occur, i.e., nucleation more easily occurs as interface energy γ in the numerator of Eq. (3.22) is low or supersaturation pressure $P_G^* - P$ in the denominator is high. Because the amount of decompression increases and the difference between the solubility and the amount of water actually dissolving increases, supersaturation pressure increases and nucleation more easily occurs.

Classical theory of nucleation rate 1: History of development

Theoretical studies of nucleation rate were performed by Becker and Döring (1935) on the formation of droplets in gas. They discussed formation and vaporization of droplets using the kinetic theory of gases (Friedlander 1977). Their study was inspired by the fact that in research on elementary particles, which was flourishing at that time, tracks of elementary particles were observed in a cloud chamber and by the elucidation of the mechanism of raindrop formation in the atmosphere in meteorology. After that, aiming at elucidating nucleation of solid crystals in liquid and using the absolute reaction rate theory (or transition state theory) developed by Eyring, which then demonstrated outstanding applicability to the kinetics of complex reactions, Turnbull and Fisher (1949) considered phase transition in which solid forms from the liquid as a kind of reaction and then formulated it. What lay behind this study seemed to be the steel industry, which was flourishing at that time. Zel'dovich (1942) established a general method to calculate nucleation rate by transforming the discontinuous cluster dynamics equation to a Fokker-Planck (F-P) equation in a continuous field and using the macroscopic growth process. Kashchiev (1969) discussed an unsteady solution of the F-P equation and the delay of the nucleation rate. Hirth et al. (1970) formulated a bubble formation process in liquid. This study was motivated by the elucidation of a boiling phenomenon in the cooling systems of nuclear reactors in atomic power plants, which had started to be practically used. Moreover, using the Einstein-Stokes relation and the transition state theory, Uhlmann (1972) rewrote the pre-exponential factor in the nucleation rate in the formulation by Turnbull and Fisher (1949) using both viscosity and diffusion. As seen above, the development of the classical theory of nucleation rate appears to be influenced by the social situation and industrial applications at that time.

3.3 Kinetics of Bubble Nucleation

How does special energy $\Delta\mathcal{F}^*$ for bubble nucleation relate to the formation rate of bubble nucleation? Nucleation rate of bubbles is defined as the number of bubbles that emerge larger in size than the critical size per unit time in unit volume (number/(m^{3}s)). Expressing this symbolically, the nucleation rate is expressed by the product of the number of bubbles with the critical size N^* and the frequency with which a molecule adds to the bubble with the critical size leading to the larger size of bubble $\mathcal{K}^*$, $J = \mathcal{K}^* \cdot N^*$, i.e.,

$$J = \underbrace{\text{Transition frequency (1/s)}}_{\mathcal{K}^*} \cdot \underbrace{\text{Critical population (number/m}^3\text{)}}_{N^*} \tag{3.23}$$

N^* is, as seen above, proportional to $\exp(-\Delta\mathcal{F}^*/kT)$, the probability of having special energy $\Delta\mathcal{F}^*$, and it is conceivable that $N^* = N_1 \exp(-\Delta\mathcal{F}^*/k_BT)$ holds (N_1 is a constant related to the monomer number density).[7] The problem of bubble nucleation is nothing other than determining the $\Delta\mathcal{F}^*$, the coefficient N_1, and transition rate $\mathcal{K}^*$. When examining from a microscopic point of view, molecules are coming in (growth) and out of (shrinkage) a bubble and the size of the bubble is always fluctuating. This bubble size fluctuation determines the coefficient (the monomer number density) and the transition rate and is related to the growth rate of bubbles. Because these factors are mutually influenced by each other, the pre-exponential factor of the nucleation rate is determined.

[7]Here it is not definitely stated whether the number of bubbles at the critical size is a value in the equilibrium size distribution N_E^* or a value in the stationary size distribution N_S^* discussed later. They are in proportion to each other, as seen later.

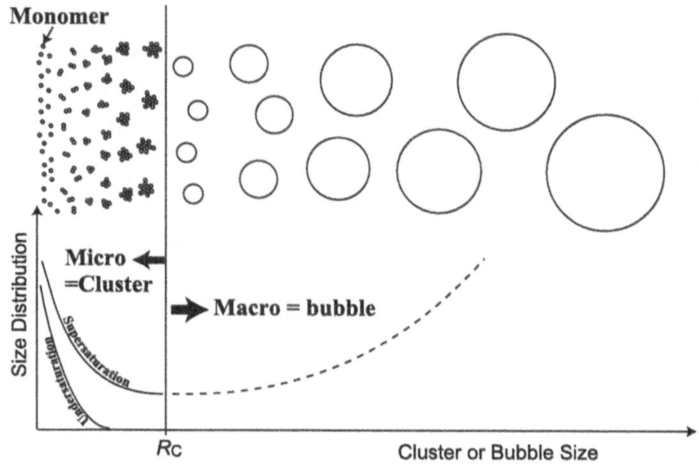

Fig. 3.6 Conceptual images of monomers, clusters, and bubbles in the equilibrium size distribution. A monomer is a cluster comprising a single molecule. R_C is the critical radius. In supersaturation conditions, size distribution larger than the R_C has no realistic meaning

For the actual calculation, putting aside an image of real space, it is useful for us to consider the size space by which we can deal with the dynamics of clusters (among them, those clusters that grow beyond the critical radius R_C into a macroscopic scale are called bubbles (Fig. 3.6)) as aggregates of molecules. In the size space, the nucleation rate can be considered as the flux flowing from the side of smaller size to the side of larger size through the critical size. This is expressed by the product of the migration velocity G_R^* (m/s) and the value of the size distribution function at the critical size F^*, $J = G_R^* \cdot F^*$, i.e.,

$$J = \underbrace{\text{Migration velocity (m/s)}}_{G_R^*} \cdot \underbrace{\text{Critical population density (number/m}^4\text{)}}_{F^*} \tag{3.24}$$

is thought to be hold.[8] In this equation F^* is a value of the size distribution function at the critical size and can be expressed as $F^* = F_0 \exp(-\Delta\mathcal{F}^*/k_B T)$ with using $\Delta\mathcal{F}^*$. Moreover, F_0 is a constant related to the number of monomers.

In any case, the nucleation rate is expressed in the form of

$$J = J_0 \exp\left(-\frac{\Delta\mathcal{F}^*}{k_B T}\right) \tag{3.25}$$

and J_0 is called a pre-exponential factor and determined by the transition probability, the growth rate, and the number density coefficient, mentioned earlier, as follows:

$$J_0 = \mathcal{K}^* \cdot N_1 = G_R^* \cdot F_0 \tag{3.26}$$

The central term $\mathcal{K}^* \cdot N_1$ is of the case where J_0 is connected with the number of monomers per unit volume N_1. The term on the right-hand side $G_R^* \cdot F_0$ is of the case where J_0 is connected with the

[8]Note that the migration velocity used here is different from the growth rate. The migration velocity used here includes the effect of fluctuation. F^* can be represented by the equilibrium size distribution F_E^*. Even if the stationary distribution F_S is assumed, the flux J of clusters with fluctuation in the size space cannot be simply expressed by the product of size distribution F_S and the growth rate A (actual rate at which clusters at the critical size grow, i.e., zero), $A \cdot F_S$. J should have the fluctuation term (B), i.e., $J = A \cdot F - B \cdot \partial F/\partial R$ (Eq. 3.34). It is rather preferable that F^* is symbolically represented by a value of the equilibrium size distribution F_E^* at the critical size.

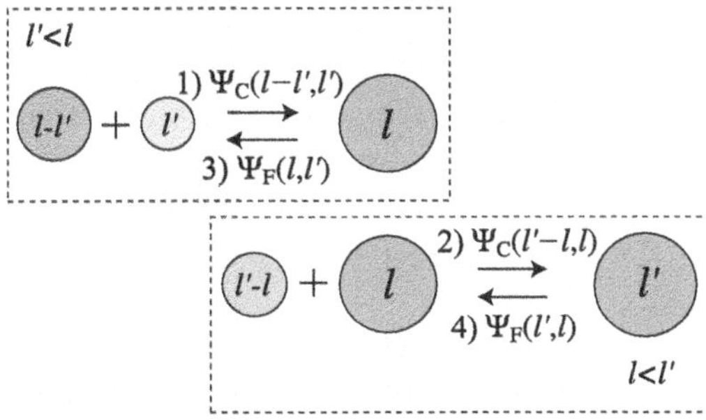

Fig. 3.7 Transition process of the bubble size. The upper process is a transition process with the size l' smaller than the size of interest l, given $l' < l$. The lower process is a transition process with the size l' larger than l, given $l' > l$

size distribution function F (the number per unit size or unit volume). Let us look at the fact that the nucleation rate can be calculated in this way of understanding.

3.3.1 Master Equation

We measure the size of a cluster by the number of contained molecules l.[9] The number of bubbles with the size l is written as $N(l)$ (number/m^3). When writing the transition probability from the size l' to the size l as $W_o(l' \to l)$, change in the number of bubbles $N(l)$ (number/m^3) is expressed as follows:

$$\frac{\partial N(l,t)}{\partial t} = \sum_{l'} \left(\underbrace{W_\circ(l' \to l) \cdot N(l',t)}_{\text{gain}} - \underbrace{W_\circ(l \to l') \cdot N(l,t)}_{\text{loss}} \right) \tag{3.27}$$

This equation is the most fundamental equation describing time-development of the number of clusters of any size l; therefore, is called a master equation (Haken 1978). This equation demonstrates that change in the population classified according to the size l can be calculated if only we are aware of the transition probability between the size, $W_\circ(l' \to l)$.[10]

Moreover, $W_\circ(l' \to l)$ is related to the following four processes (Fig. 3.7):

(1) As a bubble of the size l' $(< l)$ and a bubble of the size $l - l'$ $(< l)$ coalesce together to form a bubble of the size l, the number of bubbles of the size l increases. Assume that the transition probability at this time is $\Psi_C(l - l', l')$.
(2) As a bubble of the size $l' - l (> 0)$ and a bubble of the size l coalesce together to form a bubble of the size l' $(> l)$, the number of bubbles of the size l decreases. Assume that the transition probability at this time is $\Psi_C(l' - l, l)$.
(3) As a bubble of the size $l'(< l)$ forms from a bubble of the size l by fragmentation, the number of bubbles of the size l decreases. Assume that the transition probability at this time is $\Psi_F(l, l')$.
(4) As a bubble of the size $l' - l$ (> 0) forms from a bubble of the size $l'(> l)$ by fragmentation, the number of bubbles of the size l increases. Assume that the transition probability at this time is $\Psi_F(l', l)$.

[9]This is an expedient approach to hold change in size associated with coalescence and fragmentation of bubbles and conservation of mass and to simplify an expression of the addition and subtraction of the size. It will be converted to the radius of bubble later. Assuming that the molecular volume of gas is v_G, $l v_G = 4\pi R^3/3$.

[10]Therefore, this equation can be universally applied to other aspects such as change in populations of human beings or insects by replacing the size in Eq. (3.27) with a human being's income or the size of the insects. However, it is possible only when the transition probability can be completely described as a function of size.

Rewrite $W_{\circ}$ representing increase in number (gain) and decrease in number (loss) with phrases of the coalescence/fragmentation process. Next, we introduce an assumption of detail balance. The detailed balance indicates that in an equilibrium distribution $N_{\mathrm{E}}(l)$, the rates of coalescence and fragmentation between a given size l and another size l' is in balance.

$$\Psi_{\mathrm{C}}(l-l',l')\cdot N_{\mathrm{E}}(l-l')\cdot N_{\mathrm{E}}(l') = \Psi_{\mathrm{F}}(l,l')\cdot N_{\mathrm{E}}(l) \equiv W(l':l) \tag{3.28}$$

The assumption allows us to describe coalescence and fragmentation by a unique transition frequency $W(l':l)$. Moreover, this assumption has an important meaning of connecting a disequilibrium process of coalescence and fragmentation to an equilibrium distribution, a characteristic of an equilibrium state. Based on calculations (see Sect. 11.3.1 in the Appendix (Chap. 11)), the master equation is obtained as follows by rewriting Eq. (11.17):

$$\frac{\partial N(l,t)}{\partial t} = -\frac{\partial}{\partial l}\left(-(\Delta l)^2\cdot W(l)\cdot\frac{\partial N_{\mathrm{E}}(l)^{-1}}{\partial l}\cdot N(l,t) - \frac{(\Delta l)^2\cdot W(l)}{N_{\mathrm{E}}(l)}\frac{\partial N(l,t)}{\partial l}\right) \tag{3.29}$$

This equation describing the time evolution of the number of cluster $N(l,t)$ covers from the molecular scale to macroscopic bubbles near the critical radius.

3.3.2 Fokker-Planck Equation for Bubble (Cluster) Size Distribution

On examining Eq. (3.29) against a general conservation law (Equation below) of size distribution $F(R,t)$ that only considers macroscopic growth (growth rate G), mentioned later,

$$\frac{\partial F(R,t)}{\partial t} = -\frac{\partial}{\partial R}(G\cdot F(R,t)) \tag{3.30}$$

a coefficient before $N(l,t)$, the first term on the right-hand side, is found to correspond to the growth rate of bubbles G. This growth rate of clusters, $A(l)$, is expressed as follows:

$$A(l) = -(\Delta l)^2\cdot W(l)\cdot\frac{\partial N_{\mathrm{E}}(l)^{-1}}{\partial l} \tag{3.31}$$

and Eq. (3.29) is rewritten with $A(l)$. Using this $A(l)$, the temporal development of $N(l,t)$, the number of bubbles (clusters) of the size l, when the size of the bubble (cluster) is fluctuating by coalescence and fragmentation of a molecule (monomer), can be described by

$$\frac{\partial N(l,t)}{\partial t} = -\frac{\partial}{\partial l}\left(A(l)N(l,t) - B(l)\frac{\partial N(l)}{\partial l}\right), \tag{3.32}$$

a differential equation of the form of a Fokker-Planck equation. In this equation, $B(l)$ plays the role of a diffusion coefficient of bubbles in the size space and has a relationship

$$B(l) = A(l)(\partial \ln N_{\mathrm{E}}(l)/\partial R)^{-1} \tag{3.33}$$

with the growth rate, which is attributed to the detailed balance assumed earlier. This diffusion coefficient represents the coalescence/fragmentation of a molecule on a bubble (cluster) by molecular-scale fluctuations and is effective only on a microscopic scale around the R_{C}.

Here, a size distribution function of bubbles $F(R, t)$ is introduced. As we introduce a macroscopic scale R at this time, we will call bubbles (clusters) up to now simply bubbles. A bubble size distribution function is then defined as the number of bubbles existing in the bubble size interval $R \sim R + dR$ per unit volume, $N(l, t)dl = F(R, t)dR$. Because l is a number without a unit, the unit of $N(l, t)dl$ is number/m^3. As seen later, the moment of this size distribution function is an observable. Writing Eq. (3.32) with $F(R, t)$ (see Sect. 11.3.2 in the Appendix (Chap. 11)) yields the following Fokker-Planck equation of the bubble (cluster) size distribution function.

$$\frac{\partial F(R,t)}{\partial t} = -\frac{\partial}{\partial R}\left(A(R)F(R,t) - B(R)\frac{\partial F(R,t)}{\partial R}\right) \tag{3.34}$$

Note that in this equation the growth rate $A(R)$ covering even a molecular scale is $A(R) = -(\partial R/\partial l) \cdot W(R) \cdot (\partial(F_E(R)^{-1})/\partial R)$, which differs from Eq. (3.31) in a factor $(\partial R/\partial l)$ to convert l to R.

Relationship between $A(R)$ and $B(R)$ is

$$B(R) = A(R)\left(\frac{\partial \ln F_E(R)}{\partial R}\right)^{-1} \tag{3.35}$$

The physical meaning of this equation is that the growth of bubbles $A(R)$ and the diffusion (fluctuations of increase and decrease in size) on the size space $B(R)$ are balanced, thereby determining the equilibrium size distribution, i.e., the exponential part (extracted by taking logarithm) of the equilibrium size distribution has a meaning of potential to determine the number distribution of bubbles in the size space, thereby determining the relationship between growth and diffusion. Although this is attributed to the detailed balance, if $A(R)$ and $B(R)$ are known, this equation can be said to define the equilibrium distribution F_E. In the case of condensation from gas, described in the Box later, $A(R)$ and $B(R)$ are determined from the kinetic theory; moreover, the equilibrium distribution is also obtained. Unfortunately, in the bubble problem here, such a kinetic theory has not been known. We will now determine the equilibrium distribution from the fluctuation probability and energy required for nucleation $\Delta\mathcal{F}$ (Eq. 3.18).

All variation in the number of bubbles is described by Eq. (3.34), which represents a conservation law of the number of bubbles, in which the flow rate of bubbles in the size space, J, is

$$J = A(R)F(R,t) - B(R)\frac{\partial F(R,t)}{\partial R} \tag{3.36}$$

$$= -B(R) \cdot F_E(R)\frac{\partial}{\partial R}\left(\frac{F(R,t)}{F_E(R)}\right) \tag{3.37}$$

i.e.,

$$\frac{\partial F(R,t)}{\partial t} = -\frac{\partial J}{\partial R} \tag{3.38}$$

The nucleation rate is nothing but the number of bubbles moving to the side in which the bubble size is larger than the critical nucleus size in the size space, i.e., the flow rate of bubbles JJ. For this purpose, the equilibrium distribution of bubble size $F_E(R)$, which is included in the diffusion coefficient in the size space of bubbles $B(R)$, has to be determined first.

Population distribution of droplets condensed from gas

For droplet condensation in a gas, as the evaporation rate and the condensation rate appearing in the detail balance are known from the molecular kinetic theory, accurate analytical solutions of the equilibrium distribution [a]can be obtained (Friedlander 1977). The detailed balance in a discrete size is[a]

$$A_+(l-1)N_E(l-1) = A_-(l)N_E(l) \tag{3.39}$$

Here the size effect of the surface area is neglected.
In the condensation process of droplets, from the kinetic theory, the evaporation rate A_- and the condensation rate A_+ are known as a function of the number of molecules contained in particles l and vapor pressure (partial pressure) P_1, and these parameters are calculated as follows:

$$A_-(l) = \frac{P_R}{(2\pi m k_B T)^{1/2}} = \frac{P_s}{(2\pi m k_B T)^{1/2}} \exp\left[\frac{2\gamma v_m}{R k_B T}\right] \tag{3.40}$$

$$A_+(l-1) = \frac{P_1}{(2\pi m k_B T)^{1/2}} \tag{3.41}$$

where P_R is the saturation vapor pressure of droplets of radius R and connected with the saturation vapor pressure via a flat interface, P_s, using the Kelvin relation (Gibbs-Thomson relation). [b]The relationship between the droplet radius R and the size counted with the number of molecules l is $v_m l = 4\pi R^3/3$. In the Eq. (3.39) transposing the N_E-term and A_+ (A_-)-term yields $N_E(l-1)/N_E(l) = A_-(l)/A_+(l-1)$. Thus, multiplying each side from $l = 2$ gives

[a]Here we merely consider a number density distribution N(number/m^3), not a size distribution function.
[b]Thomson was called Lord Kelvin.

Population distribution of droplets condensed from gas

$$\text{Term}\, N_E = \frac{N_E(1)}{N_E(2)} \cdot \frac{N_E(2)}{N_E(3)} \cdots \frac{N_E(l-1)}{N_E(l)} = \frac{N_E(1)}{N_E(l)} \tag{3.42}$$

$$\text{Term}\, A = \frac{A_-(2)}{A_+(1)} \cdot \frac{A_-(3)}{A_+(2)} \cdots \frac{A_-(l)}{A_+(l-1)} \tag{3.43}$$

$$= \Delta^{-(l-1)} \exp\left[\frac{2\gamma v_m (4\pi/3v_m)^{1/3}}{k_B T} \sum_{l=2}^{l} l^{-1/3}\right] \tag{3.44}$$

In the term N_E, $N_E(l-1)$ is canceled one by one, and $N_E(1)/N_E(l)$ remains in the end. $N_E(1)$ is the number of molecules of equilibrium monomers N_1 and has a relation $P_1 = N_E(1)k_B T$ with supersaturation pressure. The degree of supersaturation Δ is defined as the ratio of supersaturation pressure P_1 to saturated vapor pressure P_s, $P_1/P_s = \Delta$. Note that the he right-hand side is a computable series, and

$$\sum_{l=2}^{l} l'^{-1/3} \approx \int_0^l l'^{-1/3} dl' = \frac{3}{2} l^{2/3} \tag{3.45}$$

is obtained, and

$$N_E(l) = N_s \Delta^l \exp\left[-\frac{3\gamma v_m (4\pi/3v_m)^{1/3} l^{2/3}}{k_B T} \right] \tag{3.46}$$

holds. In this equation, N_s is the number of molecules corresponding to the saturated vapor pressure P_s, $N_s \equiv P_s/k_B T$.

Critical population of droplets condensed from gas

The equilibrium distribution, Eq. (3.46), can be rewritten in simpler expression in the case of the critical size. The critical size R_C is obtained from the Kelvin equation as follows:

$$R_C = \frac{2\gamma v_m}{k_B T \ln \Delta} \tag{3.47}$$

Using this equation and $v_m l = 4\pi R^3/3$ gives

$$\Delta^l = \exp\left[\frac{32\pi}{3} \frac{\gamma^3 v_m^2}{(k_B T)^3 (\ln \Delta)^2} \right] \tag{3.48}$$

The inside of exp-term of Eq. (3.46) becomes $-16\pi\gamma^3 v_m^2 (k_B T)^{-3} (\ln \Delta)^{-2}/3$, and approximating N_s as the number of molecules of monomers N_1 yields[a],

$$N_E^* = N_1 \exp\left[-\frac{16\pi}{3} \frac{\gamma^3 v_m^2}{(k_B T)^3 (\ln \Delta)^2} \right] = N_1 \exp\left[-\frac{4\pi}{3} \frac{\gamma R_C^2}{k_B T} \right] \tag{3.49}$$

Note that within the last exp-term is the same expression as that in the equilibrium size distribution function (Eq. 3.77) obtained for the bubble.[a]

[a] Strictly speaking, it is unequal to the monomer number density because the case where clusters other than monomers exist is considered. See Sect. 3.3.3.

3.3.3 Equilibrium Distribution

Energy required for bubble formation is equal to change in Helmholtz free energy ($\Delta\mathcal{F}$), and the probability of bubble generation by fluctuation is expressed by Eq. (3.14). This $\Delta\mathcal{F}$ is given by Eq. (3.18) as a function of the radius of the bubble R. Because equilibrium distribution of the bubble size is proportional to the probability of bubble generation,

$$F_E = F_E^0 \exp\left(-\frac{\Delta\mathcal{F}(R)}{k_B T} \right) = F_E^0 \exp\left(-\frac{-\frac{4\pi}{3} R^3 (P_G^* - P) + 4\pi R^2 \gamma}{k_B T} \right) \tag{3.50}$$

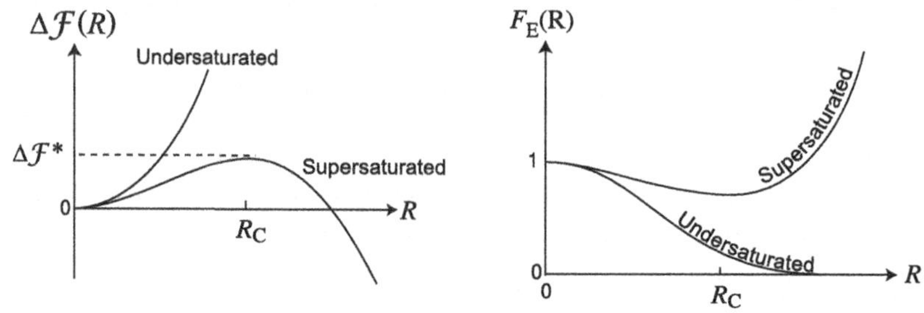

Fig. 3.8 (Left) Change in Helmholtz free energy ($\Delta\mathcal{F}$) as a function of the bubble size. A supersaturated state and an unsaturated state. (Right) The equilibrium size distribution in a supersaturated state and an unsaturated state

is obtained. From the scaling discussion (Lifshitz and Pitaevskii 1981), the coefficient will be[11]

$$F_{\mathrm{E}}^{0} \approx \frac{C R_{\mathrm{C}}{}^{2}}{v_{\mathrm{G}}} \tag{3.51}$$

For the subsequent calculation, expand the inside of the exp-term of the size distribution function (Eq. 3.50) around R, where $\Delta\mathcal{F}(R)$ gives the maximum value, i.e., expanding $\Delta\mathcal{F}(R)$ yields

$$\Delta\mathcal{F}(R) = \Delta\mathcal{F}^{*} + \frac{1}{2}\left.\frac{d^{2}\Delta\mathcal{F}}{dR^{2}}\right|_{R=R_{\mathrm{C}}}(R-R_{\mathrm{C}})^{2} + \cdots \tag{3.52}$$

where $\Delta\mathcal{F}^{*}$ is the maximum value of $\Delta\mathcal{F}(R)$ given by Eq. (3.22). Of course, fortunately, R yielding the maximum value of $\Delta\mathcal{F}$ has a physical meaning of the critical nucleus radius R_{C}. Note that the first-order term is zero by definition and, as seen from the fact that $\Delta\mathcal{F}(R)$ is a convex upward curve in the supersaturated state, $d^{2}\Delta\mathcal{F}/dR^{2}$ is calculated as follows and is negative. This relates to the Zeldovich factor described later.

$$\left.\frac{d^{2}\Delta\mathcal{F}}{dR^{2}}\right|_{R=R_{\mathrm{C}}} = -8\pi\gamma \tag{3.53}$$

Finally, the equilibrium distribution can be expressed as follows:

$$F_{\mathrm{E}} = \frac{C(R_{\mathrm{C}})^{2}}{v_{\mathrm{G}}}\exp\left(-\frac{\Delta\mathcal{F}^{*}}{k_{\mathrm{B}}T}\right)\cdot\exp\left(-\frac{\frac{1}{2}\left.\frac{d^{2}\Delta\mathcal{F}}{dR^{2}}\right|_{R=R_{\mathrm{C}}}(R-R_{\mathrm{C}})^{2}}{k_{\mathrm{B}}T}\right) \tag{3.54}$$

$$= \frac{C(R_{\mathrm{C}})^{2}}{v_{\mathrm{G}}}\exp\left(-\frac{4\pi\gamma(R_{\mathrm{C}})^{2}}{3k_{\mathrm{B}}T}\right)\cdot\exp\left(\frac{4\pi\gamma(R-R_{\mathrm{C}})^{2}}{k_{\mathrm{B}}T}\right) \tag{3.55}$$

Fig. 3.8 shows this equilibrium size distribution. Moreover, $\partial \ln F_{\mathrm{E}}/\partial R$ in the Eq. 3.35) is

$$\frac{\partial \ln F_{\mathrm{E}}(R)}{\partial R} = -\left.\frac{d^{2}\Delta\mathcal{F}}{dR^{2}}\right|_{R=R_{\mathrm{C}}}\frac{(R-R_{\mathrm{C}})}{k_{\mathrm{B}}T} = \frac{8\pi\gamma(R-R_{\mathrm{C}})}{k_{\mathrm{B}}T} \tag{3.56}$$

[11] See Sect. 11.4 in the Appendix (Chap. 11). As the ambiguity concerning F_{E}^{0} remains to the end, it should be accurately described from the microscopic point of view. Regarding homogeneous reaction of solution and gas, a partition function of the transition state is statistically defined and the transition rate is calculated by the theory of the transition state.

and $B(R)$ is given by

$$B(R) = \frac{k_B T A(R)}{\left|\frac{d^2 \Delta \mathcal{F}}{dR^2}\right| (R - R_C)} = \frac{k_B T A(R)}{8\pi\gamma(R - R_C)} \tag{3.57}$$

Thus, in the diffusion coefficient in the size space, the curvature (a second derivative: the Zeldovich factor) of the cluster generation energy $\Delta\mathcal{F}(R)$ as a function of size appears.

Flux by the molecular kinetic theory

When considering droplet condensation in gas, we calculate the number of molecules passing via the unit area using the kinetic theory. The velocity components in the x, y, and z directions of a molecule, v_x, v_y, and v_z, follow the following Maxwell distribution $f(v_x, v_y, v_z)$.

$$f(v_x, v_y, v_z) = \left(\frac{m}{2\pi k_B T}\right)^{3/2} e^{-m(v_x^2+v_y^2+v_z^2)/2k_B T} \tag{3.58}$$

To calculate the number of molecules going into a droplet, we calculate the total number of molecules $\dot{N}$, per unit time, which passes via a cross section $\mathcal{A}$ that has the normal line parallel to the x direction with a positive velocity.

$$\dot{N} = \mathcal{A} \int_0^\infty dv_x \int_{-\infty}^\infty dv_y \int_{-\infty}^\infty dv_z \cdot v_x \cdot N_0 \cdot f(\mathbf{v}) \tag{3.59}$$

where N_0 is the number of molecules per unit volume.

$$\int_{-\infty}^\infty f(\mathbf{v}) dv_y = \int_{-\infty}^\infty f(\mathbf{v}) dv_z = \int_{-\infty}^\infty e^{-\frac{m}{2k_B T}\xi^2} d\xi = \left(\frac{2\pi k_B T}{m}\right)^{1/2} \tag{3.60}$$

holds, then

$$\dot{N} = \mathcal{A} N_0 \left(\frac{m}{2\pi k_B T}\right)^{1/2} \int_0^\infty v_x e^{-\frac{m}{2k_B T} v_x^2} dv_x \tag{3.61}$$

$$= \mathcal{A} N_0 \left(\frac{m}{2\pi k_B T}\right)^{1/2} \int_0^\infty e^{-\frac{m}{k_B T}\left(\frac{v_x^2}{2}\right)} d\left(\frac{v_x^2}{2}\right) \tag{3.62}$$

$$= \mathcal{A} N_0 \left(\frac{m}{2\pi k_B T}\right)^{1/2} \left(\frac{k_B T}{m}\right) \tag{3.63}$$

is obtained. Obtaining N_0 from an equation of state for a unit volume 1 (m^3), $P = N_0 k_B T$, and substituting N_0 into the equation above yields

$$\dot{N} = \mathcal{A} \frac{P}{k_B T} \left(\frac{m}{k_B T}\right)^{1/2} \left(\frac{k_B T}{m}\right) \tag{3.64}$$

Because $\dot{N}/\mathcal{A}$ gives the flux of the molecules A_+,

$$A_+ = \frac{P}{(2\pi m k_B T)^{1/2}} \tag{3.65}$$

3.3.4 Derivation of the Steady State Nucleation Rate

Equation (3.34) can be rewritten using Eq. (3.37) as

$$\frac{\partial F(R,t)}{\partial t} = \frac{\partial}{\partial R}\left[\underbrace{B(R)\cdot F_{\mathrm{E}}(R)}_{\text{Diffusion coefficient}} \underbrace{\frac{\partial}{\partial R}\left(\frac{F(R,t)}{F_{\mathrm{E}}(R)}\right)}_{\text{Field}}\right] \tag{3.66}$$

This is the fundamental equation from now on.

Consider the case where a certain degree of supersaturation is obtained using changing conditions from an undersaturated or saturated state. Although an undersaturated state shows a monotonous equilibrium size distribution, in which the smaller the bubble radius (cluster size), the more the bubbles (clusters), a supersaturated state shows an equilibrium size distribution in which the existence probability of bubbles with the critical radius is the least; the larger the bubble radius, the higher the existence probability. In the undersaturated state, gas exists as molecular clusters in which single molecules account for the greatest majority. When the state becomes supersaturated, molecular clusters are efficiently formed and bubbles larger than the critical radius are effectively formed. The size distribution at that time evolves with time following this Eq. (3.66).

Under the situation where bubbles abundantly grow here and there, molecules (bubbles with zero size) at that time always exist in extremely large numbers (determined by $F_{\mathrm{E}}(0)$) and the process where those molecules assemble and increase in size is itself the time evolution of $F(R,t)$. At that time, the number of bubbles passing via the critical radius R_{C} can be considered as the flux in the size space, J.

Among them, in a steady state case $F_{\mathrm{S}}(R)$ in which the shape of a size distribution function does not change, steady flux is set as steady state nucleation rate J_{S}. In the case of a Fokker-Planck equation, considering microscopic fluctuations, the flux involves diffusion in the size space and differs from flux because of simple advection in the size space, i.e., it differs from (growth rate)×(size distribution function). In our problem, it can be considered as diffusion flux based on the gradient of $F(R,t)/F_{\mathrm{E}}(R)$, as shown in Eq. (3.66). Here, it is $B(R)F_{\mathrm{E}}(R)$ that plays a role of the diffusion coefficient. In the beginning of this chapter, the steady state nucleation rate was explained using the transition probability and the migration velocity in the size space. As shown in Eqs. (3.23) and (3.24), using the result of analysis shown here enables us to intuitively understand, which has been described later.

Our problem is to identify a solution that makes the flux in the size space, given as follows, constant, i.e., identify a solution $F(R)$ to

$$J_{\mathrm{S}} = -B(R)\cdot F_{\mathrm{E}}(R)\frac{\partial}{\partial R}\left(\frac{F_{\mathrm{S}}(R)}{F_{\mathrm{E}}(R)}\right) = \text{Constant} \tag{3.67}$$

and estimate J_{S}. This is a first-order differential equation, and its integral is

$$\frac{F_{\mathrm{S}}(R)}{F_{\mathrm{E}}(R)} = -J_{\mathrm{S}}\int_0^R \frac{dR'}{B(R')\cdot F_{\mathrm{E}}(R')} + \frac{F_{\mathrm{S}}(0)}{F_{\mathrm{E}}(0)} \tag{3.68}$$

Applying boundary conditions

(a) When $R \to 0$, $F_{\mathrm{S}}/F_{\mathrm{E}} \to 1$
(b) When $R \to \infty$, $F_{\mathrm{S}}/F_{\mathrm{E}} \to 0$

to Eq. (3.68) gives,

$$\frac{F_S(0)}{F_E(0)} = 1 \tag{3.69}$$

from (a), and

$$J_S = \frac{1}{\displaystyle\int_0^\infty \frac{dR'}{B(R')\cdot F_E(R')}} = \frac{1}{In} \tag{3.70}$$

from (b).

To describe $B(R)$ within the integral In, we should consider it from a molecular viewpoint. However, we avoid doing so and obtain the growth rate $A(R)$ from macroscopic consideration and use it in Eq. (3.35) to calculate $B(R)$.[12] On the assumption that macroscopic growth rate is applicable up to the vicinity of the critical radius. As seen later, $A(R)$ is given by Eq. (4.30) as

$$A(R) \equiv \frac{dR}{dt} = \frac{v_G DC}{R}\left\{1 - \left[1 + R_P^*\left(\frac{1}{R} - \frac{1}{R_C}\right)\right]^{\frac{1}{2}}\right\} \tag{3.71}$$

$$\approx \frac{v_G DC}{R} \times \frac{(R - R_C)R_P^*}{2RR_C} \tag{3.72}$$

Here because the behavior of R at the vicinity of R_C is important, an approximation of $R_P^*/R - R_P^*/R_C \ll 1$ is used. $R_P^*(C, P)$ has the size scale in which the interfacial tension is effective, and $R_P^* = 2\gamma/P_G^*$ is given. P_G^* is a supersaturation pressure, defined by Eq. (3.21) as a saturation pressure at a curved interface in a supersaturated state. Using an equation of state $v_G P_G^* = k_B T$ gives $R_P^* = 2\gamma v_G/k_B T$, and $B(R)$ is obtained from Eq. (3.57) as follows:

$$B(R) = \frac{v_G DCk_B T R_P^*}{16\pi\gamma R_C R^2} = \frac{v_G^2 DC}{8\pi R_C R^2} \tag{3.73}$$

Note that v_G in this equation is the molecular volume of gas under the supersaturation pressure P_G^* and is obtained by calculation using Eq. (3.21).[13] Parts of $R - R_C$ are conveniently canceled by the denominator and numerator of Eq. (3.57). This indicates that the values of the denominator and numerator of $B(R)$ diminish at $R = R_C$ by the same rate. Moreover, Integrating the denominator of Eq. (3.70) yields (see Sect. 11.5 in the Appendix),

$$J_S = Z \cdot B_C \cdot F_E^* \tag{3.75}$$

where B_C and F_E^* are values of $B(R)$ and $F_E(R)$ when $R = R_C$.

$$B_C \equiv B(R_C) = \frac{v_G^2 DC}{8\pi R_C^3} \tag{3.76}$$

---

[12]This is the approach of Lifshitz and Pitaevskii (1981) and is convenient because the result is determined only by well-known macroscopic physical properties. Another method is to calculate a reaction rate in accordance with the theory of the transition state in terms of statistical mechanics.

[13]Toramaru (1995) introduces $y = C_{eq}/C$ to obtain $R_0 = 2\gamma/P \cdot P/P_G^* = 2\gamma y^2/P$ and defines v_G by $v_G = k_B T/P$. In this case,

$$B(R) = \frac{v_G^2 DCy^2}{8\pi R_C R^2} \tag{3.74}$$

is obtained, which contains y^2 in its numerator. Note that v_G is the molecular volume of gas under the liquid pressure P.

$$F_E^* \equiv F_E(R_C) = \frac{CR_C^2}{v_G} \exp\left(-\frac{4\pi\gamma R_C^2}{3k_BT}\right) \tag{3.77}$$

Z is called a Zeldovich factor, is related to the second-order term (the second derivative) of the equation obtained by expanding the free energy in terms of the bubble size giving the maximum value, and is defined as follows:

$$Z = \sqrt{\frac{\left|\frac{d^2\Delta\mathcal{F}}{dR^2}\right|}{2\pi k_BT}} = \sqrt{\frac{4\gamma}{k_BT}} \tag{3.78}$$

Z has a dimension of the inverse of length (m^{-1}) and Z^{-1} gives a scale characterizing motion on the size space in the vicinity of the critical size. Moreover, Z^{-1} can be considered as a characteristic size scale in which diffusion $B(R_C)$ effectively works. Moreover, Eq. (3.75) can be seen as the product of the number of bubbles contained in the size width of the critical size, F_E^*, and the transition rate when passing through the critical size, $Z \cdot B(R_C)$.

The nucleation rate is then given by Eq. (3.75) and can be interpreted as follows:

$$J_S = \underbrace{(\text{Growth rate } G_R^*)}_{Z\cdot B_C\ (\text{m/s})} \cdot \underbrace{(\text{Critical population density } F_E^*)}_{F_E^*\ (\text{number/m}^4)} \tag{3.79}$$

$$= \underbrace{(\text{Transition frequency } \mathcal{K}^*)}_{Z^2\cdot B_C\ (\text{s}^{-1})} \cdot \underbrace{(\text{Critical population } N^*)}_{F_E^*\cdot Z^{-1}\ (\text{number/m}^3)} \tag{3.80}$$

Finally, the nucleation rate is expressed as follows[14]:

$$J_S = Z\frac{v_G DC^2}{8\pi R_C} \exp\left(-\frac{4\pi\gamma R_C^2}{3k_BT}\right) = \frac{\phi_G DC}{4\pi R_C}\sqrt{\frac{\gamma}{k_BT}} \exp\left(-\frac{4\pi\gamma R_C^2}{3k_BT}\right) \tag{3.82}$$

where $\phi_G = v_G C$ is the dissolved volumetric concentration of water in melt, which was measured at supersaturation pressure $P_G^*(C, P)$ and R_C, and is calculated using Eqs. (3.20) and (3.21). This equation differs from the nucleation rate reported by Toramaru (1995)[15] in 4π because of the difference in F_E^0 and definition of v_G. Although this difference of this factor does not exert such an important influence on the history of nucleation, the formulation described above is clearer and expresses correspondence with the nucleation rate of crystals more directly. Although it is basically the same as the nucleation rate at crystallization, this factor differs by the factor of two, which results from the fact that the activity depends on the square of the molar fraction.

---

[14] If the coefficient of the equilibrium size distribution F_E^0 is left as it is,

$$J_S = Z\frac{v_G^2 DCF_E^0}{8\pi R_C^3} \exp\left(-\frac{4\pi\gamma R_C^2}{3k_BT}\right) \tag{3.81}$$

is.

[15]

$$J_S = \frac{DC^2(1-y^2)}{8\pi}\sqrt{\frac{k_BT}{\gamma}} \exp\left(-\frac{4\pi\gamma R_C^2}{3k_BT}\right) = Z\frac{v_G DC^2 y^2}{8\pi R_C} \exp\left(-\frac{4\pi\gamma R_C^2}{3k_BT}\right) \tag{3.83}$$

where $v_G = k_BT/PD$.

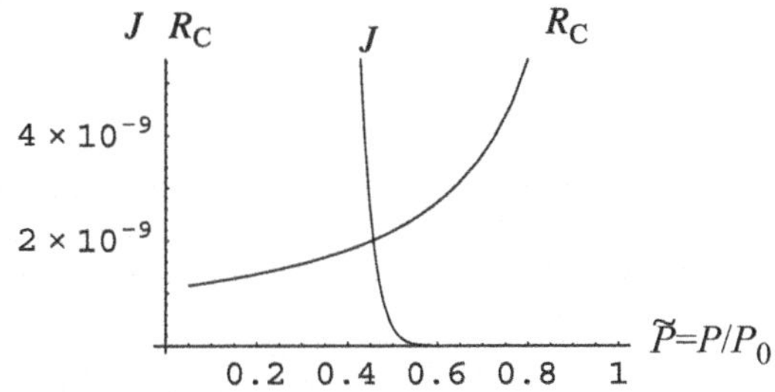

Fig. 3.9 Relationship between the scaled pressure P/P_0 and the critical nucleus radius R_C (m) and the nuleation rate J (m^{-3} s^{-1})

The nucleation rate (instantaneous nucleation rate for a given pressure and concentration) considerably depends on the degree of supersaturation $P_G^* - P$, i.e., the critical nucleus radius R_C, and a slight change in the degree of supersaturation leads to a change in nucleation rate by orders of magnitude. Figure 3.9 shows the behavior. Under a given constant pressure, the bubble nucleation rate increases exponentially as a function of the amount of decompression from the saturation pressure and the number of bubbles increases with time in the form of the integral of the bubble nucleation rate with respect to time. As seen in the next section, because formed bubbles grow by diffusion of volatiles in melt, they become depleted of volatiles and the degree of supersaturation decreases. Therefore, nucleation eventually ends and the bubble number density is determined.

In the case where the amount of decompression changes with time, the history of nucleation is generally determined by competition between the increase in the degree of saturation because of the increase in the amount of decompression and the decrease in the degree of saturation because of the growth of nucleated bubbles. Therefore, dependence of the resultant bubble number density (the number of bubbles in unit volume of melt) on the decompression rate $|dP/dt|$ is closely related to the growth law of bubbles (Toramaru 1995). For diffusion-controlled growth, the bubble number density increases in proportion to the power of 3/2 of the decompression rate. This indicates that when magma has the same chemical composition and temperature, magma under the faster decompression rate forms more bubbles compared with that under the slower decompression rate. As seen later, this has been confirmed by experiments conducted using rhyolitic melt.

Classical theory of nucleation rate 2: Difference in the pre-factor

There are various expressions for a pre-exponential factor J_0 in the steady state nucleation rate $J = J_0 \exp(-\Delta\mathcal{F}^*/(k_B T))$. The essential difference is the approach taken for calculating monomer flux into the critical nucleus. The approach used includes the kinetic theory of gases and phenomenology. In the following, J_0 is expressed using Eq. (3.26), $J_0 = \mathcal{K}^* \cdot N_1$.

Becker and Döring (1935) discussed the formation of droplets in gas using the kinetic theory of gas and drew

$$J_0 = \left[\frac{P}{(2\pi m k_B T)^{1/2}}\right]\left[v_m\left(\frac{4\gamma}{k_B T}\right)^{1/2}\right] N_1 \tag{3.84}$$

Friedlander (1977), where P is partial pressure of supersaturation of the component of droplets, N_1 is the number of monomer molecules in gas, v_m is the molecular volume of the monomer, and m is the mass of the molecules. The first square brackets are monomer flux to the critical size (number/m^2/s) and the next square brackets have a dimension of area measuring the surface area of clusters at the critical size.

Hirth et al. (1970) considered that, in critical nucleus bubbles, the monomer flux from liquid is in balance with the monomer flux in gas in bubbles obtained from the kinetic theory of gases, applied to the bubbles in liquid. Thus, Eq. (3.83) can be rewritten using the molecular volume of gas $v_G = k_B T/P$, resulting in

$$J_0 = \frac{v_{\rm m}}{v_{\rm G}}\left[\frac{\gamma}{\pi m}\right]^{1/2} N_1 \tag{3.85}$$

(The original equation includes a condensation constant as a coefficient.)
Turnbull and Fisher (1949) considered the nucleation of solid crystals in liquid, including monomer flux in liquid, as a series of reactions and applied the theory of absolute reaction rate theory to yield

$$J_0 = n_A^* \left(\frac{\gamma}{k_{\rm B}T}\right)^{1/2} \left(\frac{2v}{9\pi}\right)^{1/3} \frac{k_{\rm B}T}{\hbar} \exp\left(-\frac{Q}{k_{\rm B}T}\right) N_1 \tag{3.86}$$

where n_A^* is the number of molecules per unit surface area of critical nuclei and $\hbar$ is the Planck constant. Uhlmann (1972) made some revisions using the diffusion coefficient D and the viscosity coefficient η.

$$J_0 = \hat{J}_0 \exp\left(-\frac{Q}{k_{\rm B}T}\right) = \frac{D}{a_0^2} N_1 = \frac{k_{\rm B}T}{3\pi a_0^3 \eta} N_1 \tag{3.87}$$

where a_0 is the intermolecular distance.
In this book, formulation is conducted on the basis of the phenomenological diffusion growth rate following Lifshitz and Pitaevskii (1981) and Zel'dovich (1942). For crystallization (assuming that $N_1 = F_{\rm E}^0 \cdot \delta R = v_{\rm S} F_{\rm E}^0 / R_{\rm C}^2$),

$$J_0 = \frac{\phi_{\rm S} Z D}{4\pi R_{\rm C}} N_1 = \frac{\phi_{\rm S} D}{4\pi R_{\rm C}/Z} N_1 \tag{3.88}$$

holds, and because $1/Z$ defines the size width in the critical state, $a_0^2 \approx 4\pi R_{\rm C}/Z$ can be considered. In this equation, $\phi_{\rm S}$ is volumetric concentration of the crystallization component in liquid and $\phi_{\rm S} = v_{\rm S} C$ holds, where $v_{\rm S}$ is the molecular volume of the crystallizing component in liquid. In this book, $F_{\rm E}^0$ is determined as $N_1 = C$ holds in the end.

3.3.5 Steady State Size Distribution

Using the calculation result (Eq. 11.40) of Eq. (3.70) for Eq. (3.68) yields

$$\frac{F_{\rm S}(R)}{F_{\rm E}(R)} = 1 - Z \cdot \int_0^R \left(\frac{B(R_{\rm C})}{B(R')}\right)\left(\frac{F_{\rm E}(R_{\rm C})}{F_{\rm E}(R')}\right) dR' \tag{3.89}$$

$$= 1 - Z \cdot \int_0^R \left(\frac{R'}{R_{\rm C}}\right)^2 \exp\left[-\pi (Z R_{\rm C})^2 \left(\frac{R'}{R_{\rm C}} - 1\right)^2\right] dR' \tag{3.90}$$

Although the integration result is very complicated, its graphical representation is shown in Fig. 3.10.
The ratio at the critical size

$$\frac{F_{\rm S}(R_{\rm C})}{F_{\rm E}(R_{\rm C})} = \text{Constant} \tag{3.91}$$

is always lesser than one and approaches one as the critical size becomes small.

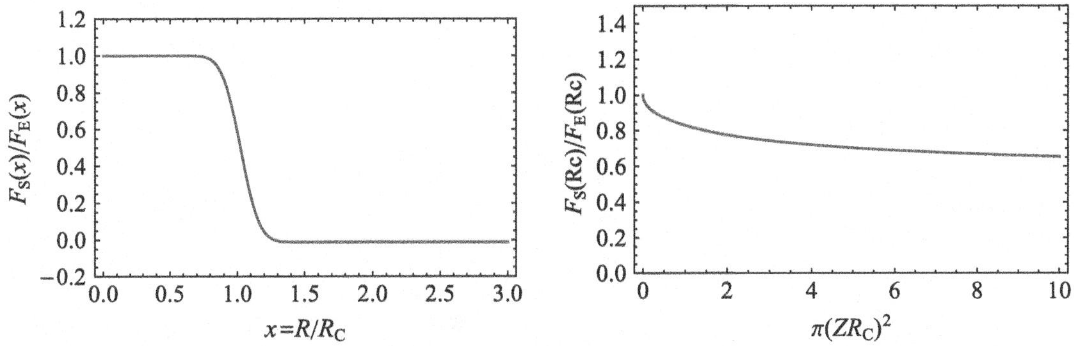

Fig. 3.10 (Left) Steady state solution normalized by the equilibrium size distribution. (Right) The ratio of values of the equilibrium size distribution to values of the steady state size distribution at the critical size as a function of the critical size. This ratio approaches one as the critical size becomes small

3.4 Heterogeneous Nucleation

In nucleation phenomena in actual magma, heterogeneous nucleation in which the surface of crystals works as nuclei is often dominant. In this case, by applying the effective interfacial energy between heterogeneous nuclei and the surrounding medium, $\gamma_{\rm eff}$, rather than the gas—liquid interfacial energy γ, the understanding about the homogeneous nucleation described above can be used as it is. As shown in Fig. 3.11, let us consider the geometry of heterogeneous nuclei after assuming that the shape of a nucleus is part of a sphere. The same consideration as the case of a sphere in the homogeneous nucleation yields changes in free energy corresponding to Eq. (3.10) as follows:

$$\Delta\mathcal{F}_{\rm hetero} = -(P_{\rm G} - P)V_{\rm CAP} + (\gamma_{\rm S,G} - \gamma_{\rm S,L})\mathcal{A}_{\rm BOTTOM} + \gamma_{\rm G,L}\mathcal{A}_{\rm TOP} \tag{3.92}$$

where $V_{\rm CAP}$ is the volume of bubbles and $\mathcal{A}_{\rm BOTTOM}$ is the area of the bottom of the nucleus. $\mathcal{A}_{\rm TOP}$ is the area of the interface between the gas phase newly formed with the nucleus formation and the liquid phase. Because the interface between a nucleating agent solid and liquid exists before the nucleus appears, change in the interfacial energy because of the appearance of the nucleus is obtained by deducting it (the second term of the right-hand side, $(\gamma_{\rm SG} - \gamma_{\rm SL})\mathcal{A}_{\rm BOTTOM}$). $V_{\rm CAP}$, $\mathcal{A}_{\rm BOTTOM}$, and $\mathcal{A}_{\rm TOP}$ can be written as a function of a contact angle θ formed by the crystal and melt sandwiching the gas phase as follows:

$$V_{\rm CAP} = \frac{\pi R^3}{3}(1 - \cos\theta)^2(2 + \cos\theta) \tag{3.93}$$

$$\mathcal{A}_{\rm BOTTOM} = \pi(R\sin\theta)^2 \tag{3.94}$$

Fig. 3.11 Schematic representation of a heterogeneous nucleus on the surface of solid (crystal)

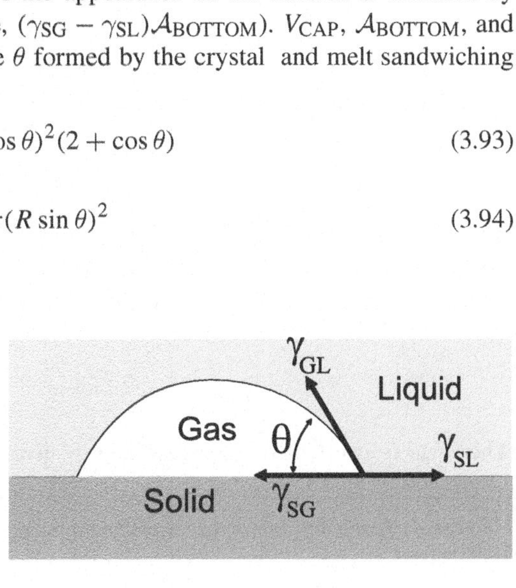

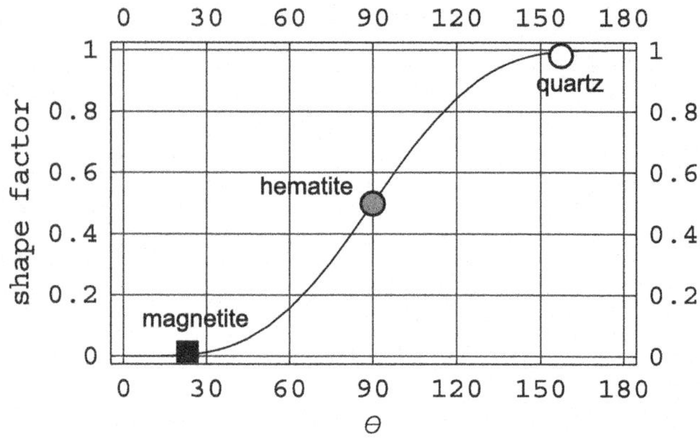

Fig. 3.12 Relationship between the shape factor and contact angles in energy required for heterogeneous nucleation. Contact angles for magnetite, hematite, and quartz in rhyolitic melt, obtained by experiments, are shown

$$\mathcal{A}_{\mathrm{TOP}} = 2\pi R^2(1 - \cos\theta) \tag{3.95}$$

where R is the radius of a sphere inscribed in the bubble nucleus. The volume of the nucleus V_{CAP} can be expressed as above.

Considering the equilibrium of the interfacial tension at the triple point in the direction along the interface between the crystal and the liquid, the relationship of

$$\gamma_{\mathrm{SG}} = \gamma_{\mathrm{SL}} - \gamma_{\mathrm{GL}} \cos\theta \tag{3.96}$$

holds.[16] Using this equation to rearrange the equation of energy change yields

$$\Delta\mathcal{F}_{\mathrm{hetero}} = -\frac{4}{3}\pi R^3 s_{\mathrm{f}}(\theta)\cdot(P_{\mathrm{G}} - P) + 4\pi R^2 \cdot s_{\mathrm{f}}(\theta)\cdot\gamma_{\mathrm{GL}} \tag{3.97}$$

where $s_{\mathrm{f}}(\theta)$ is the effect of the contact angle on the shape of the nucleus and is called the shape factor. Assuming that the nucleus is part of a sphere, $s_{\mathrm{f}}(\theta)$ can be calculated as follows (Fig. 3.12):

$$s_{\mathrm{f}}(\theta) = \frac{(2+\cos\theta)(1-\cos\theta)^2}{4} \tag{3.98}$$

Using this shape factor, change in energy associated with heterogeneous nucleation is expressed as a relation of the equation for change in energy associated with homogeneous nucleation (Eq. 3.18) $\Delta\mathcal{F}_{\mathrm{homo}}$ as

$$\Delta\mathcal{F}_{\mathrm{hetero}} = s_{\mathrm{f}}(\theta)\Delta\mathcal{F}_{\mathrm{homo}} \tag{3.99}$$

Although the critical radius R_{C}, which is important in the nucleation rate, is given as $2\gamma_{\mathrm{GL}}/(P_{\mathrm{G}}^* - P)$, if the critical radius of the heterogeneous nuclei is rewritten symbolically as $R_{\mathrm{C,hetero}} = 2\gamma_{\mathrm{eff}}/(P_{\mathrm{G}}^* - P)$, change in critical energy corresponding to Eq. (3.22) is given by

$$\Delta\mathcal{F}^*_{\mathrm{hetero}} = \frac{4\pi}{3}(R_{\mathrm{C,hetero}})^2\gamma_{\mathrm{eff}} \tag{3.100}$$

where γ_{eff} is an effective interfacial energy involved in forming the heterogeneous nuclei and given by

[16]Force balance in a direction perpendicular to the interface between the crystal and liquid is compensated by the strength of the crystal.

Table 3.1 Effective interfacial energy for crystals in rhyolitic melt. The interfacial energy between rhyolitic melt and water vapor is assumed to be 0.103 (N/m) (Cluzel et al. 2008)

Minerals	Magnetite	Hematite	Quartz
θ	≈20°	≈90°	160°–150°
Effective interfacial (N/m)	0.025	0.078	0.099

$$\gamma_{\text{eff}} = s_{\text{f}}(\theta)^{1/3}\gamma_{\text{GL}} \tag{3.101}$$

For example, $\gamma_{\text{eff}} = 0$ when $\theta = 0°$ and $\gamma_{\text{eff}} = \gamma$ when $\theta = 180°$, and γ_{eff} monotonously decreases with decrease in θ.

The contact angle defines wettability of the gas phase in the liquid phase to the crystal (non-wettability if focusing on the liquid in the gas phase), and $\theta = 0°$ indicates a completely wetted state (in which the gas phase fully spreads on the crystal surface) and yields $\gamma_{\text{eff}} = 0$, which easily causes nucleation. Although no study has been performed that directly measures γ_{eff} of actual crystals by experiments, it is known that magnetite has a tendency to be wettable and silicate minerals have a tendency to be non-wettable (Table 3.1). This is because the interface between magnetite (oxide mineral) and melt has a considerable difference in structure at a molecular level between the two phases compared with the interface between silicate minerals and melt. Furthermore, the interfacial energy between crystals and melt (= tension; to be exact, they are not equivalent but are regarded as being equivalent for simplicity.) becomes large. Because of interfacial force balance of the gas phase, crystal, and melt at the triple point, θ of magnetite becomes smaller than that of silicate minerals. Moreover, if there are clusters of H_2O molecules and uneven melt structures in melt, they seem to play a role of nucleation sites as impurities and cause heterogeneous nucleation. Whether homogeneous nucleation or heterogeneous nucleation occurs in actual magma will be described in Chap. 5.

3.5 Non-steady State Nucleation Rate

The non-steady or transient nucleation rate is related to non-steady state solutions of the Fokker-Planck equation (Eq. 3.66) and is nothing but the transient flux in the size space, J. Rewriting the Fokker-Planck equation with $\varphi(R,t) = F(R,t)/F_{\text{E}}(R)$ yields

$$\frac{\partial\varphi(R,t)}{\partial t} = \frac{1}{F_{\text{E}}(R)}\frac{\partial}{\partial R}\left[B(R)\cdot F_{\text{E}}(R)\frac{\partial\varphi(R,t)}{\partial R}\right] \tag{3.102}$$

To this equation, divide φ into a part corresponding to steady state solution φ_{st} and a non-steady term φ_{tr} as follows:

$$\varphi(R,t) = \varphi_{\text{st}}(R) + \varphi_{\text{tr}}(R,t) \tag{3.103}$$

Moreover, use the separation of variables for the non-steady term and define them as follows:

$$\varphi_{\text{tr}}(R,t) = f(R)\cdot g(t) \tag{3.104}$$

Substituting these equations above into Eq. (3.102) gives

$$\frac{1}{g(t)}\frac{dg(t)}{dt} = \frac{1}{F_{\text{E}}(R)f(R)}\frac{d}{dR}\left[B(R)\cdot F_{\text{E}}(R)\frac{df(R)}{dR}\right] = -c \tag{3.105}$$

The left-hand side is a function of only time and the central term is a function of only size. For them to be equal, they should be constants ($c = c_n$).[17] For $g(t)$,

$$g(t) = g_0 e^{-c_n t} \tag{3.106}$$

is obtained.

However, an equation for $f(R)$ is written as follows:

$$\frac{d}{dR}\left[B(R) \cdot F_{\mathrm{E}}(R) \frac{df(R)}{dR}\right] + c_n F_{\mathrm{E}}(R) f(R) = 0, \tag{3.107}$$

where c_n is an eigenvalue, and is a Sturm-Liouville equation with the weight function $FE(R)$.[18] Kashchiev (1969) solved it as an eigenvalue problem and obtained time-dependent nucleation rate defined by

$$J(t) = -B(R_{\mathrm{C}}) \cdot F_{\mathrm{E}}(R_{\mathrm{C}}) \left[\frac{\partial \varphi(R,t)}{\partial R}\right]_{R=R_{\mathrm{C}}} \tag{3.109}$$

as follows[19]:

$$J(t) = J_{\mathrm{st}} \left[1 + 2\sum_{n=1}^{\infty} (-1)^n \exp\left(-n^2 \frac{t}{\tau}\right)\right] \tag{3.110}$$

where J_{st} is a steady state nucleation rate given by Eq. (3.82). τ is a first-order eigenvalue having a dimension of time characterizing the transient process and is given by

$$\tau = \frac{4}{\pi^3 Z^2 B_{\mathrm{C}}} = \frac{8 k_{\mathrm{B}} T R_{\mathrm{C}}^3}{\pi^2 \gamma v_{\mathrm{G}}^2 D C y^2} \tag{3.111}$$

Because the inverse of the Zeldovich factor Z yields the size scale of fluctuations in the size space near the critical nucleus size and B_{C} gives diffusion in the size space, $Z^2 B_{\mathrm{C}}$ can be considered as the inverse of the relaxation time of fluctuations.

The left chart of Fig. 3.13 shows the behavior of Eq. (3.110). A series of Eq. (3.110) can be well approximated when $n > 3$. The temporal behavior of the nucleation rate in this figure demonstrated that it can be approximated as the steady state nucleation rate when $t = 5\tau$. Because τ depends on the critical nucleus size R_{C} and the critical nucleus size depends on the degree of supersaturation, time leading to the steady state nucleation should be a function of the degree of supersaturation. The right chart of Fig. 3.13 shows an instance calculated by substituting actual values. Time required to reach the steady state depends on the degree of supersaturation and the diffusion coefficient (or viscosity). When the diffusion coefficient D is 10^{-11} ($\mathrm{m^2/s}$), the time is approximately 0.1 s even if the degree of supersaturation is small, i.e., $C - C_{\mathrm{eq}} = 0.01 C_{\mathrm{eq}}$. It appears that practical homogeneous nucleation

[17] c_n is used rather than c because there can be multiple constants.

[18] This equation is a typical second-order ordinary differential equation

$$\frac{d^2 y(x)}{dx^2} + p(x)\frac{dy(x)}{dx} + q(x)y(x) = 0 \tag{3.108}$$

in which the coefficient is not a constant. Many equations appearing in mathematical physics, such as the Legendre equation and Bessel's equation, result in this form. Because the case involving constant coefficients is familiar, the case in which coefficients are not constants seems to be easily solved. However, the normal method of solving it is very complicated.

[19] This solution is expressed by the result in which solutions to an infinite number of eigenvalues are superimposed.

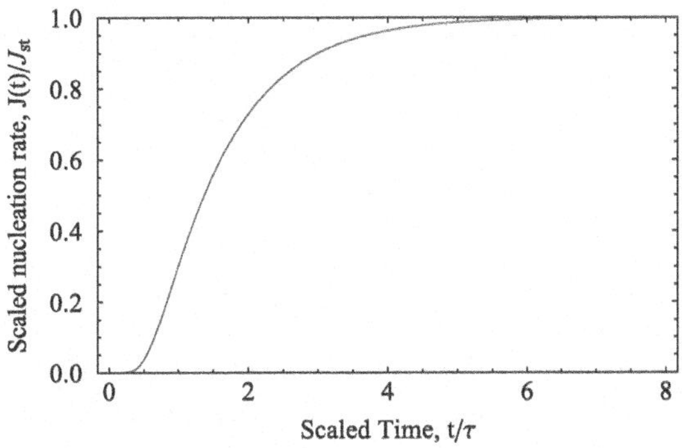

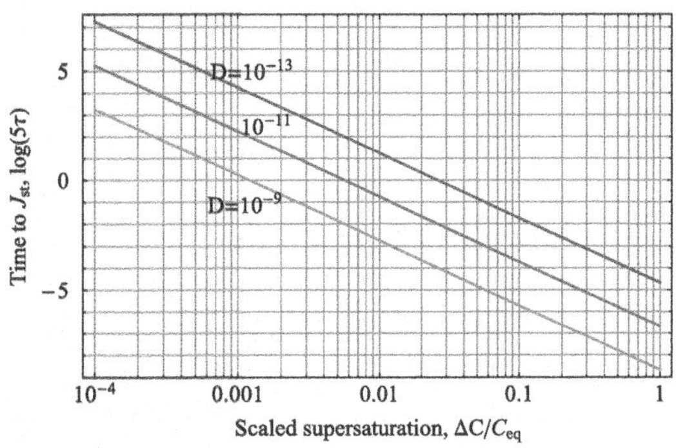

Fig. 3.13 (Left) Relationship between the transient nucleation rate and time. (Right) Relationship between the time (s) required to reach steady state nucleation and the dimensionless degree of supersaturation $(C - C_{\text{eq}})/C_{\text{eq}}$. Three lines with different diffusion coefficients $D\,(\text{m}^2/\text{s})$ are shown

can be handled as a normal steady state nucleation. However, when D is 10^{-13} (m^2/s), the time is approximately 10 s, and it may exert an influence in very drastic decompression. Moreover, this may have an effect on decompression degassing-induced crystallization.

3.6 Various Kinds of Correction for Classical Nucleation Theory

3.6.1 Tolman Correction

Correction for classical homogeneous nucleation theory includes curvature effects on the surface energy. When the curvature of the interface increases, i.e., the bubble size decreases (nanoscale), the surface energy will deviate from the normal one because the geometric effect of the curved interface works on the interaction between molecules (Tolman 1949). This is called the Tolman correction, and assuming the correction coefficient δ_{T}, the interfacial energy γ is calculated as follows:

$$\gamma = \frac{1}{1 + 2\frac{\delta_{\text{T}}}{R}}\gamma_0 \tag{3.112}$$

where γ_0 is the interfacial energy when the curvature is zero and δ_{T} for a pure substance is on the order of 10^{-8} m and may become smaller than the interfacial energy determined by macroscopic measurement (Tanaka et al. 2015). Moreover, high-order correction can be possible (Schmelzer and Baidakov 2016; Tanaka et al. 2016). However, because these studies discussed the interface between the gas phase and the liquid phase of a pure substance, i.e., a one-component system, it is unclear whether they are applicable to the binary system handled in this book.

3.6.2 Poynting Correction

In a problem of boiling of a pure substance, boiling at a constant pressure was assumed and decompression-induced vesiculation, as handled in this book, was not assumed. Blander and Katz (1975) examined the case where the ambient pressure changes and evaluated the influence of the compressibility of liquid on the equilibrium vapor pressure within bubbles. They called it the Poynting correction. For magma dealt within this book, this correction has already been considered in the exp-term of Eq. (3.21).

3.6.3 Viscosity Correction

Kagan (1960) thought that the viscosity of liquid has an effect when critical nuclei form. When he obtained the transition rate of critical nuclei in a problem of boiling of a one-component system, he simultaneously solved not only the inflow of molecules into bubbles, determined by the kinetic theory, but also the below-mentioned Rayleigh-Plesset equation and derived an equation in which a pre-exponential factor of the steady state nucleation rate includes viscosity. For magma, the inflow of molecules by diffusion and the Rayleigh-Plesset equation have to be simultaneously solved because magma is a binary (or multi-component) system. Consequently, the viscosity coefficient is included in a pre-exponential factor of the steady state nucleation rate in the form of the Péclet number, described later, and as viscosity increases infinitely, the pre-exponential factor approaches zero (Nishiwaki and Toramaru 2019).

References

Becker R, Döring W (1935) Ann d Physik [5] 24:719

Blander M, Katz JL (1975) Bubble nucleation in liquids. AIChE J 21(5):833–848

Cluzel N, Laporte D, Provost A, Kannewischer I (2008) Kinetics of heterogeneous bubble nucleation in rhyolitic melts: implications for the number density of bubbles in volcanic conduits and for pumice textures. Contrib Mineral Petrol 156:745–763

Friedlander SK (1977) Smoke, dust and haze: fundamentals of aerosol behaviour. Wiley

Haken H (1978) Synergetics: an introduction. Non-equilibrium phase transition and self-self organization in physics, chemistry and biology, 2nd ed. Springer

Hirth JP, Pound CM, Pierre GR St (1970) Bubble nucleation. Metal Trans 1:939–945

Kagan Y (1960) The kinetics of boiling of a pure liquid. Russ J Phys Chem 34:42–46

Kashchiev D (1969) Solution of the non-steady state problem in nucleation kinetics. Surface Sci 14:209–220

Landau LD, Lifshitz EM (1958) Statistical physics. Pergamon, London

Lifshitz HM, Pitaevskii LP (1981) Physical kinetics. Pergamon Press, Oxford

Nishiwaki M, Toramaru A (2019) Inclusion of viscosity into classical homogeneous nucleation theory for water bubbles in silicate melts: reexamination of bubble number density in ascending magmas. J Geophys Res Solid Earth 124(8):8250–8266

Schmelzer JW, Baidakov VG (2016) Comment on "Simple improvements to classical bubble nucleation models". Phys Rev E 94(2):026801

Tanaka KK, Tanaka H, Angèlil R, Diemand J (2016) Reply to "Comment on 'Simple improvements to classical bubble nucleation models"'. Phys Rev E 94(2):026802

Tanaka KK, Tanaka H, Angèlil R, Diemand J (2015) Simple improvements to classical bubble nucleation models. Phys Rev E 92(2):026801

Tolman RC (1949) The effect of droplet size on surface tension. J Chem Phys 17(3):333–337

Toramaru A (1995) Numerical study of nucleation and growth of bubbles in viscous magmas. J Geophys Res 100:1913–1931

Turnbull D, Fisher JC (1949) Rate of nucleation in condensed systems. J Chem Phys 17:71. https://doi.org/10.1063/1.1747055

Uhlmann DR (1972) A kinetic treatment of glass formation. J Non-Crystal Solids 7:337–348

Zel'dovich YB (1942) Zhur Eksper Teor Fiz 12:525

4 Growth and Expansion of Bubbles

Abstract

Bubbles nucleating in supersaturated liquids grow by H_2O diffusion in melts into bubbles (the number of H_2O molecules in the bubbles increases) and expand by decompression (the number of H_2O molecules in the bubbles does not change but the molecular volume increases). In this manner, "growth" ("grow") and "expansion" ("expand") are used differently for different phenomena in this chapter. Although growth and expansion simultaneously progress, their behavior will be separately understood by ignoring the other at first. Then, the two phenomena will be combined to understand the experimental results. In this chapter, important factors including time constants of various elemental processes and their dimensionless number ratios, such as the Pélclet number, which is the ratio of the diffusion to the viscosity-limited expansion rate, are discussed.

4.1 Outline of Calculation of Bubble Growth and Expansion

Bubbles nucleating in supersaturated liquids grow by H_2O diffusion in melts into bubbles (the number of H_2O molecules in the bubbles increases) and expand by decompression (the number of H_2O molecules in the bubbles does not change but the molecular volume increases). In this manner, "growth" ("grow") and "expansion" ("expand") are used differently for different phenomena in this chapter. Although growth and expansion simultaneously progress, their behavior will be separately understood by ignoring the other at first. Then, the two phenomena will be combined to understand the experimental results. In this chapter, important factors including time constants of various elemental processes and their dimensionless number ratios, such as the Pélclet number, which is the ratio of the diffusion to the viscosity-limited expansion rate, are discussed.

Bubble growth and expansion processes are understood by simultaneously solving the following equations: (1) an equation controlling the rate of increase of molecules in bubbles by diffusion flux, (2) an equation of conservation of momentum, which describes change in bubble radius (the Rayleigh-Plesset equation), (3) an equation of the diffusion field of H_2O in melt (steady-state solution is sometimes used), (4) an equation of conservation of mass of H_2O in the entire system, and (5) an equation of state of gas in bubbles.

A. Toramaru, *Vesiculation and Crystallization of Magma*, Advances in Volcanology,
https://doi.org/10.1007/978-981-16-4209-8_4

The problem of bubble growth includes two points. One problem is to obtain a concentration field surrounding spherical bubbles, the radius of which increases, i.e., to solve a diffusion equation. The diffusion equation for the concentration of a gas component in liquid, C, is as follows:

$$\frac{\partial C}{\partial t} + \underbrace{u_r \frac{\partial C}{\partial r}}_{\text{Advection}} = \underbrace{\frac{1}{r^2}\frac{\partial}{\partial r}\left(r^2 D \frac{\partial C}{\partial r}\right)}_{\text{Diffusion}} \tag{4.1}$$

where u_r is the velocity of liquid in the radial direction and r is the distance from the bubble center. Assuming that the liquid is incompressible, the conservation of mass of liquid, as described later, yields u_r as follows:

$$u_r = \frac{R^2}{r^2}\dot{R} \tag{4.2}$$

where $\dot{R} \equiv dR/dt$ is the rate of change in bubble radius. Then, the other problem is to obtain the rate of change in radius. $\dot{R}$ is then obtained by solving the Rayleigh-Plesset equation (Eq. 4.52)

$$R(t)\ddot{R}(t) + \frac{3}{2}\dot{R}(t)^2 = \frac{1}{\rho}\left(P_{\mathrm{G}} - P_{\infty} - \frac{2\gamma}{R(t)} - 4\eta\frac{\dot{R}(t)}{R(t)}\right) \tag{4.3}$$

Pressure within a bubble P_{G} is then determined by the equation of state of gas and P_{G} determines the equilibrium concentration at the interface by the relationship with solubility. The equilibrium concentration will be a boundary condition for the diffusion equation at $r = R$. These equations are for the case of a single bubble; if many bubbles form from nucleation and the amount of a gas component being present as a gas phase cannot be ignored, the conservation of mass of vesiculating gas should be considered.

$$[\text{the amount of gas as bubbles}] + [\text{the amount of gas remaining in liquid}] = [\text{the initial amount of gas}] \tag{4.4}$$

For the diffusion equation, a boundary condition at an appropriate distance from bubbles (infinity in the case of a single bubble) is required. If the volume fraction of bubbles increases as the presence of adjacent bubbles cannot be ignored in the diffusion and flow fields, a cell model in which each bubble in multi-bubbles are assigned a cell (an extended Rayleigh-Plesset equation: an effective viscosity coefficient considering the geometric effect of bubbles' arrangement is used rather than a melt viscosity coefficient) is applied.

These processes can be numerically solved; however, there are many analytical difficulties such as the coupling of the content field with the rate of change in the bubble radius (the coupling of diffusion with viscosity); the dependence of physical properties such as diffusion coefficients; viscosity coefficients; and surface tension on the content of the gas component, time-dependent behavior of diffusion; and mean/non-mean fields of the conservation of mass of gas components. Therefore, to analytically understand the bubble growth behavior, we will consider a simple case after repeated approximations.

First, ignoring the flow field and assuming that the bubble radius is in a mechanical equilibrium, we discuss growth driven only by diffusion flux to bubbles under a certain degree of supersaturation. Subsequently, we deal with the expansion problem. Here, we deal with the case where mechanical equilibrium holds and the case where it does not hold. In cases where mechanical equilibrium holds, among pressure within a bubble P_{G}, liquid pressure P, and bubble radius R is a special relationship expressed by the Laplace equation. However, for cases where mechanical equilibrium does not hold, viscous resistance and inertia of the surrounding fluid have to be evaluated. This will be performed by the Rayleigh-Plesset equation derived later.

A precondition in a series of processes of bubble growth is the idea that, at the interface between bubbles and liquid, the local chemical equilibrium always holds.[1] Hence, the equilibrium concentration at the bubble surface will be explained first.

4.2 Equilibrium Concentration at the Bubble Surface

4.2.1 General Expression

The diffusion flow of a gas component to bubbles nucleating in supersaturated solution is driven by differences between an equilibrium concentration at the gas-liquid interface and supersaturation concentration in liquids. The equilibrium concentration $C_{\text{int}}(P_G, P)$ at the interface between bubbles with gas pressure P_G and liquid (liquid pressure P) is derived by Eq. (2.35) because a saturation concentration in the case where the gas pressure of bubbles and the liquid pressure is equal to P is $C_{\text{eq}}(P)$, i.e.,

$$C_{\text{int}}(P_G, P) = \left(\frac{P_G}{P}\right)^{\frac{1}{2}} C_{\text{eq}}(P) \tag{4.5}$$

where $C_{\text{eq}}(P)$ is given by Eq. (2.25) at a reference saturated state (C_0, P_0).

4.2.2 Equilibrium Concentration in the Mechanical Equilibrium

For being in mechanical equilibrium, pressure within the bubbles P_G is connected to liquid pressure P and the bubble radius R through the Laplace equation (2.42) $P_G - P = \frac{2\gamma}{R}$, and an equilibrium concentration at the interface is given by the Gibbs-Thomson equation (2.43):

$$C_{\text{int}} = C_R(R, P) = \left(1 + \frac{1}{P}\frac{2\gamma}{R}\right)^{\frac{1}{2}} C_{\text{eq}}(P). \tag{4.6}$$

4.2.3 Expression Using the Critical Radius

When combining mechanical processes such as the Laplace equation and the Rayleigh-Plesset equation and solving them with P_G, Eq. (4.5) is sufficient to numerically solve bubble growth. However, for formulating the nucleation rate of bubbles and Ostwald ripening, it is convenient to express a supersaturated state with the critical radius R_C in an almost saturated state because the stability of a bubble with an arbitrary radius R; in other words, whether the bubble grows or contracts, can be determined.

The bubble's critical radius R_C and the saturated pressure within the bubble P_G^* are connected by the Laplace equation.

[1] It slightly differs from crystal growth. Reaction-controlled crystal growth rate is driven by difference between concentration in liquid and equilibrium concentration at interface (i.e., the degree of supersaturation).

$$P_G^* - P = \frac{2\gamma}{R_C} \tag{4.7}$$

Because the interface equilibrium concentration of a bubble having the critical radius is equal to the supersaturation concentration C ($C = C_{int} = C_R(R_C, P)$), Eq. (4.6) can be rewritten as follows:

$$\frac{C}{C_{eq}(P)} = \left(1 + \frac{1}{P}\frac{2\gamma}{R_C}\right)^{\frac{1}{2}} = \left(\frac{P_G^*}{P}\right)^{\frac{1}{2}} \tag{4.8}$$

Moreover, this equation defines the critical radius $R_C(P, C)$ as a function of liquid pressure P and concentration of the gas component in liquid C. Squaring Equation (4.6) for any bubble radius R and the corresponding equilibrium concentration C_R, and Eq. (4.8), and subtracting one from the other yields,[2]

$$C_R(R, C, P) = C\left[1 - R_P^*(C, P)\left(\frac{1}{R_C(C, P)} - \frac{1}{R}\right)\right]^{\frac{1}{2}} \tag{4.9}$$

where R_P^* is given as follows.

$$R_P^*(C, P) = \frac{2\gamma}{P\left(\dfrac{C}{C_{eq}(P)}\right)^2} = \frac{2\gamma}{P_G^*} \tag{4.10}$$

$R_C(P, C)$ is derived from Eq. (4.8) as follows

$$R_C(C, P) = \frac{2\gamma}{P\left[\left(\dfrac{C}{C_{eq}(P)}\right)^2 - 1\right]} = \frac{2\gamma}{P_G^* - P} \tag{4.11}$$

This is, as a matter of course, a condition for the gas-liquid mechanical and chemical equilibriums.[3] Here, because the equilibrium concentration at pressure P, $C_{eq}(P)$, is expressed by saturation

[2] This equation slightly differs from the equation for crystallization (Eq. 7.38). This results from (1) difference in the equation of state between bubbles and crystals (This is attributed to the equation of state of gas, whereas crystals are assumed to be linear elastic bodies) and (2) difference in activity relation (water is proportional to the square of concentration, whereas crystals are assumed to be an ideal solution). If an ideal gas $P_G^* = k_B T / v_G$ is applied to gas, the equation agrees with the corresponding definitional equation of crystallization in the following form: $R_P^* = 2\gamma v_G / k_B T$.

[3] Assuming $y = C_{eq}/C$, $y^{-2} - 1$ gives an estimate of the dimensionless degree of supersaturation, Δ, and the dimensionless degree of supersaturation Δ and the critical nucleus radius R_C generally have a relationship

$$\Delta = y^{-2} - 1 = \frac{C^2 - C_{eq}^2}{C_{eq}^2} = \frac{1}{P}\frac{2\gamma}{R_C}, \quad \text{or,} \tag{4.12}$$

$$\left(\frac{C}{C_{eq}}\right)^2 = y^{-2} = \Delta + 1$$

then

$$R_C = \frac{2\gamma}{P \cdot \Delta} \tag{4.13}$$

can be written.

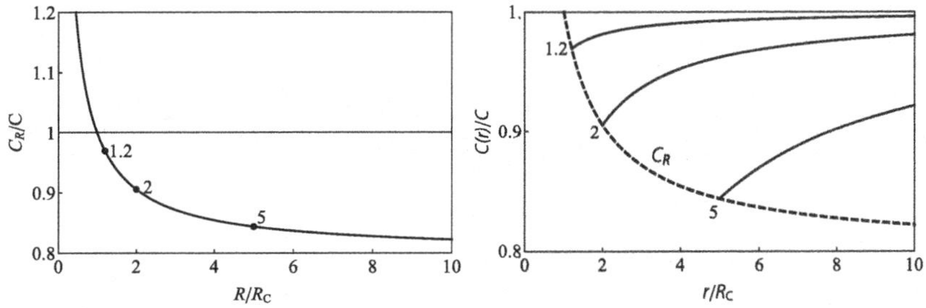

Fig. 4.1 (Left) Relationship between the bubble radius and the interface equilibrium concentration in mechanical equilibrium. (Right) Steady-state distribution of the gas component concentration around bubbles with radii of 1.2, 2, or 5 times the critical nucleus (solid lines). The curve represented by the broken line is the interface equilibrium concentration shown in the left chart. The degree of supersaturation $(C - C_{eq})/C = 0.2$ is then assumed

concentration C_0 at a reference pressure P_0 as Eq. (2.25), R_P^* and R_C can be written as follows:

$$R_P^*(C, P) = \frac{2\gamma}{P_0 \left(\frac{C}{C_0}\right)^2 \exp\left(-\frac{v_m P_0}{k_B T_0}\left(1 - \frac{P}{P_0}\right)\right)} \tag{4.14}$$

$$R_C(C, P) = \frac{2\gamma}{P_0 \left[\left(\frac{C}{C_0}\right)^2 \exp\left(-\frac{v_m P_0}{k_B T_0}\left(1 - \frac{P}{P_0}\right)\right) - \frac{P}{P_0}\right]} \tag{4.15}$$

Because a supersaturated state is considered now, $C > C_{eq}(P)$. From this equation, when a bubble is larger than the critical radius ($R > R_C$), $1/R - 1/R_C < 1$ holds and therefore $C_R < C$ holds (left chart of Fig. 4.1). This indicates that when bubbles formed after nucleation grow even slightly larger than the critical radius, the equilibrium concentration at the interface is smaller than the supersaturated concentration in the surrounding melt, C, and diffusion of the gas component to bubbles from the surrounding occurs because of the concentration gradient. This is consistent with discussion in terms of energy theory.

4.3 Steady-State Diffusion-Limited Growth

4.3.1 Case Not Including the Advection Term

A content field in liquid follows

$$\frac{\partial C(t, r)}{\partial t} = D\frac{1}{r^2}\frac{\partial}{\partial r}\left(r^2 \frac{\partial C(t, r)}{\partial r}\right), \tag{4.16}$$

a spherically symmetric diffusion equation that is obtained by ignoring the advection term in Eq. (4.1). The fact that concentration distribution around bubbles is steady means $\partial C/\partial t = 0$ in terms of mathematics. This means, in terms of physics, as the bubble radius changes with expansion of bubbles, the content field $C(t, r)$ changes in response to it; however, it immediately reaches the steady state for the bubble radius and pressure within the bubble.

A steady-state solution for this problem takes the form of

$$C(r) = c_1 + \frac{c_2}{r} \tag{4.17}$$

Lyakhovsky et al. (1996), where c_1 and c_2 are constants of integration determined from boundary conditions. Boundary conditions for the problem here are

$$\text{At } r \to R\,, \quad C(r) \to C_R$$
$$\text{At } r \to \infty\,, \quad C(r) \to C_\infty$$

and

$$C_R = c_1 + \frac{c_2}{R} \tag{4.18}$$

$$C_\infty = c_1; \tag{4.19}$$

therefore, the solution is given as follows:

$$C(r) = C_\infty - (C_\infty - C_R)\frac{R}{r} \tag{4.20}$$

(right chart of Fig. 4.1). From these equations, concentration gradient at the surface of bubbles is found to be

$$\left|\frac{\partial C(r)}{\partial r}\right|_{r=R} = \frac{C_\infty - C_R}{R} \tag{4.21}$$

assuming that diffusion flux of the gas component to bubbles in supersaturated solution (the number of molecules/unit time/unit area) is I. Note that the direction in which r is positive is set to be the direction the flux is positive; therefore, the flux to bubbles is negative. Change in the number of molecules of the gas component in the bubbles is $-4\pi R^2 I$ and the change in volume of bubbles is written as follows:

$$\frac{d}{dt}\left(\frac{4\pi}{3}R^3\right) = -\upsilon_G \times 4\pi R^2 I, \tag{4.22}$$

and the growth rate dR/dt is given by the following equation

$$\frac{dR}{dt} = -\upsilon_G I \tag{4.23}$$

where υ_G is the volume of a single molecule of the gas component of the gas phase and is assumed to be constant here[4]. Because diffusion flux I is proportional to the concentration gradient at the gas-liquid interface, it is given by

$$I = -D\left|\frac{\partial C(r)}{\partial r}\right|_{r=R} = -D\frac{C_\infty - C_R}{R} \tag{4.24}$$

Bubble growth by steady-state diffusion can be obtained by substituting the stationary diffusion flux (Eq. 4.24) into Eq. (4.23), i.e.,

$$\frac{dR}{dt} = \upsilon_G D\frac{C_\infty - C_R}{R} \tag{4.25}$$

[4]In actuality, υ_G changes in accordance with the equation of state, e.g., $\upsilon_G = k_B T/P_G$.

where v_G is generally a function of pressure within the bubbles P_G. Subsequently, the following cases in which v_G, C, and P are approximated are examined.

The Case in Which v_G = constant and $C_\infty - C_R$ = constant
Consider the case where R is sufficiently larger than R_C. Assuming that the difference in concentration $C_\infty - C_R$, which is a driving force, is a constant value ΔC,

$$\frac{dR}{dt} = v_G D \frac{\Delta C}{R} \tag{4.26}$$

holds. Integrating this equation yields

$$\frac{1}{2}\left(R^2 - R(0)^2\right) = D v_G \Delta C \cdot t \tag{4.27}$$

that is,

$$R = R(0)\sqrt{1 + \frac{2Dv_G\Delta C}{R(0)^2}t} \tag{4.28}$$

is obtained. This is an approximation in which the effect of curvature on the interface equilibrium concentration is ignored. In this equation, $R(0)$ is the bubble radius at $t = 0$. When $t \gg R(0)^2/(v_G \Delta C D)$,

$$R = \sqrt{2Dv_G\Delta C \cdot t} \tag{4.29}$$

is obtained, and a parabolic growth law, i.e., a growth law that is proportional to the square root of time, is derived.

The Case in Which $v_G = C_\infty$, and P = constant, C_R of Mechanical Equilibrium
To examine this in more detail, using Eq. (4.9), the equilibrium concentration at the bubble interface in mechanical equilibrium, for Eq. (4.25) yields the growth rate of bubbles as follows:

$$\frac{dR}{dt} = \frac{v_G D C_\infty}{R}\left\{1 - \left[1 - R_P^*(C_\infty, P)\left(\frac{1}{R_C(C_\infty, P)} - \frac{1}{R}\right)\right]^{\frac{1}{2}}\right\} \tag{4.30}$$

The diffusion driving force ΔC corresponds to the inside of the braces { }. When focusing on the nucleation stage, because $R_P^*(1/R_C - 1/R) \ll 1$, an approximation

$$\frac{dR}{dt} = \frac{v_G D C_\infty R_P^*(C_\infty, P)}{2R}\left(\frac{1}{R_C(C_\infty, P)} - \frac{1}{R}\right) \tag{4.31}$$

is obtained. This was used when Eq. (3.72) was calculated from Eq. (3.35) and the nucleation rate was obtained. In the abovementioned equations, $-1/R$ in the parentheses ( ) corresponds to the Gibbs-Thomson effect. Moreover, when $R \gg R_C$, the maximum driving force of diffusion is $\Delta C \approx C_\infty R_P^*/R_C$. Although the actual driving force approaches this value with increase in R owing to the Gibbs-Thomson effect, the initial driving force is lesser than this value and depends on the initial bubble radius $R(0)$.

Using R_P^* defined by Eq. (4.10) and bubbles' diffusion growth time t_{difG}

$$t_{difG} = \frac{(R_P^*)^2}{v_G D \Delta C} \tag{4.32}$$

Let us nondimensionalize Eq. (4.31) by a scale ($\tilde{t} = t/t_{difG}$). [5] ΔC is then the maximum driving force. Equation (4.31) can then be written without explicitly indicating the independent variable ($\tilde{t}$) in $\tilde{R}(\tilde{t})$ ($= R(t)/R_p^*$) as follows:

$$\frac{d\tilde{R}}{d\tilde{t}} = \frac{1}{2\tilde{R}}\left(1 - \frac{\tilde{R}_C}{\tilde{R}}\right) \tag{4.33}$$

When $\tilde{R}_C = R_C/R_p^* =$ constant (the degree of supersaturation is constant), Eq. (4.33) can be integrated by separation of variables and the solution is obtained as an implicit function as follows:

$$\frac{1}{2}(\tilde{R}^2 - \tilde{R}(0)^2) + \tilde{R}_C(\tilde{R} - \tilde{R}(0)) + \tilde{R}_C^2 \ln\left(\frac{\tilde{R} - \tilde{R}_C}{\tilde{R}(0) - \tilde{R}_C}\right) = \frac{\tilde{t}}{2} \tag{4.34}$$

where $\tilde{R}(0)$ is what is obtained by scaling the initial bubble radius $R(0)$ by R_p^*.

We also evaluated differences among a solution obtained by numerically integrating Eq. (4.30), which considers the Gibbs-Thomson effect; Equation (4.33) or (4.29), both of which are obtained by approximating Eq. (4.30); and Eqs. (4.28) and (4.29), both of which do not consider the Gibbs-Thomson effect. Furthermore, the influence of the initial bubble radius is evaluated. Figure 4.2 shows the result of the evaluation. Constant supersaturation concentration is assumed in this calculation, which indicates that concentration greater than the solubility under a certain pressure remains constant and equivalently corresponds to the constant critical radius ($R_C =$ constant). Furthermore, a state close to saturation at the interface (the initial bubble radius is slightly larger than the critical radius) is assumed as the initial condition of bubble growth. From this figure, the following are read in terms to modeling of diffusion growth.

(1) In the case where the effect of the bubble radius R on the interface equilibrium concentration, i.e., the Gibbs-Thomson effect, is considered (Eqs. 4.30 and 4.33), the initial delay of growth appears. Note that this occurs when the initial condition is close to the saturated state at the interface (The bubble radius is equal to the critical nucleus radius. This does not necessarily mean the time immediately after the nucleation). Furthermore, when the initial bubble radius is one and a half times as large as the critical radius, the initial delay is hardly observed. Moreover, in this case, if the square root term for the solubility is linearly approximated, growth is overestimated in the latter half of the bubble growth stage (the thick solid line and the thick broken line in the right Chart in Fig. 4.2).
(2) In the case where the Gibbs-Thomson effect is ignored, the initial delay does not appear. This is a natural result. In this case, the influence of the initial bubble radius and approximation can be ignored in the stage sufficiently after the time scale of diffusion scale.

First, let us discuss case (1), i.e., when the initial bubble radius is absolutely equal to the critical radius, a state in which there is no concentration gradient is a solution. However, this state is in an unstable equilibrium and bubbles do not grow in this state. When the bubble radius slightly increases because of fluctuation, the equilibrium concentration at the interface decreases because of slight gas pressure drop, a concentration gradient is generated, and the bubbles grow at an accelerated pace. When the initial bubble radius is close to the critical radius, the concentration gradient as a driving force is initially less (right chart of Fig. 4.1) and the growth rate is close to zero, as shown in the left chart of Fig. 4.2. Therefore, the delay in growth occurs. This corresponds to ignoring terms other than

[5] Difference from the case of crystallization lies at 2 in the denominator.

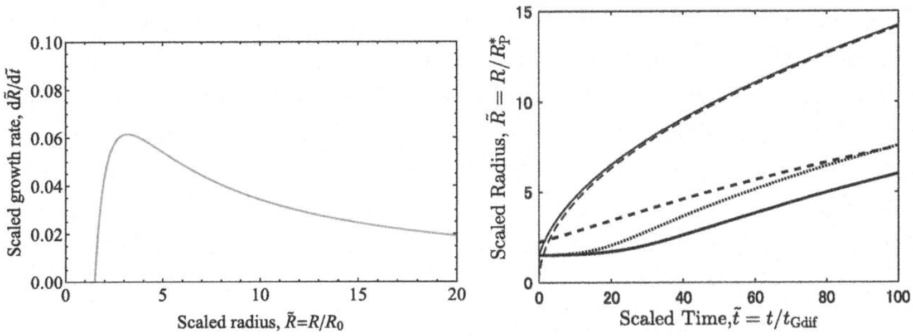

Fig. 4.2 (Left) The bubble growth rate at the constant degree of supersaturation ($\tilde{R}_C = 1.5$). $\tilde{R} = 1.5$ where the growth rate equals zero corresponds to the critical radius. (Right) Change in the bubble radius ($\tilde{R} = R/R_P^*$). The fine solid line represents Eq. (4.28) and the fine broken line represents Eq. (4.29) (two lines almost overlapping at the top), in the cases where the Gibbs-Thomson effect is not considered. $\tilde{R}(0) = 1.01\tilde{R}_C$ in both cases. The thick solid line and the thick broken line represent the results of the numerical integral of Eq. (4.30) considering the Gibbs-Thomson effect. The dotted line is an analytical solution of the approximate expression (Eq. 4.34). The thick solid line (dotted line) and the thick broken line differ in the initial bubble radius: the former is $\tilde{R}(0) = 1.01\tilde{R}_C$ and the latter is $\tilde{R}(0) = 1.5\tilde{R}_C$

ln on the left-hand side of Eq. (4.34) and $\ln((\tilde{R} - \tilde{R}_C)/(\tilde{R}(0) - \tilde{R}_C)) = \tilde{t}/\tilde{R}_C^2$ is obtained. Rewriting this equation yields

$$\tilde{R} = \tilde{R}_C + \left(\tilde{R}(0) - \tilde{R}_C\right) \exp\left(\frac{\tilde{t}}{2\tilde{R}_C^2}\right) \tag{4.35}$$

This is an expression similar to viscosity-limited growth described later.[6] To distinguish whether the initial delay is attributed to the surface tension in diffusion growth or by viscosity described later requires knowing the initial bubble radius and the critical radius and comparing both time constants (Sect. 4.9.3). In any case, exponential growth expressing the initial delay is a growth pattern characteristic of the case where growth or expansion starts from the initial bubble radius closed to chemical balance (in diffusion-limited growth) or mechanical balance (in viscosity-limited growth).

As seen from this figure, although the linear approximation ignoring the square root term is approximated while the bubbles are small, the error increases with increase in time and bubble growth. This indicates that good approximation is obtained when the bubble radius is close to the critical radius, and there will be no great inconvenience in using it to estimate the nucleation rate. Moreover, in the approximation in which the Gibbs-Thomson effect on the interface equilibrium concentration is ignored, the growth rate is overestimated when calculated at the same time. Note that the abovementioned result is of the growth rate under the condition of constant degree of supersaturation (R_C^{-1} = constant) (Eq. 4.12). Because the gas component concentration in liquid changes with the growth of bubbles, the degree of supersaturation changes.

The Case Where v_G Changes as an Ideal Gas and Mechanical Equilibrium Holds

In the previous subsections, calculation is conducted assuming that the volume of gas molecules in bubbles v_G is constant. In actuality, at a temperature higher than a gas-liquid critical point of the gas component, the volume of gas molecules follows the equation of state of an ideal gas and changes with

[6]For the initial delay in bubble growth, when Proussevitch et al. (1993) first indicated it in calculation of bubble growth by a cell model, Sparks (1994) argued that the delay was attributed to viscosity. Sparks (1978) had reviewed the problem of the initial delay in bubble growth and had recognized the problem of the critical size at which the surface tension was effective, as well as the influence of viscosity. Against this, Proussevitch et al. (1993) showed additional simulation results and were adamant that it was the effect of interfacial tension. After that, Navon et al. (1998) revealed that the delay in bubbles' viscosity-limited growth has exponential characteristics.

the pressure within the bubbles. The pressure within the bubbles is determined by the change in the number of molecules in bubbles and in the surrounding pressure, and the expansion process of bubbles in viscous fluid, explained later, has to be investigated to determine the pressure within the bubbles. Here, for the case where mechanical equilibrium holds, formulation including change in the number of molecules in bubbles and change in the volume of gas molecules is performed. For this purpose, change in the number of molecules, n_G, in bubbles associated with diffusion is described. Because the inflow of the number of gas molecules into bubbles is –4 π R2I and $I = -D(C_\infty - C_R)/R$,

$$\frac{dn_G}{dt} = 4\pi D(C_\infty - C_R(R, C_\infty, P))R \tag{4.36}$$

is obtained. In this equation, the interface equilibrium concentration C_R basically follows Eq. (2.35) and is a function of liquid pressure P, gas pressure P_G, and the equilibrium concentration at the liquid pressure P, $C_{eq}(P)$ (Eq. 4.6). $C_\infty > C_{eq}(P)$ is in a supersaturated state and the C_R that considers them is Eq. (4.9). That is expressed as $C_R(R, C_\infty, P)$.

Gas pressure P_G is related to the bubble radius R and liquid pressure P through the mechanical equilibrium (the Laplace equation (2.42)), i.e.,

$$P_G = P + \frac{2\gamma}{R} \tag{4.37}$$

holds. The equation of state is $4\pi R^3 P_G/3 = n_G k_B T$, and substituting the last Equation into P_G of this equation yields

$$\frac{4}{3}\pi R^3 \left(P + \frac{2\gamma}{R}\right) = n_G k_B T \tag{4.38}$$

Assuming that C_∞ is constant, if liquid pressure P is given as a function of time, Eqs. (4.36) and (4.38) can be considered and solved as simultaneous equations in terms of the bubble radius R and the number of gas molecules in the bubbles n_G. Consequently, an equation of rate of increase of the bubble radius

$$\frac{dR}{dt} = \frac{R^2|\dot{P}|/3 + k_B T D C_\infty \left(1 - \left[1 - R_P^* \left(\frac{1}{R_C} - \frac{1}{R}\right)\right]^{1/2}\right)}{R\left(P + \frac{4\gamma}{3R}\right)} \tag{4.39}$$

is obtained. R_P^* in this equation has been given by Eq. (4.10). Assuming that the pressure is constant ($\dot{P} = 0$), comparison with Eq. (4.30) shows that a correction term that makes effective pressure of gas in the bubbles. Note that $P + 4\gamma/3R$ is included in the denominator.

4.3.2 Case Including the Advection Term

Here, let us consider a steady state including the advection term in Eq. (4.1).

$$u_r \frac{\partial C}{\partial r} = \frac{1}{r^2}\frac{\partial}{\partial r}\left(r^2 D \frac{\partial C}{\partial r}\right) \tag{4.40}$$

As seen later, from the conservation of mass of liquid, the velocity field is related to the rate of growth or expansion of bubbles $\dot{R} = dR/dt$, and determined as follows

$$u_r = \frac{R^2}{r^2}\dot{R} \tag{4.41}$$

Substituting this into Eq. (4.40) yields

$$\frac{\partial}{\partial r}\left(R^2\dot{R}C - r^2 D\frac{\partial C}{\partial r}\right) = 0 \tag{4.42}$$

Integral of this gives

$$r^2 D\frac{dC}{dr} = R^2\dot{R}(C_\infty - c_1) \tag{4.43}$$

where c_1 corresponds to a constant of integration and is determined by a boundary condition. Moreover, integrating from r to infinity and setting a boundary condition as $C(R) = C_R$ yields,

$$C(r) = c_1 + (C_\infty - c_1)e^{-\mathrm{Pe}\frac{R}{r}} \tag{4.44}$$

$$\mathrm{Pe} = \frac{R\dot{R}}{D} \tag{4.45}$$

$$c_1 = \frac{C_R - C_\infty e^{-\mathrm{Pe}}}{1 - e^{-\mathrm{Pe}}} \tag{4.46}$$

where Pe is a dimensionless number (Péclet number) representing the ratio of diffusion D/R to the magnitude of advection by the increase in the bubble radius $\dot{R}$ and C_∞ is supersaturation concentration when $r = \infty$. Obtaining diffusion flux at the interface from these equations yields

$$I = D\left.\frac{dC}{dr}\right|_R = \dot{R}(C_\infty - C_R)\frac{e^{-\mathrm{Pe}}}{1 - e^{-\mathrm{Pe}}} \tag{4.47}$$

Although by substituting this Equation into Eq. (4.23), $\dot{R}$ in the coefficient is cancelled out, $\dot{R}$ in $e^{-\mathrm{Pe}}$ remains. Solving the equation in terms of $\dot{R}$ yields

$$\dot{R} = \frac{D}{R}\ln\left(1 + v_G(C_\infty - C_R)\right) \tag{4.48}$$

This is a growth law that can be called the steady advection-diffusion growth. When $v_G(C_\infty - C_R)$ is sufficiently small, this results in Eq. (4.25), the steady-state diffusion growth that does not include the advection term. Because realistic $v_G(C_\infty - C_R)$ under high pressure is sufficiently small, ignoring the advection term is not a problem. However, because v_G becomes large under low pressure, the advection term is expected to be non-negligible.

4.4 Non-steady State Diffusion Growth

Because C_R changes with growth of bubbles even in mechanical equilibrium (right chart of Fig. 4.1), the assumption that a concentration profile is a steady state diffusion profile is not convincing. At that time, we have to numerically solve a diffusion equation (4.1) or (4.16), but it will be numerically troublesome because the interface of bubbles that provides a boundary condition itself moves. To solve such an equation, the equation is often rewritten to a coordinate system that has $r = R(t)$ as the origin point and moves at dR/dt with the growth of bubbles. Because the bubble radius considerable changes with time, to solve it by the finite difference method may lead to a memory problem for the spatial mesh. To avoid the problem, we will solve a diffusion equation rewritten on the basis of a coordinate system that moves with $R(t)$. This approach was first proposed in the field of engineering and then applied to problems of magma. The same approach is used for crystal growth and is explained in Sect. 7.7.3.

4.5 Bubble Expansion Under Mechanical Equilibrium

4.5.1 Adiabatic Expansion Versus Isothermal Expansion and the Influence of Latent Heat

Consider the case where bubbles expand under the condition in which the number of molecules in bubbles is constant, i.e., bubbles are in chemical disequilibrium but in mechanical equilibrium with the surrounding melt. When bubbles adiabatically expand, the temperature of gas in the bubbles decreases. The rate of decline can be estimated from an equation for adiabatic change of ideal gas $PV^{\gamma_h} = \text{constant}$ and $TV^{\gamma_h - 1} = \text{constant}$. Consequently, temperature and volume are obtained at $\frac{V}{V(0)} = \left(\frac{P}{P(0)}\right)^{-\frac{1}{\gamma_h}}$ and $\frac{T}{T(0)} = \left(\frac{P}{P(0)}\right)^{\frac{\gamma_h - 1}{\gamma_h}}$, where γ_h is the specific heat ratio; it is 1.3 for water vapor. Because $V \propto R^3$, the bubble radius and temperature associated with adiabatic change will change (Fig. 4.3). Decompressing from the initial pressure $P(0)$, 100 MPa, to 1 atmospheric pressure, i.e., 0.1 MPa, will cause cooling of 20% $\approx$ 200–300 K of the initial temperature $T(0)$. In reality, bubbles are surrounded by high-temperature melt, and the exchange of heat between the bubbles and the melt is expected to be quick. The specific heat of silicate melt is about 3000 J/kg/K and isobaric specific heat of water vapor is about 2000 J/kg/K. The abundance ratio of water vapor, a gas component, in magma is several percent in weight and the degree of temperature decline is expected to be about 10 °C at the most (Sparks 1978).

Next, in cases where the number of molecules in bubbles changes, i.e., volatiles exsolve to approach chemical equilibrium, latent heat is absorbed and therefore the temperature of the system decreases. Considering this influence, Sahagian and Proussevitch (1996) estimated that magma would be cooled by about 30 °C. As seen in Sect. 2.7, assuming that latent heat of water separation $\Delta\mathcal{H}_{\text{water}}$ is 20 kJ/mol $\approx 10^6$ J/kg, if all water dissolving in melt at 5 wt% vesiculates and exsolves from melt and then is in thermal equilibrium with the melt, the temperature decreases by about 19 °C. Although a process of water exsolution and expansion of steam during magma vesiculation can be considered, in a first approximation, as an isothermal process, only about 30 °C can affect crystallinity in crystallization of microlites, as seen later. Furthermore, because the crystallization of microlites emits latent heat of crystallization, temperature decrease by water exsolution and gas expansion will be suppressed or, on the contrary, temperature will increase (Blundy et al. 2006).

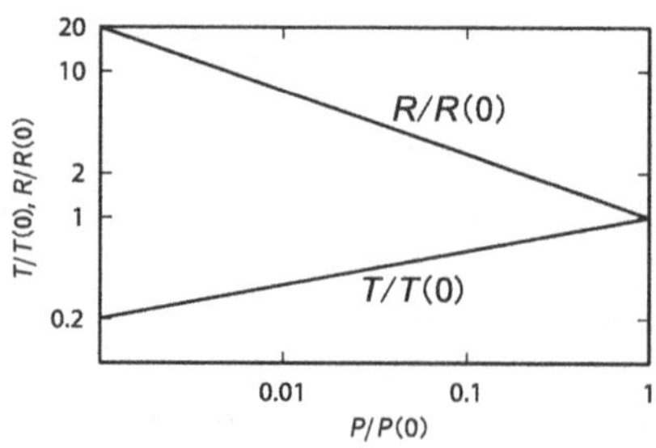

Fig. 4.3 Cooling by adiabatic expansion of bubbles

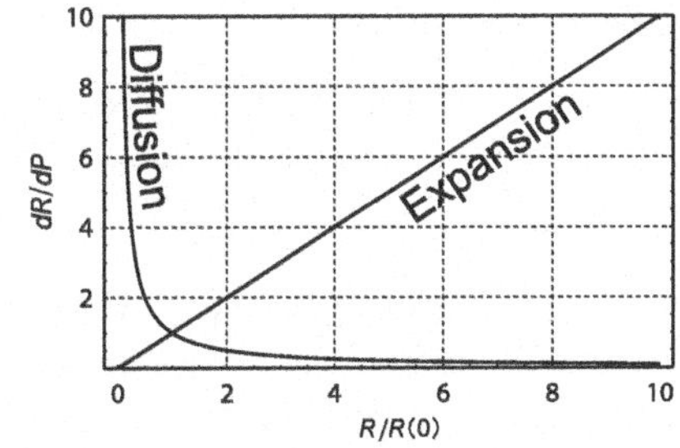

Fig. 4.4 Bubbles' isothermal expansion coefficient and diffusion-limited growth rate

4.5.2 Expansion Rate of Bubbles as a Function of the Bubble Radius

Let us consider the case where the number of molecules in bubbles is constant, bubbles are in chemical disequilibrium with the surrounding melt, bubble expansion overcomes viscous resistance, and bubbles expand under a relatively low pressure as being in equilibrium with the pressure of melt. In this case, we can assume isothermal change as the first approximation, as seen in the previous subsection. In this case, the expansion coefficient of bubbles can be written from the equation of state of an ideal gas as follows:

$$PV \approx PR^3 = \text{constant} \tag{4.49}$$

and then

$$\left.\frac{dR}{dt}\right|_T = \frac{1}{3}\frac{R}{P}\left|\frac{dP}{dt}\right| \tag{4.50}$$

is obtained. For an adiabatic process,

$$\left.\frac{dR}{dt}\right|_S = \frac{1}{3\gamma_h}\frac{R}{P}\left|\frac{dP}{dt}\right| \tag{4.51}$$

holds. In any case, the expansion rate of bubbles proportionally increases with the bubble radius when the decompression rate is constant.

For diffusion-limited growth, however, bubble growth rate dR/dt is inversely proportional to bubble size, according to Eq. (4.26). This fact reveals that while diffusion-limited growth is dominant at the time immediately after nucleation when the bubble size is small, expansion due to decompression becomes predominant when the bubble size increases under low pressure (Fig. 4.4).

4.6 An Equation Describing Change in the Bubble Radius in a Viscous Fluid: The Rayleigh-Plesset Equation

Because diffusion of a gas component into bubbles causes an increase in gas pressure in the bubbles, mechanical equilibrium expressed by Laplace's equation will break down. It is the so-called Rayleigh-Plesset equation, shown below[7] that describes change in the bubble radius with respect to time under mechanical disequilibrium.

[7] An inviscid liquid case had been derived by Rayleigh (1917); after that, Poritsky (1952) introduced the viscosity term and then Plesset and Prosperetti (1977) widely spread the use of the equation.

$$\underbrace{R(t)\ddot{R}(t) + \frac{3}{2}\dot{R}(t)^2}_{\text{Inertia term}} = \frac{1}{\rho}\left(\underbrace{P_{\mathrm{G}} - P_{\infty} - \frac{2\gamma}{R(t)}}_{\text{Driving force}} - \underbrace{4\eta\frac{\dot{R}(t)}{R(t)}}_{\text{Viscous resistance}}\right) \tag{4.52}$$

where ρ is the density of liquid, $R(t)$ is the bubble radius, P_{G} is the pressure within the bubbles, P_{∞} is the pressure of liquid at a place distant from the bubbles, and η is the viscosity of liquid. This equation satisfies the conservation of mass and the conservation of momentum around bubbles that expand or contract. Let us derive this equation in the following to clarify the physics behind the equation and meanings of each term.

Considering a spherical coordinate system, spherical symmetry, and a nongravitational field, the conservation law of mass and momentum of liquid surrounding bubbles that expand or contract while keeping a spherical shape is expressed as follows (Bird et al. 1960) (See Box):

Conservation of mass

$$\frac{\partial \rho}{\partial t} + \frac{1}{r^2}\frac{\partial}{\partial r}\left(r^2 \cdot \rho u_r\right) = 0 \tag{4.53}$$

Conservation of momentum in the radial direction

$$\rho\left(\frac{\partial u_r}{\partial t} + u_r\frac{\partial u_r}{\partial r}\right) = -\frac{\partial P}{\partial r} + \frac{1}{r^2}\frac{\partial}{\partial r}\left(r^2 \cdot \sigma_{rr}\right) - \frac{\sigma_{\theta\theta} + \sigma_{\phi\phi}}{r} \tag{4.54}$$

where u_r is the velocity of liquid in the radial direction and P is the pressure of the liquid.

σ_{rr}, $\sigma_{\theta\theta}$, and $\sigma_{\phi\phi}$ are normal stress tensors in the radial, longitudinal, and latitudinal directions, respectively, and are given generally as follows:

$$\sigma_{rr} = \eta\left[2\frac{\partial u_r}{\partial r} - \frac{2}{3}(\nabla \cdot \boldsymbol{u})\right] \tag{4.55}$$

$$\sigma_{\theta\theta} = \eta\left[2\left(\frac{1}{r}\frac{\partial u_\theta}{\partial \theta} + \frac{u_r}{r}\right) - \frac{2}{3}(\nabla \cdot \boldsymbol{u})\right] \tag{4.56}$$

$$\sigma_{\phi\phi} = \eta\left[2\left(\frac{1}{r\sin\theta}\frac{\partial u_\phi}{\partial \phi} + \frac{u_r}{r} + \frac{u_\theta \cot\theta}{r}\right) - \frac{2}{3}(\nabla \cdot \boldsymbol{u})\right] \tag{4.57}$$

The left-hand side of Eq. (4.54) is the inertia term. The first term of the right-hand side

$$-\frac{\partial P}{\partial r} \tag{4.58}$$

is a force by radial pressure gradient; the second term of the right-hand side

$$\frac{1}{r^2}\frac{\partial}{\partial r}\left(r^2 \cdot \sigma_{rr}\right) \tag{4.59}$$

is radial viscous resistance that directly occurs by radial movement of the interface; and the third term

$$-\frac{\sigma_{\theta\theta} + \sigma_{\phi\phi}}{r} \tag{4.60}$$

is a force originating from stress introduced by compression or extension in the latitudinal and longitudinal directions and acting on the radial direction because of a geometric effect. Although, similar to the conservation of mass $r^{-2}\partial(r^2\rho u_r)/\partial r$, radial momentum is considered to be related only to the second term occurring only by radial stress σ_{rr}, tangential stress $\sigma_{\theta\theta}$ and $\sigma_{\phi\phi}$ get involved and their contribution is expressed in the third term. This can be understood by recalling Laplace's equation in which the surface tension, which is tangential force, is added to the external pressure and the force balances with the internal pressure of bubbles.

The coordinate transformation of the equation of motion: I

The derivation of Eq. (4.54) is very complicated and follows the two principles below (Bird et al. 1960). An equation of motion in a Cartesian coordinate system will be mathematically transformed into an equation of motion in a curvilinear coordinate system. Derivation from geometry of force balance in a curvilinear coordinate system is not advisable.

Principle 1. Because each side of an equation of motion is a vector quantity, follow a rule of coordinate transformation of vector, i.e., transformation from a vector (F_x, F_y, F_z) in a Cartesian coordinate system into a vector $F_{k'}$ (components (F_r, F_θ, F_ϕ)) in a spherical coordinate system can be given by

$$F_x = l_{xk} F_{k'}$$

(the rule of Einstein summation convention was used), and the transformation matrix l_{ij} is

$$\begin{pmatrix} l_{xr} & l_{x\theta} & l_{x\phi} \\ l_{yr} & l_{y\theta} & l_{y\phi} \\ l_{zr} & l_{z\theta} & l_{z\phi} \end{pmatrix} = \begin{pmatrix} \sin(\theta)\cos(\phi) & \cos(\theta)\cos(\phi) & -\sin(\phi) \\ \sin(\theta)\sin(\phi) & \cos(\theta)\sin(\phi) & \cos(\phi) \\ \cos(\theta) & -\sin(\theta) & 0 \end{pmatrix}$$

For a tensor, transformation from τ_{ij} ($\tau_{xx}, \tau_{xy}, \tau_{xz}$ etc.) into $\tau_{i'j'}$ ($\tau_{rr}, \tau_{\theta\theta}, \tau_{\phi\phi}$, etc.) in a spherical coordinate system is given by

$$\tau_{ij} = l_{ii'} l_{jj'} \tau_{i'j'}$$

Principle 2. Following a chain rule, for example,

$$\frac{\partial}{\partial x} = \frac{\partial}{\partial r}\frac{\partial r}{\partial x} + \frac{\partial}{\partial \theta}\frac{\partial \theta}{\partial x} + \frac{\partial}{\partial \phi}\frac{\partial \phi}{\partial x}$$

differentiation in a Cartesian coordinate system will be transformed into differentiation in a spherical coordinate system. In this equation,

$$\frac{\partial r}{\partial x} = \sin(\theta)\cos(\phi), \quad \frac{\partial \theta}{\partial x} = \frac{\cos(\theta)\cos(\phi)}{r},$$
$$\frac{\partial \phi}{\partial x} = \frac{-\sin(\phi)}{r\sin(\theta)}$$

hold.

The coordinate transformation of the equation of motion: II

A fundamental equation is a Navier-Stokes equation in a Cartesian coordinate system that neglects gravity

$$\underbrace{\rho\left(\frac{\partial u_i}{\partial t}+u_j\frac{\partial u_i}{\partial x_j}\right)}_{\text{Inertia term}}=\underbrace{-\frac{\partial P}{\partial x_i}+\frac{\partial \sigma_{ij}}{\partial x_j}}_{\text{Force}}$$

This equation is transformed into an equation in a spherical coordinate system. First, because the both sides are vector quantity, according to Principle 1 n Box above, a r component of an inertia term of a spherical coordinate is

$$\begin{aligned}[\text{Inertia term}]_r &= \hat{l}_{rx}\,[\text{L.H.S.}]_x+\hat{l}_{ry}\,[\text{L.H.S.}]_y\\ &\quad+\hat{l}_{rz}\,[\text{L.H.S.}]_z\end{aligned}$$

where $\hat{l}_{rx}$, etc. are components of an inverse matrix of l_{ij}. Each vector component, for example, $[\text{L.H.S.}]_x$ is

$$\begin{aligned}&[\text{Inertia term}]_x\\ &=\rho\left[\frac{\partial u_x}{\partial t}+u_x\frac{\partial u_x}{\partial x}+u_y\frac{\partial u_x}{\partial y}+u_z\frac{\partial u_x}{\partial z}\right]\end{aligned}$$

Vector quantity ux and so on follow Principle 1 and differentiation $\partial/\partial x$ and so on follow Principle 2. Many terms are cancelled and

$$\begin{aligned}&[\text{Inertia term}]_r\\ &=\left(\frac{\partial u_r}{\partial t}+u_r\frac{\partial u_r}{\partial r}+\frac{u_\theta}{r}\frac{\partial u_r}{\partial \theta}+\frac{u_\phi}{r\sin(\theta)}\frac{\partial u_r}{\partial \phi}-\frac{u_\theta^2+u_\phi^2}{r^2}\right)\end{aligned}$$

is obtained. Although [Force] is calculated by the same method, transformation of stress is calculated, for example, as follows:

$$\begin{aligned}\sigma_{xx}&=l_{xi'}l_{xj'}\sigma_{i'j'}\\ &=\cos^2(\phi)\sin^2(\theta)\sigma_{rr}+\cos(\theta)\cos^2(\phi)\sin(\theta)\sigma_{r\theta}\\ &-\cos(\phi)\sin(\theta)\sin(\phi)\sigma_{r\phi}+\cos(\theta)\cos^2(\phi)\sin(\theta)\sigma_{\theta r}\\ &+\cos^2(\theta)\cos^2(\phi)\sigma_{\theta\theta}-\cos(\theta)\cos(\phi)\sin(\phi)\sigma_{\theta\phi}\\ &-\cos(\phi)\sin(\theta)\sin(\phi)\sigma_{\phi r}-\cos(\theta)\cos(\phi)\sin(\phi)\sigma_{\phi\theta}+\sin^2(\phi)\sigma_{\phi\phi}\end{aligned}$$

The coordinate transformation: II continued

On the other hand, when writing a stress component in the r direction in a spherical coordinate system as $\left[\partial\sigma_{ij}/\partial x_j\right]_r$,

$$\left[\frac{\partial\sigma_{ij}}{\partial x_j}\right]_r = \hat{l}_{rx}\left[\frac{\partial\sigma_{ij}}{\partial x_j}\right]_x + \hat{l}_{ry}\left[\frac{\partial\sigma_{ij}}{\partial x_j}\right]_y + \hat{l}_{rz}\left[\frac{\partial\sigma_{ij}}{\partial x_j}\right]_z$$

holds and for example,

$$\left[\frac{\partial\sigma_{ij}}{\partial x_j}\right]_x = \frac{\partial\sigma_{xx}}{\partial x} + \frac{\partial\sigma_{xy}}{\partial y} + \frac{\partial\sigma_{xz}}{\partial z}$$

holds. Substituting σ_{xx} expressed by σ_{rr} and so on into σ_{xx} and so on in the above equation, transforming it into differentiation in the spherical coordinate, following Principle 2, substituting the differentiation into the equation of $\left[\partial\sigma_{ij}/\partial x_j\right]_r$, substituting them into $[Inertiaterm]_r = [Force]_r$ in the spherical coordinate system, and rearranging the obtained equation yields Eq. (4.54). A θ component and a ϕ component are calculated by the same method.

Moreover, this is similar to force balance in the rubber membrane of a rubber balloon. In this case, rubber is an elastic body and stress in the latitudinal and longitudinal directions corresponds to the tension of rubber. The state where the rubber balloon maintains a certain size indicates that radial stress by internal pressure of gas and elastic stress in the latitudinal and longitudinal directions in the rubber membrane is balanced. This situation works in both elastic membrane similar to the rubber balloon and in fluids in the same manner.

If the incompressible fluid is assumed in Eq. (4.53); $\partial\rho/\partial t = 0$,

$$\frac{\partial}{\partial r}\left(r^2 \cdot \rho u_r\right) = 0 \tag{4.61}$$

holds, and integrating this equation from the bubble surface $r = R$ to an arbitrary position $r = r$ gives

$$r^2 \cdot u_r - R^2 \cdot u_R = \text{const}, \tag{4.62}$$

and because at $r = R$, $u_R = \dot{R}$ is obtained, and therefore const = 0, i.e., from the conservation of mass, a velocity field in the radial direction that is attenuated by r^{-2} is obtained.

$$u_r(t, r) = \frac{R(t)^2}{r^2}\dot{R}(t) \tag{4.63}$$

Because stress tensor is in spherical symmetry ($\sigma_{\theta\theta} = \sigma_{\phi\phi}$) and incompressible ($\nabla \cdot \mathbf{u} = 0$), for the Newtonian viscous fluid

$$\sigma_{rr} = 2\eta\frac{\partial u_r}{\partial r} \tag{4.64}$$

$$\sigma_{\theta\theta} = \sigma_{\phi\phi} = 2\eta\frac{u_r}{r} \tag{4.65}$$

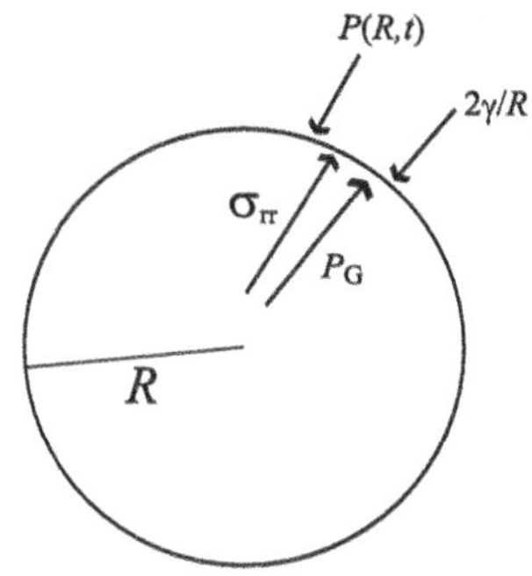

Fig. 4.5 Mechanical balance at the interface of a bubble expanding in viscous fluid (dynamic mechanical equilibrium)

hold. Substituting Eq. (4.63) into them yields

$$\sigma_{rr} = -4\eta \frac{R(t)^2 \dot{R}}{r^3} \tag{4.66}$$

$$\sigma_{\theta\theta} = \sigma_{\phi\phi} = 2\eta \frac{R(t)^2 \dot{R}}{r^3}, \tag{4.67}$$

i.e., $\sigma_{rr} = -(\sigma_{\theta\theta} + \sigma_{\phi\phi})$ holds. By using this, the conservation of momentum, Eq. (4.54), can be rewritten as follows:

$$\rho \left(\frac{\partial u_r}{\partial t} + u_r \frac{\partial u_r}{\partial r} \right) = -\frac{\partial P}{\partial r} + \left(\frac{\partial \sigma_{rr}}{\partial r} + 3\frac{\sigma_{rr}}{r} \right) \tag{4.68}$$

Substituting Eq. (4.63) of the velocity field into the left-hand side, taking

$$\frac{\partial u_r}{\partial t} = \frac{1}{r^2} \frac{\partial}{\partial t} \left(R^2 \dot{R}(t) \right) \tag{4.69}$$

and

$$u_r \frac{\partial u_r}{\partial r} = \frac{\partial}{\partial r} \left(\frac{1}{2} u_r^2 \right) \tag{4.70}$$

into consideration, and integrating from $r = R$ to ∞ gives

$$\text{L.H.S.} = \rho \left(\frac{3}{2} \dot{R}(t)^2 + R\ddot{R} \right) \tag{4.71}$$

$$\text{R.H.S.} = -(P_\infty - P(R,t)) - \sigma_{rr}(R,t) + 3 \int_R^\infty \frac{\sigma_{rr}}{r} dr \tag{4.72}$$

The reason why the integral for stress is left in these equations is that these are general expressions applicable to viscoelastic liquid. At the bubble surface, radial force balance (which can be called as dynamic mechanical equilibrium) is

$$P_G + \sigma_{rr}(R,t) = P(R,t) + \frac{2\gamma}{R} \tag{4.73}$$

(Figure 4.5).

By substituting this equation into Eq. (4.72) and rearranging it, the conservation of momentum can be written as follows:

$$\rho\left(\frac{3}{2}\dot{R}^2 + R\ddot{R}\right) = P_{\mathrm{G}} - P_\infty - \frac{2\gamma}{R} + 3\int_R^\infty \frac{\sigma_{rr}}{r}dr \tag{4.74}$$

This is a general form of an equation of motion for a spherical bubble with the radius that changes in liquid with any rheological properties. σ_{rr} in Eq. (4.72) is the result of the integral of $\partial\sigma_{rr}/\partial r$ that is independent of rheology. Therefore, Eq. (4.73), dynamic mechanical balance at the interface, always holds independent of rheology; however, if rheology is not a simple Newtonian viscosity, the integral of Eq. (4.74) will result in a different form because of the different rheology. This behavior can be concretely understood when the equation is extended to a viscoelastic liquid later.

Here, using Eq. (4.66) for Newtonian viscosity for σ_{rr} gives

$$\int_R^\infty \frac{\sigma_{rr}}{r}dr = -\frac{4}{3}\eta\frac{\dot{R}}{R}, \tag{4.75}$$

and substituting this equation into Eq. (4.74) yields the Rayleigh-Plesset equation (4.52). Moreover, the mechanical equilibrium at $\dot{R} = 0$ results in Laplace's equation. The above liquid is dealt with as an incompressible fluid. However, formulation after considering compressibility of liquid was conducted (Keller and Miksis 1980).

4.7 Time Change of Bubble Expansion: Inertial Expansion

Section 4.5 describes the case where bubbles expand always keeping mechanical equilibrium with the surrounding pressure. This section describes changes with time until equilibrium is attained on the basis of the Rayleigh-Plesset equation derived in the previous section. Note that the collapse and oscillation of bubbles will be dealt with in Sects. 6.6 and 6.7.

4.7.1 Bubble Expansion in Inviscid Liquid

Assuming liquid viscosity, which is sufficiently small and negligible, the Rayleigh-Plesset equation (4.52) can be written as follows:

$$\rho\left(R(t)\ddot{R}(t) + \frac{3}{2}\dot{R}(t)^2\right) = P_{\mathrm{G}} - P_\infty - \frac{2\gamma}{R(t)} \tag{4.76}$$

(Rayleigh equation).

After examining this equation, when the initial bubble radius $R(0)$ is in mechanical equilibrium satisfying Laplace's equation, i.e., the right-hand side is 0, the left-hand side is also 0, and therefore both $\dot{R}$ and $\ddot{R}$ are 0. Let us define the bubble radius in the mechanical equilibrium R_{B} as follows:

$$R_{\mathrm{B}} = \frac{2\gamma}{P_{\mathrm{G}} - P_\infty} \tag{4.77}$$

This radius in equilibrium has a meaning similar to the critical radius at nucleation; however, this mechanical equilibrium is an unstable equilibrium. If the pressure in bubbles does not change and the bubble radius even slightly exceeds R_B, pressure by the surface tension decreases, mechanical equilibrium breaks, the right-hand side becomes positive, both $\dot{R}$ and $\ddot{R}$ become non-zero, positive values, and bubbles start expanding. Moreover, if the bubble radius becomes lesser than R_B, the pressure by the surface tension, which acts on the bubbles, increases, and the right-hand side becomes negative. Because $\dot{R}^2$ is positive, $\ddot{R}$ becomes negative, i.e., in space of the bubble radius versus time, the bubble radius shows a convex-upward curve and the bubbles contract. In reality, if the number of molecules of gas in bubbles does not change, the internal pressure of bubbles decreases by bubble expansion and increases by bubble contraction; therefore, the bubble radius remains stable or oscillates.

4.7.2 The Case Where the Pressure in Bubbles is Constant: Simple Inertial Expansion

Equation (4.76) can be analytically solved with respect to the rate of change in the bubble radius $\dot{R}$ under the condition where the pressure of gas P_G is constant and under the initial condition where $\Delta P_G = P_G - P_\infty = 2\gamma/R(0)$ and when $R = R(0)$, $\dot{R} = 0$.[8] The solution will then be

$$\dot{R} = \frac{dR}{dt} = \left[\frac{2}{3}\frac{\Delta P_G}{\rho}\left(1-\left(\frac{R(0)}{R}\right)^3\right) - \frac{2\gamma}{\rho}\frac{1}{R}\left(1-\left(\frac{R(0)}{R}\right)^2\right)\right]^{\frac{1}{2}} \tag{4.79}$$

For Eq. (4.79), assuming $R(0) = R_B$ when $t = 0$, $\ddot{R} = \dot{R} = 0$ is obtained, and then expansion and contraction do not occur. To examine unstable equilibrium, as described above, the case where the bubble radius is instantaneously changed from the equilibrium radius is shown in Fig. 4.6. If the initial bubble radius is larger than R_B even a little, the bubbles expand at an accelerated pace.

The behavior of bubble expansion systematically changes according to the abovementioned deviation from the equilibrium radius, i.e., the magnitude of driving force. As the driving force is large, change in the bubble radius shows more linearity. As the driving force decreases, more time is required to approach this linear expansion. In any case, when sufficient time has passed, it asymptotically approaches a constant expansion rate

$$\frac{dR}{dt} = \left(\frac{2}{3}\frac{\Delta P_G}{\rho}\right)^{\frac{1}{2}} \tag{4.80}$$

The timescale characterizing the transient behavior of inertial expansion of these bubbles in the inviscid liquid is as follows:

---

[8]Taking $\ddot{R} = d\dot{R}/dt = d\dot{R}/dR \cdot dR/dt = \dot{R} \cdot d\dot{R}/dR$ into consideration, Eq. (4.76) can be rewritten as

$$\rho\left(R(t)\frac{d}{dR}\left(\frac{\dot{R(t)}^2}{2}\right) + \frac{3}{2}\dot{R(t)}^2\right) = \Delta P_G - \frac{2\gamma}{R(t)} \tag{4.78}$$

Assuming $y = \dot{R(t)}^2/2$,

$$\frac{dy}{dR} + \frac{3}{R}y = \frac{1}{R}\left(\frac{\Delta P_G}{\rho} - \frac{2\gamma}{\rho}\frac{1}{R}\right)$$

is obtained, which is an inhomogeneous first-order linear differential equation with respect to y as a function of R and can be solved by the variation of constants method. Assuming $y = 0$ when $R = R(0) = R_B$, the constants of integration (0) can be determined.

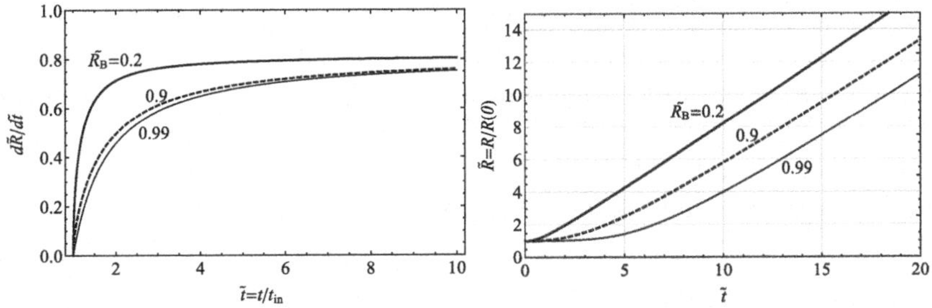

Fig. 4.6 Bubble expansion under the condition of constant bubble gas pressure in an inviscid liquid. The case where the initial radius $R(0)$ instantaneously increases from the equilibrium radius R_B. (Left) Rate of bubble expansion as a function of the bubble radius (Eq. (4.79)). Expansion rates differ depending on the magnitude of deviation from the equilibrium radius $\tilde{R}_B = R_B/R(0)$. When the value of this ratio is 1, the pressure in bubbles equals the surface tension, i.e., the system is in mechanical equilibrium as the initial condition. When the ratio is <1, the pressure in bubbles is larger than the surface tension. (Right) The bubble radius as a function of time, obtained by numerically integrating the Rayleigh equation (Eq. (4.76)) (the same condition as the left chart). The scale of the bubble radius R is the initial bubble radius $R(0)$ and the scale of time t is $R(0)(\rho/\Delta P_G)^{1/2}$. Consequently, the scale of the rate of change in the bubble radius $\dot{R} = dR/dt$ is the timescale of bubble oscillation $(\Delta P/\rho)^{1/2}$

$$t_{in} = R(0)\left(\frac{\rho}{\Delta P_G}\right)^{\frac{1}{2}} \tag{4.81}$$

and depends on overpressure ΔP_G and the initial bubble radius $R(0)$ (Fig. 4.6). This is the timescale of bubble oscillation and is generally longer than the timescale $R(0)(\rho_G/P_G)^{1/2}$, in which pressure travels at the velocity of sound of gas in bubbles. If ΔP_G becomes small, while the time required to reach the steady rate becomes long, the degree of overpressure is small in the end. [9] However, overpressure in bubbles can also be affected by other factors such as bubble oscillation, ascent rate of magma, and viscosity of magma.

When the number of molecules of gas in bubbles is constant, pressure in the bubble decreases with bubble expansion and therefore the bubbles oscillate. As seen later, the generation of bubble oscillation is considerably affected by the viscosity of surrounding liquid. Moreover, when vacuum bubbles are present under liquid pressure P, the bubbles collapse but the timescale at that time is called the Rayleigh collapse time t_{col}, which is roughly equivalent to the case of $\Delta P_G = P$. This is extensively used in cavitation-related studies in engineering (Sect. 6.6.1). The case where the number of molecules of gas in bubbles is constant will be explained in the next section along with the influence of viscosity.

4.8 The Influence of Viscosity on Bubble Expansion: Viscosity-Limited Expansion

To investigate the influence of viscosity, let us consider the Rayleigh-Plesset equation after neglecting inertia. Liquid pressure around the gas-liquid interface is determined by an equation of the conservation of momentum, Eq. (4.68), on which the Rayleigh-Plesset equation is based, and is generally different

[9] Scaling of the above equation by this timescale ($\tilde{R} = R/R(0)$, $\tilde{t} = t/t_c$) gives

$$\frac{d\tilde{R}}{d\tilde{t}} = \left[\frac{2}{3}\left(1 - \tilde{R}^{-3}\right) - \frac{\tilde{R}_C}{\tilde{R}}\left(1 - \tilde{R}^{-2}\right)\right]^{\frac{1}{2}} \tag{4.82}$$

from liquid pressure at infinity far from the bubble, which is expressed by P_∞ in the Rayleigh-Plesset equation. However, for slow bubble expansion that can neglect the inertia term, because the pressure gradient is 0 and $\sigma_{rr}(R,t) = -4\eta\dot{R}/R$ holds in an equation of dynamic mechanical equilibrium at the interface, Eq. (4.73), P_∞ in the Rayleigh-Plesset equation and liquid pressure at the gas-liquid interface become equal to the liquid pressure at infinity. Therefore, unless otherwise specified, the liquid pressure at infinity P_∞ will be written as simply P. Then, the Rayleigh-Plesset equation can be written as follows:

$$P_{\mathrm{G}}(t) - P(t) - \frac{2\gamma}{R(t)} - 4\eta\frac{\dot{R}(t)}{R(t)} = 0 \tag{4.83}$$

that is,

$$\dot{R}(t) \equiv \frac{dR(t)}{dt} = \frac{R}{4\eta}\left(P_{\mathrm{G}}(t) - P(t) - \frac{2\gamma}{R(t)}\right) \tag{4.84}$$

is obtained.

4.8.1 The Case Where Pressure in Bubbles and Overpressure in Bubbles are Constant

The Case Where Overpressure in Bubbles is Constant ($P_{\mathrm{G}}(t) - P(t) = \Delta P_{\mathrm{G}} =$ **constant)**
Consider the case where overpressure in bubbles $P_{\mathrm{G}} - P$ is constant (ΔP_{G}). This equation can be integrated and

$$R(t) = R_{\mathrm{B}} + (R(0) - R_{\mathrm{B}})\exp\left(\frac{\Delta P_{\mathrm{G}}}{4\eta}t\right) \tag{4.85}$$

is obtained (Lyakhovsky et al. 1996). In this equation, R(0) is the initial bubble radius and R_{B} is the bubble radius in mechanical balance, given by Eq. (4.77). Furthermore, $4\eta/\Delta P_{\mathrm{G}}$ gives the timescale of viscous deformation t_{vis} as follows:

$$t_{\mathrm{vis}} = \frac{4\eta}{\Delta P_{\mathrm{G}}} \tag{4.86}$$

where ΔP_{G} has a meaning of pressure difference that drives viscous deformation.

When $\tilde{R} = R/R(0)$, $\tilde{R}_{\mathrm{B}} = R_{\mathrm{B}}/R(0)$, and $\tilde{t} = t/t_{\mathrm{vis}}$ is defined using the initial bubble radius $R(0)$, Eq. (4.84) and its solution can be written by dimensionless expression as follows:

$$\frac{d\tilde{R}}{d\tilde{t}} = \tilde{R}\left(1 - \frac{\tilde{R}_{\mathrm{B}}}{\tilde{R}}\right) \tag{4.87}$$

$$\tilde{R}(\tilde{t}) = \tilde{R}_{\mathrm{B}} + \left(1 - \tilde{R}_{\mathrm{B}}\right)\exp(\tilde{t}) \tag{4.88}$$

When $R(0) \gg R_{\mathrm{B}}$,

$$R(t) \approx R(0)\exp\left(\frac{\Delta P_{\mathrm{G}}}{4\eta}t\right) \tag{4.89}$$

is obtained, which is exponential expansion, i.e., the so-called viscosity-limited growth (Toramaru 1995).

Figure 4.7 shows how the viscous expansion occurs. When the timescale of viscosity is equivalent to or shorter than the timescale of inertia, the viscosity exerts almost no influence and inertia that

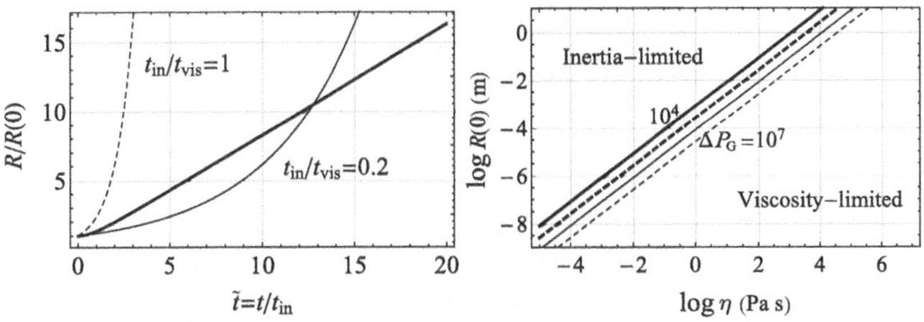

Fig. 4.7 Comparison between inertial expansion and viscosity-limited expansion of bubble expansion under the condition where gas pressure in bubbles is constant ($\tilde{R}_B = R_B/R_p^* = 2\gamma/\Delta P_G R_p^* = 0.5$). (Left) Change in the bubble radius as a function of time. Time is scaled by the timescale of inertial expansion. The thick line represents a case only of inertial expansion. The fine lines represent a case only of viscous expansion and correspond to Eq. (4.85). The broken fine line and the solid fine lines correspond to the ratio of timescale $t_{in}/t_{vis} = 1$ and 0.2, respectively. (Right) Relationship between the viscosity and the bubble radius, which determine viscosity-limited and inertia-limited (the viscosity and the bubble radius when $t_{in}/t_{vis} = 1$). The lines are of $\Delta P_G = 10^4$, 10^5, 10^6, and 10^7. Although the density of liquid is assumed to be 2500 kg/m^3, there is no great difference if the density of water is used. The region in which the viscosity is <1 Pa s is used to evaluate the influence of the viscosity of water

moves the mass of liquid around bubbles determines the rate of bubble expansion. On the other hand, as the timescale of viscosity becomes larger, the initial delay of bubble expansion increases. This fact indicates that inertia can be substantially neglected and viscosity controls bubble expansion.

The ratio of timescale of inertial expansion to that of viscous expansion

$$\frac{t_{in}}{t_{vis}} = \frac{R(0)(\rho\Delta P_G)^{1/2}}{4\eta} \tag{4.90}$$

is largely dependent on the bubble radius and viscosity of liquid. Qualitatively, for low viscosity liquids such as water and a relatively large bubble radius, the viscosity does not affect expansion; note that bubble expansion is inertial expansion. When the viscosity increases and the bubble radius decreases, bubble expansion will be controlled by the viscosity of the liquid (Fig. 4.7).

4.8.2 Expansion Under Constant Amount of Decompression

This subsection examines the case where the pressure in bubbles changes with change in the size of bubbles. Assuming the liquid pressure before decompression is P_0, we deal with the liquid pressure as a step function with respect to time, i.e.,

$$P(t) = P_0 \quad t = 0 \tag{4.91}$$

$$P(t) = P_f \quad t > 0 \tag{4.92}$$

where P_f is the liquid pressure after decompression.

When $\tilde{R} = R/R(0)$, $\tilde{t} = t/t_{vis}$, and $\tilde{P}_G = P_G/P_G(0)$ are defined using timescale of viscosity-limited growth $t_{vis} = 4\eta/P_{G0}$ and the initial bubble radius $R(0)$, Eq. (4.84) is then rewritten as follows:

$$\frac{d\tilde{R}}{d\tilde{t}} = \tilde{R}\left(\tilde{P}_\mathrm{G} - \tilde{P}_\mathrm{f} - \frac{b}{\tilde{R}}\right) \tag{4.93}$$

$$\tilde{P}_\mathrm{f} = \frac{P_\mathrm{f}}{P_\mathrm{G}(0)} \tag{4.94}$$

$$b = \frac{2\gamma}{R(0)P_\mathrm{G}(0)} \tag{4.95}$$

In these equations, $\tilde{P}_\mathrm{f}$ and b are connected with the final equilibrium bubble radius $\tilde{R}_\mathrm{f}$.

Constant Number of Molecules in Bubbles, Isothermal

The state where the number of molecules in bubbles is constant, corresponds to the case where gas molecules diffuse infinitely slow in liquids. In an isothermal condition, $P_\mathrm{G} = P_\mathrm{G}(0)(R(0)/R)^3$ ($\tilde{P}_\mathrm{G} = \tilde{R}^{-3}$ is obtained from an equation of state and neglecting the term of the interfacial tension ($b = 0$) yields

$$\tilde{P}_\mathrm{G} = \frac{1}{\tilde{R}^3} \tag{4.96}$$

$$b = 0 \tag{4.97}$$

and Eq. (4.93) will be

$$\frac{d\tilde{R}}{d\tilde{t}} = \tilde{R}\left(\frac{1}{\tilde{R}^3} - \tilde{P}_\mathrm{f}\right) \tag{4.98}$$

This equation can be integrated, thus giving

$$\tilde{R} = \tilde{P}_\mathrm{f}^{-1/3}\left[1 - \left(1 - \tilde{P}_\mathrm{f}\tilde{R(0)}^3\right)e^{-3\tilde{P}_\mathrm{f}\tilde{t}}\right]^{1/3} \tag{4.99}$$

(Figure 4.8).

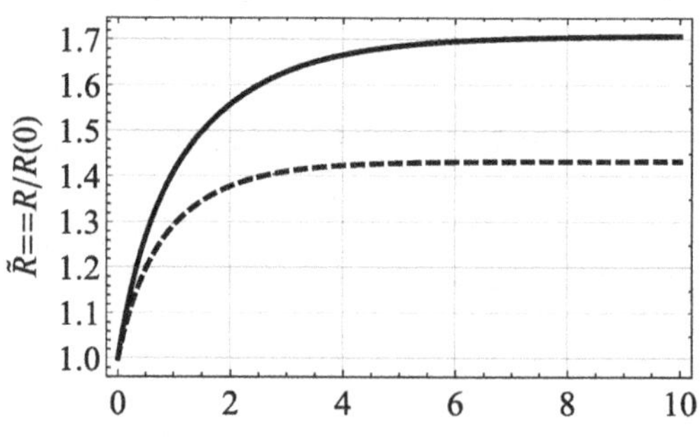

Fig. 4.8 The behavior of bubble expansion under the constant amount of decompression when the amount of gas is constant and the influence of the surface tension on the behavior. Melt pressure P_f is then assumed to be $\tilde{P}_\mathrm{f} = P_\mathrm{f}/P_\mathrm{G}(0) = 0.2$, where $P_\mathrm{G}(0)$ is the initial bubble pressure. $\tilde{P}_\mathrm{f} = 0.2$ indicates that the initial pressure in bubbles is five times the liquid pressure. A solid line: without surface tension ($b = 0$), Eq. (4.99). A broken line: with surface tension ($b = 0.2$), Eq. (4.100). $b = 0.2$ indicates that the pressure of the surface tension to the initial bubble radius $2\gamma/R(0)$ is one-fifth of the pressure in bubbles $P_\mathrm{G}(0)$. The equilibrium radius of the case where the surface tension is neglected becomes slightly larger

For Eq. (4.99) for a single bubble and for a cell model equation (4.156) for foamy bubbles, Barclay et al. (1995) obtained analytical solutions for instantaneous decompression and the case of linear decompression at a constant decompression rate. Although bubble oscillation occurs when the inertia term is involved, it will be dealt with in Sect. 6.7. Barclay et al. (1995)

If the surface tension is not negligible,

$$\frac{d\tilde{R}}{d\tilde{t}} = \tilde{R}\left[\frac{1}{\tilde{R}^3} - \tilde{P}_{\mathrm{f}} - \frac{b}{\tilde{R}}\right] \tag{4.100}$$

is obtained. Figure 4.8 shows results that were numerically solved; the surface tension affects the final equilibrium radius and the equilibrium radius diminishes in size equivalent to the surface tension. When the number of molecules in bubbles is constant, a bubble expansion curve is convex upward and the radius reaches the equivalent radius in several times the timescale of viscosity.

4.8.3 The Case of Decompression at a Constant Rate

Change in Pressure

We consider the case where liquid pressure linearly decreases with time from the pre-decompression saturated pressure P_0.

$$P(t) = P_0\left(1 - \frac{t}{t_{\mathrm{dec}}}\right) \tag{4.101}$$

where decompression time t_{dec} is defined by

$$t_{\mathrm{dec}} = \frac{P_0}{|dP/dt|} \tag{4.102}$$

That is,

$$\frac{dP(t)}{dt} = -\frac{P_0}{t_{\mathrm{dec}}} \tag{4.103}$$

The Case of Constant Pressure in Bubbles (Constant Molecular Volume of Gas)

For cases where pressure in bubbles is constant ($P_{\mathrm{G}} = P_{\mathrm{G}}(0)$), the interfacial tension can be neglected ($P_{\mathrm{G}}(0) \approx P_0$) and pressure is linearly decompressed. Note that the Rayleigh-Plesset equation (4.84) can be written as follows:

$$\frac{dR(t)}{dt} = \frac{P_0}{4\eta}\frac{t}{t_{\mathrm{dec}}}R(t) \tag{4.104}$$

and its solution is given as an exponential function of the square of time as follows (Lensky et al. 2004):

$$R(t) = R(0)\exp\left(\frac{t^2}{2t_{\mathrm{dec}}t_{\mathrm{vis}}}\right) \tag{4.105}$$

This case is different from the case of constant amount of decompression in the point that this timescale of expansion is given by the square root of the product of the decompression time and the timescale of viscosity-limited growth. This corresponds to Eq. (4.89) under a constant pressure, and the timescale of viscosity-limited growth here, t_{vis}, is defined as follows:

$$t_{\rm vis} = \frac{4\eta}{P_0} \tag{4.106}$$

which is different from the timescale of viscosity (Eq. (4.86)) described above because of difference in the scale of the amount of decompression. The assumed state where pressure in bubbles is constant is the state where gas pressure is in equilibrium with supersaturated liquid, such as a nucleation stage, which is maintained by efficient diffusion.

4.8.4 Bubble Growth Calculation with a Combination of Diffusion and Viscosity

The above discussion is for the case where the number of molecules in bubbles does not change or the case where overpressure in bubbles remains constant because of infinitely fast diffusion. In reality, the number of molecules in bubbles is controlled by diffusion of water in liquid, and the efficiency of diffusion is controlled by the diffusivity, and the difference between the equilibrium content at the bubble-liquid interface and the content in liquid at a point distant from bubbles. When the number of molecules in bubbles increases owing to diffusion, gas pressure in bubbles increases and defeats viscous resistance, leading to bubble expansion. Therefore, to know change in the bubble radius, molecular diffusion in liquid, the equation of state of gas in bubbles, and the Rayleigh-Plesset equation have to be simultaneously solved.

Derivation of a Governing Equation

The equilibrium concentration at the bubble-liquid interface, generally given by Eq. (4.5), is a function of gas pressure in bubbles. Substituting gas pressure in bubbles $P_{\rm G}$ determined by Eq. (4.83), the Rayleigh-Plesset equation in which the inertia term is neglected, into Eq. (4.5) and using Eqs. (4.7) and (4.8) yield the following equation corresponding to Eq. (4.9),

$$C_R(R, P, C_\infty) = C_\infty \left[1 - R_P^*(C_\infty, P)\left(\frac{1}{R_{\rm C}(C_\infty, P)} - \frac{1}{R} - \frac{2\eta}{\gamma}\frac{\dot{R}}{R}\right)\right]^{\frac{1}{2}} \tag{4.107}$$

in which the term representing increase in the pressure in bubbles owing to viscous resistance, $-2\eta/\gamma \cdot \dot{R}/R$, is added. When bubbles expand, this term becomes negative, the inside of the parentheses decreases and the equilibrium concentration becomes large, and a driving force of diffusion flux decreases. Moreover, $R_P^*(C_\infty, P)$ and $R_{\rm C}(C_\infty, P)$ are parameters that define a supersaturated state and, if neglecting the dependence of the solubility on liquid pressure (Poynting factor), are influenced only by a constant saturation pressure ($P_{\rm G}^* = P_{\rm G}(C_\infty)$) and variable liquid pressure $P(t)$ under a decompression state if C_∞ does not change.

$$R_{\rm C}(C_\infty, P) = \frac{2\gamma}{P_{\rm G}^* - P(t)} \tag{4.108}$$

Because $R_P^*(C_\infty, P)$ is given by Eq. (4.10), it is determined only by saturation pressure corresponding to water content at infinity (i.e., pre-decompression saturation concentration), $P_{\rm G}^*$.

$$R_P^*(C_\infty, P) = \frac{2\gamma}{P_{\rm G}^*} \tag{4.109}$$

Assume that the pre-decompression state involves liquid pressure P_0 and homogeneous concentration in liquid C_∞. Considering the case where bubbles with a finite size exist in equilibrium in liquid, the liquid should be a supersaturated state (gas and liquid having common pressure P_0 are not in equilibrium via a flat interface). Pressure in bubbles $P_G(0)$ is increased by the surface tension as compared with liquid pressure. Because the gas pressure is equal to P_G^*, which corresponds to a concentration in equilibrium with liquid, and the bubble radius $R(0)$, which is equal to the critical radius $R_C(0)$, the following equation holds:

$$P_G(0) - P_0 = \frac{2\gamma}{R(0)} \tag{4.110}$$

$$P_G^* - P_0 = \frac{2\gamma}{R_C(0)} \tag{4.111}$$

After decompression, the liquid pressure $P(t)$, the bubble radius $R(t)$, the gas pressure in the bubbles $P_G(t)$, and the rate of change in the bubble radius $\dot{R}(t)$ are determined using the Rayleigh-Plesset equation.

$$P_G(t) - P(t) - \frac{2\gamma}{R(t)} - 4\eta\frac{\dot{R}(t)}{R(t)} = 0 \tag{4.112}$$

When the number of molecules in bubbles increases by diffusion, the number of molecules in bubbles n_G changes after Eq. (4.36). We then substituted $P_G(t)$ of the Rayleigh-Plesset equation (Eq. (4.112)) in which the inertia term is neglected into the equation of state of an ideal gas $4\pi R^3 P_G/3 = n_G k_B T$ and differentiated it with respect to time. Finally, in the term dn_G/dt in the equation, we substituted the equation of diffusion growth (Eq. (4.36)). For C_R of the Eq. (4.36), we use Eq. (4.107). For constant decompression rate (Eq. (4.101)), the obtained equation of change in the bubble radius will be

$$\underbrace{\tilde{R}\ddot{\tilde{R}} + 2\dot{\tilde{R}}^2}_{\text{Viscous resistance}} = \underbrace{\frac{t_{\text{vis}}}{t_{\text{dec}}}\tilde{R}^2}_{\text{Decompression}} + \underbrace{\left[3\frac{t_{\text{vis}}}{t_{\text{difG}}}\tilde{P}_G^*\Delta\tilde{C} - (3\tilde{R}\tilde{P} + 2b\tilde{P}_G^*)\dot{\tilde{R}}\right]}_{\text{Growth}} \tag{4.113}$$

after scaling time by characteristic time of viscosity t_{vis}, the bubbles radius by the initial bubble radius $R(0)$, and pressure by pre-decompression saturation pressure P_0 and rearranging. In this equation, b is defined by Eq. (4.95) and because $b = 1 - \tilde{P}_G(0)^{-1} = 1 - 1/\tilde{P}_G^*$ holds from Eq. (4.110) in this case if b is given, $\tilde{P}_G^*$ will be determined. The physical meaning of the parameter b is the excess pressure worth of surface tension in the pressure in bubbles that is in equilibrium with a supersaturated liquid. The characteristic time of viscosity is $t_{\text{vis}} = 4\eta/P_0$ and characteristic time of diffusion growth is $t_{\text{difG}} = R(0)^2/(\phi_{G0}D)$, where $\phi_{G0} = v_G C_\infty = k_B T C_\infty/P_G(0)$.

$\Delta\tilde{C}$ in this equation corresponds to concentration difference between infinity and the bubble interface, $C_\infty - C_R$, which drives diffusion and is expressed with Eq. (4.107) as follows:

$$\Delta\tilde{C} = 1 - \left[1 - \frac{1}{\tilde{P}_G^*}\left(\tilde{P}_G^* - \tilde{P} - \frac{b/(1-b)}{\tilde{R}} - \frac{\dot{\tilde{R}}}{\tilde{R}}\right)\right]^{1/2} \tag{4.114}$$

Equations (4.113) and (4.114) describe the diffusion growth of a single bubble in liquid decompressed from the equilibrium pressure under the condition where water content in liquid at infinity is constant and represent balance between viscous resistance in liquid (the left-hand side) and a driving force (the right-hand side). As seen from these equations, the behavior of bubble growth is determined only by the ratios of characteristic time t_{vis}/t_{dec} and t_{vis}/t_{difG} and the parameter b (or P_G^*) defining the state before decompression. The inverse of t_{vis}/t_{difG} is the Péclet number (Pe) defined later. In cases where pressure is instantaneously decompressed to constant pressure, $\tilde{P} = \tilde{P}_f < P_0$ is used and the first term of the right-hand side of Eq. (4.113) will be 0 because of $t_{dec} = \infty$. For linear continuous decompression, $\tilde{P} = 1 - t_{vis}\tilde{t}/t_{dec}$ is used as per Eq. (4.101).

Although the above formulation is performed under the condition where gas concentration in liquid is constant, the concentration will decrease with bubble growth because of the conservation of the gas component. In that case, a system that comprises bubbles having the same size and has the bubble number density N has to be considered; moreover, the decrease in the concentration at infinity has to be considered based on the conservation of mass of water $C_\infty(t) = C_0 - 4\pi R^3 N/3$. Using the cell model explained later (Sect. 4.10.2), Proussevitch et al. (1993a), Blower et al. (2001), and Lensky et al. (2004) considered the conservation of the gas component and examined bubble growth by competition between diffusion and viscosity.

Bubble Growth Curves

Using Eqs. (4.113) and (4.114) which describe diffusion growth of bubbles in a viscous fluid, derived above, let us examine the competition among diffusion, viscosity, and decompression rate, taking an example from bubble growth at constant gas concentration in liquid (C_∞). We assume that a bubble with the initial radius $R(0)$ is in equilibrium at liquid pressure P_0 before decompression, i.e., although the bubble radius in this condition is equal to the critical radius R_C in terms of physics, it does not play a role of the critical nucleus radius in the nucleation process because the problem of nucleation is not dealt with here. Because the pressure in the bubble is increased by the surface tension as compared with liquid pressure, the water concentration in liquid being in equilibrium with the bubble is higher than the saturated concentration at pressure P_0, $C_{eq}(P_0)$, i.e., the liquid is in a supersaturated state. This overpressure in the bubble due to the surface tension $(2\gamma/R(0))$ and the water concentration in liquid depend on the bubble radius. Furthermore, when the bubble radius is infinity, the overpressure will be zero and the water content in liquid will be equal to the saturated content $C_{eq}(P_0)$.

A bubble coexisting with liquid in equilibrium (either mechanical or chemical) with a finite size is, so to say, in an unstable equilibrium. If the bubble radius is smaller than the equilibrium bubble radius maintaining mechanical equilibrium, the pressure in the bubble increases, the equilibrium interface concentration increases, the chemical equilibrium at the interface breaks, water in the bubble dissolves in the liquid, and the concentration approaches the equilibrium content at the interface. Furthermore, the number of molecules in bubbles decrease, the bubble radius decreases, the interface concentration of bubbles increases, and diffusion from the interface to the inside of liquid occurs. Therefore, the interface concentration decreases and the dissolution and diffusion of water in the liquid and decrease in the bubble radius continuously occur. Consequently, the concentration gradient in the liquid is maintained and diffusion and dissolution of water from the bubble will continue until the bubble disappears. For cases where the bubble radius is larger than the equilibrium bubble radius, the opposite process occurs and a bubble continues to grow. In the case where the bubble radius changes under the chemical equilibrium, although the same process occurs, pressure difference at the interface drives the change in the bubble radius. In the case where liquid pressure changes, the bubble grows or disappears in a similar manner.

For the constant amount of decompression, the decompression rate is 0; moreover, in this case, $t_{vis}/t_{dec} = 0$ because $t_{dec} = \infty$. Figure 4.9 shows differences by the difference in Pe $= t_{difG}/t_{vis}$. As the Péclet number is high, diffusion takes more time and growth is suppressed. Moreover, when the Péclet number is large, the time required for diffusion is more and the behavior is represented by convex upward curves, similar to the behavior at an initial stage in the case of the constant number of molecules in bubbles, as relaxing toward the equilibrium radius (Fig. 4.9c and d). These curves are steeper than a

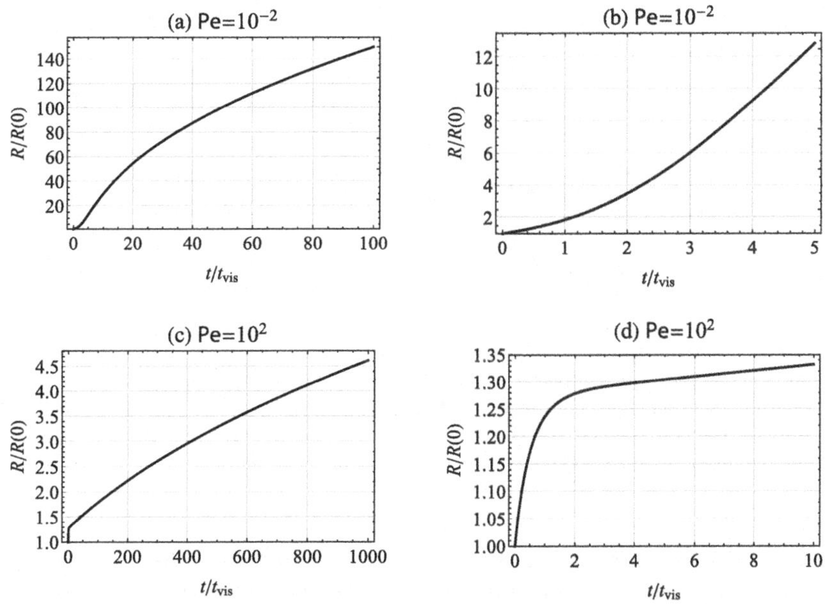

Fig. 4.9 Comparison of the behavior of bubble growth associated with gas exsolution at a constant pressure (Eq. (4.113), $P_f/P_0 = 0.4$, $b = 0.2$). **a** and **b** Pe $= t_{difG}/t_{vis} = 10^{-2}$. **c** and **d** Pe $= t_{difG}/t_{vis} = 10^2$. **b** and **d** show the behavior in the initial stage. When the Péclet number is small, time domains where convex downward curves appear fall within the characteristic timescale of viscosity ($t/t_{vis} < 10$). When Pe $= 10^2$, the behavior shows convex upward curves, obeying an expansion law at the constant number of bubbles (Fig. 4.8) at the initial stage

diffusion-limited growth curve, which is proportional to the square root of time. On a long-term basis, because the influence of viscosity can be neglected, the bubble growth curves will demonstrate a gentle convex upward curve of the diffusion-limited growth. When Pe $= t_{difG}/t_{vis}$ is < 1, the behavior shows convex downward curves characteristic of viscosity-limited growth at an initial stage shorter than the viscosity timescale; furthermore, at the later stage longer than the viscosity timescale, the behavior is convex upward curves characteristic of diffusion growth (Fig. 4.9a and b).

Figure 4.9 shows that such behavior of growth curves characteristic of viscosity-limited growth dependent on the Péclet numberis is observed in the time regime within the timescale characteristic of viscosity, t_{vis}, at most. An absolute quantity of bubble growth becomes large as the value of the Péclet number is reduced, i.e., diffusion works more effectively. These facts hold for continuous decompression. To summarize, judgement whether viscosity-controlled or diffusion-controlled will be based on the behavior at an initial stage of bubble growth, and the absolute quantity of bubble growth within the characteristic time of viscosity.

When the decompression rate is not 0, the behavior changes according to a value of t_{vis}/t_{dec} or t_{difG}/t_{dec}. Figure 4.10 shows bubble growth curves when the values of these parameters are changed. Before decompression, a bubble and liquid are in equilibrium, i.e., chemical and mechanical equilibria. However, if liquid pressure infinitesimally decreases, force balance at the interface breaks and the bubble expands under the condition of constant number of molecules in the bubble. The expansion rate at that time is influenced by viscosity. When the bubble radius increases, the pressure in a bubble decreases as per the equation of state and the equilibrium concentration at the interface decreases. Water then moves from the interface to the bubble to maintain the chemical equilibrium at the interface, and the concentration at the interface decreases and a concentration gradient is produced in the liquid. This concentration gradient drives water diffusion from the liquid to the bubble, the number of water molecules in the bubble then increases, and the bubble grows.

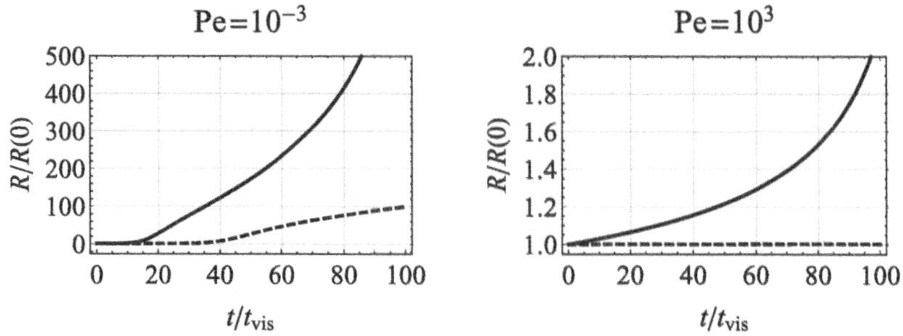

Fig. 4.10 Comparison of the behavior of bubble growth associated with gas exsolution at a constant decompression rate ($b = 0.2$). (Left) Pe $= 10^{-3}$. (Right) Pe $= 10^{3}$. Note that the scales of the vertical axes differ. Broken lines: small constant decompression rate ($t_{vis}/t_{dec} = 0.0001$). Solid lines: large constant decompression rate ($t_{vis}/t_{dec} = 0.01$). For relationship between time and pressure, multiplying a value on the horizontal axis by tvis/tdec yields the amount of decompression at that time, i.e., a value of 100 on the horizontal axis for the broken lines is equivalent to 1% of decompression and that for the solid lines is equivalent to 100% of decompression

Because pressure difference between a bubble and liquid, which drives bubble expansion, is small at an initial stage of continuous decompression, the higher the viscosity is (i.e., t_{vis} is larger and the Pe is smaller), the greater the delay of bubble expansion is. Therefore, when the Pe number is small, diffusion and entry of water into the bubble is delayed and the initial delay of bubble growth becomes clearly apparent (Fig. 4.10a). Furthermore, in the case where the Péclet number is small, concentration gradient associated with the initial delay is large and growth curves will be convex upward curves characteristic of diffusion-limited growth. On the other hand, when the Péclet number is large, the bubble expands by almost maintaining mechanical equilibrium with the liquid and shows convex downward growth curves. This can be inferred from $R \propto P^{-1/3} \propto (1 - t/t_{dec})^{-1/3}$ obtained from the equation of state of gas ($R^3 P =$ constant).

To examine the influence of continuous decompression on growth curves, we have rewritten Eq. (4.113) as follows:

$$\ddot{\tilde{R}} = \frac{1}{\tilde{R}} \left\{ \underbrace{\frac{t_{vis}}{t_{dec}} \tilde{R}^2}_{\text{Decompression}} + \underbrace{\left(3\frac{t_{vis}}{t_{dif}} \tilde{P}_G^* \Delta\tilde{C} - (3\tilde{R}\tilde{P} + 2b\tilde{P}_G^*)\dot{\tilde{R}} \right)}_{\text{Growth}} - \underbrace{2\dot{\tilde{R}}^2}_{\text{Inertia}} \right\} \tag{4.115}$$

This equation shows that the curvature of growth curves $\ddot{\tilde{R}}$, which is determined by interplay among continuous decompression, diffusion, and increase in the bubble radius by viscous resistance. To the curvature, continuous decompression contributes positively and a part of viscous resistance contributes negatively. Figure 4.10 shows growth curves when the decompression rate changes. Regardless of the Péclet number, as the decompression rate is high, the growth curves tend to show convex downward curves; moreover, as the decompression rate is low, they tend to show convex upward curves. This results from competition between growth by diffusion and expansion by continuous decompression.

Including the case of constant liquid pressure, the characteristics of growth curves do not completely agree with the simple classification of growth rate-determining processes based on the Péclet number, i.e., diffusion-limited growth = convex upward (parabolic) and viscosity-limited growth = convex downward (exponential). Therefore, to identify the growth rate-limiting process, the comparison of the amount of bubble growth for each time constant and of the shape of growth curves within the time regime of the characteristic timescale of viscosity is important.

4.9 Outline of Bubble Growth and Experimental Results

4.9.1 Characteristic Timescale in Bubble Growth

Timescale related to bubble growth and expansion processes will be summarized in this section. Note that the boundary conditions of timescale somewhat vary as per the situation. Here, neglecting the influence of boundary conditions, we see the orders of magnitude of the timescale and the parameters on which timescale depends.

With reference to Eq. (4.32), the timescale of diffusion-limited growth to the bubble radius R can be given as follows:

$$t_{\mathrm{difG}} = \frac{R^2}{\phi_{\mathrm{G0}} D} \tag{4.116}$$

where $\phi_{\mathrm{G0}} = v_{\mathrm{G}} C_0$, which is conversion volume corresponding to the initial water concentration, and plays a role of a correction term using the initial concentration C_0, which is a driving force of diffusion growth. Assuming that the molecular volume of the gas obeys the equation of state of an ideal gas, we obtain $\phi_{\mathrm{G0}} = k_{\mathrm{B}} T C_0 / P_0$. If we assume the saturated state (this is not a bad assumption), concentration is determined by pressure and ϕ_{G0} is determined only by pressure. The concentration can then be estimated using the simplified Burnham model and $C(P(\mathrm{Pa})) = C(10^8)(P/10^8)^{1/2}$. $C(10^8\mathrm{Pa})$ represents the saturation concentration of basaltic melt at 100 MPa, 3.0 wt%, or $3.0 \times 10^{27}\,\mathrm{m}^{-3}$ of the number of molecules (4.2 wt% or $4.2 \times 10^{27}\,\mathrm{m}^{-3}$ is used for andesitic, granitic, and albite melt in common). When these values are used, ϕ_{G0} will be a value ranging from 0.2 to 6.0 when the pressure ranges from 1000 to 1 MPa. Therefore, there is no considerable difference in the orders of magnitude if the timescale of simple diffusion

$$t_{\mathrm{dif}} = \frac{R^2}{D} \tag{4.117}$$

is used as the timescale of diffusion growth. The timescale is shown in the left chart of Figure 4.11 as a function of the bubble radius and diffusion coefficient. Note that it depends on the bubble radius and is in the range from 10^{-6} to 10^{10} s.

Using a reference pressure P_0, the timescale of viscosity-limited growth can be given from Eq. (4.86) as follows:

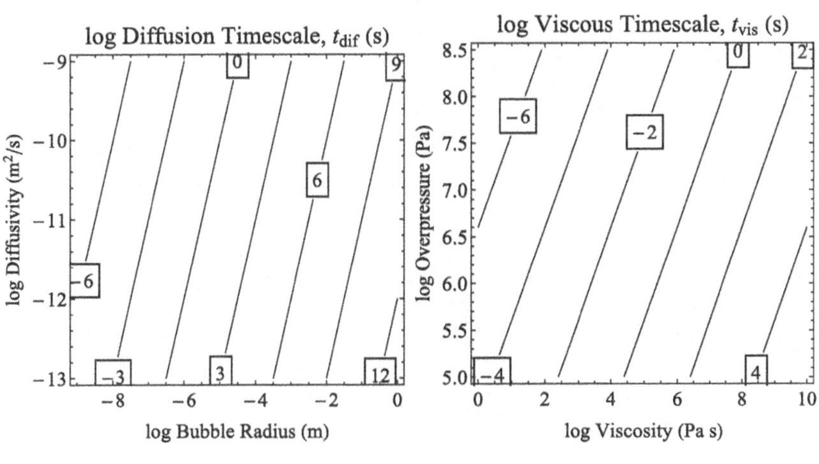

Fig. 4.11 The timescale of diffusion growth (left) and the timescale of viscosity growth (right)

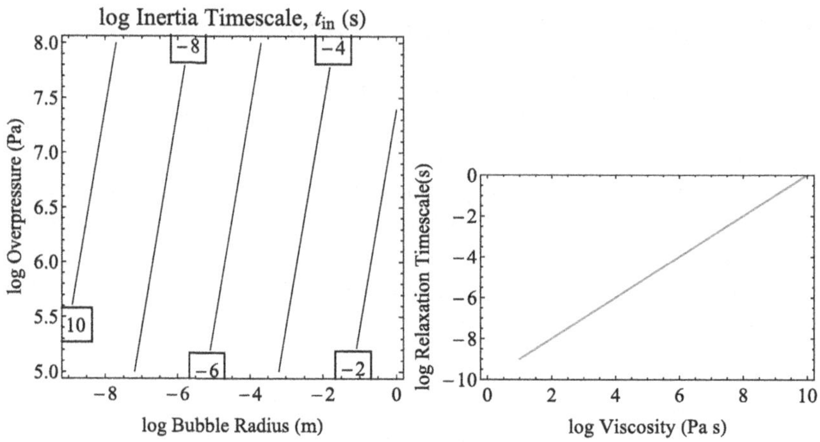

Fig. 4.12 The timescale of inertial expansion (left) and the timescale of viscoelastic relaxation (right)

$$t_{vis} = \frac{4\eta}{P_0} \tag{4.118}$$

The timescale of viscosity-limited growth depends on the viscosity coefficient and the reference pressure P_0, which serves as a rough equivalent of overpressure; note that it is in the range from 10^{-6} to 10^5 s (right chart of Fig. 4.11) and does not depend on the bubble radius. In discussion, as the reference pressure, liquid pressure at the saturated state, gas pressure P_G^*, and the initial pressure in bubbles $P_G(0)$ can be used.

On the basis of Eq. (4.81), the timescale of inertial expansion is given as

$$t_{in} = R\left(\frac{\rho}{\Delta P_G}\right)^{1/2} \tag{4.119}$$

and is the same with the timescale of the Rayleigh collapse t_{col} (Eq. (6.109)) and the timescale of bubble oscillation in terms of physics. Because vacuum bubbles are assumed in the Rayleigh collapse, ΔP_G is equal to liquid pressure P. The timescale of inertial expansion is extremely short, which is on the order of microseconds to milliseconds (left chart of Fig. 4.12).

Although magma has linear (Newtonian) viscosity in some situations, complex rheology becomes important in other situations. When magma is approximated as a Maxwell viscoelastic fluid, the following timescale of viscoelastic relaxation becomes important (see Appendix in 11.7 of Chap. 11).

$$t_{re} = \frac{\eta}{k_E} \tag{4.120}$$

Because shear modulus of silicate melt k_E is of the order of 10 GPa and does not considerably change, viscoelastic relaxation time is almost determined by the viscosity η of melt and extensively varies between 10^{-9} and 1 s (right chart of Fig. 4.12).

Because melt pressure changes with the motion of magma, the timescale of decompression below is important.

$$t_{dec} = \frac{P}{|dP/dt|} \tag{4.121}$$

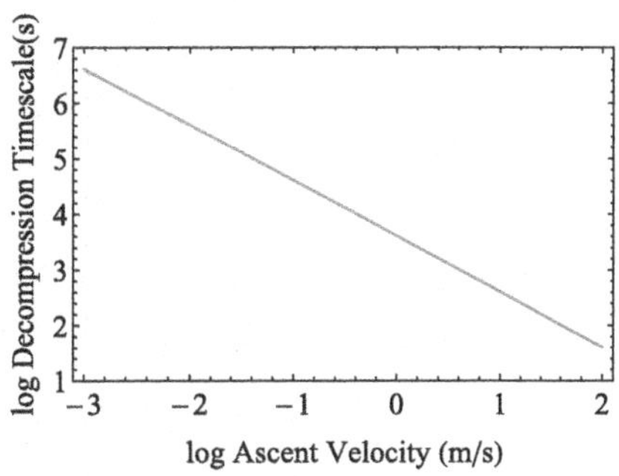

Fig. 4.13 Relationship between the timescale of decompression and the ascent velocity. $P = 100$ MPa, $\rho = 2500$ kg/m^3 is assumed in the case of steady state ascent from a depth of about 4 km

When the pressure of magma is determined by the hydrostatic pressure of magma, decompression is related to the discharge velocity of magma, i.e., the ascent velocity of magma U,

$$t_{\text{dec}} = \frac{P}{\rho g U} \tag{4.122}$$

holds (see Box in Sect. 5.4, decompression rate, for details). In this equation, ρ is the density of magma. According to the ascent velocity of magma, the timescale of decompression considerably varies on a scale between 10 s and 1 year (Fig. 4.13).

4.9.2 Dimensionless Parameters Controlling Bubble Growth

The timescales shown above are the approximate equivalent speed of phenomena of each elemental process. Growth and expansion of bubbles under the disequilibrium state (under decompression) is a complex process in which such elemental processes simultaneously proceed. Therefore, a process having a longer timescale limits the rate of the whole complex process. To evaluate which process is dominant, it is useful to examine the dimensionless numbers which are the ratios of timescales.

Originally, the Péclet number is defined as the ratio of the magnitude of advective transfer to the magnitude of diffusional transfer, as follows:

$$\text{Pe} = \frac{[\text{Advection}]}{[\text{Diffusion}]} \tag{4.123}$$

For bubble growth, the magnitude of diffusion can be estimated by the velocity of diffusion and a smaller time constant of diffusion indicates the magnitude of diffusion is larger because diffusion progresses faster, i.e., [Diffusion] = $1/t_{\text{difG}}$.

On the other hand, the magnitude of advection can be estimated by the rate of increase in the bubble radius, $\dot{R}$, and its time constant is $(\dot{R}/R)^{-1}$. From the Rayleigh-Plesset equation in which inertia is neglected (Eq. (4.83)), $\dot{R}/R = 4\eta/\Delta P_{\text{G}} \approx 4\eta/P_0$ is obtained. Therefore, the magnitude of advective transfer for the bubble growth in viscous fluid can be measured by the time constant of viscosity and will be [Advection] = $1/t_{\text{vis}}$. Thus, the Péclet number is given as

$$\text{Pe} = \frac{t_{\text{difG}}}{t_{\text{vis}}} = \frac{P_0 R^2}{4\phi_{\text{g0}}\eta D} \tag{4.124}$$

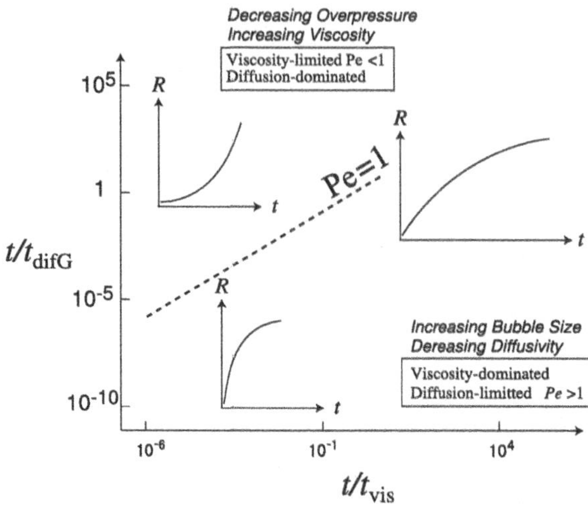

Fig. 4.14 Features of bubble growth curves under constant decompression and relationship of timescales. t_{difG} is the timescale of diffusion growth, t_{vis} is the characteristic time of viscosity, and t is the timescale of interest. $\mathrm{Pe} = t_{\mathrm{difG}}/t_{\mathrm{vis}} = 1$ roughly divides dominant regions of diffusion and viscosity. In the region where the time is longer than the timescale of viscosity, the influence of the Péclet number is small. "-dominated" indicates that viscosity or diffusion effectively works in a short timescale and "-limited" indicates that viscosity or diffusion relatively suppresses growth

(Navon et al. 1998).[10] This Péclet number indicates which factor, i.e., diffusion or viscosity, controls the rate of the bubble growth; moreover, when $\mathrm{Pe} > 1$, the time constant of viscosity is lesser than that of diffusion, indicating diffusion is dominant. On the other hand, when $\mathrm{Pe} < 1$, the time constant of viscosity is larger, indicating viscosity is dominant. Furthermore, $\mathrm{Pe} = 1$ can serve a rough guide for distinguishing the diffusion-controlled region from the viscosity-controlled region (Fig. 4.14). This has been confirmed by experiments described later.

The Péclet number is influenced not only by the diffusion coefficient and the viscosity coefficient but also by the bubble radius, the degree of supersaturation, and the overpressure in bubbles. In silicate melt, the diffusion coefficient of water varies in the range from 10^{-9} to 10^{-13} according to composition and temperature (Appendix in 11.6.2 of Chap. 11). Moreover, the viscosity coefficient varies in the range from 10^{1} to 10^{10} (see Appendix in 11.6.3 of Chap. 11).[11] Therefore, even if other parameters are equal, the Péclet number changes by several orders of magnitude. When the bubble radius increases or the overpressure in bubbles and the degree of supersaturation change with growth of bubbles, the Péclet number changes. This fact will be conducive to examine the modes of controlling the growth rate change with bubble growth.

Although the Péclet number is directly effective for examining bubble growth after instantaneous decompression, in which pressure of melt does not change, the balance with characteristic time of

[10]The Péclet number defined here does not represent the ratio of advection to diffusion conventionally used in matter transport. The conventionally used ratio is the ratio of the advection term of the diffusion equation (the second term on the left-hand side of Eq. (4.1)) to the diffusion term (the right-hand side of Eq. (4.1)), such as the Péclet number described in Sect. 4.3.2. However, note that the advection term is not included in the discussion here.

[11]Between diffusivity and viscosity, the classical Stokes-Einstein relation holds

$$D = \frac{k_{\mathrm{B}} T}{6\pi \eta a_{\mathrm{M}}} \tag{4.125}$$

in some cases. In this equation, a_{M} is the effective radius of a molecule; however, this relation does not hold between the viscosity and the molecular diffusion of silicate melt (Liang et al. 1997).

decompression will be important for continuous decompression (e.g., ascending magma). The ratios of characteristic time of decompression to the timescale of diffusion growth and that of viscosity growth are

$$\frac{t_{\mathrm{dif}}}{t_{\mathrm{dec}}} = \Theta_{\mathrm{D}} = \alpha_3^{-1} \tag{4.126}$$

$$\frac{t_{\mathrm{vis}}}{t_{\mathrm{dec}}} = \Theta_{\mathrm{V}} = \alpha_4^{-1} \tag{4.127}$$

respectively. These equations will play an essentially important role in the problem of bubble nucleation in a continuous decompression field dealt with later. Lensky et al. (2004) expressed them as symbols Θ_{D} and Θ_{V}. Both α_3 and α_4 defined by Toramaru (1995) are the inverses of them, and these three dimensionless numbers, the Péclet number, Θ_{D}, and Θ_{V}, control the behavior of bubble growth.

$$\mathrm{Pe} = \frac{\Theta_{\mathrm{D}}}{\Theta_{\mathrm{V}}} = \frac{\alpha_4}{\alpha_3} \tag{4.128}$$

From this relationship, if two of the three are given, the remaining one will be determined as a matter of course.

4.9.3 Comparison with Experiments

As seen in Sect. 4.8.4, the growth and expansion of bubbles is influenced by diffusion of the gas component into bubbles and viscous flow of melt associated with bubble expansion. Moreover, whether it occurs under constant pressure or under continuous decompression affects the growth and expansion of bubbles and is related to whether $t_{\mathrm{vis}}/t_{\mathrm{dec}}$ in Eq. (4.113) will become zero or not. Continuous decompression causes increase in supersaturation in the background, whereas constant pressure does not.

Using a diamond anvil cell (DAC), Martel and Bureau (2001) conducted an in situ observation experiment under high pressure and high temperature and observed the growth of bubbles. Although temperature and pressure change simultaneously and continuously in many cases because of the properties of DAC, let us take an example of the result of an experiment conducted under almost constant temperature and pressure. The left chart of Fig. 4.15 shows the result of their experiment conducted under the conditions in which the initial temperature is 879 °C, the final temperature is 869 °C, the initial pressure is 1.26 GPa (12.6 kbar), and the final pressure is 1.24 GPa (12.4 kbar). The starting material is haplogranite (haplo means artificial granite synthesized after natural granite from an oxide commercially available as a chemical) containing 78.6 wt% of SiO_2 and considerable amount of water, 23.9 wt% of the initial water content.

Although the diffusion coefficient in this substance estimated from the experimental result of Shaw (1974) is about 10^{-9} m²/s and the viscosity coefficient is about 10^{-4} Pa s, because there are no experimental data on physical properties for such high water content, these estimates are doubtful. The diffusion coefficient reaches a limit at about 10^{-9} and the viscosity coefficient reaches a limit if bonds of SiO_4 networks are cut. Thus, we estimate the viscosity coefficient at 10 Pa s at the most. Using these values, we estimated the time constants of diffusion and viscosity when the radius is 1 μm; moreover, the overpressure in bubbles is 10^6 Pa at tdif = $t_{\mathrm{dif}} = 10^{-3}$ s and $t_{\mathrm{vis}} = 10^{-5}$ s, which yield 10^2 of the Péclet number. The left chart of Figure includes a curve obtained by numerically solving Eq. (4.113) and calculated under the condition where $t_{\mathrm{vis}}/t_{\mathrm{dec}} = 0$ and $\mathrm{Pe} = 10^2$. The curve appears to approximately explain the experimental result. Subsequently, Martel et al. improved the DAC device,

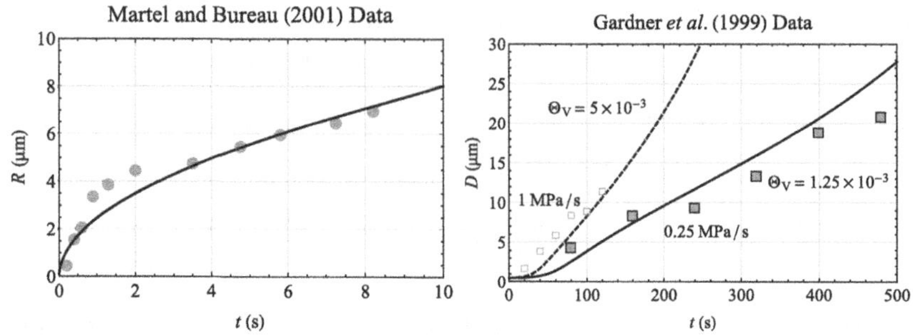

Fig. 4.15 Comparison with bubble growth experiments. (Left) Comparison between the experimental data (black circles) by Martel and Bureau (2001) and theory (curve). (Right) Comparison between the experimental data (squares) by Gardner et al. (1999) and theory. Difference in symbols indicates difference in decompression rate. Difference in line types corresponds to difference in parameters explained in the main text. The simplified Burnham model was used for water solubility. $\tilde{R}_0 = \text{constant}$, $\tilde{R}_C = b/(1 - \tilde{P})$. The water content in liquid is assumed to be a constant. Note that the timescales of the horizontal axes differ

conducted a similar experiment and reported an almost identical bubble growth curve (Gondé et al. 2011).

Using a gas-medium high pressure apparatus, Gardner et al. (1999) conducted an experiment in which obsidian containing 77 wt% of SiO_2 was used as a starting material and the samples (the initial temperature: 825 °C, the initial water content: 5.5 wt%) were continuously decompressed and then rapidly quenched at various pressures to observe vesicular texture. In their experiment, the initial pressure was 200 MPa and the supersaturation pressure of bubbles was estimated at about 40 MPa from bubble formation pressure. Their results are shown in the right chart of Fig. 4.15.

The diffusion coefficient and the viscosity coefficient corresponding to the condition of their experiment are about 10^{-10} m^2/s and 10^6 Pa s, respectively. When we estimate time constants of diffusion and viscosity on the basis of the values above, t_{dif} is in the order of 10^{-4} s to 10^{-2} s, t_{vis} is in the order of 10^{-1} s to 10^1 s; therefore, the Péclet number is of the order of 10^{-5} to 10^{-1}, although they depend on the bubble radius. Moreover, decompression time tdec is estimated at 200 s for 1 MPa/s of the decompression rate and at 800 s for 0.25 MPa/s of the decompression rate. When the Péclet number corresponding to this experiment is assumed to be in the order of 10^{-3}, we obtain $\Theta_V = t_{\text{vis}}/t_{\text{dec}} = 5 \times 10^{-3}$ and 1.25×10^{-3}, respectively.

Using these parameters, we solved Eq. (4.113) numerically and showed the obtained growth curves in the right chart of Fig. 4.15. Gardner et al. indicated that the bubble growth would be almost linear. These linear growth curves suggest that the Péclet number is <1. However, the Péclet number seems not to be extremely small (see Fig. 4.10) because the experiment time is as long as 50–100 s, while the characteristic time of viscosityt_{vis} is 10^{-1} s, and the bubble radius grows 30 times.

Change in the bubble radius in the decompression process has been studied relatively well elsewhere (e.g., Liu et al. 2004). In an experiment under a constant amount of decompression, the bubble radius often describes S-shaped curves with the lapse of time. The initial convex-downward part of the curves represents exponential growth in which diffusion is inefficient under the condition where the Péclet number is large and overpressure in bubbles is large and almost constant. In this initial stage, overpressure in bubbles is affected by viscous resistance of melt and the mechanical process that expands bubbles governs the bubble growth. In the next stage, the mechanical process become negligible, transport of H_2O to bubbles controls the bubble growth (diffusion-limited region), and change in the bubble radius describes parabolic curves (convex upward curves) (Navon et al. 1998). Convex-downward growth curves at the initial stage in an experiment under a constant liquid pressure are attributed to a combination of the following two factors: (1) Overpressure in initial bubbles conserves saturation pressure at nucleation under high pressure and difference in pressure between the pressure in bubbles and melt drives bubble expansion as of the point in time when melt is decompressed. (2) Diffusion of water molecules in bubbles can be substantially neglected at the initial stage because H_2O equilibrium concentration at the bubble interface is large because of high gas pressure in bubbles. These experiments were conducted under a special situation where the amount of decompression is constant. Moreover, we have to pay attention when we apply the experimental results obtained to nature.

The fact that the dependence of the bubble number density on the decompression rate, which is obtained in the experiment under a constant decompression rate, can be explained well by the dependence in the case where the diffusion growth of bubbles is assumed ($BND \propto |dP/dt|^{3/2}$). This suggests that bubble growth at the nucleation stage is diffusion-limited growth. Unfortunately, we have not obtained sufficiently dense data on change in the bubble radius at the nucleation stage obtained in the experiment under a constant decompression rate to discuss a growth mode under the present circumstances.

4.10 Extension of the Rayleigh-Plesset Equation

4.10.1 Extension to Viscoelastic Liquid

In cases where liquid is a viscoelastic medium, the term of stress is described using constitutive equations describing the relationship between stress and shear rate. Because there are various stress-shear rate constitutive equations according to the viscoelastic model, there are as many Rayleigh-Plesset equations for viscoelastic media as there are constitutive equations. Here, following Fogler andGoddard (1970), let us consider a model in which a Maxwell body and a Newtonian viscous body are connected in parallel. This model is called an Oldroyds model. This model was applied to an experiment of dynamics of bubbles in analog magma by Ichihara et al. (2004).

In the model, a memory function $N(t)$ in

$$\sigma_{rr} = 2\int_0^t N(t-t')\cdot\dot{\epsilon}_{rr}(r',t')\cdot dt', \tag{4.129}$$

which corresponds to the equation of the Maxwell model (11.65) (see Appendix in Sect. 11.7 of Chap. 11), is given by

$$N(t) = \eta\delta(t) + k\exp\left(-\frac{t}{t_{\mathrm{re}}}\right) \tag{4.130}$$

The first term $\eta\delta(t)$ is the term of Newtonian viscosity and corresponds to the term of viscous resistance in the Rayleigh-Plesset equation derived earlier. The second term reflects a viscoelastic property of a Maxwell body.

Strain rate is given by

$$\dot{\epsilon}_{rr}(r,t) = \frac{\partial u_r}{\partial r} = -2\frac{R^2(t)\dot{R}(t)}{r^3} \tag{4.131}$$

Note that strain rate $\dot{\epsilon}_{rr}$ in Eq. (4.129) is that at the past time t'. Therefore, we have to bring the position r in the denominator of the shear rate into correspondence with the position r' at the past time t'. Assuming that the position of a fluid particle is r, $u_r = dr/dt$ holds. Substituting it into $u_r = R^2(t)\dot{R}(t)/r^2$, a velocity field derived from the conservation of mass, yields $r^2dr = R^2(t)\dot{R}(t)dt = R^2(t)dR$. Integrating it gives $r^3 - (r')^3 = R^3(t) - R^3(t')$. The shear rate of a substance at position r

at the time t underwent at the position r' at the past time t', which is expressed by,

$$\dot{\epsilon}_{rr}(r', t') = -\frac{2R^2(t')\dot{R}(t')}{(r')^3} = -\frac{2R^2(t')\dot{R}(t')}{r^3 + R^3(t') - R^3(t)} \tag{4.132}$$

The integral of Eq. (4.75) related to stress is

$$\int_R^\infty \frac{\sigma_{rr}}{r} dr = -4 \int_R^\infty \int_0^t \frac{N(t - t')R^2(t')\dot{R}(t')}{r(r^3 + R^3(t') - R^3(t))} dt' dr \tag{4.133}$$

and integrating it with respect to r gives

$$\int_R^\infty \frac{\sigma_{rr}}{r} dr = -4 \int_0^t \frac{N(t - t')R^2(t')\dot{R}(t') \ln\left(\frac{R(t')}{R(t)}\right)}{R^3(t') - R^3(t)} dt' \tag{4.134}$$

Finally, the Rayleigh-Plesset equation extended to the Oldroyds viscoelastic fluid is given in a dimensionless form by

$$\begin{aligned} \tilde{R}\ddot{\tilde{R}} + \frac{3}{2}\dot{\tilde{R}}^2 &= \frac{\Delta P_G}{P_\infty} - \frac{P_\gamma}{P_0 \cdot \tilde{R}} - \frac{t_{\text{vis}}}{t_c} \frac{\dot{\tilde{R}}}{\tilde{R}} \\ &- 12 \frac{k}{P_\infty} \int_0^{\tilde{t}} \left[\exp\left(-\frac{\tilde{t} - \tilde{t}_1}{t_{\text{re}}/t_c} \right) \right] \frac{\dot{\tilde{R}}_1 \tilde{R}_1^2 \ln(\tilde{R}_1/\tilde{R})}{\tilde{R}_1^3 - \tilde{R}^3} d\tilde{t}_1 \end{aligned} \tag{4.135}$$

where $\tilde{R} = R/R(0)$, $\tilde{R}_1 = \tilde{R}(t_1)$, $\tilde{t} = t/t_c$, $t_c = R(0)(\rho/P_0)^{1/2}$, $P_\gamma = 2\gamma/P_0$, and $t_{\text{vis}} = 4\eta/P_0$. The ratio between bubble collapse time and viscoelastic relaxation time, t_{re}/t_c, is a sort of the Deborah number (De) defined as the ratio between the characteristic time of a phenomenon and viscoelastic relaxation time.

$$\text{De} = \frac{t_{\text{re}}}{t_c} \tag{4.136}$$

4.10.2 Extension to Multi-bubbles: A Cell Model

Geometry of a Cell Model: Conservation of Mass of Melt

As vesiculation progresses and the volume fraction of bubbles increases, interaction between bubbles becomes nonnegligible. In such a case, we use a model describing bubble expansion of multi-bubbles, which is called a cell model (Fig. 4.16). This model was developed by Denson (1984) in the field of engineering and was applied to magma by Proussevitch et al. (1993b). In this model, different boundary conditions will be used when the diffusion equation and the Navier-Stokes equation are integrated.

The relationship between the volume fraction and characteristic scale in this geometry is as follows. In the initial state, there are no bubbles or extremely small bubbles (if any). That means the initial volume fraction of gas $\phi(0) = 0$. Assuming that the initial effective radius of a melt cell containing one bubble is l_0, the effective radius of the melt cell in the bubble expansion process is l, and the radius of the bubble contained in the melt cell in the bubble expansion process is R. Because the volume of melt is conserved in the expansion process,

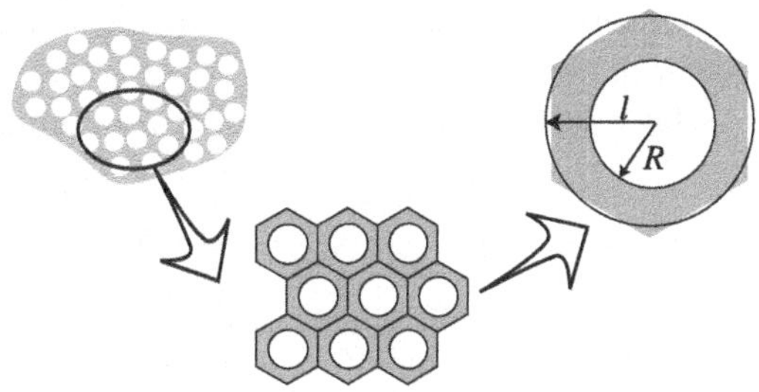

Fig. 4.16 Geometry of the cell model

$$l_0^3 = l^3 - R^3 \tag{4.137}$$

holds. The volume fraction of the bubble ϕ is given by

$$\phi = \frac{R^3}{l^3} \tag{4.138}$$

From the two equations above, we obtain

$$\frac{l_0^3}{l^3} = 1 - \phi \tag{4.139}$$

If the initial bubble radius is sufficiently small, the initial effective radius of the cell l_0 can be related to the bubble number density N (the number of bubbles per unit melt volume). Because $4\pi l_0^3 N/3 = 1$, we obtain

$$l_0 = \left(\frac{3}{4\pi N}\right)^{1/3} \tag{4.140}$$

Conservation of Mass of Volatiles

Conservation of mass of volatiles can be given by

$$\frac{4}{3}\pi R^3 \rho_G + 4\pi\rho \int_R^l \tilde{C}(r) r^2 dr = \frac{4}{3}\pi l_0^3 \rho \tilde{C}_0 \tag{4.141}$$

where the concentration $\tilde{C}$ is defined in units of wt%. In this equation, ρ_G is the density of gas and ρ is the density of liquid. To use this equation, we have to know the concentration profile $\tilde{C}(r)$.

For simplicity, we apply an approximation in which $\tilde{C}(r) = \tilde{C} =$ constant, so-called mean field approximation, to the above equation and can easily integrate it, thus obtaining the conservation of mass of volatiles as follows:

$$\rho_G\phi + \rho\tilde{C}(1 - \phi) = (1 - \phi)\rho\tilde{C}_0 \tag{4.142}$$

in which we used the geometry of the cell model (Eqs. 4.137 and 4.138). From this equation, the concentration of volatiles in melt $\tilde{C}$ can be calculated by

$$\tilde{C} = \tilde{C}_0 - \frac{\phi}{1 - \phi}\frac{\rho_G}{\rho} \tag{4.143}$$

using the volume fraction of the gas phase ϕ and the density of the gas phase ρ_{G}. Because $\phi(0) = 0$ is assumed, $\tilde{C}_0$ when $\phi = 0$.[12]

This is equal to the relationship of concentration derived from

$$\frac{4}{3}\pi R^3 N \rho_{\mathrm{G}} + \tilde{C}\rho = \tilde{C}_0\rho \tag{4.144}$$

which is conservation of mass in the case where the cell model is not followed. Because N is the bubble number density per unit melt volume, N has the relationship of $V_{\mathrm{G}} = 4\pi R^3 N/3 = \phi/(1-\phi)$ with the total volume of bubbles V_{G}.

Flow Field: The Rayleigh-Plesset Equation

In this geometry, we integrate Eq. (4.54) from $r = R$ to $r = l$.

$$\text{L.H.S.} = \rho\left[\left(2R\dot{R}^2 + R^2\ddot{R}\right)\left(\frac{1}{R} - \frac{1}{l}\right) - \frac{1}{2}\dot{R}^2\right] \tag{4.145}$$

$$\text{R.H.S.} = -(P(l,t) - P(R,t)) - \sigma_{rr}(l,t) + 3\int_R^l \frac{\sigma_{rr}}{r}dr \tag{4.146}$$

Force balance at the bubble wall is expressed by Eq. (4.73) and

$$P(R,t) - \sigma_{rr}(R,t) = P_{\mathrm{G}} - \frac{2\gamma}{R} \tag{4.147}$$

$$P(l,t) = P_{\mathrm{m}} \tag{4.148}$$

are obtained, where P_{m} is the pressure of liquid at the center of the bubble wall.

$$\text{R.H.S.} = P_{\mathrm{G}} - P_{\mathrm{m}} - \frac{2\gamma}{R} + 3\int_R^l \frac{\sigma_{rr}}{r}dr \tag{4.149}$$

Using the equation of a velocity field of a single bubble (Eq. (4.63)) for the integral part in Eq. (4.149) gives

$$\int_R^l \frac{\sigma_{rr}}{r}dr = -\frac{4}{3}\eta R^2\dot{R}\left(\frac{1}{R^3} - \frac{1}{l^3}\right) \tag{4.150}$$

and extending the Rayleigh-Plesset equation to the cell model yields

$$\rho\left[\left(2R\dot{R}^2 + R^2\ddot{R}\right)\left(\frac{1}{R} - \frac{1}{l}\right) - \frac{1}{2}\dot{R}^2\right] = P_{\mathrm{G}} - P_{\mathrm{m}} - \frac{2\gamma}{R} - 4\eta\frac{\dot{R}}{R}\left(1 - \frac{R^3}{l^3}\right) \tag{4.151}$$

From the geometry, the relationship among the volume fraction of bubbles ϕ, the spacing between bubbles l, and the bubble radius R is

$$\phi = \frac{R^3}{l^3} \tag{4.152}$$

and substituting this into Eq. (4.151) yields

[12]You can convert the unit of concentration from wt% to (number of molecules/m^3) by replacing $\rho_{\mathrm{G}}/\rho \quad P_{\mathrm{G}}/k_{\mathrm{B}}T = 1/v_{\mathrm{G}}(P_{\mathrm{G}})$, where $v_{\mathrm{G}}(P_{\mathrm{G}})$ is the volume of one water molecule in gas at the pressure P_{G}.

$$\rho\left[\left(2\dot{R}^2+R\ddot{R}\right)\left(1-\phi^{1/3}\right)-\frac{1}{2}\dot{R}^2\right]=P_\mathrm{G}-P_\mathrm{m}-\frac{2\gamma}{R}-4\eta\left(1-\phi\right)\frac{\dot{R}}{R} \tag{4.153}$$

Neglecting the inertia term gives

$$P_\mathrm{G}-P_\mathrm{m}-\frac{2\gamma}{R}-4\eta(1-\phi)\frac{\dot{R}}{R}=0 \tag{4.154}$$

This is equivalent to an equation in which the viscosity in Eq. (4.84) for a single bubble is replaced with the following effective viscosity

$$\eta_\mathrm{eff}=\eta(1-\phi) \tag{4.155}$$

Using the half length of the spherical shell thickness of the bubble wall, $h=l-R$, Barclay et al. (1995) expanded the term $1-r^3/l^3$ in Eq. (4.151) and used the following equation to discuss overpressure caused by viscosity at the time of bubble expansion.

$$P_\mathrm{G}-P_\mathrm{m}-\frac{2\gamma}{R}-4\eta h\frac{\dot{R}}{R^2}=0 \tag{4.156}$$

Change in Concentration Profile Around Bubbles with Time

Whether bubbles are multi-bubbles or single bubbles, change of the concentration in liquid follows the diffusion equation (4.1) or (4.16). The case in which moving of the bubble wall is considered, follows Eq. (4.1), whereas the case in which moving of the bubble wall is not considered, follows Eq. (4.16). In the current problem, the boundary conditions will be

$$\begin{aligned}\text{At } r\to R\,,&\quad \tilde{C}(r)\to\tilde{C}_R\\ \text{At } r\to l\,,&\quad d\tilde{C}(r)/dr=0\end{aligned}$$

The boundary conditions at the middle point between bubbles ($r=l$) are geometric requirements and the concentration there, $\tilde{C}_l$, is determined as a solution of the governing equation system. $\tilde{C}_R$ is the equilibrium concentration at the interface of the bubble radius R and is calculated from C_R given by Eq. (4.9) or (4.107). These boundary conditions constitute the governing equation system together with the Rayleigh-Plesset equation of the cell model (Eq. (4.153) or (4.154)), the equation for diffusion flux into bubbles and increase in the number of molecules in bubbles, the equation of state for pressure in bubbles, and conservation of mass of volatiles (Eq. (4.141) or the mean-field approximated Eq. (4.143)). Because the thickness of the bubble wall $2h$ changes with bubble expansion, some contrivance for calculating it is required.

In the cell model, the influence of viscosity and diffusion on the behavior of bubble growth is qualitatively the same as the case of a single bubble. However, because space scale of concentration gradient, which drives diffusion, is given by the thickness of the bubble wall in the cell model, the bubble growth rate becomes large as compared with the case of a single bubble. In the cell model, the concentration profile around the bubbles will finally be homogeneous at the equilibrium concentration at the pressure.

Figure 4.17 shows change of the concentration profile with time associated with bubble growth under a constant amount of decompression. As seen from this figure, the concentration profile temporally develops from an initial homogeneous concentration distribution to a diffusion profile with concentration gradient to a final homogeneous equilibrium concentration profile. Thus, assuming that

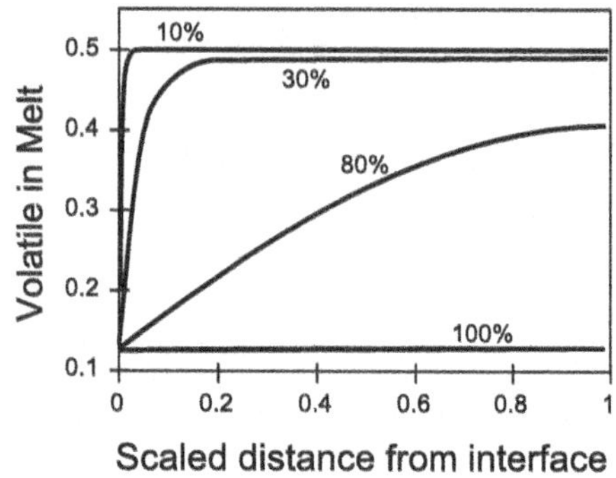

Fig. 4.17 Change of water concentration profile at the bubble wall with respect to time associated with bubble growth in the cell model (Adapted from Fig. 17 in Proussevitch et al. 1993b). As a rough guide to time, percentages of the bubble size to the final bubble radius are shown

the equilibrium concentration at the interface C_R is constant, the cell model has similarity with the problem of cooling of a dike by heat conduction in terms of mathematics.

References

Amon M, Denson CD (1984) A study of the dynamics of foam growth: analysis of the growth of closely spaced spherical bubbles. Polymer Eng Sci 24:1026–1034

Barclay J, Riley DS, Sparks RSJ (1995) Analytical models for bubble growth during decompression of high viscosity magmas. Bull Volcanol 57:422–431

Bird RB, Stewart WE, Lightfoot EN (1960) Transport phenomena. Wiley, New York, p 780

Blower JD, Madera HM, Wilson SDR (2001) Coupling of viscous and diffusive controls on bubble growth during explosive volcanic eruptions. Earth Planet Sci Lett 193:47–56

Blundy J, Cashman K, Humphreys M (2006) Magma heating by decompression-driven crystallization beneath andesite volcanoes. Nature 443(7107):76

Fogler HS, Goddard JD (1970) Collapse of spherical cavities in viscoelastic fluids. Phys Fluid 13:1135–1141

Gardner JE, Hilton M, Carroll MR (1999) Experimental constraints on degassing of magma: isothermal bubble growth during continuous decompression from high pressure. Earth Planet Sci Lett 168:201–218

Gondé C, Martel C, Pichavant M, Bureau H (2011) In situ bubble vesiculation in silicic magmas. Am Mineral 96(1):111–124

Ichihara M, Okunitani H, Ida Y, Kameda M (2004) Dynamics of bubble oscillation and wave propagation in viscoelastic liquids. J Volcanol Geotherm Res 129:37–60

Keller JB, Miksis M (1980) Bubble oscillations of large amplitude. J Acoustical Soc Am 68(2):628–633

Lensky NG, Navon O, Lyakhovsky V (2004) Bubble growth during decompression of magma: experimental and theoretical investigation. J Volcanol Geotherm Res 129:7–22

Liang Y, Richter FM, Chmberlin L (1997) Diffusion in silicate melts: III. Emprical models for multicomponent diffusion. Geochim Cosmochim Acta 61:5295–5312

Liu Y, Zhang Y, Behrens H (2004) H_2O diffusion in dacitic melts. Chem Geol 209:327–340

Lyakhovsky V, Hurwitz S, Navon O (1996) Bubble growth in rhyolitic melts: experimental and numerical investigation. Bull Volcanol 58:19–32

Martel C, Bureau H (2001) In situ high-pressure and high-temperature bubble growth in silicic melts. Earth Planet Sci Lett 191:115–127

Navon O, Chekhmir A, Lyakhovsky V (1998) Bubble growth in highly viscous melts: Theory, experiments, and autoexplosivity of dome lavas. Earth Planet Sci Lett 160:763–776

Plesset MS, Prosperetti A (1977) Bubble dynamics and cavitation. Ann Rev Fluid Mech 9:145–185

Poritsky H (1952) The collapse or growth of a spherical bubble or cabity in a viscous fluid. Proc First Nat Cong Appl Mech 813–821-D

Proussevitch AA, Sahagian DL, Kutolin VA (1993) Stability of foams in silicate melts. J Volcanol Geotherm Res 59:161–178

Proussevitch AA, Sahagian DL, Anderson VAT (1993) Dynamics of diffusive bubble growth in magmas: isothermal case. J Geophys Res 98:22283–22307

Proussevitch AA, Sahagian DL, Anderson VAT (1994) Reply to comments by sparks. J Geophys Res 99:17829–17832

Rayleigh L (1917) On the pressure developed in a liquid during the collapse of a spherical cavity. Philosophical Mag J Sci 34:94–98

Sahagian DL, Proussevitch AA (1996) Thermal effects of magma degassing. J Volcanol Geotherm Res 74:19–38
Shaw H (1974) Diffusion of H_2O in granitic liquids: part I. experimental data; part II. Mass transfer in magma chambers. In: Geochemical transport and kinetics, vol 634. Carnegie I, Washington, pp 139–170
Sparks RSJ (1994) Comments on "Dynamics of diffusive bubble growth in magmas: Isothermal case" by A. A. Proussevitch, D. L. Sahagian, and V. A. T. Anderson. J Geophys Res 99:17827–17828
Sparks RSJ (1978) The dynamics of bubble formation and growth in magmas: a review and analysis. J Volcanol Geotherm Res 3:1–37
Toramaru A (1995) Numerical study of nucleation and growth of bubbles in viscous magmas. J Geophys Res 100:1913–1931

Temporal Development of Vesiculation 5

Abstract

In previous chapters, we learned how simultaneous nucleation rate and growth rate of bubbles at instantaneous decompression are determined when temperature, pressure, and concentration of volatiles are obtained. However, to understand the entire image of the vesiculation process and vesicular texture observed in eruptive materials, such instantaneous nucleation rate and growth rate are insufficient. For this purpose, we should understand the history of the vesiculation process, i.e., the temporal development of bubble nucleation and growth processes because the fluid mechanical behavior of vesiculating magma and the vesicular texture of eruptive materials because of the nucleation rate and the growth rate temporally developed in response to temporally changing temperature, pressure, and volatile content in succession and from a mutual influence among each other. In this chapter, based on the assumption that pressure linearly decreases with time, a methodology to understand the temporal development of the vesiculation process will be explained. This helps understand the temporal development of the bubble size distribution function, and the result can be systematically understood using a small number of dimensionless controlling parameters. Moreover, the result will be applied to laboratory experiments to establish an empirical understanding of the vesiculation process.

5.1 Overall Scheme

The overall flow of calculation is as follows: temperature, pressure, and volatile concentration are derived and then bubble nucleation rate and growth rate are successively calculated (Fig. 5.1). For decompression-induced vesiculation associated with magma ascent, an isothermal process is assumed and calculation starts from a saturated condition (C_0, P_0). Pressure is then obtained as a function of time (or given from calculation of conduit flow), the critical radius is calculated, and the bubble nucleation rate J corresponding to it is calculated using steady nucleation rate J_S from the following Eqs. (3.82), (3.20), and (3.21).

A. Toramaru, *Vesiculation and Crystallization of Magma*, Advances in Volcanology,
https://doi.org/10.1007/978-981-16-4209-8_5

$$J = \frac{\phi_G DC}{R_C}\sqrt{\frac{\gamma}{k_B T}}\exp\left(-\frac{4\pi\gamma R_C^2}{3k_B T}\right) \tag{5.1}$$

$$\phi_G = v_G^* C \tag{5.2}$$

$$v_G^* = \frac{k_B T}{P_G^*} \tag{5.3}$$

$$R_C = \frac{2\gamma}{P_G^* - P} \tag{5.4}$$

$$P_G^* = P_0\left(\frac{C}{C_0}\right)^2 \exp\left[-\frac{v_m P_0}{k_B T_0}\left(1-\frac{P}{P_0}\right)\right] \tag{5.5}$$

where P_G^* and v_G^* represent the pressure in a bubble having the critical radius and molecular volume of the gas, respectively. Next, the number of molecules in the bubble n_G and the pressure in the bubble are calculated from the flux of diffusive inflow of volatiles (Eq. (4.36) and the equation of state of ideal gas).

$$\frac{dn_G}{dt} = 4\pi D(C - C_R)R \tag{5.6}$$

For C in this equation, the concentration at infinity C_∞, which drives diffusion, is assumed to be equal to content in a mean field, C (mean field approximation). C_R is the equilibrium concentration at the interface of the bubble having radius R and determined by the pressure in bubble P_G. Generally, it is determined by Eq. (2.35) or Eq. (2.34) or alternatively calculated by Eq. (4.9) when the system is in mechanical equilibrium or by Eq. (4.107) when the system is not in mechanical equilibrium. The pressure in the bubble P_G is related to the bubble radius R and the number of molecules n_G through the equation of state of ideal gas.

$$P_G = \frac{n_G k_B T}{4\pi R^3/3} \tag{5.7}$$

Bubble growth rate (expansion rate) is calculated from the Rayleigh–Plesset equation (the inertia term is ignored for sufficiently slow change such as diffusion growth) using the pressure in bubble P_G.

$$\frac{dR}{dt} = \frac{R}{4\eta}\left(P_G - P - \frac{2\gamma}{R}\right) \tag{5.8}$$

Fig. 5.1 Scheme of calculation

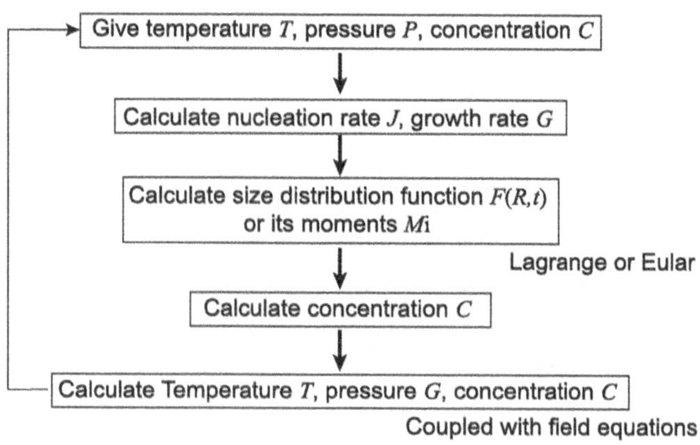

Note that the bubble growth rate is generally a function of the bubble radius. The change of the bubble size distribution function with respect to time is calculated using the bubble nucleation rate J, the number of molecules in the bubble n_{G}, and the bubble growth rate $\dot{R} = dR/dt$. Furthermore, the calculation is roughly classified into two approaches toward understanding it: the Euler and Lagrange approaches. In any case, because of the calculation of the size distribution function $F(R, t)$, the amount of concentration of the volatile exsolved C_{exs} is calculated as follows:

$$C_{\mathrm{exs}} = \int_{R_{\mathrm{C}}}^{\infty} \frac{4\pi R^3}{3v_{\mathrm{G}}} F(R, t) dR \tag{5.9}$$

$$v_{\mathrm{G}} = \frac{k_{\mathrm{B}}T}{P_{\mathrm{G}}} \text{ or } \frac{4\pi R^3}{3n_{\mathrm{G}}} \tag{5.10}$$

where P_{G} is given by Eq. (5.7). From the conservation of mass, the volatile concentration remaining in melt, C, can be calculated as follows.

$$C = C_0 - C_{\mathrm{exs}} \tag{5.11}$$

Under the conditions of another temperature and pressure at the new next step, the bubble nucleation rate, the number of molecules in bubbles, and the bubble growth rate are obtained using the new value for the volatile concentration C and liquid pressure P and change of bubble size distribution with respect to time is calculated in the same way. In this manner, the temporal development of the vesiculation process associated with change in temperature and pressure is obtained by calculating the moment of bubble size distribution or distribution function.

5.2 Temporal Development of Vesiculation Using the Eulerian Approach

5.2.1 Partial Differential Equation Representing the Conservation of the Number of Bubbles

Generally, the Eulerian approach defines absolute space and describes balance, or conservation, of a physical quantity in a unit cell in the absolute space. The absolute space here is size space and the physical quantity here is the bubble number density BND (the number of bubbles in unit melt volume). Indeed, although the size distribution function is a function of positions in real space, it is assumed here to be homogeneous and independent of positions. Let us understand change with time of the bubble size distribution associated with nucleation and growth of bubbles by explicitly dealing with it as a function of size and time $F(R, t)$. As described in the previous chapter, $F(R, t)$ is defined as the BND between the bubble size R and $R + dR$ will be $F(R, t)dR$. This bubble size distribution $F(R, t)$ is the number density in size space and is the number density in real space, and has a dimension of (number/m^4). The temporal development of the size distribution function in the size space obeys the conservation law for the number of bubbles in the size space (Fig. 5.2). Assuming that change in $F(R, t)dR$, the number of bubbles in a size interval between R and $R + dR$ in a certain time interval dt is

$$d(F(R, t)dR) \tag{5.12}$$

where change in the number of bubbles is determined by deduction of flux of the number of bubbles in the size interval $I_R(R, t)$ (number/(m^{3}s)) as follows:

$$[I_R(R, t) - I_R(R + dR, t)]\, dt \tag{5.13}$$

Expanding this equation by the Taylor expansion, ignoring the higher order terms in it, and rearranging it by regarding it as being equal to $d(F(R,t)dR)$ yields the following:

$$\frac{\partial F(R,t)}{\partial t} = -\frac{\partial I_R}{\partial R} \tag{5.14}$$

Because the bubble flux in the size space, I_R, is the product of the flow of bubbles in the size space, i.e., the bubble growth rate $G(R,t)$, and the number of bubbles having the size $F(R,t)$, it is expressed by the following equation:

$$I_R = G(R,t)F(R,t) \tag{5.15}$$

where $G(R,t)$ is $\dot{R} = dR/dt$. The conservation law of the number of bubbles is then expressed by the following:

$$\frac{\partial F(R,t)}{\partial t} = -\frac{\partial}{\partial R}[GF(R,t)] \tag{5.16}$$

Because bubbles are formed following the nucleation rate J at the critical nucleus radius R_C and infinitely large bubbles do not exist, the boundary conditions will be

$$\begin{aligned} G(R,t)F(R,t) &= J \text{ at } R = R_C \\ F(R,t) &= 0 \text{ at } R = \infty \end{aligned} \tag{5.17}$$

The temporal development of the bubble size distribution function can be obtained by directly solving a partial differential equation of the conservation law of size distribution (Eq. (5.16)) under the boundary condition (5.17) (Fig. 5.3). However, it is generally difficult to analytically solve a partial differential equation and a conservation law having a simple form $\partial f(x,t)/\partial t + \partial(f(x,t)v(x,t))/\partial x = 0$ has a numerical instability.[1] Moreover, to solve a coupling with field variables such as temperature and pressure requires complicated programming. Therefore, there is a method of solving the partial differential equation by integrating it in the size space and solving moment of the size distribution and not by directly solving it. This method will be explained in the following section.

5.2.2 Derivation of a Moment Equation

The moment of a size distribution M_i is defined by the following equation:

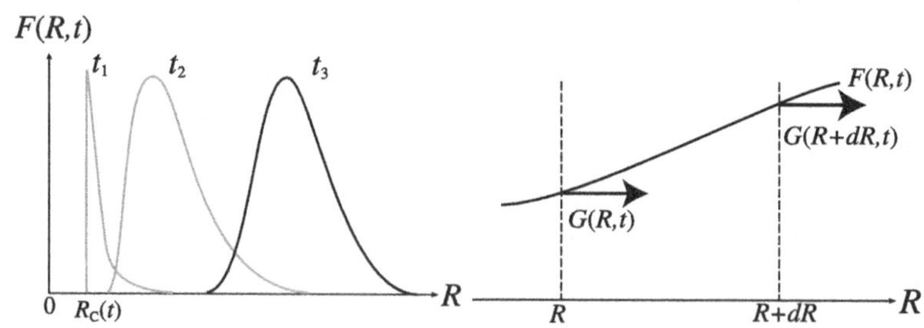

Fig. 5.2 Understanding of temporal development of size distribution by the Eulerian approach

[1] The problem of numerical instability is mostly solved using cubic interpolation (CIP).

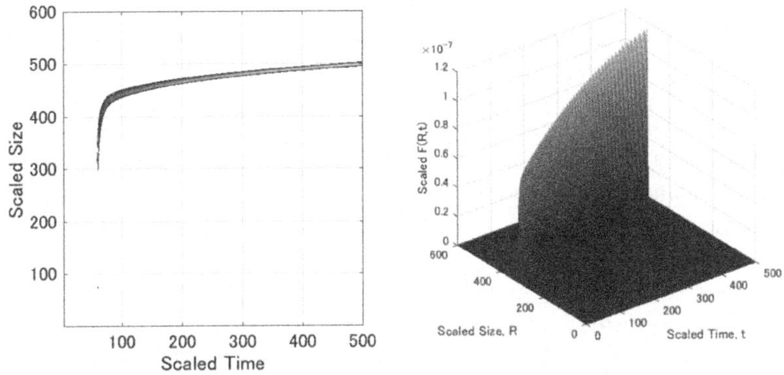

Fig. 5.3 Calculation example of temporal development of the bubble size distribution by the Eulerian approach. Numerical solution of Eq. (5.16). (Left) Contour plot of size distribution function $F(R, t)$ plotted with time on the horizontal axis and size on the vertical axis (logarithmic scale). The plot shows that bubbles forming at a certain time grow instantly and then grow slowly. (Right) 3-D graph of time—size—$F(R, t)$

$$M_i(R_C(t), t) = \int_{R_C}^{\infty} R^i F(R, t) dR \tag{5.18}$$

where i is an integer representing the order of the moment. This moment is of petrographic observables and the zero-th order moment

$$M_0(t) = \int_{R_C}^{\infty} F(R, t) dR \tag{5.19}$$

is the BND. The first order moment M_1 is the total of the bubble radius per unit melt volume

$$M_1(R_C(t), t) = \int_{R_C}^{\infty} R F(R, t) dR \tag{5.20}$$

and the mean bubble radius $\langle R \rangle$ is given by

$$\langle R \rangle = \frac{M_1}{M_0} \tag{5.21}$$

The second order moment

$$M_2(R_C(t), t) = \int_{R_C}^{\infty} R^2 F(R, t) dR \tag{5.22}$$

is related to the total surface area of bubbles per unit melt volume, and the third order moment

$$M_3(R_C(t), t) = \int_{R_C}^{\infty} R^3 F(R, t) dR \tag{5.23}$$

is related to the total volume of bubbles per unit melt volume. In particular, the third-degree moment M_3 has a relationship

$$V_G = \frac{4\pi}{3} M_3 \tag{5.24}$$

with the total volume of bubbles V_G, and the volume fraction of bubbles ϕ is calculated as

$$\phi = \frac{\frac{4\pi}{3}M_3}{1+\frac{4\pi}{3}M_3} \tag{5.25}$$

We derive the temporal development of these equations of the moment. Although the moment is a function of time, it will be a function of time and of the critical nucleus radius $R_C(t)$, i.e., as a function of time, for the sake of mathematical expression. We use a chain rule while remembering and obtaining a time derivative of the i-order moment

$$\frac{dM_i(R_C,t)}{dt} = \frac{\partial M_i(R_C,t)}{\partial t} + \frac{\partial M_i(R_C,t)}{\partial R_C}\frac{dR_C}{dt} \tag{5.26}$$

From the equation defining the moment (Eq. (5.18)),

$$\frac{\partial M_i}{\partial R_C} = -(R_C)^i F(R_C) \tag{5.27}$$

is obtained. Moreover, remembering that the growth rate $G(R,t)$ and the size distribution $F(R,t)$ are a function of R and t, we obtain

$$\frac{\partial M_i(R_C,t)}{\partial t} = \frac{\partial}{\partial t}\left(\int_{R_C}^{\infty} R^i F(R,t)dR\right) \tag{5.28}$$

$$= -\int_{R_C}^{\infty} R^i \frac{\partial}{\partial R}(G(R,t)F(R,t))dR \tag{5.29}$$

$$= (R_C)^i G(R_C,t)F(R_C,t) + i\int_{R_C}^{\infty} R^{i-1}G(R,t)F(R,t)dR \tag{5.30}$$

$$= (R_C)^i G(R_C,t)F(R_C,t) + iM_0\langle R^{i-1}G(R,t)\rangle \tag{5.31}$$

Here, we define

$$\langle R^{i-1}G(R,t)\rangle = \frac{1}{M_0}\int_{R_C}^{\infty} R^{i-1}G(R,t)F(R,t)dR \tag{5.32}$$

While remembering that in Eq. (5.31) there is a symbolic relationship between the nucleation rate, the size distribution function, and the growth rate,

$$J = G(R_C,t)F(R_C,t) \tag{5.33}$$

Equation (5.31) will be

$$\frac{\partial M_0(R_C,t)}{\partial t} = J \tag{5.34}$$

when $i = 0$.

Ultimately, the equation of motion of the moment (Eq. (5.26)) will be

$$\frac{dM_0(R_C,t)}{dt} = J - F(R_C)\frac{dR_C}{dt} \tag{5.35}$$

when i = 0, and

$$\frac{dM_i(R_C, t)}{dt} = (R_C)^i J - (R_C)^i F(R_C)\frac{dR_C}{dt} + iM_0\langle R^{i-1}G(R, t)\rangle \tag{5.36}$$

when i > 0.

The first term of the right-hand side of Equations (5.35) and (5.36) is contribution from nucleation and the second term represents the influence of change in the critical nucleus radius on the moment. The third term of Equation (5.36) represents contribution from growth. Although the second term of Equations (5.35) and (5.36) will be a correction term of nucleation at a phase where supersaturation is progressing, it can be ignored because $F(R_C)$ is small. For a supersaturated, or rather nearly saturated state, it will represent the disappearance or dissolution of bubbles at the critical size or smaller into melt and describe the effect of Ostwald ripening (Chap. 6). Although it is difficult to describe the value of $F(R_C)$ in macroscopic quantity, Langer and Schwartz (1980) approximated it as follows:

$$F(R_C) = b_{LS}\frac{M_0}{\langle R\rangle - R_C} \tag{5.37}$$

In this equation, b_{LS} is determined such that the exponent of change of the number density with time at Ostwald ripening will be -1; i.e. $b_{LS} = 0.317$ when $\langle R\rangle - R_C < 0.5R_C$ and $b_{LS} = 0$ when $\langle R\rangle - R_C > 0.5R_C$. When evaluating Equation (5.36), the integral of the third term of the right-hand side of Equation (5.32), approximation

$$\langle R^{i-1}G\rangle = G(\langle R\rangle) \cdot M_{i-1}/M_0 \tag{5.38}$$

is made using the mean value of R, $\langle R\rangle = M_1/M_0$. When G is independent of R, this equation strictly holds; if not, examination is required. Because the influence of Ostwald ripening and nucleation on the moment of the first-order or higher can be ignored, actual calculation is very easy. The only thing we have to do is to solve a set of hierarchical equations comprising the following ordinary differential equations.

$$\frac{dM_0}{dt} = J \tag{5.39}$$

$$\frac{dM_1}{dt} = G(\langle R\rangle)M_0 \tag{5.40}$$

$$\frac{dM_2}{dt} = 2G(\langle R\rangle)M_1 \tag{5.41}$$

$$\frac{dM_3}{dt} = 3G(\langle R\rangle)M_2 \tag{5.42}$$

$$\vdots$$

$$\frac{dM_i}{dt} = iG(\langle R\rangle)M_{i-1} \tag{5.43}$$

where i is an integer and $G(\langle R\rangle)$ calculated by combining the equation for expansion of a bubble with the mean bubble radius $\langle R\rangle$ (Eq. (4.52) obtained by ignoring the inertia term of the Rayleigh–Plesset equation (4.84)) and Eq. (4.30)) for diffusion growth. Actually, it is sufficient to calculate up to $i = 3$.

5.3 Temporal Development of Vesiculation Using the Lagrangian Approach

This approach was first devised in a study of formation of dust in the primordial solar nebula (Yamamoto and Hasegawa 1977). In the Lagrangian approach, the motion of a particle in time—size space itself is tracked. Here, a particle is a bubble and a bubble moves in a time—size space immediately driven by growth after nucleation (Fig. 5.4). The number of bubbles present in a given size range R and $R + dR$ is related to the nucleation rate at the time when they nucleated. Assuming that bubbles forming at the same time will be the same size after a given time, the number of bubbles will be conserved. Because the nucleation rate at time t' is $J(t')$, the number of the bubbles that nucleated at the time interval between the time t' and the time $t' + \delta t'$ can be approximated as follows:[2]

$$J(t')\delta t' \tag{5.44}$$

The reason why symbol δ is used here for an infinitesimal time difference is that this is not normal differentiation but has an implication of variation[3]. The size of a bubble at time t which nucleated at time t', $R(t', t)$, is given by the following:

$$R(t', t) = R_C + \int_{t'}^{t} G(R(t', \tau), \tau)d\tau \tag{5.45}$$

and the size of a bubble at time t which nucleated at time$t' + \delta t'$ is given by the following:

$$R(t' + \delta t', t) = R_C + \int_{t'+\delta t'}^{t} G(R(t' + \delta t', \tau), \tau)d\tau \tag{5.46}$$

where

$$G(R(t', \tau), \tau) = \frac{dR(t', \tau)}{d\tau} \tag{5.47}$$

holds. Because the bubble that nucleated $\delta t'$ after, $R(t' + \delta t', \tau)$, is smaller in size, the size interval is $R(t', t) - R(t' + \delta t', t)$ and will be written as $\delta R(t', t)$ hereafter. The number of bubbles present in this size interval can be approximated as follows:

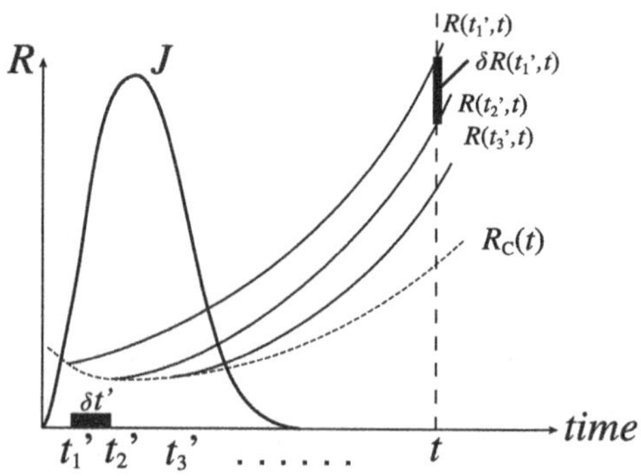

Fig. 5.4 The figure showing temporal development of bubble size distribution in the Lagrangian approach. In actual numerical calculation, sufficiently small δt will be used. $R(t_1', t)$ is the growth trajectory of bubbles nucleated at time t_1' and others are the same

[2] Contribution of changing R_C to the number density is ignored.

[3] Differentiation is the rate of change of y, or dy, when x is changed infinitesimally in a function $y = f(x)$, while variation is the rate of change of a function f when f itself changes. In this case, the curve of bubble growth $R(t', t)$ itself changes because of change of nucleation time t'.

$$F(R(t', t), t)\delta R(t', t) \tag{5.48}$$

using a size distribution function. Because this is equal to the number of the bubbles at nucleation described above,

$$J(t')\delta t' = F(R(t', t), t)\delta R(t', t) \tag{5.49}$$

holds.

Therefore, size distribution can be written as follows:

$$F(R(t', t), t) = \frac{J(t')}{\frac{\delta R(t', t)}{\delta t'}} \tag{5.50}$$

Note that $\delta R(t', t)/\delta t'$ in the denominator is what is obtained by dividing size difference at time t between two bubbles (that are generated at time t' and time $t' + \delta t'$) by time differences $\delta t'$ and is not the ordinary so-called growth rate of bubbles. Numerical calculation is made by tracking the growth of bubbles forming at time t' and the slightly later time $t' + \delta t'$, recording their sizes in the PC memory, calculating the difference at time t

$$R(t', t) - R(t' + \delta t', t) \tag{5.51}$$

and dividing it by the difference of nucleating time $\delta t'$.

The nucleation rate sensitively changes, responding to change in the degree of supersaturation associated with change in liquid pressure and water concentration. Liquid pressure is determined by fluid-mechanical dynamics of magma in simulation of eruptions and follows experimental conditions to interpret the experiment. Water concentration follows the conservation law (Eqs. (5.9)–(5.11)). Thus,

$$C = C_0 - \int_{R_C}^{\infty} \frac{4\pi R^3}{3v_G} F(R, t) dR \tag{5.52}$$

or expressed as

$$C = C_0 - \int_0^t \frac{4\pi R^3(t', t)}{3v_G} J(t') dt' \tag{5.53}$$

using nucleation rates. Numerical calculations are performed by individually inserting $F(R, t)$ or $J(t')$ into this equation to calculate change in the nucleation rate and water concentration with time. At that time, the history of J up to the time, $J(t')$, and the size of bubbles at the time t which nucleated at time t' have to be recorded. Figure 5.5 shows an example of the calculation. The greatest merit of this method is that if the nucleation rate and the growth rate are given as functions of time, the size distribution function can be analytically obtained relatively easily. We will explain this method in the chapter on crystal size distribution (CSD).

Scaling of equations and dimensionless numbers

When we want to identify how boundary conditions, initial conditions, and various properties influence the behavior of a solution, scaling is an effective approach. "Scaling" is making a variable dimensionless using an appropriate scale. The scale can be arbitrarily obtained or determined in the equation system. Let us consider a one-dimensional Navier-Stokes equation without body force as an example.

$$\rho \frac{Du}{Dt} = -\frac{\partial P}{\partial x} + \eta \frac{\partial^2 u}{\partial x^2} \tag{5.54}$$

where u is the flow velocity and D is the Lagrange derivative. Scaling this equation with $u = u_0 \tilde{u}$, $t = t_0 \tilde{t}$, $P = P_0 \tilde{P}$, $x = x_0 \tilde{x}$, and $u_0 = x_0/t_0$ yields

$$\frac{D\tilde{u}}{D\tilde{t}} = -\frac{P_0}{\rho u_0^2} \frac{\partial \tilde{P}}{\partial \tilde{x}} + \frac{1}{Re} \frac{\partial^2 \tilde{u}}{\partial \tilde{x}^2} \tag{5.55}$$

Reynolds number (Re) is then defined as follows:

$$\mathrm{Re} = \frac{u_0 x_0}{\nu},$$

which is the ratio of inertia and viscosity. Here, ν is kinematic viscosity η/ρ that has a dimension of diffusion coefficient (momentum diffusion by viscosity).

When the geometry of a system (e.g., the diameter of a cylinder) yields x_0, as in the case of a Kármán vortex, conditions of a vortex formation can be organized using Reynolds number Re. When we want to identify potential properties of a system of equations without geometric constraints, we can assume $\mathrm{Re} = 1$ and take $x_0 = \nu/u_0$ as a space scale to identify that $P_0/(\rho u_0^2)$ remains as the only dimensionless parameter appearing in the dimensionless Navier- Stokes equation and controls the fluid-mechanical behavior of the system. In this case, however, note that the space scale x_0 is related to a scale for viscosity relaxation ν/u_0 and the time scale t_0 is related to the timescale for viscosity diffusion x_0^2/ν. If x_0 is determined by the geometry of the system beforehand, the Reynolds number will remain as a variable parameter. In this case, assuming that $P_0/(\rho u_0^2)$, a coefficient ahead of $\partial \tilde{P}/\partial \tilde{x}$ is 1 yields $u_0 = (P_0/\rho)^{1/2}$, which indicates that we can see the flow velocity with a scale for the velocity of sound. In the case where a velocity field u_0 is also given, two dimensionless numbers are identified to control the behavior of the system.

Moreover, for a system of equations solved by combining multiple equations as dealt with in this book, physical properties and physical conditions, such as γ, η, P_0, D, v_m, t_{dec}, T, and C_0, etc., can independently vary. Although the influence on the behavior of a solution can be identified by independently and systematically changing these parameters, identifying the relationships between the parameters is difficult and discovering the essential parameters that determine the behavior of the solution became complicated. Scaling allows us to extract dimensionless groups that are determined by the special combinations of parameters and to understand the roles of the parameters and the dimensionless numbers.

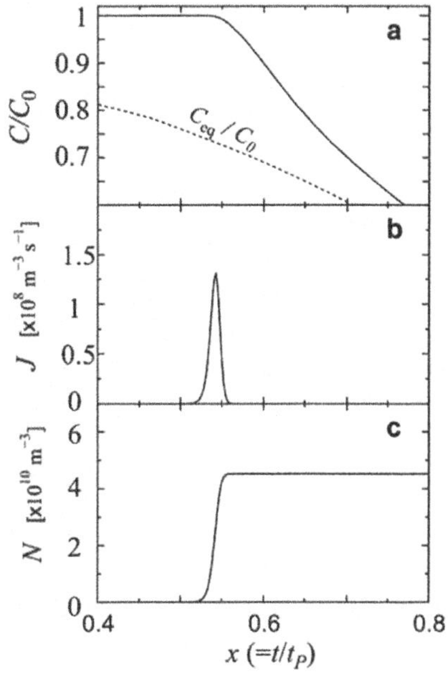

Fig. 5.5 An example of calculation by Lagrangian formulation (Yamada et al. 2005). **a** Water concentration and solubility in melt, **b** nucleation rate, and **c** BND with time

5.4 Controlling Parameters in Decompression-Induced Vesiculation

Scaling (see Box above) of a problem in the case where pressure changes at a constant rate demonstrates that the following four dimensionless groups control the behavior of solution. We call these dimensionless groups the controlling parameters of vesiculation.

$$\alpha_1 = \frac{4\pi\gamma(R_P^*)^2}{3k_B T} \tag{5.56}$$

where R_P^* is the scale of the critical radius defined by the following:

$$R_P^* = 2\gamma/P_0 \tag{5.57}$$

and will be 2×10^{-10} when $\gamma = 10^{-2}$ N/m and $P_0 = 100$ MPa. α_1, which indicates that the difficulty of nucleating (nucleation barrier), largely depends on the interfacial tension and changes by orders of magnitude. It changes from 10^{-4} to 10^2, which corresponds to values from $\gamma = 0.001$ (N/m), corresponding to heterogeneous nucleation, to $\gamma =$ c.a.0.1 (N/m), which corresponds to homogeneous nucleation (Fig. 5.5).

$$\alpha_2 = \frac{v_m P_0}{k_B T} \tag{5.58}$$

This equation represents the influence of the initial saturation pressure and appears in the Poynting correction for solubility (Eq. (2.25)). It takes values from about 0.2 to 1.

$$\alpha_3 = \frac{\phi_{G0} D t_{dec}}{(R_P^*)^2} = \frac{t_{dec}}{t_{difG}} \propto \left|\frac{dP}{dt}\right|^{-1} \tag{5.59}$$

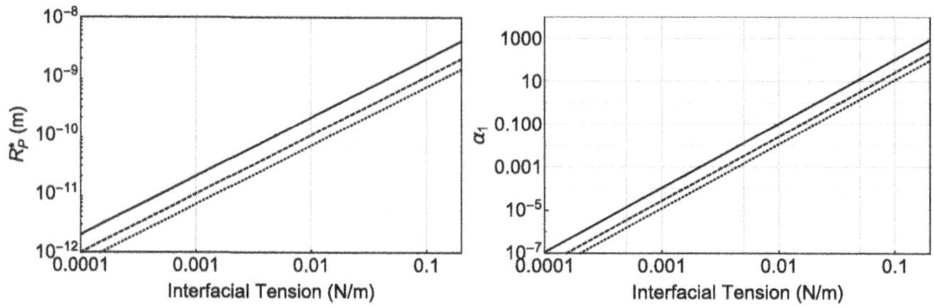

Fig. 5.6 (Left) Relationship between R_p^* and surface tension. (Right) Relationship between α_1 and surface tension. Different line types represent different initial saturation pressure. Solid lines: 10^8 Pa, broken lines: 2×10^8 Pa, dotted lines: 3×10^8 Pa

About the decompression rate

The decompression rate dealt with here is the decompression that magma experiences in motion of magma and is the Lagrangian decompression rate. Generally, in the conduit flow model, change in pressure is calculated as a function of time in fixed spatial coordinates from Eulerian description, i.e., a Navier-Stokes equation. Let us connect these two. We consider magma ascending in a 1D vertical conduit. The Lagrangian decompression rate $(DP/Dt)_{z_n}$ that magma particles experience at a certain depth z_n with magma ascent is given as follows:

$$\left(\frac{DP}{Dt}\right)_{z_n} = \left(\frac{\partial P}{\partial z}\right)_{z_n} \cdot \left(\frac{dz}{dt}\right)_{z_n} + \left(\frac{\partial P}{\partial t}\right)_{z_n} \tag{5.60}$$

$$= \left(\frac{\partial P}{\partial z}\right)_{z_n} \cdot U(z_n) + \left(\frac{\partial P}{\partial t}\right)_{z_n} \tag{5.61}$$

The first term of the right-hand side is contribution from pressure gradient caused by the movement of the position and the second term is Eulerian pressure change at position z_n. The pressure gradient of the first term is then determined by the momentum conservation law, i.e., a Navier-Stokes equation. For a steady flow, the pressure gradient will be as follows:

$$\left(\frac{\partial P}{\partial z}\right)_{z_n} = -\rho g - \rho U \left(\frac{\partial U}{\partial z}\right)_{z_n} - F_{\text{ric}} \tag{5.62}$$

and the velocity gradient in the second term of the right-hand side is given as follows:

$$\left(\frac{\partial U}{\partial z}\right) = -\frac{U}{A}\left(\frac{\partial A}{\partial z}\right) - \frac{M^2}{\rho U}\left(\frac{\partial P}{\partial z}\right) \tag{5.63}$$

where M is a Mach number. In the case where the velocity of magma is lesser than the velocity of sound and change in the cross-sectional area of a volcanic conduit (A) or friction with a wall (F_{ric}) can be ignored because the contribution of velocity gradient $\partial U/\partial z$ to pressure gradient $\partial P/\partial z$ is small, the pressure gradient is determined by hydrostatic gradient $-\rho g$ and the Lagrangian decompression rate will be Eq. (5.67). On the other hand, Eulerian pressure change $\partial P/\partial t$ is 0 in a steady flow and is not 0 in an unsteady flow. It is caused by, for example, volume change associated with phase change or unsteady change in velocity.

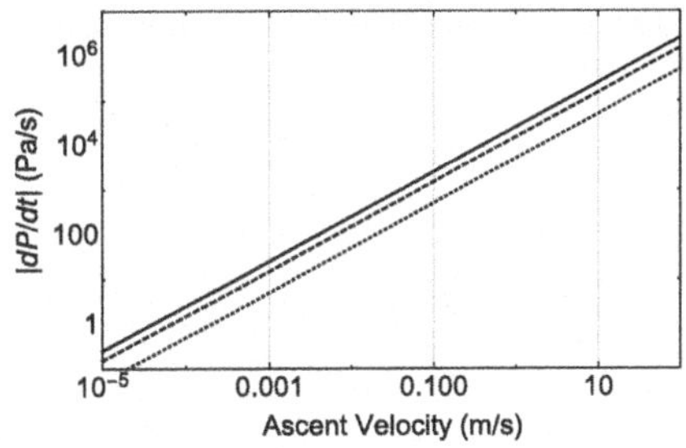

Fig. 5.7 Relationship between decompression rate and magma ascent velocity. Line types represent mean density of magma column. Solid lines: 2500 kg/m^3, broken lines: 1500 kg/m^3, dotted lines: 500 kg/m^3

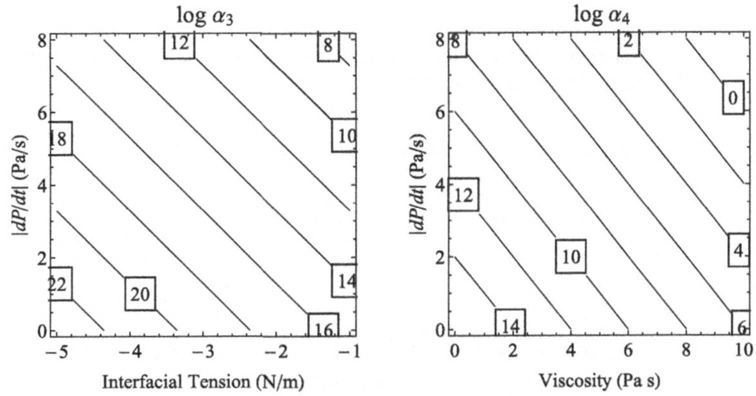

Fig. 5.8 (Left) Values of α_3 as a function of surface tension and decompression rate. (Right) Values of α_4 as a function of viscosity and decompression rate. Numbers on the contours represent a power of 10

This is the ratio of the decompression time to the diffusion growth time, which represents how effectively diffusion works. Moreover, it represents the influence of the decompression rate in the diffusion region. For the decompression time t_{dec}, if linear decompression from the initial pressure P_0 is assumed, P_0/t_{dec} is equal to the decompression rate $|dP/dt|$. The last equation is based on that. Decompression time scale t_{dec} can be expressed as $P_0/\rho g U$ with ascent velocity U in the case where hydrostatic pressure of magma changes with magma ascent (Fig. 5.7). t_{difG} is a timescale of the bubbles' diffusion-limited growth, $(R_p^*)^2/\phi_{G0} D$. ϕ_{G0} is conversion volume corresponding to the initial water content, $k_B T C_0/P_0$ (= 0.33, 100 MPa). This is not so sensitive to the initial saturation pressure P_0 because C_0 and P_0 are somewhat canceled. The value of α_3 varies from 10^8 to 10^{22} depending on the surface tension, the diffusion coefficient, and the decompression rate.

$$\alpha_4 = \frac{P_0 t_{dec}}{4\eta} = \frac{t_{dec}}{t_{vis}} \tag{5.64}$$

This is the ratio of decompression time to timescale of viscosity-limited growth (($t_{vis} = 4\eta/P_0$)), representing the influence of viscosity. Moreover, it represents the influence of the decompression rate of the viscosity regime.

Timescales and parameters introduced above are strongly dependent on the circumstances. Their representative values are shown in Fig. 5.8 and Table 5.1. Because t_{dec} cannot be defined for instantaneous decompression, the ratio of t_{difG} tot_{vis}, i.e., Péclet number (Pe), is defined as follows:

$$\mathrm{Pe} = \frac{t_{difG}}{t_{vis}} = \frac{\alpha_4}{\alpha_3} \tag{5.65}$$

Table 5.1 Examples of physical properties and controlling parameters. Other parameters and constants are assumed as follows: $\alpha_1 = 1, \alpha_2 = 0.17, R_P^* = 4.3 \times 10^{-10}$ m, $\gamma = 0.0215$ N/m, $P_0 = 10^8$ Pa, $C_0 = 4.2(\text{or } 3) \times 10^{27}$ number m^{-3}, $k_B = 1.38 \times 10^{-23}$ J/K, $T = 1200$ K, $v_m = 3 \times 10^{-29}$ m^3, $\rho = 10^3$ kg/m^3.

α_3	α_4	t_{dec}, s	t_{difG}, s	t_{vis}, s	D, m^2/s	η, Pa s	U, m/s
2.15×10^{12}	2.5×10^6	10^4	4.6×10^{-9}	4×10^{-3}	10^{-10}	10^5	1
2.15×10^{11}	2.5×10^6	10^4	4.6×10^{-8}	4×10^{-3}	10^{-11}	10^5	1
2.15×10^{12}	2.5×10^3	10^4	4.6×10^{-9}	4	10^{-10}	10^8	1
2.15×10^{12}	2.5×10^2	10^3	4.6×10^{-9}	4	10^{-10}	10^8	10

5.5 Temporal Development of Vesiculation Under the Condition of Constant Decompression Rate

5.5.1 Decompression Rate

When decompression rate is constant and pressure linearly decreases from the initial saturation pressure P_0 with time.

$$P = P_0 - |\dot{P}|t \tag{5.66}$$

where $|\dot{P}| = |dP/dt|$. In a natural conduit flow, assuming that the ascent rate of magma is U, such decrease in pressure holds in regions where the flow is steady without large velocity gradient ($U \cdot \partial U/\partial z = 0$) and in regions on which the influence of viscosity dissipation and friction with a wall is small ($F_{ric} = 0$),

$$|\dot{P}| = \rho \cdot g \cdot U \tag{5.67}$$

holds (see Box).

5.5.2 Solution by Moment Equations Based on Eulerian Description

Toramaru (1989) attempted to numerically solve the conservation law of the number of bubbles in size space (Eq. (5.16)) using the growth law of bubbles in mechanical equilibrium, which ignores the effect of viscosity and qualitatively obtained the same understanding as the understanding by moment equations shown here. However, he failed to derive an empirical formula for a realistic range of parameters because of numerical instability. Toramaru (1995) incorporated the effect of viscosity as the Rayleigh–Plesset equation (Eq. (5.8)) and quantitatively understood the behavior of vesiculation process for a range of parameters corresponding to actual physical properties of silicate melt. An outline of the result is given below.

5.5.2.1 Outline of Nucleation and Growth

Figure 5.9 shows the calculation examples by moment equations. The calculation result revealed that bubble nucleation occurs only once in a narrow range of supersaturation and then bubbles grow and expand. Although bubble growth in a sufficiently supersaturated state is diffusion-limited growth, expansion is predominant at low pressure near the Earth's surface (isothermal expansion because isothermal process is assumed in this case). This is the case where diffusion-limited growth controls bubble growth and we call it diffusion-controlled bubble nucleation process (or diffusion nucleation regime).

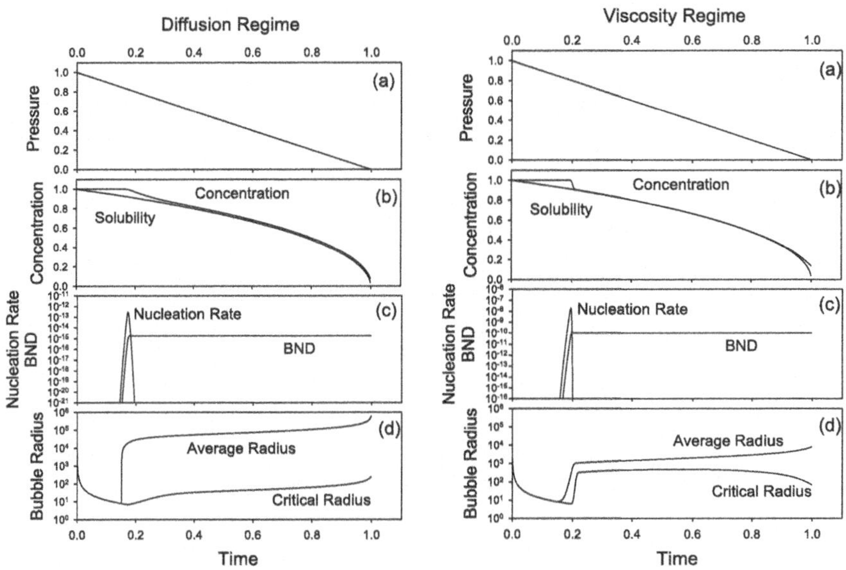

Fig. 5.9 Calculation examples of temporal development of vesiculation by the moment equations. Variables are scaled by the following scales: time t_{dec}, bubble radius R_P^*, content C_0, bubble number density BND $C_0/2$, pressure P_0, nucleation rate $C_0/2t_{\text{dec}}$ (see Table 5.1). (Left) Diffusion regime $\alpha_4 = 10^5$. (Right) Viscosity regime $\alpha_4 = 10^3$. Change with time in (a) pressure, (b) water concentration in melt and saturation concentration, (c) nucleation rate and BND, and (d) average bubble radius and critical radius, are shown in order from top to bottom. Other parameters are $\alpha_1 = 1$, $\alpha_2 = 0.17$, and $\alpha_3 = 10^{11}$ for both. The value of α_1 is obtained from Fig. 5.6 (corresponding to about 0.02 N/m of the interfacial tension). The value of α_3 is obtained from Fig. 5.8. The decompression rate is 1 MPa/s corresponding to about 0.02 N/m of the interfacial tension (from Fig. 5.7. Corresponding to several dozen meters per second of steady-flow ascent rate). Viscosity corresponds to 10^6 Pa s in the diffusion regime and 10^8 Pa s in the viscosity regime. The difference in the magnitude of the nucleation rate and of BND between the diffusion regime and the viscosity regime should be noted

On the other hand, in the case where viscosity is predominant over diffusion and decompression (the case where α_4 is small), although one nucleation event is followed by bubble growth and the expansion process is the same, the quantitative aspects of bubble nucleation and growth considerably differ. The difference is shown in Fig. 5.10. In the viscosity-controlled case, the degree of supersaturation required for nucleation becomes large and bubble growth at the nucleation stage and subsequent supersaturation relaxation stage shows viscosity-limited growth, which follows an exponential law under constant bubble overpressure. This is called viscosity-controlled bubble nucleation process (or viscosity nucleation regime).

To quantitatively understand the behavior of such entire vesiculation process, we have to identify how the parameters characterizing the temporal development of vesiculation (vesiculation parameters), including nucleation pressure P_{n}, timescale in which nucleation occurs δt_{n}, pressure scale in which nucleation occurs ΔP_{n}, the maximum value of nucleation rate $J_{\max}$, and BND N, are influenced by the controlling parameters $\alpha_1 \sim \alpha_4$.[4] Among the parameters, BND is representative of them because it is the final product of the vesiculation process. A parametric study revealed that the difference between the diffusion regime and the viscosity regime obviously appears in BND among vesiculation parameters (Fig. 5.11). There is a negative correlation between α_3, the ratio of diffusion to decompression rate, and BND, and a relationship BND $\approx \alpha_3^{-1.5}$ quantitatively holds (left chart of Fig. 5.11). This means that BND increases in proportion to the power of 3/2 of the decompression rate.

[4] Numerical calculation with systematically changed controlling parameters is called a parametric study.

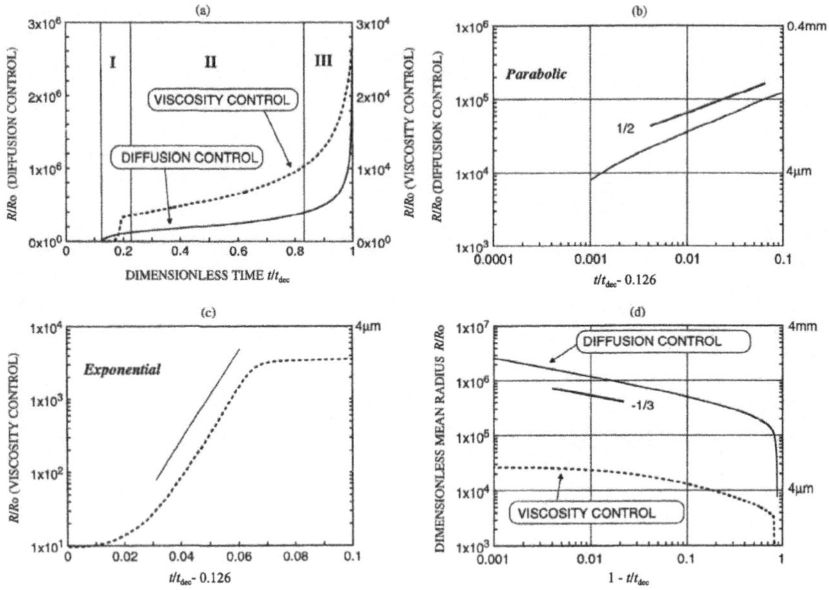

Fig. 5.10 Bubble growth at constant decompression rate (Toramaru 1995). The vertical axis is the bubble radius and the horizontal axis is time. Variables are scaled as in Fig. 5.9. The solid lines correspond to $\alpha_4 = 10^7$ in diffusion-limited growth and the broken lines correspond to $\alpha_4 = 10^3$ in viscosity-limited growth. $\alpha_1 = 1$, $\alpha_2 = 0.17$, and $\alpha_3 = 10^{12}2$ are used for both cases. Approximate values of the bubble radius are on the right vertical axis. **a** The bubble radius and the time are linear. I, II, and III represent bubble growth in the relaxation regime, the quasi-equilibrium regime, and the expansion regime, respectively. **b** Log-log plot. Growth with a slope 1/2 represents diffusion growth. Note that time is what has elapsed from the nucleation time. **c** Semi-logarithmic plot (the bubble radius is logarithmic and time is linear). Linearity means exponential growth. Time is what has elapsed from the nucleation time. **d** Log-log plot. Note that time is that required to reach the surface and corresponds to dimensionless pressure. The line with a slope $-1/3$ represents isothermal expansion

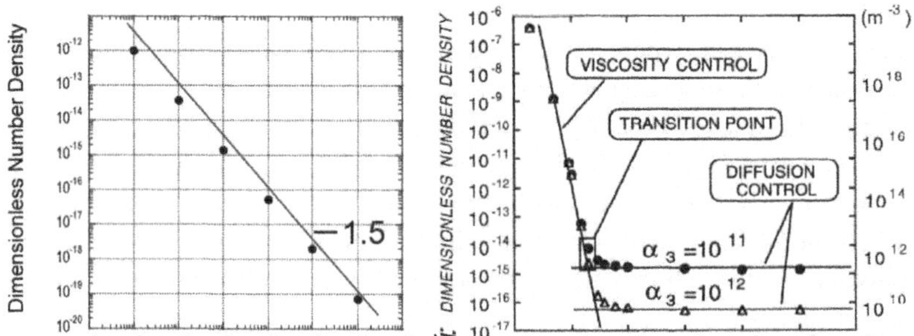

Fig. 5.11 Relationship between dimensionless BND and controlling parameters obtained by numerical experiments. The marks are the bubble number densities obtained by giving controlling parameters on the horizontal axis at the point after the decompression from the initial saturation pressure to 1 atm. (Left) BND vs α_3, (Right) BND vs α_4. Plots like these are made for other vesiculation parameters by examining dependence of all controlling parameters and empirical formulae are made. The right vertical axis is approximate real scale (Toramaru 1995)

Next, let us examine the relationship between α_4, the ratio of viscosity to decompression rate, and BND (right chart of Fig. 5.11). This figure shows that dependence of BND on α_4 is clearly divided into two regimes: a diffusion-controlled regime in which BND depends only on α_3 and a viscosity-controlled regime in which BND depends on α_4, one of the controlling parameters. The transition point is around$\alpha_4 = 2 \times 10^3$. In the case where α_4 is larger than the value of the transition point, i.e., viscosity and decompression rate is small and BND is affected only byα_3, the ratio of diffusion to decompression, changes following BND $\approx \alpha_3^{-1.5}$, and is not related to viscosity. In the case where α_4 is lesser than the value of the transition point, BND is not completely affected by diffusion but affected by viscosity and decompression rate and BND exponentially increases as viscosity and decompression rate increase. The dependence of vesiculation parameters on controlling parameters in the diffusion nucleation regime and in the viscosity nucleation regime, obtained by numerical experiments, is separately described below.

5.5.2.2 Diffusion Nucleation Regime

The pressure at which nucleation occurs, scaled by the initial saturation pressure P_0, is expressed as follows:

$$\Delta \tilde{P}_n = 1 - \frac{P_n}{P_0} \approx 10^{-1} \times \alpha_1^{1/2} \alpha_2^{0.06} \alpha_3^{-0.03} \tag{5.68}$$

$1 - P_n/P_0$ is the degree of supersaturation required for nucleation (scaled by the initial saturation pressure). Because exponents of α_2 and α_3 are small, nucleation pressure is almost only determined by α_1, i.e., interfacial energy (tension). Depending on the interfacial energy, magma can ascend to the surface while maintaining a supersaturated state without the occurrence of nucleation. This will be described again in Sect. 5.8, on the limits of nucleation.

The maximum nucleation rate is then determined by

$$J_{max} \approx 10^4 \times C_0 t_{dec}^{-1} \alpha_1^{-5/2} \alpha_2^{-1/4-0.06} \alpha_3^{-3/2+0.03} \tag{5.69}$$

The maximum nucleation rate depends on interfacial energy (α_1) and the ratio of diffusion to decompression rate $\alpha_3 = t_{dec}/t_{difG}$. As the interfacial energy increases, a nucleation barrier increases; moreover, as the diffusion coefficient increases, α_3 increases and J_{max} reduces. Moreover, as the decompression rate increases, α_3 decreases and Jmax becomes large. Although J_{max} in a dimensionless form is proportional to $\alpha_3^{-3/2}$, i.e., the power of 3/2 of the decompression rate $|dP/dt|$, it is divided by the timescale of the decompression rate tdec when converted to a quantity with a dimension and therefore J_{max} itself is proportional to the power of 5/2 of the decompression rate.

Nucleation duration is a time giving a half-width of the maximum nucleation rate.

$$\delta t_n \approx 10^{-3} \times t_{dec} \alpha_1^{1/2} \alpha_2^{0.06} \alpha_3^{-0.03} \tag{5.70}$$

This timescale is proportional to the power of 3/2 of the surface tension. As the decompression time becomes longer, the nucleation duration also becomes longer in proportion. Eventually, the nucleation duration is proportional to $\alpha_1^{1/2}$ in the corresponding pressure range δP_n. The BND is the number of bubbles forming in a unit melt volume.

$$N \approx 10 \times C_0 \alpha_1^{-2} \alpha_2^{-1/4} \alpha_3^{-3/2} \tag{5.71}$$

Regardless of diffusion-controlled or viscosity-controlled regime, the following relationship basically holds among maximum nucleation rate, nucleation duration, and BND:

$$N \approx J_{\max} \times \delta t_{\mathrm{n}} \tag{5.72}$$

5.5.2.3 Viscosity Nucleation Regime

In the viscosity-limited regime, α_4 is involved in vesiculation parameters rather than α_3. An equation corresponding to Eq. (5.68) will be determined as described below. Basically, while dependence of α_1 and α_2 is the same as for the diffusion regime, dependence of α_4 will be in a form of an exponent. This is related to the growth behavior of bubbles in the nucleation regime (Sect. 5.5.4)

$$\Delta \tilde{P}_{\mathrm{n}} \approx 10^{-1} \times \alpha_1^{1/2} \alpha_2^{0.06} \alpha_4^{-0.1} \tag{5.73}$$

$$J_{\max} \approx 10^2 \times C_0 t_{\mathrm{dec}}^{-1} \alpha_1^{-5/2} \alpha_2^{-1/4-0.06} \alpha_4^{-1} e^{-0.0128\alpha_4} \tag{5.74}$$

$$\delta t_{\mathrm{n}} \approx 10^{-1} \times t_{\mathrm{dec}} \alpha_1^{1/2} \alpha_2^{0.06} e^{0.0052\alpha_4} \tag{5.75}$$

$$N \approx 10 \times C_0 \alpha_1^{-2} \alpha_2^{-1/4} \alpha_4^{-1} e^{-0.0075\alpha_4} \tag{5.76}$$

These are rough relations and the exponents require further examination.

5.5.3 Solution by Lagrangian Description

Assuming a diffusion-limited growth law that ignores viscosity, Yamada et al. (2005) analytically solved a problem based on Lagrangian description and obtained analytical expressions of the BND, the maximum nucleation rate, and nucleation pressure (Fig. 5.5). Because the whole process of the calculation is complicated, only the outline will be described below.

Nucleation rate $J(t)$, which is a function of water concentration and pressure, will be a function of time through change of them with time. Because pressure and time have a linear relationship ($P = P_0 - |\dot{P}|t$), pressure can be considered as time. To clarify this history is to solve the problem. Because bubble growth by diffusion is proportional to the square root of time from nucleation time t', $(t - t')$, water concentration given by Eq. (5.53) is analytically given as a function of time $C(t)$.

Next, nucleation pressure P_{n} (time t_{n}) is determined. For that, the fact that a degree of supersaturation just reaches the maximum value at the time of maximum nucleation is used. The degree of supersaturation ΔC is given by $C(P(t)) - C_{\mathrm{eq}}(P(t))$, and a derivative of it with respect to time will be 0, i.e.,

$$\frac{d\Delta C}{dt} = \underbrace{\frac{dC}{dt}}_{\text{Decrease}} - \underbrace{\frac{dC_{\mathrm{eq}}}{dt}}_{\text{Increase}} = 0 \tag{5.77}$$

The central part of this equation represents competition between decrease in the degree of supersaturation caused by water exsolution associated with nucleation and growth and increase in the degree of supersaturation associated with decrease in pressure. From this equation, the maximum nucleation time t_{n} can be obtained, and the exponential part of the nucleation rate is expanded by Taylor expansion around t_{n} with respect to time (i.e., pressure) and up to second-order terms are taken (the first-order terms equal zero by definition). Using this relation, BND is obtained as a function of time by calculating the following integral:

$$N = \int_0^\infty J(t')dt' \tag{5.78}$$

The result of this integration almost agrees with the temporal development of vesiculation in the diffusion regime based on the moment equations. Although dependence of BND on diffusion coefficient and decompression rate almost agrees, the maximum nucleation rate slightly reduces and the nucleation duration increases by a factor of < 10. Although Lagrangian description is considered to more explicitly include the form of the size distribution function and to provide more accurate parameter dependence, transcendental equations[5] have to be numerically solved to obtain the actual value of BND using analytical expressions.

Yamada et al. (2008) expanded the method of Yamada et al. (2005) to the bubble growth law including the effect of viscosity and numerically solved the system of equations. Consequently, they reported that the transition point between the viscosity regime and the diffusion regime slightly shifts toward lower viscosity compared with Toramaru (1995). Moreover, they demonstrated that in the viscosity regime, Ostwald ripening, i.e., redissolution of bubbles, occurs and yields smaller BND and that bubble nucleation continues at a constant rate under high viscosity conditions. However, whether such behavior in a high viscosity regime actually occurs or not has to be experimentally confirmed in future.

5.5.4 Bubble Growth as a Factor Determining BND

An important point established by calculating the temporal development of vesiculation process is that the number density of bubbles formed by nucleation is basically controlled by bubble growth, i.e., bubble growth consumes volatiles in melt and reduces the degree of supersaturation of volatiles in melt. This effect acts during the active bubble nucleation and determines the history of nucleation.

Let us examine the relationship between nucleation and growth, which determines the BND. The maximum value of nucleation determining the BND corresponds to an extreme value of a degree of supersaturation. The fundamental relationship that connects nucleation with growth can be clarified by examining the conditions for the extreme value of the degree of supersaturation. As a rough guide for the degree of supersaturation, we use the inverse of the critical nucleus radius

$$\frac{2\gamma}{R_C} = P_G^* - P = P_0\left(\left(\frac{C}{C_0}\right)^m \times \exp\left[-\alpha_2\left(1-\frac{P}{P_0}\right)\right] - \frac{P}{P_0}\right) \tag{5.79}$$

Using dimensionless variables such as $\tilde{t} = t/t_{\mathrm{dec}}$, $\tilde{R}_C = R_C P_0/2\gamma$, $\tilde{P} = P/P_0$, and $\tilde{C} = C/C_0$, the condition for the extreme value $(d(\tilde{R}_C^{-1})/d\tilde{t} = 0)$ is expressed as follows:

$$\underbrace{m\tilde{C}^{m-1}\exp\left\{-\alpha_2(1-\tilde{P})\right\}\frac{d\tilde{C}}{d\tilde{t}}}_{\text{Decrease by bubble growth: negative}} + \underbrace{\left\{\alpha_2\tilde{C}^m\exp[-\alpha_2(1-\tilde{P})] - 1\right\}\frac{d\tilde{P}}{d\tilde{t}}}_{\text{Increase by decompression: positive}} = 0 \tag{5.80}$$

This equation corresponds to Eq. (5.77). In this equation, m is an index that determines dependence of solubility on pressure and $m = 2$ for H_2O and $m = 1$ for CO_2. Equation (5.80) shows that the maximum value nucleation rate reaches is determined by competition between decrease in the content of volatiles in melt by bubble growth, and increase in the degree of supersaturation by decompression.

Thus, not absolute values of the concentration of volatiles but the amount of infinitesimal change from the initial saturation concentration is a problem at the nucleation stage. Therefore, the concentration of volatiles can be approximated by the initial value $\tilde{C} = 1$. Change in the concentration of volatiles is connected with bubble growth, as seen earlier, and is expressed using the bubble radius R and the number density N as $C = 1 - 4\pi R^3 N/(3v_G)$, where v_G is the volume of gas molecules and

[5] Equations that include exponential function and/or logarithmic function and are algebraically unsolvable.

follows the equation of state of ideal gas. Scaling this equation with $R = R_P^* \tilde{R}$ and the number density $N = C_0\tilde{N}/2$ gives $\tilde{C} = 1 - \alpha_1 \tilde{P}_G \tilde{N} \tilde{R}^3$, and differentiating it (all but $\tilde{R}$ are constant) yields

$$\frac{d\tilde{C}}{d\tilde{t}} = -3\alpha_1 \tilde{P}_G \tilde{N} \tilde{R}^2 \frac{d\tilde{R}}{d\tilde{t}} \tag{5.81}$$

At the nucleation stage, pressure in bubbles is close to the initial saturation pressure and can be approximated as $\tilde{P}_G = 1$. Moreover, $d\tilde{P}/d\tilde{t}$ is 1 by definition. From Eq. (5.80), the BND $\tilde{N}$ can be finally obtained as follows:

$$\tilde{N} \approx \frac{1 - \alpha_2}{3m\alpha_1 \tilde{R}^2 d\tilde{R}/d\tilde{t}} \tag{5.82}$$

This equation connects the bubble growth rate $d\tilde{R}/d\tilde{t}$ and the BND.

Because $\tilde{R} = \tilde{R}_0(\alpha_3\tilde{t})^{1/2}$ holds under diffusion-limited growth, approximating the bubble growth time by the nucleation duration ($\tilde{t} = \delta\tilde{t}_n$) gives the BND

$$\tilde{N} \approx \frac{1 - \alpha_2}{3m\alpha_1 \tilde{R}_0^3 \alpha_3^{3/2} \delta\tilde{t}_n^{1/2}} \tag{5.83}$$

which shows that BND is proportional to the power of $-3/2$ of α_3, a parameter for diffusion and decompression.

On the other hand, in the case of the viscosity-limited growth,

$$\tilde{R} = \exp(\alpha_4 \Delta\tilde{P}\delta\tau) \tag{5.84}$$

$$\tilde{N} = \alpha_4^{-1} \exp(-3\alpha_4 \Delta\tilde{P}\delta\tilde{t}_n) \tag{5.85}$$

hold and they are found to exponentially depend on α_4, a parameter for viscosity and decompression. $\Delta\tilde{P}$ in these equations is the supersaturation pressure at nucleation pressure.

5.6 Temporal Development of Vesiculation Process Under the Condition of a Constant Amount of Decompression

5.6.1 Outline of Nucleation and Growth

The condition of a constant amount of decompression is useful for understanding vesiculation in laboratories. Moreover, as some experiments in decompression-induced crystallization experiments are conducted under a constant amount of decompression, the condition of a constant amount of decompression is helpful in considering crystallization processes of microlites. Moreover, for a constant amount of decompression, there are two types of vesiculation processes: the diffusion-controlled type and the viscosity-controlled type. Figure 5.12 shows examples of the temporal development of each type.

Nucleation starts with the maximum nucleation rate corresponding to the amount of decompression and then the degree of supersaturation decreases as bubbles grow. While the nucleation rate slowly decreases from the start in the diffusion regime, it remains at the maximum nucleation rate for a certain time in the viscosity regime because bubble growth is suppressed in the viscosity regime.

As seen in the subsections of the constant decompression rate, we finally organize the BND of bubbles produced as a function of a parameter. Figure 5.13 shows a plot in which the Péclet number

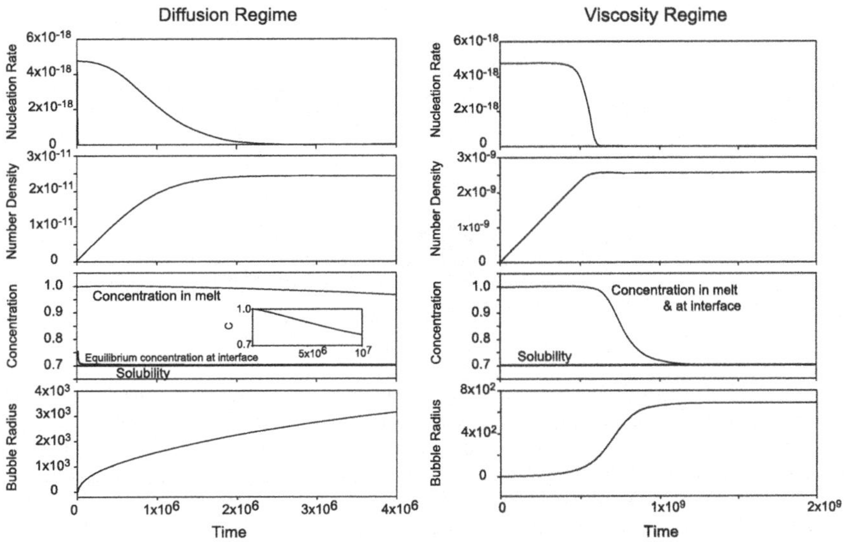

Fig. 5.12 Change with time of magma vesiculation under a constant amount of decompression (Toramaru and Miwa 2008). Change with time in the nucleation rate, the bubble number density, the H_2O concentration in melt, and the bubble radius are shown in order from top to bottom. Variables are scaled by the same scales as in Fig. 5.9. Parameters used for both include $\alpha_1 = 8.0, \alpha_2 = 0.17, \Delta\tilde{P} = 0.55, \alpha_1/\Delta\tilde{P}^2 = 26.4$. (Left) Diffusion regime. $\mathrm{Pe} = 10^{-2}$. (Right) Viscosity regime. $\mathrm{Pe} = 10^{-7}$. Curvature of bubble growth curves and difference in timescale should be paid attention. Reprinted from Toramaru and Miwa (2008), Copyright 2008, with permission from Elsevier

(Pe) is used as the parameter. The Péclet number is the ratio of characteristic time of diffusion growth to the characteristic time of viscosity growth, as defined by Eq. (4.124); it is proportional to the amount of decompression and the square of the bubble radius and is inversely proportional to the product of the diffusion and viscosity coefficients. Data in this figure is put in order of values of $\alpha_1/\Delta\tilde{P}^2$, which are included in the exponential part of the nucleation rate referred to later. This figure showed that when the Péclet number is large, the BND is independent of the Péclet number and determined only by the parameter $\alpha_1/\Delta\tilde{P}^2$. Furthermore, when the Péclet number is lesser than a certain value, the BND depends on the Péclet number and the dependence is inversely proportional to the Péclet number ($\tilde{N} \propto \mathrm{Pe}^{-1}$).

Although it is seemingly difficult to understand that BND is independent of the Péclet number, this is because the influence of the diffusion coefficient is canceled by the nucleation rate and the nucleation duration (the product of them gives the BND). In other words, the regime independent of the Péclet number can be called the diffusion-controlled regime for the behavior of the BND, determined by competition between bubble nucleation and growth.[6]

On the other hand, in the regime dependent on the Péclet number, viscosity in addition to diffusion is found to be effective. Actually, the curvature of bubble growth curves at the initial phase and difference in timescale of bubble growth in Fig. 5.12 showed that this regime can be considered as the viscosity-controlled regime.

The result of the numerical experiments reveals that the boundary between the diffusion regime and the viscosity regime is expressed by the following equation using the Péclet number, the amount of decompression $\Delta\tilde{P}$, and nucleation barrier (interfacial tension) α_1.

[6] Note that the subject differs from that for the diffusion-limited regime and the viscosity-limited regime discussed in the sections on bubble growth.

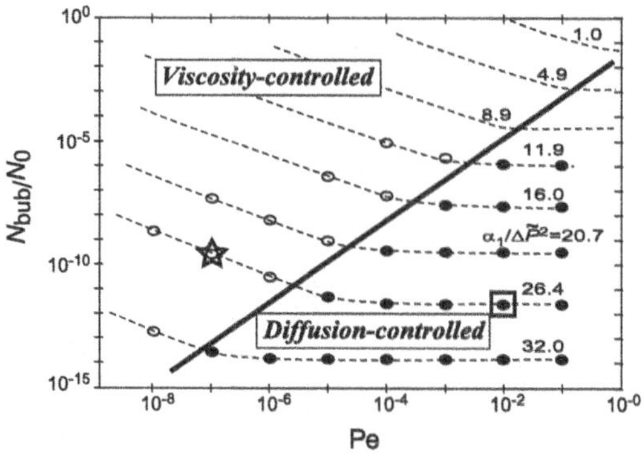

Fig. 5.13 Diffusion-controlled regime and viscosity-controlled regime of the bubble number density under a constant amount of decompression (Toramaru and Miwa 2008). The vertical axis is the bubble number density scaled by the number of water molecules or by the number of water clusters and the horizontal axis is the Péclet number. The back circles and white circles are results of numerical experiments and correspond to the diffusion-controlled regime and viscosity-controlled regime, respectively. The broken lines connect the results of the numerical experiments along values of parameter $\alpha_1/\Delta\tilde{P}^2$. The square and the star correspond to calculation conditions for diffusion-controlled and viscosity-controlled regimes in Fig. 5.12, respectively. Reprinted from Toramaru and Miwa (2008), Copyright 2008, with permission from Elsevier

$$\ln\left(\frac{\mathrm{Pe}}{\alpha_1^{3/5}}\right) = \epsilon - 0.55\frac{\alpha_1}{\Delta\tilde{P}^2} \tag{5.86}$$

where the Péclet number is defined by Eq. (5.65), and ϵ is a small value close to zero.

Because surface tension will considerably vary according to whether nucleation is homogeneous or heterogeneous, the dimensionless number α_1 representing it can take a wide-ranging value (Fig. 5.6). This fact will be important when interpreting laboratory experiments.

Using $\alpha_1 \approx 10$ (an approximate value at 300 MPa of saturation pressure, obtained from Fig. 5.6), which corresponds to 0.1 N/m of the interfacial tension, and $\Delta\tilde{P} = 0.5$ for a parameter $\alpha_1/\Delta\tilde{P}^2$, the parameter will be 40, which is in the diffusion-controlled regime in Fig. 5.13 when $\mathrm{Pe} = 10^{-2}$. If heterogeneous nucleation is assumed and $\alpha_1 \approx 0.1$, which corresponds to 0.03 N/m of the interfacial tension, is used for the parameter, it is found from Fig. 5.13 that the parameter will be 0.4 and obvious in the viscosity-controlled regime. Thus, diffusion-controlled or viscosity-controlled regime is closely related to the interfacial tension or whether nucleation is homogeneous or heterogeneous.

5.6.2 Maximum Nucleation Rate

In any regime, the maximum nucleation rate determined by the amount of decompression and the surface tension is an important factor determining the BND. Dimensionless maximum nucleation rate is approximately obtained from the equation of the nucleation rate as follows:

$$\tilde{J}_{\max} = \tilde{J}_0 \exp\left(-\varepsilon_{\mathrm{P}}\frac{\alpha_1}{\Delta\tilde{P}^2}\right) \tag{5.87}$$

$$\tilde{J}_0 = \frac{(3\alpha_1)^{1/2}}{4\pi^{3/2}} \tag{5.88}$$

where time is scaled by $t_{\mathrm{difG}} = (R_P^*)^2/(\phi_{\mathrm{G0}} D)$ and J_0 is scaled by $C_0/(2t_{\mathrm{difG}})$. ε_P is the influence of difference between the initial saturation pressure and pressure under constant decompression on chemical potential of water in liquid (Poynting correction. Ignoring it gives $\varepsilon_\mathrm{P} = 1$) and is given as follows:

$$\varepsilon_\mathrm{P} = \left(\frac{\Delta \tilde{P}}{\exp(-\alpha_2 \Delta \tilde{P}) - 1 + \Delta \tilde{P}} \right)^2 \tag{5.89}$$

This value is sensitive to the initial saturation pressure P_0 through α_2, and ε_P takes values from 1.5 to 2.5 when P_0 takes values from 100 to 250 MPa.

5.6.3 Bubble Number Density

Because the BND N is basically determined by the product of the maximum nucleation rate $J_{\max}$ and the nucleation duration δt_n, if parameter dependence of the nucleation duration is given, dependence of the BND is clarified. In the diffusion-controlled regime, the BND does not depend on the diffusion coefficient and is determined only by the amount of decompression and the surface tension (α_1) at the nucleation rate. This is because when the diffusion coefficient changes, the nucleation rate changes and the diffusion growth rate of bubbles simultaneously changes, i.e., they cancel each other out. This is significantly different from the case of constant decompression rate. On the other hand, in the viscosity-controlled regime, the BND depends on both the nucleation rate and on time for the viscosity-limited growth of bubbles. Time for viscosity-limited growth is inversely proportional to the Péclet number and therefore the BND is inversely proportional to the Péclet number.

5.6.3.1 Diffusion Regime

In the diffusion regime, the nucleation duration is determined by the maximum nucleation rate and the relationship

$$\delta \tilde{t}_\mathrm{n}^{\mathrm{dif}} = \left(\alpha_1 \tilde{J}_{\max} \right)^{-2/5} \tag{5.90}$$

holds. Substituting this into Eq. (5.87) yields

$$\tilde{N}_{\mathrm{dif}} = \alpha_1^{-2/5} \tilde{J}_{\max}^{3/5} = \alpha_1^{-2/5} \tilde{J}_0^{3/5} \exp\left(-\frac{3}{5} \varepsilon_\mathrm{P} \frac{\alpha_1}{\Delta \tilde{P}^2} \right) \tag{5.91}$$

5.6.3.2 Viscosity Regime

In the viscosity regime, the nucleation duration is inversely proportional to the Péclet number.

$$\delta \tilde{t}_\mathrm{n}^{\mathrm{vis}} = \mathrm{Pe}^{-1} \tag{5.92}$$

Thus, the BND will be

$$\tilde{N}_{\mathrm{vis}} = \tilde{J}_0 \exp\left(-\varepsilon_\mathrm{P} \frac{\alpha_1}{\Delta \tilde{P}^2} \right) \mathrm{Pe}^{-1} \tag{5.93}$$

5.6.4 The Rate of Decrease in Water Concentration in Melt

Because decrease in water content, or water exsolution, causes a rise in the melting point of crystals, the rate of the decrease corresponds to the effective cooling rate and influences nucleation and growth processes of microlites formed by decompression-induced crystallization. Generally, from the conservation of mass $C_0 = C + 4\pi R^3/(3v_G)N$, decrease in water content in melt by diffusion of water into bubbles changes as follows:

$$\left|\frac{dC}{dt}\right| = \frac{4\pi R^2}{v_G}\frac{dR}{dt}N \tag{5.94}$$

As this shows, the rate of water exsolution, i.e., the rate of decrease in water concentration is proportional to the BND (Navon and Lyakhovsky 1998). In other words, the greater the number of bubbles, the higher the effective cooling rate (the ascent rate of liquids).

This fact suggests that to interpret the texture of microlites generated by decompression-induced crystallization, textural analysis of only microlites is insufficient and textural analysis including vesicular texture is necessary. The bubble growth rate dR/dt can be calculated by the method dealt with in Chap. 4. In the following, we examine the dependence of the water exsolution rate on parameters using a simple model. In the diffusion-controlled regime, substituting the law of diffusion-limited growth (Eq. (5.94)) into Eq. (4.26) yields the following:

$$\left|\frac{dC}{dt}\right| = (48\pi^2\phi_e)^{1/3} D\Delta C_e N^{2/3} \tag{5.95}$$

(this is basically the same as Navon and Lyakhovsky 1998). In this equation, ΔC_e is effective concentration difference that drives diffusion, and ϕ_e is the volume fraction of gas at the time when water exsolution efficiently progress (ϕ_e can be approximated as $4\pi R^3 N/3$ when the gas volume is small). It is found from this that water exsolution progresses more efficiently when the BND is high.

On the other hand, in the viscosity-controlled regime, $dR/dt = \Delta P \cdot R/(4\eta)$ is obtained by substituting $\Delta P = P_G - P_\infty - 2\gamma/R$ into Eq. (4.84) and substituting $dR/dt = \Delta P \cdot R/(4\eta)$ into Eq. (5.94) yields

$$\left|\frac{dC}{dt}\right| = \frac{3\phi_e}{4\eta v_G}\Delta P \tag{5.96}$$

which linearly depends on the amount of decompression and is independent of the BND. BND is incorporated into the effective bubble volume fraction ϕ_e, and considering the dependence of ϕ_e and ΔC_e on the amount of decompression, the above examination coincides with the result of parametric study of the numerical experiments (Toramaru and Miwa 2008).

5.7 Vesiculation Experiments

Shimozuru et al. (1957) and Sakuma and Murase (1957) are probably the first studies of vesiculation experiments with silicate melt in the field of volcanology. Both experiments were conducted at a pressure of 1 atm. Shimozuru et al. (1957) conducted an in-situ observation experiment with obsidian from Imari, Saga Prefecture, under high temperature conditions and demonstrated that temperature and heating rate influence generation rate of bubbles and growth rate of bubbles. Sakuma and Murase (1957) used obsidian from the Wada Pass, Nagano Prefecture, as samples and conducted an experiment observing samples maintained at high temperature for a certain time and then quenched. They observed the process of growth of numerous microscopic (5–150 μm) bubbles contained in the samples in the initial condition and experimentally demonstrated the temporal development of size distribution. Moreover, they attempted to derive an equation corresponding to the Rayleigh–Plesset equation (but unfortunately, they could not correctly derive it because a base equation of motion contained errors) and attempted to formulate the efficiency of coalescence of bubbles in a form containing a size distribution function and to explain their experimental results.

Such vesiculation experiments under a pressure of 1 atm were hardly ever conducted for nearly 30 years. Sakuyama and Kushiro (1979) conducted a decompression-induced vesiculation experiment with piston cylinder-type high pressure apparatus and experimentally demonstrated that alkali elements were transported by generation and movement of bubbles. In addition to the above experiments, research dealing with the vesiculation process itself by decompression experiments was hardly ever conducted until the 1990s.

5.7.1 Experiment Under Constant Decompression Rates

Since about the middle of the 1990s, experiments on decompression-induced vesiculation with silicate melt have been performed on numerous occasions. They are basically conducted with gas-medium high pressure apparatus. Haplogranite, which has a composition similar to granite, and obsidian glass were used as starting materials in the early years. A starting material is heated for about 6 h at a temperature higher than the melting point (in the case where heterogeneous nucleation due to microlites should be avoided). To saturate the sample with water, the sample and water are enclosed in a Pt or Au tube and the tube is maintained at the desired temperature and pressure for several days. The sample is sometimes taken out of the tube to measure the water content or else the tube containing the sample is decompressed. The tube is then decompressed at a constant decompression rate or instantaneously (by opening the pressure valve) to a desired pressure and kept constant. Decompression is then performed by controlling the opening and closing of the pressure valve by computer or by manually opening the valve carefully. In the case where decompression is controlled by computer, a constant decompression rate is achieved by multiple decompression steps (0.1–1 MPa) with time steps. After reaching the desired pressure, the sample is quenched by moving it to the lower temperature place. The vesicularity, bubble radius distribution, number density, and water content in melt of the quenched sample are measured. The water concentration is measured by Fourier transform infrared (FTIR) spectroscopy.

Figure 5.14 shows examples of measured vesicularity and water content in melt of the samples that were saturated at 200 MPa (saturated water concentration: 6.05 wt%), decompressed at two decompression rates (1 MPa/s and 27.8 kPa/s), and then quenched at different target pressures. These measured quantities recorded in each sample were considered to be snapshots in the case where the samples were continuously decompressed to a pressure of 1 atm. This figure shows that in the case where the samples are decompressed at a constant rate from 200 MPa, bubbles do not form until about 80 MPa (the vesicularity is zero), and at around 70 MPa the water concentration in the sample glass suddenly approaches the saturation concentration and the amount of bubbles increases. The figure shows that when the decompression rate is high, the total volume of bubbles slowly increases at first and then drastically increases. The reason why a lower decompression rate caused a slower increase in the bubble volume was that the resultant BND is reduced, the effective diffusion of water in total bubbles is reduced, and the rate of decrease in water in melt reduces (Eq. (5.94)). The difference in the rate of increase in the bubble volume reflects this.

The left picture of Fig. 5.15 shows the vesicular texture of the sample obtained by the decompression experiment. Bubbles with almost the same size are seen to be homogeneously distributed. The right chart of Fig. 5.15 shows how BND changes with quenched (final) pressure for the two decompression rates. BND has jumped to a certain value at a certain pressure. This indicates that only one nucleation

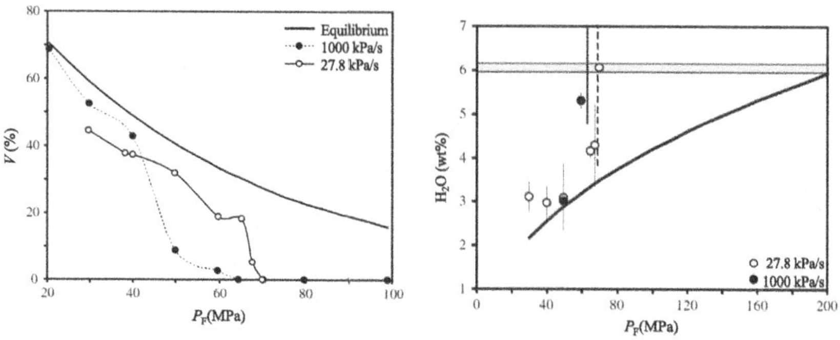

Fig. 5.14 Vesicularity associated with the bubble nucleation and growth (left) and change in the water concentration in melt (right) (Cluzel et al. 2008). Solid lines represent the solubility (right) and the vesicularity expected from equilibrium vesiculation (left). The horizontal band in the right graph represents the initial value. Reprinted by permission from Springer Nature Cluzel et al. (2008), COPYRIGHT (2008)

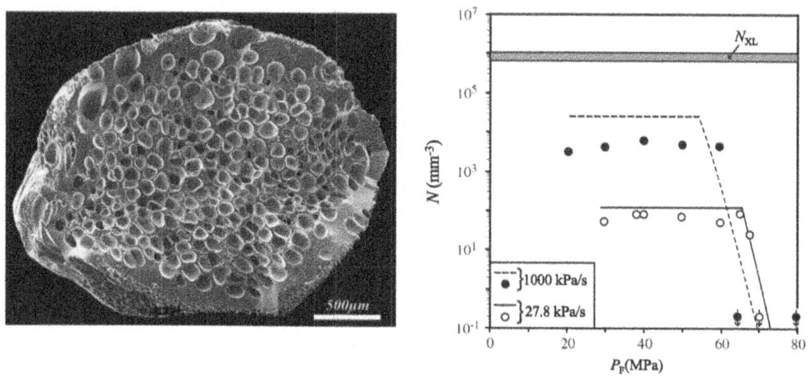

Fig. 5.15 (Left) Vesicular texture obtained by an experiment on decompression-induced vesiculation with rhyolitic melt. A sample was saturated at 200 MPa, decompressed at 1 MPa/s, and then quenched at 20.5 MPa. (Right) BND as a function of the decompression rate and quenching pressure (Cluzel et al. 2008). The horizontal bar indicated by N_{XL} is the range of the number density of hematite. The bending lines comprise rising edges (oblique line) predicted by classical nucleation theory and BND (the horizontal lines were calculated by a model equation (10.1) in Toramaru (2006) (precise version of Equation (5.71))). Reprinted by permission from Springer Nature Cluzel et al. (2008), COPYRIGHT (2008)

event occurs in a certain narrow range of pressure and it determines the number of bubbles. Moreover, after the formation of bubbles, the number of bubbles is reported to remain constant, which demonstrates that bubbles do not coalesce in these experiments.

Assuming that bubble nucleation pressure P_n is 70 MPa, $\Delta\tilde{P}_n/P_0 = (P_0 - P_n)/P_0 = (200 - 70)/200 = 0.65$ is obtained and substituting this into Eq. (5.68) gives $\alpha_1^{1/2} = 6.5$ (while ignoring the effect of α_2 and α_3) and $\alpha_1 = 42.25$. From Fig. 5.6, this value of α_1 indicates that the interfacial tension is a value between 0.1 and 0.2 N/m. To obtain an accurate value, the influence of other parameters and coefficients in Eq. (5.68) have to be accurately determined. In fact, in the experiment by Cluzel et al. (2008), the initial melt contains an infinite number of microcrystals of hematite (Fig. 5.15) and there is a high probability of heterogeneous nucleation. However, even if there are many microcrystals compared with the resultant BND, effectively correcting the surface tension $((\alpha_1))$ allows us to use Eq. (5.71) for homogeneous nucleation.

As seen above, for continuous decompression, the decompression rate governs the number of bubbles formed. Theoretically predicted dependence on decompression rate is proportional to the power of 3/2 of the decompression rate, as given by Eq. (5.71). In early experiments (e.g. Mangan and Sisson 2000),

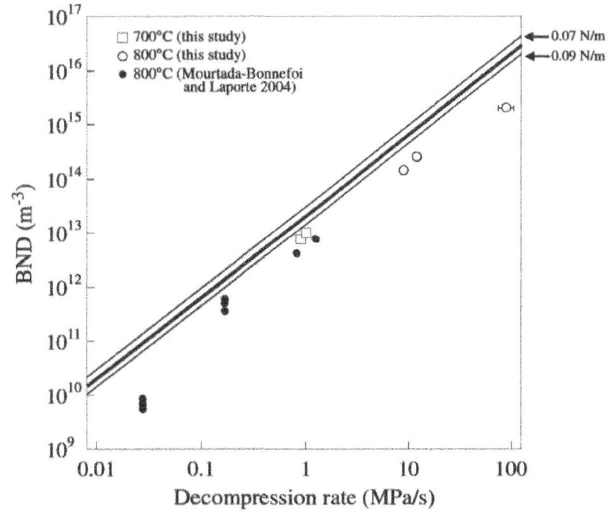

Fig. 5.16 Relationship between the BND and the decompression rate obtained by experiments with rhyolitic melt under constant decompression rate (Hamada et al. 2010). The lines represent equations by Toramaru (2006) (basically Eq. (5.71)) with different values of the surface tension. Reprinted by permission from Springer Nature Hamada et al. (2010), COPYRIGHT (2010)

although a tendency of BND to increase with increase in the decompression rate was observed, it did not exactly agree with the theoretical prediction. These experiments did not completely satisfy the conditions for homogeneous nucleation; moreover, there is the possibility of them being considerably affected by heterogeneity in the samples and heterogeneous nucleation (with insufficient heterogeneous nucleation sites) on the wall of containers. After that, the results of Mourtada-Bonnefoi and Laporte (2004) and of a subsequent experiment (Hamada et al. 2010) showed that dependence of BND on the decompression rate and approximate absolute values agreed with the theoretical prediction (5.16).

In recent years, decompression-induced vesiculation experiments with melt having composition other than rhyolite composition have been conducted. Mangan and Sisson (2005) conducted a decompression-induced vesiculation experiment with dacitic melt and demonstrated that dacitic melt shows higher BND than rhyolitic melt. They considered that dacitic melt has a lower surface energy because of difference in nucleation pressure. Similar to previous studies, Fiege (2014) and Fiege and Cichy (2015) conducted decompression-induced vesiculation experiments with andesitic melt and demonstrated that BND is higher than the theoretically predicted values. Pichavant et al. (2013) conducted a decompression-induced vesiculation experiment with basaltic melt, resulting in high BND. For melt with lower viscosity than rhyolitic melt, whether bubbles maintained the original positions at which nucleation occurs becomes a problem because of the flowability of melt. These experiments paid attention to the point, and it is probable that low-viscosity melt results in higher BND. As the cause, Fiege and Cichy (2015) discussed correlation between viscosity and BND and concluded that viscosity in one form or another influences bubble nucleation (Fig. 5.17).

Attempts to experimentally study the influence of viscosity of melt on bubble nucleation have been conducted. Gardner and Denis (2004) and Gardner (2007b) conducted experiments using rhyolite in the viscosity nucleation regime in which α_4 is 2000 or less but did not obtain high BND as predicted in Toramaru (1995). To summarize, for the homogeneous nucleation of rhyolitic melt, experiments demonstrate good consistency with theory. For the melt with this composition, the fact that the surface tension was directly measured by Bagdassarov et al. (2000) (Appendix (Chap. 11) 11.6.1) facilitates comparison between theory and experiment. However, for low-viscosity melt with lower SiO_2 content than rhyolitic melt, the fact that experiments directly measuring the surface tension have not been conducted gives more room for discussion compared with theory; in fact, a sufficiently strong understanding has not been reached. This will be a task for the future.

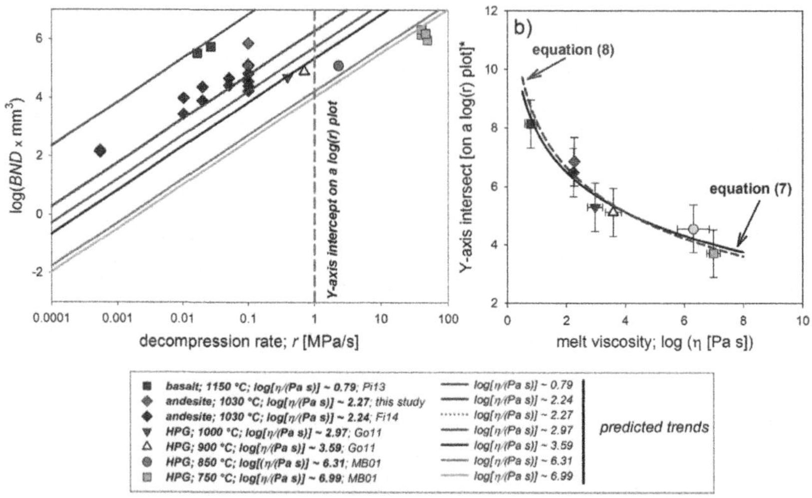

Fig. 5.17 (Left) Relationship between the BND and the decompression rate, obtained by experiments under constant decompression rate for melt with various compositions. The lines represent different viscosities, and the slope, the power of 3/2, was obtained from Eq. (5.71). (Right) Plot of the intercept of the lines at 1 MPa/s of the decompression rate against the viscosity. Reprinted from Fiege and Cichy (2015) with permission from the Mineralogical Society of America

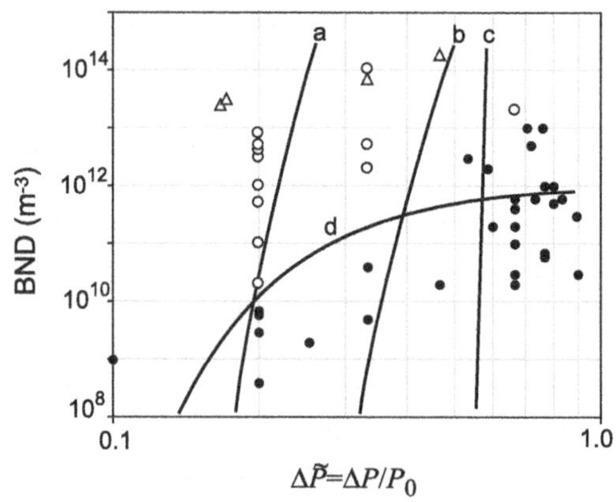

Fig. 5.18 Relationship between the number density of bubbles formed by decompression-induced vesiculation under constant amount of decompression, BND, and the amount of decompression. The black circles ● represent the case where oxide crystals were not contained initially (Hurwitz and Navon 1994) and the white circles ○ and white triangles △ represent the case where oxide crystals were contained (Hurwitz and Navon 1994 and Gardner 2007b). Curves a, b, c, and d are those expected by Eq. (5.91) for the initial number of water molecular clusters (and the surface tension) 10^{20} (0.027), 10^{19} (0.039), 10^{70} (0.1), and 10^{12} (0.016), respectively. Adapted from Toramaru and Miwa (2008), Copyright 2012, with permission from Elsevier

5.7.2 Experiment Under Constant Amount of Decompression

A study leading to vesiculation experiments is an experiment conducted by Hurwitz and Navon (1994) with rhyolitic melt under a constant amount of decompression. They investigated the relationship between the amount of decompression and BND while focusing attention on either the presence or absence of crystals. Figure 5.18 shows the result of their study. In the case where microlites of oxides are initially contained, the BND of forming bubbles does not largely depend on the amount of decompression and takes on large values. This suggests a possibility that the number of forming bubbles is determined by the number of preexisting heterogeneous nucleation sites. The interfacial tension between oxide crystals and silicate melt becomes large and consequently it becomes easy for steam bubbles to nucleate on the surface of the oxide crystals (Sect. 3.4).

In the case where microlites are not contained, the number density changes by about four orders of magnitude when the amount of decompression increases by an order of magnitude. This suggest that the number of bubbles is not governed by the number of preexisting nucleation sites but governed by a degree of supersaturation at nucleation which is determined by the amount of decompression. As seen above, because BND depends on exponential functions for the degree of supersaturation (Eqs. (5.87) and (5.91)) under constant amount of decompression, the relationship obtained in this experiment seems to be explained to some extent. However, quantitative examination revealed that when the number of water molecules is assumed, BND drastically increases by more than ten orders of magnitude with the amount of decompression less than one order of magnitude, and the experimental result cannot be explained. If the number of molecules moving in melt for bubble nucleation is 10^{12}, data shown in the figure is found to be explained. It seems from the value of 10^{12} that because the actual number of H_2O molecules is in the order of 10^{23}, water molecules exist as clusters in which some water molecules have already gathered, and the clusters govern the nucleation process. However, molecular diffusion at that time is not appropriately considered. Furthermore, for heterogeneous nucleation, this regime is the viscosity-controlled regime and the influence of the regime remains unclear. The condition of a constant amount of decompression corresponds to the condition of constant temperature in crystallization experiments. In a crystallization process, the steady state nucleation rate is determined from change of the number of crystals with time under a constant temperature in a supercooling state. On this basis, fundamentally important classical nucleation theory has been closely examined. However, as far as bubble formation is concerned, the behavior of change in the number of bubbles under a constant amount of decompression has not been sufficiently clarified by experiment and it is a task for the future.

5.8 The Limits of Homogeneous Nucleation

Supersaturation pressure required for nucleation in the case of decompression at a constant rate does not largely depend on the decompression rate and the diffusion coefficient but is almost completely determined by the magnitude of the surface tension. As the surface tension increases, the supersaturation pressure required for nucleation increases. This is expressed by Eq. (5.68) and represented in Fig. 5.19. The magnitude of the surface tension under which nucleation does not occur in the case of decompression to the possible maximum decompression, or a vacuum, can be defined as 0.13 N/m for 240 MPa of the initial saturation pressure. This initial saturation pressure corresponds to the conditions of the experiment conducted by Mourtada-Bonnefoi and Laporte (2004). Furthermore, the supersaturation pressure required for nucleation at that time was 0.625×240 MPa and the corresponding surface tension was about 0.1 N/m, which agreed well with the experimental result of the surface tension.

Let us apply it to the value of the surface tension of rhyolitic melt by Bagdassarov et al. (2000) (Fig. 11.2). The surface tension between melt and steam in saturation drastically decreases with increase in saturation pressure. The value of 0.27 N/m for 1 atm decreases to 0.14, 0.1, and about 0.06, for 100, 200, and 300 MPa, respectively. This fact means that homogeneous nucleation does not occur at about 100 MPa or less of the initial supersaturation pressure. The value of 100 MPa corresponds to 3–4 wt% of the initial water content, and melt with the initial water content equal to or less than it does not cause homogeneous nucleation; however, it causes bubble generation only by heterogeneous nucleation. On that occasion, heterogeneous nucleation will occur on the surface of magnetite, the shape factor of

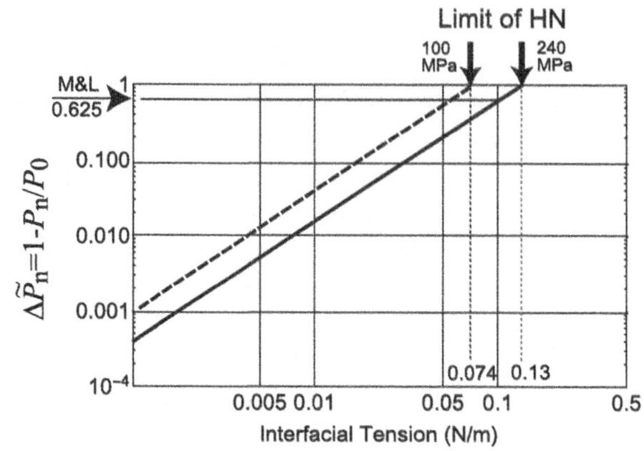

Fig. 5.19 Relationship between the supersaturation pressure required for homogeneous nucleation for decompression at a constant rate, ΔP_n, and surface tension. Calculated by using Eq. (5.68). Homogeneous nucleation does not occur even at decompression to an atmospheric pressure of 0 atm ($\Delta \tilde{P}_n = 1$) when the surface tension is 0.074 N/m or higher for 100 MPa of saturation pressure (broken line) or when the surface tension is 0.13 N/m or higher for 240 MPa of saturation pressure. M&L indicates the experimental result obtained by Mourtada-Bonnefoi and Laporte (2004)

which is small and which is used as nucleation sites. Note that the above holds true for the case where melt is quenched and solidified at a pressure of 1 atm; if melt is maintained at high temperature for a relatively extended time, nucleation may gradually progress by homogeneous nucleation. In this case, considering cooling time to solidification, the understanding of the constant amount of decompression can be applied to BND. The bubble size is then expected to be relatively large because BND is reduced.

5.9 Second Nucleation

The second nucleation of bubbles is a phenomenon in which bubbles newly nucleate when magma that already contains bubbles is decompressed further.[7] Second nucleation can occur if decompression history becomes complicated. In one case, after saturated magma is decompressed and bubbles nucleate in the magma, the magma stays at depth for a certain time and the bubble-containing magma can be rapidly decompressed by fracturing surrounding rocks. In another case, bubble nucleation occurs in cooling magma chambers, magma chambers containing bubbles form, and then the magma can be rapidly decompressed by eruption. The evidence of such second nucleation may be observed in eruptive materials such as pumice and volcanic ash as bimodal bubble size distribution. Massol and Koyaguchi (2005) numerically solved the problem of nucleation in magma ascending in volcanic conduits and demonstrated that second nucleation occurs in a low-pressure condition where decompression is accelerated. Sparks and Brazier (1982) measured bubble size distribution by the gas adsorption method and reported eruptive materials that underwent nucleation three times (Askja 1875, Mount St. Helens 1980); however, this measurement was questioned later (Whitham and Sparks 1986). By observing the texture of pyroclastic rocks to date, there are some studies that suggest the possibility of second nucleation (plinian pumice sample CP15 from Mount Mazama (Fig. 8 in Klug et al. 2002) and vulcanian pumice from Montserrat (Fig. 5 in Formenti and Druitt 2003, Fig. 4 in Giachetti et al. 2011).

As BND and the size of preexisting bubbles are reduced, the total surface area reduces, and the rate of decrease of volatiles in melt associated with the bubble growth reduces. Consequently, when

[7]Note that second nucleation is different from second boiling.

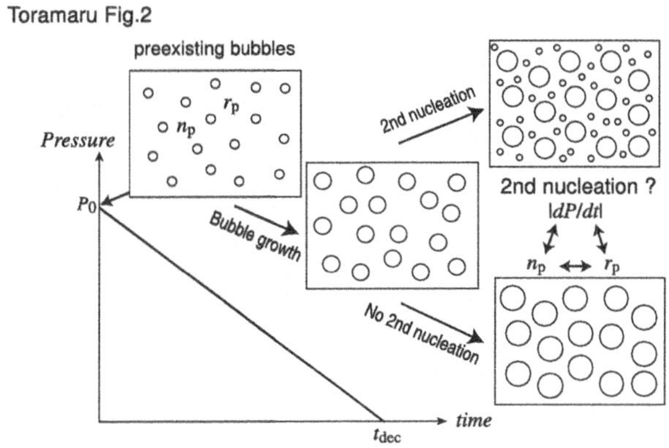

Fig. 5.20 A conceptual diagram to study the second nucleation of bubbles (Toramaru 2014). Reprinted from Toramaru (2014), Copyright 2014, with permission from Elsevier

magma containing such preexisting bubbles is decompressed, it will be easily supersaturated and easily undergoes second nucleation. Moreover, when the decompression rate is high, second nucleation is expected to easily occur (Fig. 5.20).

The manner in which BND and the size of preexisting bubbles and the decompression rate of magma influence second nucleation can be numerically investigated with the moment equation (Toramaru 2014). From the result and analysis of numerical experiments, the critical condition at which second nucleation occurs is given as follows as a function of BND of preexisting bubbles N_{PE}, the size of preexisting bubbles R_{PE}, and physical properties of magma such as the surface tension γ and the diffusion coefficient D:

$$\left|\frac{dP}{dt}\right|_{\text{limit}} = 4\pi D P_0 N_{PE} R_{PE} \tau_* \left[\mathcal{Z}(\zeta)^{-1} + \mathcal{Z}(\zeta)/3\right] \tag{5.97}$$

where $|dP/dt|_{\text{limit}}$ is the critical decompression rate, a threshold below which second nucleation does not occur. Then, functions $\mathcal{Z}$ and ζ are defined by the following equations:

$$\mathcal{Z}(\zeta) = 3\left[\frac{1}{2}\left(\zeta^{\frac{1}{2}} + \sqrt{\zeta - \frac{4}{27}}\right)\right]^{\frac{1}{3}} \tag{5.98}$$

$$\zeta = \left(\frac{\phi_{G0}\tau_*}{8\pi N_{PE} R_{PE}^3}\right)^2 = \left(\frac{\tau_*}{6}\frac{\phi_{G0}}{\phi_P}\right)^2 \tag{5.99}$$

The parameter ζ is the ratio of the volume equivalent to dissolved water content, $\phi_{G0} = k_B T C_0/P_0$, to the volume of water as preexisting bubbles, $\phi_{PE} = 4\pi N_{PE} R_{PE}^3/3$. τ_* is a numerical parameter dependent on the interfacial tension and is given by $\tau_* = \tau_0 \alpha_1^{1/2}$ ($\tau_0 = 0.15$).

If information on magma (such as physical properties and the initial water content) is given, the critical decompression rate of second nucleation is determined by giving BND and the size of preexisting bubbles. This can be represented as a contour plot (Fig. 5.21) on the plane of the number density N_{PE} and the size R_{PE}. From this figure, the dependence of the critical decompression rate on the number density and the size of the preexisting bubbles is found to be divided into two regimes: the regime in which the critical decompression rate depends on the number density and the size (high ϕ_{PE} regime) and the regime in which it depends only on the number density (low ϕ_{PE} regime), according to the water content as the preexisting bubbles ϕ_{PE}. For example, preexisting bubbles represented by point A are in a condition where the bubble radius is $R_{PE} = 1\,\mu$m and the BND is $N_{PE} = 10^{11}\,\text{m}^{-3}$. ϕ_{PE} is

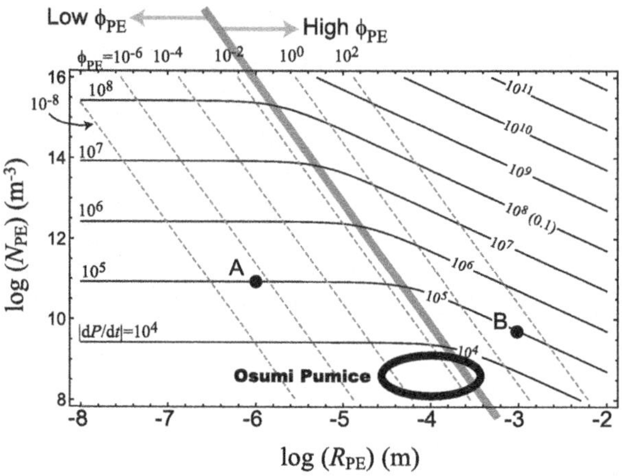

Fig. 5.21 A contour plot of the critical decompression rate $|dP/dt|$ at which second nucleation occurs, as a function of the number density N_{PE} and the size R_{PE} of preexisting bubbles. Second nucleation occurs if the decompression rate is larger than the values represented by the contours. $C_0 = 4.2 \times 10^{27}$, $k_B = 1.38 \times 10^{-23}$, $T = 1100$ (K), $D = 10^{-11}$ (m²/s), P_0= 150 (MPa), $\phi_{G0} = 0.425$, $\gamma = 0.1$ (N/m), and $\tau_0 = 0.15$ are assumed. The influence of the interfacial tension is that when γ decreases by an order of magnitude, the isograms of $|dP/dt|$ in the low ϕ_{PE} regime shift by three orders of magnitude on the vertical axis and those in the high ϕ_{PE} regime shift upward by 3/2 orders of magnitude. (Modified from Toramaru 2014)

4×10^{-7} m³ and point A will be in the low ϕ_{PE} regime where the water content as preexisting bubbles is relatively small. Assuming that the diffusion coefficient of water is 10^{-11} m²/s and the interfacial tension is 0.1 N/m, the critical decompression rate will be 10^5 Pa/s, which corresponds to an ascent rate of 4 m/s. If the decompression rate exceeds this, second nucleation occurs in the magma and if the decompression rate does not exceed it, second nucleation does not occur.

On the other hand, pre-existing bubbles represented by point B are in a condition where $N_{PE} = 5 \times 10^9\ \mathrm{m}^{-3}$, $R_{PE} = 1$ mm, and $\phi_{PE} \approx 21\ \mathrm{m}^3$, and will be in the high ϕ_{PE} regime, where a lower number density of preexisting bubbles gives the same critical decompression rate. The water content of pre-existing bubbles equivalent to $\phi_{PE} = 1$ under the conditions of temperature and pressure given in the figure can be calculated at about 12 wt% from $0.018 \times P \times \phi_P/(6 \times 10^{23} \times 2500 k_B T) \times 100$ wt%. Whether vesiculated magma with such high water content exists underground or not is unclear.

When the amount of volatiles as preexisting bubbles is relatively small, that is, when ζ is large, $\mathcal{Z}(\zeta)^{-1} + \mathcal{Z}(\zeta)/3 \approx \mathcal{Z}(\zeta)/3 = \zeta^{1/6}$ holds and the critical decompression rate can be approximated as follows:

$$\left|\frac{dP}{dt}\right|_{\mathrm{limit}} = 2\pi^{1/3} \cdot D \cdot P_0 \cdot \phi_{G0}^{1/3} \cdot (\tau_0 \alpha_1^{1/2})^{4/3} \cdot N_{PE}^{2/3} \propto N_{PE}^{2/3} \cdot \gamma^2 \tag{5.100}$$

The critical decompression rate is proportional to the power of 2/3 of the number density of preexisting bubbles and to the square of the surface tension; however, it does not depend on the bubble size. The fact that, regardless of the size of preexisting bubbles, it does not depends on the size of preexisting bubbles reflects the fact that as the size becomes equal to or smaller than a certain size, the efficiency of decrease in the water concentration by diffusion growth is constant. Furthermore, the critical decompression rate is determined only by the number density of preexisting bubbles. However, when the volume of preexisting bubbles is high, $\mathcal{Z}(\zeta)^{-1} + \mathcal{Z}(\zeta)/3 \approx 1$ holds and the critical decompression rate can be approximated as follows:

$$|dP/dt|_{\mathrm{limit}} = 4\pi \cdot D \cdot P_0 \cdot N_{PE} \cdot R_{PE} \cdot (\tau_0 \alpha_1^{1/2}) \propto N_{PE} \cdot R_{PE} \cdot \gamma^{3/2} \tag{5.101}$$

The critical decompression rate is proportional to BND and the size of preexisting bubbles and to the power of 3/2 of surface tension. The fact that it is proportional to the size of preexisting bubbles shows that the size of preexisting bubbles influences the behavior of decrease in the water concentration. The results of the numerical experiments on second nucleation can be qualitatively summarized as follows:

1. When the volume fraction of preexisting bubbles is reduced, the critical decompression rate above which second nucleation occurs is determined only by the number density of preexisting bubbles. In this case, the critical decompression rate must be about one order of magnitude higher than the decompression rate virtually required to form the BND of preexisting bubbles.
2. When the volume fraction of preexisting bubbles is high, the critical decompression rate above which second nucleation occurs is proportional to the number density and the size of preexisting bubbles.
3. If the second nucleation of bubbles occurs, the BND of secondary bubbles becomes one order of magnitude or more higher than the number density of first-generation bubbles.
4. The BND decompression rate meter is applicable to the number density of secondary bubbles.

References

Bagdassarov N, Dorfman A, Dingwell D (2000) Effect of alkalis, phosphorus, and water on the surface tension of haplogranite melt. Am Mineral 85:33–40

Cluzel N, Laporte D, Provost A, Kannewischer I (2008) Kinetics of heterogeneous bubble nucleation in rhyolitic melts: implications for the number density of bubbles in volcanic conduits and for pumice textures. Contrib Mineral Petrol 156:745–763

Fiege A, Cichy SB (2015) Experimental constraints on bubble formation and growth during magma ascent: a review. Am Mineral 100(11–12):2426–2442

Fiege A, Holtz F, Cichy SB (2014) Bubble formation during decompression of andesitic melts. Am Mineral 99(5–6):1052–1062

Formenti Y, Druitt TH (2003) Vesicle connectivity in pyroclasts and implications for the fluidisation of fountain-collapse pyroclastic flows, Montserrat (West Indies). Earth Planet Sci Lett 214:561–574

Gardner JE (2007) Heterogeneous bubble nucleation in highly viscous silicate melts during instantaneous decompression from high pressure. Chem Geol 236:1–12

Gardner JE, Denis M-H (2004) Heterogeneous bubble nucleation on Fe-Ti oxide crystals in high-silica rhyolitic melts. Geochim Cosmochim Acta 68:3587–3597

Giachetti T, Burgisser A, Arbaret L, Druitt TH, Kelfoun K (2011) Quantitative textural analysis of Vulcanian pyroclasts (Montserrat) using multi-scale X-ray computed microtomography: comparison with results from 2D image analysis. Bull Volcano 73(9):1295–1309

Hamada M, Laporte D, Cluzel N, Koga KT, Kawamoto T (2010) Simulating bubble number density of rhyolitic pumices from Plinian eruptions: constraints from fast decompression experiments. Bull Volcanol 72(6):735–746

Hurwitz S, Navon O (1994) Bubble nucleation in rhyolitic melts: experiments at high pressure, temperature and water content, Earth Planet. Sci Lett 122:267–280

Klug C, Cashman K, Bacon C (2002) Structure and physical characteristics of pumice from the climactic eruption of Mount Mazama (Crater Lake). Oregon Bull Volcanol 64:486–501

Langer JS, Schwartz AJ (1980) Kinetics of nucleation in near-critical fluids. Phys Rev A 21:948–958

Mangan MT, Sisson TW (2000) Delayed, disequilibrium degassing in rhyolite magma. Decompression experiments and implications for explosive volcanism. Earth Planet Sci Lett 183:441–455

Mangan MT, Sisson TW (2005) Evolution of melt-vapor surface tension in silicic volcanic systems: Experiments with hydrous melts. J Geophys Res 110:B01202. https://doi.org/10.1029/2004JB00321523-36

Massol H, Koyaguchi T (2005) The effect of magma flow on nucleation of gas bubbles in a volcanic conduit. J Volcanol Geotherm Res 143:69–88

Mourtada-Bonnefoi CC, Laporte D (2004) Kinetics of bubble nucleation in a rhyolitic melt: an experimental study of the effect of ascent rate. Earth Planet Sci Lett 218:521–537

Navon O, Lyakhovsky V (1998) Vesiculation processes in silicic magmas. In: Gilbert JS, Sparks RSJ (eds) The physics of explosive volcanic eruptions, vol 145. Geological Society, London, Special Publications, pp 27–50

Pichavant M, Di Carlo I, Rotolo SG, Scaillet B, Burgisser A, Le Gall N, Martel C (2013) Generation of CO_2-rich melts during basalt magma ascent and degassing. Contrib Mineral Petrol 166(2):545–561
Sakuma S, Murase T (1957) Some properties of natural glasses at high temperatures: viscosity and the growth of bubbles. Kazan 2(1):6–16, 09–30. (in Japanese with English abstract)
Sakuyama M, Kushiro I (1979) Vesiculation of hydrous andesitic melt and transport of alkalies by separated vapor phase. Contrib Mineral Petrol 71(1):61–66
Shimozuru D, Nakamuda O, Seno H, Noda H, Taneda S (1957) Mechanism of Pumice Formation, Kazan 2(1):17–25, 1957-09-30. (in Japanese with English abstract)
Sparks RSJ, Brazier S (1982) New evidence for degassing processes during explosive eruptions. Nature 295:218–220
Toramaru A (1989) Vesiculation process and bubble size distributions in ascending magmas with constant velocities. J Geophys Res 94:17523–17542
Toramaru A (1995) Numerical study of nucleation and growth of bubbles in viscous magmas. J Geophys Res 100:1913–1931
Toramaru A (2006) BND (bubble number density) decompression rate meter for explosive volcanic eruptions. J Volcanol Geotherm Res 154:303–316
Toramaru A (2014) On the second nucleation of bubbles in magmas under sudden decompression. Earth Planet Sci Lett 404:190–199
Toramaru A, Miwa T (2008) Vesiculation and crystallization under instantaneous decompression: numerical study and comparison with laboratory experiments. J Volcanol Geotherm Res 177:983–996
Whitham AG, Sparks RSJ (1986) Pumice. Bull Volcanol 48:209–223
Yamada K, Tanaka H, Nakazawa K, Emori H (2005) A new theory of bubble formation in magma. J Geophys Res 110:B02203. https://doi.org/10.1029/2004JB003113
Yamada K, Emori H, Nakazawa K (2008) Time-evolution of bubble formation in a viscous liquid. Earth Planets Space 60:661–679
Yamamoto T, Hasegawa H (1977) Grain formation through nucleation process in astrophysical environment. Prog Theor Phys 58:816–828. https://doi.org/10.1143/PTP.58.816

6 Other Bubble-Related Processes

Abstract

In previous chapters, thermodynamic equilibrium conditions for vesiculation of magma and kinetics of nucleation and growth of bubbles have been explained. Bubbles in magma experience both simple elemental processes of nucleation and growth in supersaturated liquid and various processes before reaching the surface of the Earth. Among various processes bubbles experience, this chapter explains secondary growth, deformation, coalescence, detachment, collapse, and oscillation, which are relatively well studied. In addition to the previous chapters, theoretical examination is first conducted, followed by application of the theoretical examination to the experimental results.

6.1 Secondary Growth of Bubbles: Ostwald Ripening

6.1.1 Mechanism of Secondary Growth

When a new phase forms after nucleation and growth and the phase is in a weak supersaturated state, the volume of the phase (bubbles or crystals) does not change but particles grow on an average. This phenomenon is called secondary growth or Ostwald ripening. Secondary growth of bubbles or crystals is driven by an action of lessening the surface energy. The mechanism of the secondary growth is a combination of heterogeneity of local equilibrium concentration by the Gibbs-Thomson effect at the interface and thereby driven mass transfer in a liquid phase.

Figure 6.1 is a schematic of the mechanism of secondary growth. As a mechanical analogy of the Gibbs-Thomson effect, let us consider the case in which two differently sized rubber balloons are connected by a straw. Assume that, regardless of the size of the balloons, the tensional coefficient of rubber is uniform. Following a mechanical equilibrium, i.e., the Laplace equation (3.16), the internal pressure of a balloon decreases as the size of the balloon increases. Therefore, if the two balloons are connected by a straw, a flow of gas is generated, driven by the pressure difference of gas between the balloons from the smaller balloon with higher internal pressure to the larger balloon with lower internal pressure. The size difference between the balloons increases because of the gas flow, the driving force and the gas flow becomes increasingly large, and finally the smaller balloon deflates and only the larger balloon remains.

A. Toramaru, *Vesiculation and Crystallization of Magma*, Advances in Volcanology,
https://doi.org/10.1007/978-981-16-4209-8_6

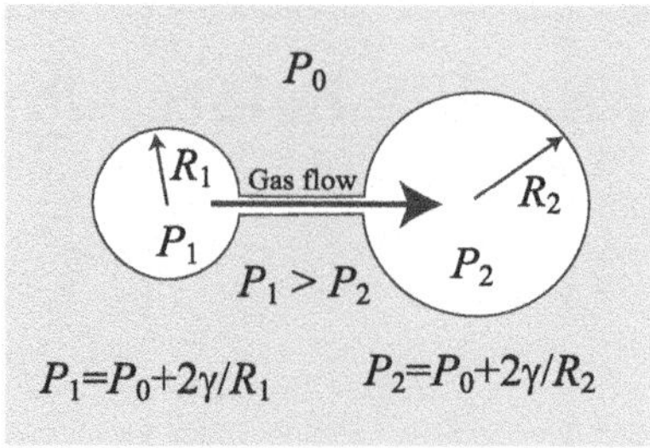

Fig. 6.1 A schematic of two different-sized rubber balloons connected by a straw. An analogy of secondary growth (Ostwald ripening) of bubbles. Internal pressures of balloons (P_1 and P_2) that are in mechanical equilibrium (the Laplace equation) with the surrounding fluid (liquid pressure: P_0) become higher as the size of the balloon R decreases; therefore, the gas flows from the smaller balloon to the larger balloon. For secondary growth of bubbles, the interfacial concentration (chemical potential) of bubbles that are in chemical equilibrium with liquid becomes higher as the bubble radius becomes smaller: $\mu_R = \mu_\infty + 2\gamma v_G/R$. Note that μ_R and μ_∞ are chemical potential at the interface of a bubble with the radius of curvature R and at the flat interface (the radius of curvature is infinite), respectively, and v_G is the molecular volume of the gas. Therefore, a concentration gradient is generated in the liquid and the gas component moves from the vicinity of the smaller bubble to the vicinity of the larger bubble by diffusion

Moreover, the same process works in the chemical equilibrium between bubbles or crystals, the interface of which has curvature, and liquids. In this case, difference in pressure in the bubbles due to the interfacial tension produces differences in chemical potential of the gas component in the liquid which is in equilibrium with the gas in the bubbles and high-internal pressure and small-radius bubbles with large curvature will have larger interfacial equilibrium concentration. In other words, concentration of the gas component in liquid becomes nonuniform and consequently mass transfer by diffusion will occur. The mass transfer, which is diffusion in liquid following the concentration gradient, occurs from the vicinity of smaller bubbles with larger curvature at the interface to that of the larger bubbles with a smaller curvature. That is, the concentration at the interface of the smaller bubbles decreases and becomes undersaturated and the vicinity of the larger bubbles becomes supersaturated. In the undersaturated region, gas that has exsolved in bubbles re-dissolves in liquid to maintain the interfacial equilibrium; in the supersaturated region, exsolution of gas to bubbles further progresses. Consequently, smaller bubbles become increasingly small and larger bubbles become increasingly large. Because of this process, smaller-sized bubbles disappear and the number of bubbles decreases and the mean size increases.

6.1.2 Solution of the Size Distribution by Lifshitz and Slyozov (The LS Theory)

For spherical crystals in solution, Lifshitz and Slyozov (1961) provided analytical solutions. For spherical crystals in solution, because the solubility linearly depends on pressure, the square root, as in the case of the Gibbs-Thomson relation (Eq. 2.43) in bubbles in silicate melt, does not appear.

For this problem, Lifshitz and Slyozov (1961) obtained an analytic solution of the temporal development of the size distribution function. Figure 6.2 shows their solution. The point to be focused on here is the conclusion that domains one and a half times as large as the critical size do not exist. The peak (the most dominant size) is positioned at $1.2/1.5 = 0.8$ in the overall size distribution. Smaller-sized spherical crystals are distributed among larger-sized ones, and the smaller the size of crystals, the smaller the number of crystals.

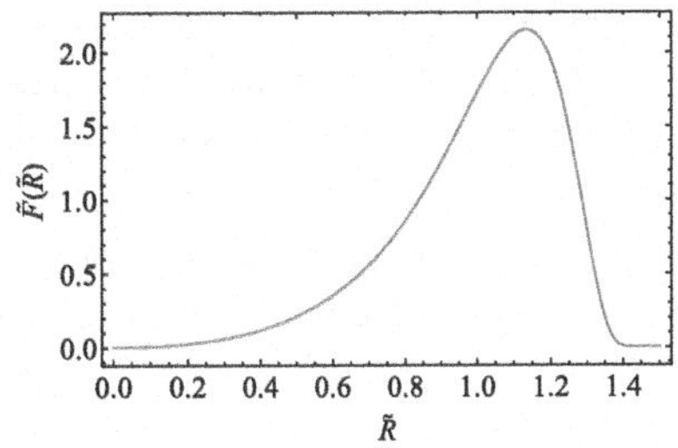

Fig. 6.2 Size distribution of a solution by Lifshitz and Slyozov (1961). The size $\tilde{R}$ (scaled by Eq. (6.2)) is the mean particle size. Distribution $\tilde{F}(\tilde{R})$ is scaled appropriately

The number of particles N m^{-3} then decreases in inverse proportion to time.

$$N \approx \frac{0.5}{DR_X^* t} \tag{6.1}$$

Moreover, the mean particle size $\langle R \rangle$ becomes the same as the critical radius and increases in proportion to the cube root of time.

$$\langle R \rangle \approx \left(\frac{4R_X^* D}{9} t \right)^{\frac{1}{3}} \tag{6.2}$$

where R_X^* is the particle size on which the surface tension is effective, which is defined by Eq. (7.27). The growth law of the LS theory is interesting compared with the growth law of diffusion growth in a relatively high degree of supersaturation (proportional to the square root of time).

6.1.3 Qualitative Understanding of the Growth Law

The growth law of Ostwald ripening is proportional to the cube root of time. However, the diffusion growth law is proportional to the square of time. Where does this difference between them arise from? In Chap. 4, a basic equation for diffusion growth of bubbles was given by Eq. (4.25). In the equation, the particle size R in the denominator of the right-hand side corresponds to diffusion length and $C_\infty - C_R$ in the numerator of the right-hand side can be considered as the concentration difference driving diffusion. For Ostwald ripening, diffusion length is half of the distance between the centers of two bubbles (in diluted solution) and difference in concentration is difference in equilibrium concentration ΔC caused by difference in pressure in bubbles. Therefore, bubble growth can be rewritten as follows:

$$\frac{dR}{dt} = v_G D \frac{\Delta C}{l} \tag{6.3}$$

Difference in equilibrium concentration ΔC produced by difference in pressure in bubbles ΔP_G can be approximated as follows:

$$\Delta C \approx \frac{\partial C}{\partial P} \Delta P_G \tag{6.4}$$

Moreover, because a scale of difference in pressure in bubbles, ΔP_G is difference in pressure by the interfacial tension, it can be approximated by the Laplace equation as follows:

$$\Delta P_G \approx \frac{2\gamma}{R} \tag{6.5}$$

On the other hand, the distance between the centers of two bubbles $2l$ increases with decrease in the number of bubbles N, in accordance with $l \approx N^{-1/3}$. In the case where the volume fraction ϕ of bubbles does not change, the number of bubbles N changes in accordance $N \propto R^{-3}$ (independent of the volume fraction of bubbles), as seen in the section of the cell model. Therefore, the distance between bubbles increases in proportion to the representative bubble radius in the system ($l \propto R$). Consequently, the bubble growth rate can be written as follows:

$$\frac{dR}{dt} = v_G D \frac{(\partial C/\partial P)2\gamma}{R^2} \tag{6.6}$$

and the integral of this equation shows that the bubble size is proportional to the cube root of time.

6.1.4 Comparison with Experiments

The fact that the growth law of Ostwald ripening of droplets or crystals in liquid or of solid phase in solid phase is proportional to the cube root of time, i.e., coarsening, has been confirmed by experiments (Kingery et al. 1976). The case of foam with large volume fraction has been confirmed to be in proportion to the square root of time (Weaire and Hutzler 2000; Proussevitch et al. 1993a). Moreover, it was reported in experiments with silicate melt (Larsen and Gardner 2000; Larsen et al. 2004; Lautze et al. 2011) that the growth law is not always in proportion to the cube root of time. In particular, the rate of decrease in the bubble number density is small compared with the LS theory in which it is inversely proportional to time. Larsen et al. (2004) conducted an experiment on constant amount of decompression with Mono Craters rhyolite and demonstrated that the bubble number density N obeys the following power-law rule on time t (minute).

$$N \propto t^{\chi_{OS}} \tag{6.7}$$

where the exponent χ_{OS} takes on values from -0.32 to -0.43, thus considerably departing from -1, a value expected from the LS theory. Furthermore, Lautze et al. (2011) conducted experiments with rhyolitic and basaltic andesite melt and reported that the exponent takes on values from -0.59 to-0.94, which are not in agreement with the LS theory.

This suggests that for steam bubbles in silicate melt, the LS theory cannot be applied as it is. As primary causes for disagreement with the theory, the following two points can be considered. One is that the solubility of water in silicate melt is proportional to the square root of pressure of bubbles. In the LS theory, equilibrium concentration is assumed to be in proportion to pressure. Another is that for bubbles, the molecular volume of gas molecules in bubbles is mutually related to pressure in bubbles through an equation of state for an ideal gas. The LS theory assumes that the molecular volume in particles is constant. Thus, to understand Ostwald ripening in silicate melt requires theoretical changes.

6.2 Deformation of Bubbles

6.2.1 Theoretical Study

Deformation of droplets of Newtonian fluid in Newtonian viscous fluid has been examined for a relatively long time. In general, the deformation of droplets is governed by the velocity field of (strain rate $\dot{\epsilon}$) the surrounding fluid, surfactants at the surface of the droplets, the ratio of viscosity of the surrounding fluid η to viscosity of the droplets η', the surface tension γ, and the Reynolds number (Re) of the surrounding fluid. The case of bubbles can be dealt with as a case in which the viscosity ratio η'/η is close to zero. Moreover, the effect of surface will be ignored.

The deformation of droplets is evaluated by determining a deformation field of the fluid surrounding the droplets and then by determining resultant stress in the fluid and a flow field of the inside of the droplets. When considering the surface tension, it will be included in the boundary conditions at the boundary between the droplets and the surrounding fluid, and then determined by calculating an equation of motion at the surface of the droplets. For the case where Re is small, Taylor (1934) used a deformation field surrounding nearly spherical droplets, obtained by Taylor (1932), to calculate the balance of forces and determined the shape at steady state. Consequently, he demonstrated that the following Capillary number (Ca) is important.

$$\mathrm{Ca} = \frac{\eta\dot{\epsilon}R}{\gamma} \tag{6.8}$$

Ca is the ratio of a force shifting a bubble from a sphere by deformation of the surrounding fluid ($\eta\dot{\epsilon}$) to a force restoring the bubble to a sphere by the surface tension (γ/R). When Ca is low, the effect of the surface tension is dominant and a spherical shape is maintained. When Ca becomes high, deformation is dominant and the shape of droplets transitions out of the spherical shape. Moreover, it was found that when Ca increases and exceeds a threshold, bubbles will be fragmented.

For the case where Re and the viscosity ratio are <1 and simple shear and pure shear are involved, Taylor investigated a degree of deformation (deformation index) DF defined by the following equation:

$$DF = \frac{L_1 - L_3}{L_1 + L_3} \tag{6.9}$$

where the major axis of a deformed droplet and the minor (shortest) axis are assumed to be L_1 and L_3, respectively. Consequently, when deformation is small ($0 \ll L_3/L_1 < 1$), DF was found to be in proportion to Ca in a steady state of simple shear.

$$DF = \mathrm{Ca} \tag{6.10}$$

For pure shear, 2 is multiplied as a coefficient.

When deformation becomes very large ($L_1/L_3 \gg 1$), tips of a bubble will become pointed (the bubble at the upper left in the left chart of Fig. 10.15). This shape is called a slender-body and influences the surrounding deformation field. Hinch and Acrivos (1980) analytically solved the slender-body and obtained the following relation of a steady-state shape:

$$\frac{L_1}{R} = 3.45\mathrm{Ca}^{1/2} \tag{6.11}$$

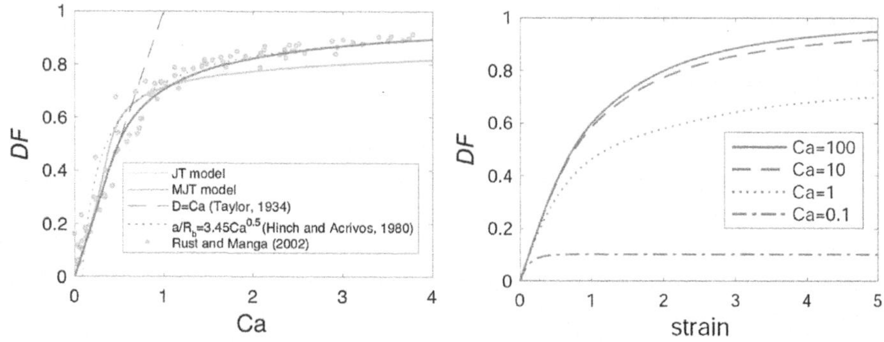

Fig. 6.3 (Left) Relationship between the shape of bubble (the degree of deformation: DF) under a steady state and Ca (Rust and Manga 2002). Different curves represent different models. (Right) Temporal development of the shape of bubble (the degree of deformation: DF) under simple shear with a parameter Ca (strain can be regarded as time) (Ohashi et al. 2018). Reprinted from (Ohashi et al. 2018), Copyright 2018, with permission from Elsevier

For pure shear (Acrivos and Lo 1978),

$$\frac{L_1}{R} = 20\text{Ca}^2 \tag{6.12}$$

was obtained.

When a droplet was slightly deformed from a sphere and can be approximated by an ellipsoid, the case without surface tension (Wetzel and Tucker 2001) and the case where surface tension works (Jackson and Tucker 2003) were analyzed by applying Eshelby's theory of an elastic body (Eshelby 1957). In these studies, Jackson and Tucker (2003) developed a model that can consistently handle a droplet, from a slightly extended droplet to an ellipsoid to a greatly extended slender-body. Note that in these recent models they allow us to calculate both the shape of droplets in a steady state and the shape of droplets in a transient state to the steady state.

These studies focused on deformation of droplets of viscous fluid in Newtonian fluid and did not include the case where the viscosity ratio η'/η of bubbles is extremely low and close to 0. In recent years, improving a model by Jackson and Tucker (2003) (the JT model), Ohashi et al. (2018) established a calculation method for temporal development of bubble shape in a transient process (the MJT model) and investigated factors controlling bubble shape in processes including a transient process. Consequently, they succeeded in expressing the bubble shape as a function of Ca (shear rate) and total strain (that is, time) (Fig. 6.3). By applying it to natural deformed bubbles, Ca contributing to deformation, i.e., shear rate and the amount of strain, can be simultaneously estimated (Chap. 10).

6.2.2 Experimental Study

Rust and Manga (2002) conducted experiments for air bubbles in a region covering the intermediate area between large deformation and small deformation with a parallel band device (two parallel bands that move to the opposite directions) developed by Taylor and confirmed that transition of a steady shape occurs when Ca $= 1$ (left chart of Fig. 6.3). The degree of deformation DF is proportional to Ca in accordance with Eq. (6.10). When Ca is small, and when Ca exceeds 1, L_1/R increases in proportion to Ca$^{1/2}$ in accordance with Eq. (6.11). An important point here is that when Ca > 1, the degree of deformation DF in the steady state is around 1 and takes on values of >0.8. This corresponds to the fact that in Fig. 6.3 curves of DF in the steady state in large Ca are close at around 1.

Furthermore, they investigated a process in which a deformation field disappears and bubbles are relaxed by the surface tension and demonstrated that the degree of deformation of the major axis $L_1 - R$ is relaxed as follows:

$$\frac{L_1(t) - R}{L_1(0) - R} = \exp(-0.67t/\tau) \tag{6.13}$$

with a time constant $\tau = R\eta/\gamma$. Thus, an important point is that bubble shape in a steady state can be described only by Ca.

6.3 Coalescence of Bubbles

As nucleated bubbles grow, the distance between bubbles shortens and the bubbles can coalesce. The efficiency of bubble coalescence is then governed by a process of the inter-bubble distance shortening and a process of drainage of melt film between approaching bubbles (Fig. 6.4). Furthermore, a process in which the shape of coalesced bubbles relaxes into a sphere influences final vesicular texture. First, as a fundamental aspect, to understand the temporal development of bubble size distribution in the bubble coalescence process, we formulate it by a discrete size distribution function and explain the concept of coalescence efficiency. Generally, coalescence efficiency depends on elemental processes and the size of bubbles to coalesce. To increase our understanding of coalescence efficiency, we shall formulate coalescence efficiency for several elemental processes. Using coalescence efficiency independent of the size of bubbles as an example for simplicity, we shall quantitatively understand the temporal development of size distribution.

6.3.1 Coalescence Frequency

We define discrete size distribution as n_i. i is consecutive integers and defines the size. A relationship in which when values of integers change as 2, 3, 4, ..., the size will change as twofold, threefold, fourfold... in proportion to them is convenient. Therefore, if the minimum size is not a single molecule but any other scale, the size is defined by an integral multiple of the other scale.

Time change of bubble number density in a size bin k can be expressed as follows:

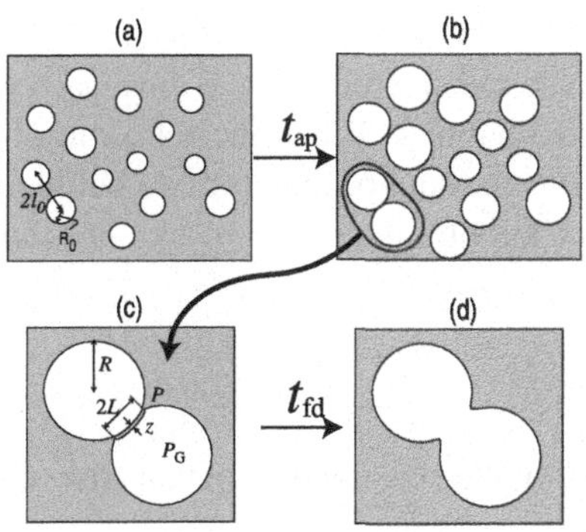

Fig. 6.4 Schematic of the coalescence process of bubbles. For bubble approach by bubble growth, $t_{ap} = t_G$ holds. (**a**, **b**) Bubble approach by bubble expansion. The centers of the bubbles do not move. (**c**, **d**) Coalescence of bubbles by drainage of liquid film between bubbles. In the Reynolds lubrication theory, liquid film between bubbles is considered as a disc with a radius L and thickness z

$$\frac{dn_k}{dt} = \underbrace{\frac{1}{2}\sum_{i+j=k}^{\infty} K_{ij}n_in_j}_{\text{Increase}} - \underbrace{n_k\sum_{i=1}^{\infty} K_{ik}n_i}_{\text{Decrease}} \tag{6.14}$$

where K_{ij} is the frequency of coalescence (coalescence frequency) of two bubbles (the size of which are i and j, respectively) among n_i bubbles and n_j bubbles and has a dimension of ^{3}/(number s). The coalescence frequency K_{ij} depends on the mechanism of coalescence.

The first term on the right-hand side is the rate of increase of bubbles with the size $k = i + j$ by coalescence of two small-sized bubbles (size i and j). When we add from the smaller size of i, we count size two times (for $i < j$ and $i > j$). Therefore, we multiply the equation by a factor of 1/2 to cancel it. The second term on the right-hand side represents that the number of bubbles with the size k decreases because a bubble with the size k and a different-sized bubble coalesce into a bubble with larger size $k + j$. The meaning of coalescence frequency $K_{ij}n_i$ is that $K_{ij}n_i$ gives the number of encounters and coalescences of a bubble of interest with the size j and a bubble with the size i per unit time.

In the case where bubbles can freely move in fluid like dispersed particles, conventional collision frequency in particle's aggregation process can be used. In that case, for $K_{ij}n_i$, we can consider the number of a population of candidates the size of interest j will encounter as the particle number density n_i and the probability of encountering the size j in the population as K_{ij}. This case will be explained later.

In the case where bubbles cannot freely move because of closely spaced bubbles or viscosity, the number of a population of bubbles with the size i that can encounter and coalesce with the size j of interest is the number of the nearest bubbles and determined by the geometric arrangement of bubbles. The number of the nearest bubbles is 1 for random distribution of equal spheres and 14 for regular close packing of equal spheres. Although this number varies according to the volume fraction of and the size distribution of bubbles, it is assumed to be 1 in this book. In this case, for $K_{ij}n_i$, the number of the population is 1 and $K_{ij}n_i$ can be seen as the probability of discovering the size i in that population per unit time. Then, 1 (the number of the population) will be replaced by the probability of selecting n_i bubbles with the size i from among the total number of bubbles N, which include all sizes and

$$K_{ij}n_i \propto \frac{n_i}{N} \tag{6.15}$$

holds.

Moreover, in this case, the coalescence frequency is inversely proportional to the total time t^* required for the coalescence of two bubbles. When the relaxation time of the shape is ignored, the total time required for coalescence is the sum of time for two bubbles to approach t_{ap} and the time for drainage of liquid film between the bubbles t_{fd}, i.e., $t^* = t_{\text{ap}} + t_{\text{fd}}$, and

$$K_{ij}n_i \propto \frac{1}{t_{\text{ap}} + t_{\text{fd}}} \tag{6.16}$$

holds. t_{ap} and t_{fd} generally depend on the size i and j. In that case, coalescence frequency K_{ij} will be

$$K_{ij} = \frac{1}{N(t_{\text{ap}} + t_{\text{fd}})} = \frac{1}{Nt^*} \tag{6.17}$$

where N is the total number of bubbles and is given by

$$N = \sum_{i=1}^{\infty} n_i \tag{6.18}$$

In the case where bubbles cannot freely move, assuming that t^* is independent of the size, substituting Eq. (6.17) into Eq. (6.14) yields

$$\frac{dn_k}{dt} = \frac{1}{2Nt^*} \sum_{i+j=k}^{\infty} n_i n_j - \frac{n_k}{Nt^*} \sum_{i=1}^{\infty} n_i \tag{6.19}$$

Operating on this equation by $\sum_{k=1}^{\infty}$ to obtain an equation for the rate of change in the total number of bubbles (the bubble number density) N yields

$$\frac{dN}{dt} = -\frac{1}{2t^*} N \tag{6.20}$$

If we consider t^* as a constant independent of N and t, a solution to this equation can be easily obtained.

$$N = N(0) \exp\left(-\frac{t}{2t^*}\right) \tag{6.21}$$

where $N(0)$ is the initial total number of bubbles. Thus, the number of bubbles exponentially decreases and its time constant is $2t^*$. We call its inverse coalescence efficiency (CE).

$$\text{CE} = \frac{1}{2t^*} \tag{6.22}$$

Moreover, in the case where bubbles can freely move in a liquid by turbulent flow, similar to particles making Brownian motion in gas,

$$N = \frac{N(0)}{1 + N(0)t/(2t^*)} \tag{6.23}$$

is obtained (Friedlander 2000).

6.3.2 Shortening Process of Interbubble Distance: An Elemental Process of Coalescence

Bubble Approach Due to Growth and Expansion

This is a mechanism characteristic of the coalescence process of bubbles in liquid (Fig. 6.4a, b). In the aggregation process of aerosols and dust in interplanetary space, particles move around by, for example, Brownian motion to encounter the other particles. However, the same does not hold for bubbles in magma.

The distance between the centers of bubbles $2l$ is related to the number density of bubbles N and shortens as the number density increases: $l \propto N^{-1/3}$. When the volume fraction is constant, an approximation $N \propto R^{-3}$ holds and thus $l \approx R$ is assumed. Assuming that the rate of growth or expansion of bubbles is dR/dt, the timescale of bubble approach tap will be $t_{\text{ap}} = l/dR/dt \approx R/dR/dt$ by zeroth order approximation. Because in the bubble coalescence process, the volume of bubbles v is used as a unit of size, using $v = 4\pi R^3/3$ yields

$$t_{ap} \approx \frac{R}{\frac{dR}{dt}} \approx \frac{v}{\frac{dv}{dt}} \tag{6.24}$$

This equation can be rewritten as follows:

$$\frac{dv}{dt} = \frac{v}{t_{ap}} \tag{6.25}$$

In the case of simple diffusion growth, because $dR/dt \propto 1/R$ holds, $dv/dt \approx v^{1/3}$ holds, and

$$\frac{dv}{dt} \approx v^{1/3} \tag{6.26}$$

$$t_{ap} \approx v^{2/3} \tag{6.27}$$

are obtained.

For the isothermal expansion of bubbles under continuous decompression at a constant rate, when the number of molecules in bubbles is constant, from an equation of state Pv = constant,

$$\frac{dv}{dt} = \frac{v}{t_{dec}} \tag{6.28}$$

$$t_{ap} = t_{dec} \tag{6.29}$$

are obtained. The bubble approach time will be equal to the decompression time $t_{dec} = P/|dP/dt|$ and does not depend on bubble radius. These results are based on a mean-field approach and the size of and the distance from the nearest bubble around a bubble of the size v is not dependent on the size of the nearest bubble but dependent on v (spatial distribution of bubbles is random and probability is equal for all sizes on average).

Bubble Approach Due to Buoyancy-Driven Relative Motion

If the ascent of bubbles in melt depends on the size of the bubbles, the difference in ascent rate between the bubbles can cause approach of bubbles of different sizes and coalescence of them (Fig. 6.5a, b). In general, when we evaluate an efficiency of coalescence of particles floating in viscous fluid by buoyancy-driven up-and-down relative motion, we begin by estimating an effective collision cross section. The effective collision cross section is obtained by correcting a geometric collision cross section while considering fluid-mechanical interactions between particles.

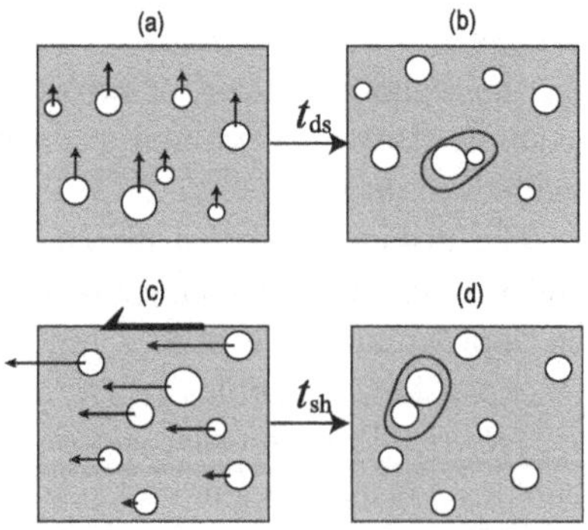

Fig. 6.5 (**a**, **b**) Approach due to difference in the ascent rate by buoyancy. (**c**, **d**) Approach due to shear

While bubbles ascend by buoyancy, particles heavier than liquid sediment. Moreover, sedimenting particles undergo a similar coalescence process (coagulation). Assuming that bubbles are spheres that are not deformed, discussion on the process of coagulation of particles by difference in rate of sedimentation (differential sedimentation) is applicable for bubbles.

For spherical particles that are not deformed, we first determine a flow field around a particle of a certain size R_2 which ascends at the Stokes velocity by buoyancy. Next, at a point lower than the particle, not fluid-mechanically influenced by the particle, and a certain horizontal distance distant from the center of the particle, another particle of a larger size R_1 is positioned and the ascending locus of the particle of the size R_1 is calculated. It is expected that when the initial horizontal distance between the two particles is sufficiently less than the sum of radii of the two particles $R_1 + R_2$, the lower particle will collide with the higher particle, and when the initial horizontal distance is sufficiently greater, the particles will not collide. If the fluid-mechanical interaction is not considered, the geometric collision cross section will obviously be $\pi(R_1 + R_2)^2$. When the fluid-mechanical interaction is considered, a correction term will be multiplied to it (Friedlander 2000). Moreover, because bubbles are deformed by interaction, its influence is fed back to the flow field. Therefore, the problem becomes more complicated and will be difficult to analyze. Formally, introducing a correction term considering a fluid-mechanical interaction and bubble deformation, β_{ij}, into coalescence efficiency yields

$$K_{ij} = \beta_{ij}\pi(R_i + R_j)^2|U_i - U_j| \tag{6.30}$$

Rewriting this equation using the Stokes velocity $U_{\mathrm{St}} = \Delta\rho g R^2/(3\eta)$ for ascending rate U and the particle volume $v = 4\pi R^3/3$ yields

$$K_{ij} = \beta_{ij}\pi(\Delta\rho g/(3\eta))(4\pi/3)^{-4/3}(v_i^{1/3} + v_j^{1/3})^2|v_i^{2/3} - v_j^{2/3}| \tag{6.31}$$

Because a dimension of factors other than the terms of β and v is (m^{-1} s^{-1}), a dimension of the term of v is (m^4), and a dimension of K_{ij} is (m^3 s^{-1}), as seen above, β is dimensionless.

Manga and Stone (1994) performed an experiment using corn syrup as viscous fluid (Fig. 6.6) and evaluated the various influences described above by the following Bond number (Bo) (also called the Eötvös number) when the Reynolds number (Re) is low:

$$\mathrm{Bo} = \frac{\Delta\rho g R^2}{\gamma} \tag{6.32}$$

Bo is (twice) the ratio of the pressure by buoyancy ($\Delta\rho g R$)) to the surface tension ($2\gamma/R$). When Bo is small, bubbles maintain their spherical shape and are less influenced by deformation. When Bo is large, bubbles deform and fluid-mechanical interaction becomes complicated. Moreover, if Re

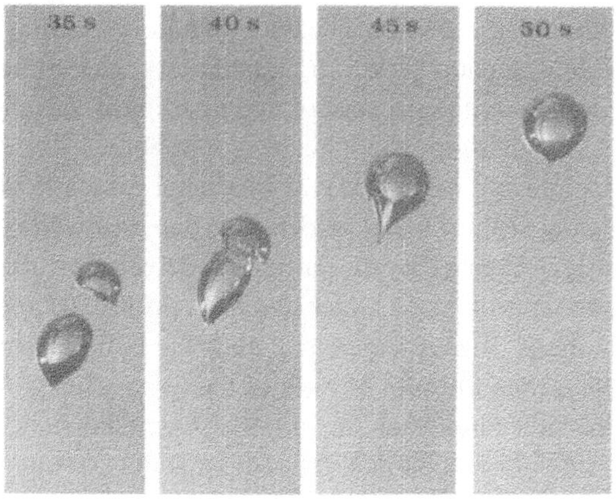

Fig. 6.6 Coalescence of two bubbles in viscous fluid (corn syrup: Re $\approx$ 0.005, Bo $\approx$ 20), which start to ascend from two points where the distance between the two centers are farther than the sum of radii (Manga and Stone 1994). The larger bubble obliquely approaches the smaller bubble from below, collides with the smaller bubble, and then they coalesce into one bubble. Under other conditions, the larger bubble overtakes the smaller bubble once and sucks up the smaller bubble to coalesce. The horizontal size of the upper bubble is $\approx$ 2 cm. Reprinted from (Manga and Stone 1994), Copyright 1994, with permission from Elsevier

becomes large, an ascending bubble can deform even solely and make complex motions (Sect. 6.5.2). Therefore, the coalescence efficiency is considered to be a function of Re.[1]

Values of Bo seen as a function of bubble radius are shown in Fig. 6.7. For magma, when the bubble radius is 1 mm or less, Bo will be roughly 1 or less and bubble deformation can be neglected. When bubbles grow to about 10 cm, Bo will exceed 10 and the influence of deviation from a sphere has to be considered.

They determined the correction term β_{ij}, the ratio of effective collision cross section to geometric cross section, as follows:

$$\beta_{ij} = 0.3\left(\frac{R_2}{R_1}\right)^{1/2} + 0.5\mathrm{Bo}\left(\frac{R_2}{R_1}\right)^{6} \tag{6.34}$$

from their experimental result. Thus, it is found from this equation and $R_1 > R_2$ and $0 < R_2/R_1 < 1$ that for Bo = 0, where bubbles do not deform, the coalescence efficiency will be 0.3 times as small as the geometric collision cross section. This indicates that coalescence is more difficult than simple collision because of the presence of liquid film between bubbles. Moreover, when Bo is > 10, the coalescence efficiency increases in a certain range of a particle size ratio R_2/R_1. Although actual

[1] As seen in Sect. 6.2, deformation of bubbles can be evaluated by the Capillary number (Ca, Eq. 6.8). As for Ca, assuming that what is obtained by dividing the Stokes velocity V_S by the bubble radius R is the shear rate $\dot{\epsilon}$

$$\mathrm{Ca} = \frac{\eta R V_S}{\gamma R} \approx \frac{\Delta\rho g R^2}{\gamma} = \mathrm{Bo} \tag{6.33}$$

is obtained, which is equal to Bo. Therefore, the coalescence efficiency is considered to be well expressed by considering Bo and Re. The Weber number (We), the ratio of the surface tension $(2\gamma/R)$ to the inertial force (ρV_S^2):

$$\mathrm{We} = \frac{\rho V_S^2 R}{\gamma}$$

may influence bubbles ascending at high velocity..

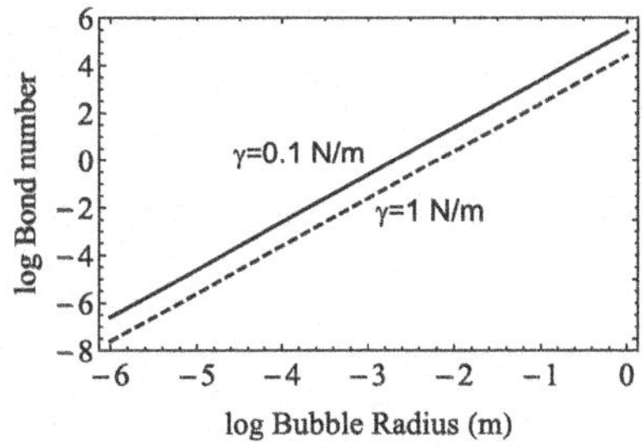

Fig. 6.7 Relationship between the bubble radius and the Bond number. Difference in the density is assumed to be 2500 kg/m^3. The figure shows two examples where the surface tension is 0.1 N/m or 1 N/m

fluid-mechanical interaction between bubbles is complicated, it is interesting that there are some cases where two bubbles do not coalesce, although they approach very closely.

Experiments by Manga and Stone (1994) and Stewart (1995) demonstrated that coalescence caused by collision of ascending bubbles does not occur simultaneously with collision but occurs at entanglement of bubbles after collision. Also in the case of an ascending single bubble, the surrounding fluid is entangled into the lower stagnation point below the bubble. Thus, to understand coalescence of two ascending bubbles, we have to investigate the entanglement process and the behavior of liquid film formed between the bubbles.

Bubble Approach Due to Shearing

In addition, for the approach and coalescence of bubbles by shear flow (Fig. 6.5c, d), we consider collision of particles by shearing by the same procedure as for differential sedimentation of bubbles. We then begin by ignoring fluid-mechanical interaction and deformation of bubbles and assuming ballistic movement of a material particle placed in a shear flow field (shear rate $\dot{\epsilon}$). We now assume that a direction perpendicular to the shear plane of the shear flow is x and place a sphere with a radius of $R_i + R_j$ as the center of the sphere is situated at $x = 0$ to consider a collision of a point mass with the sphere. Note that the size i and the size j are exchangeable. The number of the ballistic particles that are parallel to the shear flow and flow into the sphere (that is, $K_{ij} n_i$) can be geometrically calculated using the following equation (Friedlander 2000):

$$K_{ij} n_i = 4n_i \int_0^{\pi/2} (R_i + R_j)^3 \dot{\epsilon} \sin^2 \theta cos\theta d\theta \tag{6.35}$$

Integration of the equation results in

$$K_{ij} = \beta_{ij} \frac{4}{3} (R_i + R_j)^3 \dot{\epsilon} \tag{6.36}$$

This equation can be seen as the product of the geometric collision cross section $(R_i + R_j)^2$ and the rate of inflow of the particle by shear $(R_i + R_j)\dot{\epsilon}$. A factor 4/3 is a geometric correction produced by integration. Moreover, β_{ij} expresses the fluid-mechanical interaction of particles (bubbles) in the shear flow field and deformation of bubbles and will be 1 in the above calculation. Because deformation of bubbles can be evaluated by Ca, as explained in Sect. 6.2, β_{ij} can be evaluated as a function of Ca: $\beta_{ij}(\text{Ca})$.

Okumura et al. (2008) produced a quasi-simple shear using a double cylindrical rotary apparatus and made in-situ observation of time change of bubble texture of rhyolitic melt at 9755 °C by Spring-8. In this apparatus shear rate becomes large on the outer side of the apparatus. Consequently, it was found that because coalescence of bubbles and deformation of bubbles occurs together, the amount of strain increases and the shape of bubbles becomes complicated. This is closely related to a permeability problem described later (Sect. 6.4.3).

Process of Drainage of Melt Film Between Bubbles

A state where bubbles are closely packed as they cannot move solely is called foam. In this state, the pressure difference that drives motion of liquid in melt film between bubbles results from the surface tension. In a static state where whole foam does not move, the only force causing motion of melt is surface tension. In this case, the process of drainage of melt film is most simply dealt with using Reynolds lubrication theory (Batchelor 1967). In this case, liquid film is assumed to be a thin disk with a radius of L and a thickness of z (Figs. 6.4c, d and 6.9). The thickness of the bubble film z changes using the following equation:

$$\frac{d(z^{-2})}{dt} = \frac{4\Delta P}{3\eta \cdot L^2} \tag{6.37}$$

where driving pressure ΔP is the difference between pressure at the vicinity of the center of the liquid film disk (at which the disk intersects with a line connecting the centers of two bubbles) and that at the edge of the disk (a Plateau border: a prismatic portion formed by three bubbles). Pressure in the prismatic portion then becomes negative by bubble's curvature K_C worth of pressure, i.e.,

$$\Delta P = K_C \cdot \gamma \tag{6.38}$$

holds.

Note that although we assume foamy bubbles, such change of the shape of bubbles is attributed to the state in which some force is acting on the bubbles. For example, in the case where bubbles gather upward by buoyancy, buoyancy acts as a driving force for deformation of bubbles and flow of melt film. Moreover, in the case where magma vesiculates in a finite volume and bubbles increase in quantity, confining the pressure of the vessel and acts as the driving force. Furthermore, for the case where bubbles perform a relative motion, a driving force that moves the surrounding liquid acts as the driving force for draining the melt film. In the case where spherical bubbles grow and expand in a situation where the volume fraction of bubbles is relatively small and foam is not formed, because the radius of the disk-shaped liquid film between the bubbles L at a moment of contact of two bubbles can be effectively regarded as 0, the time required for draining liquid is infinitely reduced and can be neglected compared with the time required for approach of bubbles. It is theoretically demonstrated for draining melt film that in a region where the volume fraction is large and bubbles form foam; moreover, melt film thins the most not at the center of the melt liquid disk but a site distant from there (Traykov et al. 1977). This can be certainly recognized in scoria produced by a Strombolian eruption with low-viscosity basaltic magma (Fig. 1.4).

Integrating Eq. (6.37) with respect to z from the initial thickness z_0 to the minimum thickness $z_{\min}$ at which melt film becomes unstable because of intermolecular force and ruptures and assuming $z_0 \gg z_{\min}$, we obtain the time required for draining liquid until rupture of the film t_{fd} as follows:

$$t_{\mathrm{fd}} = \frac{3\eta \cdot L^2}{4\Delta P \cdot z_{\min}^2} \tag{6.39}$$

where L is the effective radius of liquid film disks. Moreover, because the pressure difference driving the flow of liquid in the liquid film is equal to the difference between the pressure at the Plateau border and that at a flat region of contact between two bubbles (curvature is 0), we can estimate $\Delta P = 2\gamma/L$.

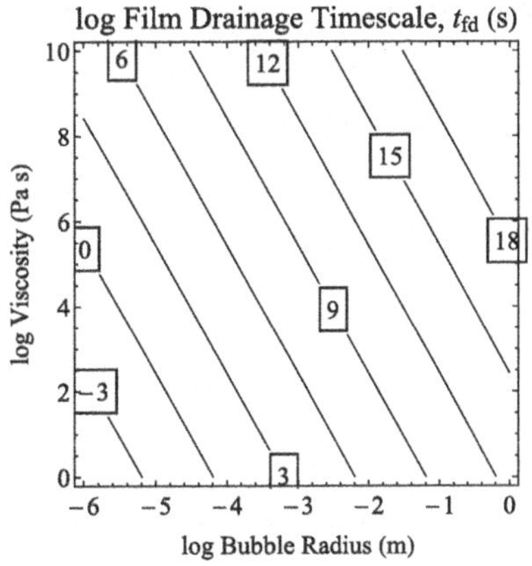

Fig. 6.8 Timescale of drainage of liquid film in foamy bubble swarm the bubble volume fraction of which is large. This is in the case where the curvature radius of the Plateau border and the radius of the liquid disk are assumed to be equal. In the case of contact of spherical bubbles, $t_{fd} = 0$ can be assumed regardless of viscosity or radius

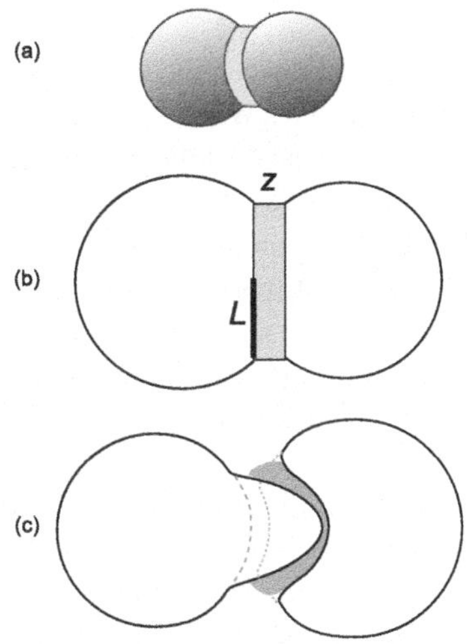

Fig. 6.9 Change of the shape of liquid film between bubbles into rupture (Adapted from Fig. 6 in Castro et al. 2012). **a** Liquid film between two bubbles. **b** Disk-like shape. **c** Dimple-like shape

We assume for simplicity that the radius of the liquid film disk and the curvature radius of the Plateau border are equal to L. Consequently, the timescale of liquid drainage is

$$t_{fd} = \frac{3\eta \cdot L^3}{4\gamma \cdot z_{min}^2} \tag{6.40}$$

t_{fd} seen as a function of L and liquid viscosity η is shown in Fig. 6.8.

In a region where the volume fraction of bubbles is small, a small hollow called a dimple is formed between the bubbles (Fig. 6.9). The dimple is formed when the difference in internal pressure between two bubbles overcomes shape preservation due to the surface tension of the liquid film. The growth of dimples is considered to be self-accelerating growth. Although this is experimentally confirmed for various materials, theoretical handling is difficult and a computational approach has been developed.

6.3.3 Shape Relaxation of Bubbles

Deformed or coalesced bubbles vary spatially in the curvature of the gas-liquid interface. Mechanical equilibrium requires that the curvature of the gas-liquid interface is constant and that the pressure

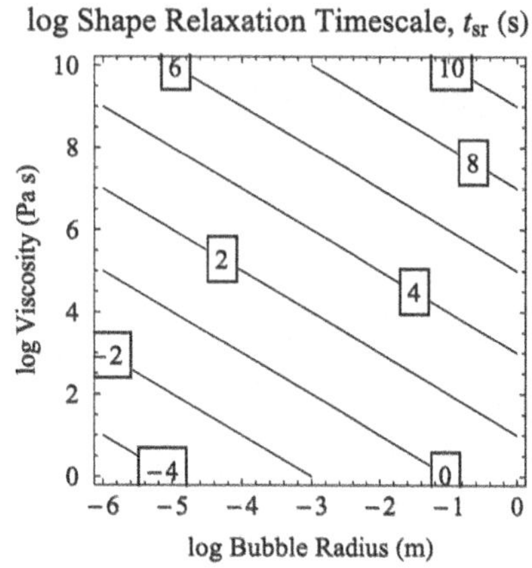

Fig. 6.10 Timescale of shape relaxation. Bubble size is taken as a scale of heterogeneity of the interface curvature. The interfacial tension is assumed to be 0.1 N/m

difference between the liquid phase and the gas phase is in balance with pressure due to the interfacial tension $2\gamma/R$. When the curvature of the interface is not spatially uniform, heterogeneous liquid pressure because of the interfacial tension (the pressure of the gas phase is considered to be quasistatic constant) drives the liquid flow. Moreover, the liquid flow is generated extremely locally, driven by local heterogeneity of the curvature.

Let us estimate the timescale of shape relaxation (Toramaru 1988). A Navier-Stokes equation in which gravity and the inertia term are ignored is $-\partial P/\partial x + \eta \partial^2 U/\partial x^2 = 0$, where x is coordinates along the interface and U is a representative flow velocity in a relaxation process. This equation represents a balance between the pressure gradient caused by heterogeneity of the interfacial tension $\partial P/\partial x$ and viscous resistance $\eta \partial^2 U/\partial x^2$. The pressure gradient is about $2\gamma/R^2$, where R is a local curvature radius of the shape of the irregular gas-liquid interface. When we assume that the timescale of shape relaxation (a process by which the curvature of the interface becomes constant) is t_{sr}, the velocity U can be written as R/t_{sr} and viscous resistance will be $\eta R/t_{sr} \cdot 1/R^2$. Consequently, we obtain

$$t_{sr} = \frac{\eta R}{\gamma} \tag{6.41}$$

as the timescale of shape relaxation.

Figure 6.10 shows the relaxation time as a function of viscosity and the bubble size R of a typical spatial scale. While in the case where heterogeneous scale is small and viscosity is small, the relaxation time is 1 s or less, in the case where the bubble size exceeds 1 mm and the viscosity exceeds 10^3 Pa s, which is assumed for basaltic melt, the shape relaxation time exceeds 1 s. A comparison between timescale of coalescence and timescale of drainage of liquid film demonstrated that drainage of liquid film is always several orders of magnitude larger. Therefore, the shape of coalesced bubbles can be considered to immediately relax and return to the sphere. However, this does not hold for the case in which an external force that causes continuous shape deformation of bubbles or gas phase acts. For example, on developing permeability of gas phase by deforming magma, coalescence and shape relaxation progress by similar timescales.

6.3.4 Temporal Development of Size Distribution in the Case Where the Initial Size Distribution Is Monodisperse

Analysis by Discrete Size Distribution

Monodisperse bubbles as the initial condition are expressed as follows:

$$n_1(0) = N_0 \tag{6.42}$$

$$n_i(0) = 0, \quad i > 1 \tag{6.43}$$

Following Friedlander (2000), let us obtain the temporal development of size distribution in the case where the initial size distribution is monodisperse. We then substitute $k = 1, 2, 3, \cdots ...$ into Eq. (6.19) as required.

When $k = 1$, the first term of the right-hand side of Eq. (6.19) is 0, and we obtain

$$\frac{dn_1}{dt} = -\frac{n_1}{t^*} \tag{6.44}$$

and solve it using the monodisperse initial condition $n_1(0) = N_0$, thus obtaining

$$n_1 = N_0 \exp\left(-\frac{t}{t^*}\right) \tag{6.45}$$

When k = 2,

$$\frac{dn_2}{dt} = \frac{n_1^2}{2\,Nt^*} - \frac{n_2}{t^*} = \frac{N_0}{2t^*}\exp\left(-\frac{3t}{2t^*}\right) - \frac{n_2}{t^*} \tag{6.46}$$

is obtained. Here, we used the solution for the total bubble number, N, (Eq. 6.21) and the solution for $k = 1$, n_1, (Eq. 6.45). The solution for $k = 2$ is

$$n_2 = N_0 \exp\left(-\frac{t}{t^*}\right)\left[1 - \exp\left(-\frac{t}{2t^*}\right)\right] \tag{6.47}$$

Solving for $k = 3, 4, 5, \cdots ..$ in the same way yields

$$n_k = N_0 \exp\left(-\frac{t}{t^*}\right)\left[1 - \exp\left(-\frac{t}{2t^*}\right)\right]^{k-1} \tag{6.48}$$

for general k.

For bubbles, as described on page 171, when we measure bubble size in units of the minimum sized bubbles, discrete size bin is defined as k (integer) multiple of the unit. Using bubbles formed in a single event of homogeneous nucleation as the minimum bubbles (that is the critical nucleus size theoretically) and the volume v as the bubble size (the initial bubble volume is assumed to be v_0) yields

$$k = \frac{v}{v_0} \tag{6.49}$$

Substituting this into Eq. (6.48) yields

$$n_v = N_0 \exp\left(-\frac{t}{t^*}\right)\left[1 - \exp\left(-\frac{t}{2t^*}\right)\right]^{(v/v_0)-1} \tag{6.50}$$

Moreover, this equation can be expressed using a frequently used exponential function as follows:

$$n_v = N_0 \frac{\exp\left(-\frac{t}{t^*}\right)}{1 - \exp\left(-\frac{t}{2t^*}\right)} \exp\left(-\frac{v}{v^*}\right) \tag{6.51}$$

where v^* is a characteristic size that defines the slope of the distribution function by its inverse number,

$$\frac{1}{v^*} = \frac{\ln\left[1 - \exp\left(-\frac{t}{2t^*}\right)\right]^{-1}}{v_0} \tag{6.52}$$

v^* increases with time.[2] I.e., it shows the characteristic that bubble size increases as coalescence progress. t^* is time characterizing progress of coalescence in terms of decrease in bubble numbers. Moreover, when $t \gg t^*$,

$$n_v = N_0 \exp\left(-\frac{t}{t^*}\right) \exp\left(-\frac{v}{v^*}\right) \tag{6.53}$$

is obtained.

These calculations of discontinuous distribution have the advantage of making it possible to easily calculate the temporal development of size distribution by assuming the initial monodispersity. However, they have the disadvantage of not including bubble growth and bubble expansion in temporal development of the size distribution despite bubbles approaching due to bubble growth and bubble expansion are assumed as elemental processes of bubble approaching. This problem can be solved to some extent by developing a theory of continuous size distribution described in the following section.

6.3.5 Analysis by Continuous Size Distribution

Smoluchowski Equation

Temporal development of size distribution function by growth and coalescence of particles has been studied in the fields of aerosol science and chemical engineering and has been given a general equation including the effect of advection by motion of the surrounding fluid (Friedlander 1977). In the field of magma, Sahagian (1985) and Sahagian et al. (1989) discussed temporal development of size distribution of bubbles in lava flow in the case where bubbles coalesce due to difference in ascent rate resulting from bubble size. In the field of physics, Meakin (1990) examined the coalescence of ascending bubble swarm and spatial distribution of bubbles by computer simulation. In a decompression field associated with magma ascent, Mancini et al. (2016) separated growth (increase in the number of molecules in bubbles) and expansion (increase in the size under invariant number of molecules in bubbles), introduced a distribution function as a function of mass and volume of bubbles, and presented a numerical model that can handle three coalescence kernels (collision because of difference in ascent rate Fig. 6.5a, b, drainage of liquid film between bubbles, and deformation of liquid film between bubbles).

[2] Although this v^* appears to diverge at $t = 0$, because v = v0 at t = 0 is obtained from examination of the original Eq. (6.48), $(1 - \exp(-t/2t^*))^{(v/v_0-1)}$ converges to 1 in the limit.

A general equation ignoring the advection effect and describing temporal development of particle size distribution function by coalescence of particles, not including particle growth, with respect to continuous size distribution is called the Smoluchowski coagulation equation[3]:

$$\frac{\partial n(v,t)}{\partial t} = \underbrace{-n(v,t)\int_0^\infty K(v,v')n(v',t)dv'}_{\text{Decrease by coalescence}} + \underbrace{\frac{1}{2}\int_0^v K(u,v-v')n(v',t)n(v-v',t)dv'}_{\text{Increase by coalescence}} \tag{6.54}$$

where $n(v,t)$ is a distribution function $[\frac{\text{number}}{\text{m}^6}]$ of particles present in the range of time from t to $t+dt$ and in the range of volume from v to$v+dv$. $K(v,v')$ is coagulation kernel $[\frac{\text{m}^3}{\text{s}}]$ and depends on elemental processes of coalescence, and in the case where it is a function of particle size (assuming that the size of interest is v and the size of particles that coalesce into it is v'), it will be included in the integral. From the coagulation kernel seen in the section on discrete size distribution, Eq. (6.17), in the case of continuous distribution, coagulation kernel can be rewritten as follows:

$$K(v,v') = \frac{1}{t^* N(t)} \tag{6.55}$$

Solution to a Model Including Growth of the Initial Monodisperse Bubbles

Substituting Eq. (6.55) into Eq. (6.54) yields the integration of the first term of the right-hand side $1/t^*$. By adding the advection term by bubble growth, the following coalescence equation is obtained.

$$\frac{dn(v,t)}{dt} + \frac{\partial}{\partial v}[G_v(v)n(v,t)] = -\frac{n(v,t)}{t^*} + \frac{1}{2N(t)t^*}\int_0^v n(v-v')n(v')dv' \tag{6.56}$$

where $G_v(v)$ is the velocity of volume growth and expansion of a single bubble (m^3/s).

Integrating this equation with respect to v from 0 to ∞ yields a differential equation with respect to the total bubble number density $N(t)$ as a function of time, as follows:

$$\frac{dN(t)}{dt} = -\frac{1}{2t^*}N(t) \tag{6.57}$$

Integrate this equation under the initial condition of monodisperse

$$n(v,0) = N_0\delta(v-v_0) \tag{6.58}$$

where N_0 is the total bubble number density (number/m^3) at $t=0$ and v_0 is the bubble size (m^3) at that time. Consequently, the following law of decrease, which is the same as a law of decrease in the number density (Eq. 6.21) obtained from analysis under the conditions of the initial monodisperse, no growth, and discontinuous distribution, is obtained.

---

[3] His is a continuous size distribution version of an equation describing temporal development of cluster size distribution describing coalescence and fragmentation, which is found in the derivation of the Fokker-Planck equation from the master equation.

$$N(t) = N_0 \exp\left(-\frac{1}{2t^*}\right) \tag{6.59}$$

Using the method of Ramabhadran et al. (1976), Ohashi obtained the following analytical solution for the case where the volume growth rate is proportional to the bubble volume ($G_v(v) = v/t_g$).

$$n(v,t) = N_0 \exp\left(-\frac{t}{t^*}\right) \sum_{i=0}^{\infty} \left\{1 - \exp\left(-\frac{t}{2t^*}\right)\right\}^i \delta\left(v - (i+1)v_0 \exp\left(\tilde{t}^* \frac{t}{t^*}\right)\right) \tag{6.60}$$

where $\tilde{t}^*$ is the ratio of timescale of coalescence t^* to timescale of growth t_g.

$$\tilde{t}^* = \frac{t^*}{t_g} \tag{6.61}$$

Equation (6.60) shows that this ratio of timescales turns out to be the only controlling parameter, and the temporal development of size distribution is found to be influenced by this value. Although this equation appears to be complicated in the term of the infinite series, in the case where the delta function is focused on, only a special i will be selected for the number of bubbles $n(v, t)$ of a certain size v at a certain time t by

$$v - (i+1)v_0 \exp\left(\tilde{t}^* \frac{t}{t^*}\right) = 0 \tag{6.62}$$

The infinite series starts from $i = 0$ in the case of this solution for continuous distribution.[4] Ultimately, we have to consider the size v as $v = kv_0 \exp(t/t_g)$ and to consider it as what results by considering extension of size scale by growth. That is,

$$n(v,t) = n(kv_0 e^{t/t_g}, t) = N_0 \exp\left(-\frac{t}{t^*}\right) \left[1 - \exp\left(-\frac{t}{2t^*}\right)\right]^{\frac{v}{v_0}\exp(-t/t_g)-1} \tag{6.63}$$

holds and measured size v is a result that bubbles experience growth and expansion. A discussion in the case without growth can be used as it is when the plot is converted into the size indicator $\hat{v}(= kv_0) = ve^{-t/t_g}$, which is obtained by multiplying by a factor e^{-t/t_g} that considers both growth and expansion, and the initial bubble number density N_0 formed by nucleation before the coalescence process can be estimated. Note that for this purpose, the volume growth time t_g has to be found in some way and it mathematically holds precisely only when the volume growth rate is proportional to the volume. This holds in the case of isothermal expansion of bubbles associated with magma decompression (Eq. 6.28).

Scaling Theory in Regard to Power Law Distribution

Gaonac'h et al. (1996a) developed a scaling argument of power law distribution often observed in nature. Power law distribution is often observed in the size distribution of bubbles in lava flow and scoria (Gaonac'h et al. 1996b, 2005). Now, we yield power law distribution as follows:

$$n(v,t) = n_0 v^{-\chi_v - 1} t^{-\chi_t} \tag{6.64}$$

For the kernel $K(v, v')$ of the coalescence equation without growth, Eq. (6.54), assuming an elemental process of coalescence caused by bubble ascending yields

[4] In the case of the discrete initial monodisperse without growth (Eq. 6.48), the infinite series starts from $k = 1$.

$$K(v, v') = \left(v^{1/3} + v'^{1/3}\right)^2 |U(v) - U(v')| \tag{6.65}$$

where U is the ascent rate of bubbles as a function of bubble size and its dependence is

$$U(v) \propto v^{\chi_U} \tag{6.66}$$

For the Stokes velocity, $\chi_U = 2/3$.

The left-hand side of the coalescence equation (6.54) will be

$$\frac{\partial n(v,t)}{\partial t} \propto v^{-\chi_v - 1} t^{-\chi_t - 1} \tag{6.67}$$

Let us evaluate the right-hand side by the same way. For this purpose, when we assume that the same power law holds for bubbles of the volume λv, in which λ is a multiplier indicating the different volume size, we obtain

$$n(\lambda v, t) = \lambda^{-\chi_v - 1} n(v, t) \tag{6.68}$$

The integral of the first term of the right-hand side I_1 is as follows:

$$I_1(\lambda v, t) = \int_0^\infty K(\lambda v, v') n(v', t) dv' \tag{6.69}$$

Introducing $v^\dagger = v'/\lambda$ for a dummy variable v' yields

$$K(\lambda v, v') = K(\lambda v, \lambda v^\dagger) \propto \lambda^{\chi_U + 2/3} K(v, v^\dagger) \tag{6.70}$$

and taking notice of $n(v', t) = n(\lambda v^\dagger, t)$ yields

$$I_1(\lambda v, t) \propto \lambda^{\chi_U - \chi_v + 2/3} I_1(v, t) \tag{6.71}$$

This indicates that $I_1(v, t) \propto v^{\chi_U - \chi_v + 2/3} t^{-\chi_t}$. Given $n(v, t)$ before the integral I_1, the first term of the right-hand side of Eq. (6.54) and the second term have the same dimension

$$v^{\chi_U - 2\chi_v - 1/3} t^{-2\chi_t} \tag{6.72}$$

Comparing this to exponent portions of dimensions of v and t in the left-hand side yields

$$\chi_v = \frac{2}{3} + \chi_U \tag{6.73}$$

$$\chi_t = 1 \tag{6.74}$$

It is found from this that the exponent of power law distribution can be related to the elemental process (differential sedimentation in this case) of coagulation kernel. If the ascent of bubbles at the Stokes velocity (which is proportional to the square of the bubble radius) is assumed, the exponent of the volume is $\chi_U = 2/3$, the exponent of the accumulated power law distribution is $-\chi_v = -4/3$, and the exponent of the differential power law distribution is $-\chi_v - 1 = -7/4$. This has been confirmed by computational simulation. Note that the above discussion assumes that power law distribution with the same exponent steadily holds. For this purpose, the bubbles of the smallest size have to be always supplied at a constant rate.

Blower et al. (2002) demonstrated using computational simulation that in the case where bubble nucleation duration is equal to or longer than the characteristic time of bubble growth, power law distribution is produced when nucleation occurs successively in melt in spaces between bubbles that have already nucleated and grown large. This can occur for heterogeneous nucleation and future experimental studies are required.

Other Distribution

It has been known that when particles of the smallest size are not supplied, multiple cases considered to undergo a coalescence process approach logarithmic normal distribution (Friedlander 2000). In this case, because the number of particles rapidly decreases because of coalescence in the size region smaller than the mean value, a bell-shaped distribution function in which the maximum value is found at around the mean value is expected. Furthermore, it be optimistically considered that in the size region larger than the mean value, steady power law distribution holds. Actually, for logarithmic normal distribution, distribution in the size region larger than the mean size can be approximated using power law distribution.

Friedlander and Wang (1966) assumed coagulation by Brownian motion and required

$$\frac{n(v,t)dv}{N(t)} = \psi\left(\frac{v}{\langle v\rangle}\right) d\left(\frac{v}{\langle v\rangle}\right) \tag{6.75}$$

for a solution to a coagulation equation without growth. The idea is that a function ψ exists by which normalized number of particles in each size width dv by the total number $N(t)$(the left-hand side) becomes equal to the number in the size width $d(v/\langle v\rangle)$ (the right-hand side) using distribution function $\psi(v/\langle v\rangle)$ to normalized scale of the size by its mean value $\langle v\rangle$ (self-preserving). They demonstrated that distribution close to logarithmic normal distribution can be obtained by obtaining analytical expression for the solution and calculating it numerically. Moreover, logarithmic normal distribution is derived by assuming proportionate growth (the law of proportionate growth, or Gibrat's law) (for coalescence, coalescence efficiency proportional to the size) (Eberl et al. 1998).

6.3.6 Comparison with Experiments

Outline of Experimental Results

Using Lipari obsidian, Larsen et al. (2004) conducted an experiment at a constant decompression rate 0.5 MPa/s and demonstrated that the number density of bubbles N decreases exponentially with respect to time (note that in the experiment investigating Ostwald ripening at a constant amount of decompression, N decreased in power function with respect to time).

$$N \propto \exp(-0.018t) \tag{6.76}$$

In this equation, the unit of time t is second. The experiment was conducted under the condition where decompression rate is constant and assuming that monodisperse initial bubbles are generated in one nucleation event, Eq. (6.59), a result of a theory of coalescence including growth using continuous distribution function, can be applied. Because 0.018, a value in front of time indicating the decrease in the bubble number density is equivalent to $1/2t^*$, timescale of coalescence of bubbles t^* is estimated at $1/0.036 = 27.8$ s.

Gardner (2007a) conducted an experiment using melt with a viscosity from 10^5 to 10^7 Pa s at a constant decompression rate in the range from 0.0064 to 0.025 MPa/s and demonstrated that the viscosity and the decompression rate influence decrease in the bubble number density. Using a moissanite cell apparatus, Masotta et al. (2014) conducted in situ observations of a coalescence process of CO_2 bubbles in the melt of different compositions (basalt, andesite, and rhyodacite) under a constant pressure and reported that the bubble number density exponentially decreases over time. Furthermore, they showed that the size distribution is exponential distribution and time constant of coalescence changes by two orders of magnitude with composition (Table 6.1).

Application of Theory to Experiments

Let us theoretically interpret the experimental result of Masotta et al. (2014). In the condition such as the experiment, growth or expansion of bubbles is considered to control approach of bubbles. As seen in Chap. 4, factors controlling growth and expansion of bubbles can be evaluated by the Péclet number (Pe), the ratio of timescale of diffusion to timescale of viscosity. In the present case, Pe is given as

$$\mathrm{Pe} = \frac{t_{\mathrm{Gdif}}}{t_{\mathrm{vis}}} = \frac{R^2 \Delta P}{4 v_{\mathrm{G}} \Delta C D \eta} \tag{6.77}$$

Although Pe may be calculated by t_{Gdif} and t_{vis} roughly estimated from Fig. 4.11, more precise estimation as they fit with the experimental condition yields values in Table 6.2. The table shows that Pe becomes larger than 1 and diffusion limits the rate of the bubble growth. In this experiment, diffusion plays an important role for approaching bubbles and actually diffusivity changes by an order of magnitude with composition and coalescence efficiency changes by an order of magnitude.

Table 6.1 Time constants of coalescence of bubbles obtained by experiments

Composition	Viscosity(Pa s)	Time constant(1/s)	Condition	References
Obsidian	$9.5\times10^4 \sim 3.2\times10^6$	1.8×10^{-2}	Constant $\lvert dP/dt \rvert$	Larsen et al. (2004)
Rhyodacite	2.51×10^5	1.02×10^{-4}	Constant P	Masotta et al. (2014)
Andesite	1.32×10^3	5.08×10^{-4}	Constant P	Masotta et al. (2014)
Andesite	3.98×10^1	1.06×10^{-3}	Constant P	Masotta et al. (2014)

Table 6.2 Estimation of the Péclet number in the experiment by Masotta et al. (2014). The solubility at 1 atm is 5.00×10^{-5} (wt%). The molecular volume v_{G} is calculated at 2×10^{-25} (m^3) from the equation of state of an ideal gas when temperature is assumed to be 1473 K. a are values given in Masotta et al. (2014). b is calculated using an equation for CO_2 solubility $C = 5 \times 10^{-10} P$ (see Sect. 2.10.1). c is calculated using Giordano et al. (2008). e is calculated using Spickenbom et al. (2010). f is calculated by assuming $R = 10^{-5}$ (m) and using an estimation for ΔC from the slope of the solubility curve

Composition	Basalt	Andesite	Rhyodacite	Remarks
Initial CO_2 (ppm)	101	289	1135	a
Initial pressure (Pa)	2.02×10^7	5.78×10^7	2.27×10^8	b
Temperature T (K)	1513	1473	1373	a
SiO_2 (wt%)	50.64	57.90	68.65	a
Viscosity η (Pa s)	3.98×10^1	1.32×10^3	2.51×10^5	c
Diffusivity D (m^2/s)	5.43×10^{-11}	1.37×10^{-11}	3.17×10^{-12}	e
Time constant (1/s)	1.06×10^{-3}	5.08×10^{-4}	1.02×10^{-4}	a
Pe	3.32×10^5	4.08×10^4	9.95×10^2	f

The timescale of coalescence by growth and expansion of bubbles, t^*, is the addition of time required for bubble approach by their growth and expansion and time required for drainage of liquid film: $t^* = t_{fd} + t_g$. As seen in the section of elemental processes of coalescence, where the volume fraction in this experiment is small and foam is not formed, the time required for drainage of liquid film can be ignored and $t^* = t_g$ is assumed. Moreover, this can be understood from the fact that coalescence efficiency is almost determined by timescale of bubble growth. Note that comparison between theory and experiments examined here is of the case where a process of relaxation from monodisperse initial bubble size distribution holds in a relatively ideal situation.

6.4 Development of Gas Permeability

6.4.1 Importance of Gas Permeability

Gas permeability has been receiving considerable attention as a factor determining whether eruption will be explosive or nonexplosive. Eichelberger et al. (1986) measured porosity and permeability of drilling core samples of Obsidian Dome, which is the largest one in Inyo rhyolite in Eastern California, and demonstrated a relationship in which permeability increases steeply with increase in porosity. Following the observation result, Jaupart and Allègre (1991) revealed that the magnitude of permeability and the ascent rate of magma at depths control the ascent rate of magma at the surface using an (open-system) gas-liquid two-phase steady flow model considering out-gassing in a 1D volcanic conduit.

First, as shown in Fig. 1.22 of Chap. 1, if magma reaches the surface without out-gassing, the vesicularity exceed 99% (Sect. 10.2.4), i.e., the magma should swell up 99 times. The expansion will be converted into the velocity to the vertical direction in a quasi-one-dimensional flow in a volcanic conduit, and the velocity will reach the velocity of sound at the surface according to a choked condition as compressible fluid. If magma behaves as a closed system for volatiles, an explosive eruption cannot avoid occurring in the range of probable initial amounts of volatiles (3–6 wt%). For nonexplosive eruptions such as lava flows and lava domes to occur, underground magma should behave as an open system and out-gassing must effectively occur.

As demonstrated by calculation by Jaupart and Allègre (1991), a parameter of permeability controls out-gassing efficiency. The velocity U of gas flowing in a certain material to the 1D x direction is proportional to ΔP, pressure gradient in its direction or the difference in pressure between both ends (the interval: Δx) (Darcy's law).[5]

$$U = \frac{k_p}{\eta}\frac{\Delta P}{\Delta x} \tag{6.78}$$

Its constant of proportionality includes permeability and gas viscosity. The dimension of permeability k_p is m^2, thus having a meaning of effective cross-sectional area of gas flow. The velocity of gas flow becomes large as the permeability becomes large, and the velocity of gas flow becomes small as the viscosity becomes large because gas becomes difficult to flow. Then, by what mechanism is the permeability of gas in vesiculating magma acquired?

In the observation by Eichelberger et al. (1986), they analyzed samples after outgassing and rock texture is a result of complex history magma experienced. Therefore, there has been uncertainty about how bubbles are connected and how permeability develops with progress of magma vesiculation and that is among the most examined topics in the current volcanology.

[5] Multiplying the velocity U, which has a dimension m/s, by the cross-sectional area gives m^3/s, which means the flux.

6.4.2 Bubble Connection in an Isotropic Field Without Flow

In a series of processes where magma is continuously decompressed and bubble nucleation, growth, and expansion progress, a process in which bubbles are connected and network of a gas phase is formed is a fundamental process for permeability development. This subsection summarizes the case without magma flow.

When bubbles of the same size are randomly distributed in a 3D space, we simply assume that geometrically overlapped bubbles are connected. This progresses in a situation where large-scale bubble coalescence occurs in space or shape relaxation of coalesced bubbles does not effectively occur. Therefore, elemental processes of connectivity development include bubble coalescence and shape (un)relaxation process.

Then, where is the threshold of the volume fraction of bubbles above which bubbles form a network connecting over the whole system? This is a problem called continuum percolation,[6] which has been investigated in the field of the percolation theory in condensed matter physics; moreover, the answer is theoretically found to be about 30% (e.g. Shant and Kirkpatrick 1971). A volume fraction at which infinite size network is formed is called a critical volume fraction and is often written as ϕ_C. It is uncertain that bubbles are actually randomly distributed, and bubbles formed by vesiculation grow into the state of nearly close packing without coalescing.

In the observation by Eichelberger et al. (1986), permeability increases steeply when the volume fraction was about 60%. Therefore, we can consider $\phi_C = 0.6$ as 0.6. A pressure at which magma that is saturated with water at 200 MPa (in which 6 wt% water is dissolved) and undergoes decompression-induced vesiculation in a closed system in equilibrium reaches the critical value of 0.6 is about 30 MPa, which is equivalent to the depth of about 1.2 km. A pressure at which the magma reaches the critical value $\phi_C = 0.3$ is 70 MPa, equivalent to the depth of about 2.8 km.

Saar and Manga (2002) conducted numerical simulation in which the shape of objects distributed in a continuous space is variously changed, such as plate-like ones and needle-like ones, and demonstrated that flat objects and needle-like objects show extremely small critical volume fraction. This means that effective volume increased because of the effect of the object shape.

Permeability is geometrically determined by connectivity, a topological property, and the minimum aperture of a path that influences the flow (a_p), and is empirically given by

$$k_p = \frac{a_p^2 \phi^n (1-\phi)^m}{b_p} \tag{6.79}$$

in general. The topological property is expressed by exponents of the term of ϕ and the constant in the denominator b_p, and the aperture size is expressed by a_p^2. Where $n = 5.5$ and $m = 0$, the equation will be the Rumpf-Gupte relation (Rumpf and Gupte 1971) and where $n = 3$ and $m = 2$, the equation will be the Carman-Kozeny relation (Rumpf and Gupte 1971). Klug and Cashman (1996) and Klug et al. (2002) analyzed pumice samples, compared relationship between gas permeability and vesicularity ϕ with existing relations of permeability, and demonstrated that permeability can be explained by the critical phenomena-like behavior predicted by a percolation theory in which $n = 2$, $m = 0$, and $\phi_C = 0.3$ and is influenced by the aperture parts.

Because these natural samples have been altered by various physical processes after the development of permeability, such as compaction by out-gassing and microfracture by cooling or impacts, it is

[6] Continuum percolation is so named because the percolation theory originally handles vertices on a discrete lattice and connection of bonds linking the vertices, but "continuum" indicates the connection in a continuous medium (Stauffer and Aharony 2014).

dangerous to directly understand how permeability increases with increase in vesicularity from the relationship between permeability and vesicularity. To resolve this difficulty, Takeuchi et al. (2008, 2009) measured the permeability of products of a decompression-induced vesiculation experiment immediately after the experiment. Consequently, the critical volume fraction was $\phi_C = 0.8$, showing that the development of permeability is more difficult than expected for natural samples. This indicates that even if the volume fraction increases, coalescence of bubbles does not progress and connection of bubbles does not develop. Note that this experimental result was of the experiment conducted under continuous decompression.

6.4.3 Bubble Connection in Shear Flow

As described in Sect. 6.3.2, bubble connection in shear flow is closely related to the bubble coalescence process. Moreover, it is closely related to bubble deformation, handled in Sect. 6.2. Because theoretical handling of such complex process is almost impossible, some researchers has started to study it by numerical simulation. Huber et al. (2014) studied rupture of liquid film between bubbles and deformation of a single bubble under simple shear using the lattice Boltzmann method, determined calculation parameters because they would agree with theoretical and experimental results, and demonstrated the effectiveness of this calculation method. Using these parameters, they investigated the behavior of multi-bubbles in a deformation field. Other limitations including the development of bubble connection will be issues to be addressed in the future. Moreover, using this calculation method, Gupta and Kumar (2008) simulated the shape of a single bubble ascending in viscous fluid (Fig. 6.15), seen later, to a certain extent.

In the experiment mentioned in Sect. 6.2, Okumura et al. (2008) investigated the connectivity of bubbles and demonstrated that the connectivity of bubbles increases as the amount of deformation becomes large. Because brittle fracture of magma occurs in a shear field because of viscoelastic characteristics of magma, the problem of bubble connection becomes more complicated. To investigate such phenomena in laboratory experiments is a future issue to be addressed.

6.5 Detachment and Ascending of Bubbles

6.5.1 Detachment of Bubbles

As seen in Sect. 3.4 for heterogeneous nucleation of bubbles, on the surface of crystals of different molecular structure from silicate melt, such as magnetite, the contact angle becomes small because of the interfacial tension and heterogeneous nucleation is easy to occur. Because heterogeneous nucleation occurs in a low degree of supersaturation, it is related to overpressurization by cooling and vesiculation in magma chambers. We then examine a mechanism and conditions by which bubbles that have heterogeneously nucleated on the horizontal surface of solid phase are detached from the surface by buoyancy.[7]

First, we calculated buoyancy acting on bubbles adhering to the horizontal surface of solid phase. Figure 6.11 shows a spherical bubble adhering to the horizontal substrate with a dihedral angle of θ'. The dihedral angle θ' has the following relationship with the gas-liquid (γ_{GL}), liquid-solid (γ_{SL}), and gas-solid (γ_{SG}) interfacial tension:

[7] Although enormous knowledge of this research has been accumulated in the field of mechanical engineering, to review it is far out of the scope of this book. Please refer to, for example, Carey (2007) if necessary.

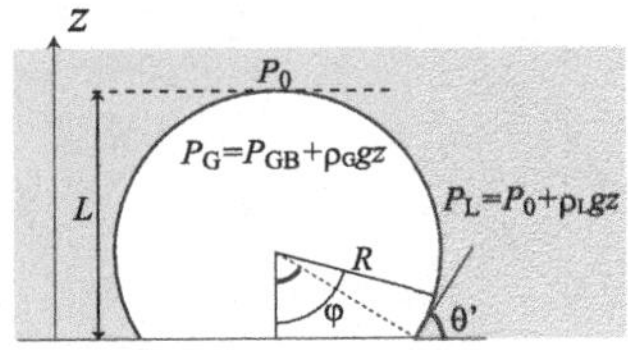

Fig. 6.11 Geometry of a bubble heterogeneously nucleating. The bubble is assumed to be a sphere with a radius of R

$$\cos\theta' = \frac{\gamma_{SG} - \gamma_{SL}}{\gamma_{GL}} \tag{6.80}$$

This is derived from horizontal force balance of the interfacial tension at the triple point (Sect. 3.4. θ in Eq. (3.95) is $\pi - \theta$). The vertical component is balanced with the strength of the solid phase. Buoyancy F_B is the integral of difference in pressure between the gas phase and the liquid phase, which acts via the gas-liquid interface, over the whole gas-liquid interface. We now set a vertical upward direction from the surface of the solid phase as the z-axis and set an angle between a line drawn from the center of a spherical bubble to the gas-liquid interface and a vertical line drawn from the center as φ, as shown in the figure. An infinitesimal area dA made by a zone along the latitude of an infinitesimal angle $d\varphi$ will be

$$dA = 2\pi R^2 \sin\varphi d\varphi \tag{6.81}$$

A force acting on the zone vertically upward at the interface from the liquid phase to the gas phase is $(P_L(z) - P_G(z))dA\cos\varphi$. Here, P_L and P_G denote liquid pressure and gas pressure, respectively, and because $P_L(z) = P_0 + \rho_L g(L-z)$ and $P_G = P_0 + \rho_G g(L-z)$ hold, $P_L(z) - P_G(z) = \Delta\rho g(L-z)$. Moreover, $L - z = R(1+\cos\varphi)$ holds. Buoyancy F_B is the integral of the vertical upward component $(P_L(z) - P_G(z))\cos\varphi dA$ from the triple point (θ') to the top of the bubble (π), that is,

$$F_B = 2\pi R^3 \Delta\rho g \int_{\theta'}^{\pi} (1+\cos\varphi)\cos\varphi\sin\varphi d\varphi \tag{6.82}$$

$$= \frac{\pi}{3} R^3 \Delta\rho g (2\cos^3\theta' + 3\cos^2\theta' - 1) \tag{6.83}$$

Linde et al. (1994). The buoyancy becomes negative when $\theta' > 60°$, i.e., when θ' is $60°$ or larger, buoyancy is not produced. This is because a force acting downward from the upward gas-liquid interface is larger than a force acting upward from the downward gas-liquid interface. For buoyancy to act on bubbles, the downward interface of bubbles has to occupy a sufficiently large proportion. It is the case where θ' is $60°$ or smaller; however, actual bubble detachment can occur when the dihedral angle θ' is $60°$ or larger.

What becomes important in practical problems is the size of bubbles at detachment. Because it is obvious that large bubbles cannot remain on the substrate, we have to know how the size of bubbles at detachment or the maximum size at which bubbles can adhere to is determined and what parameter controls it. For this purpose, let us think a detachment mechanism that the following deformation of bubbles is considered.

When a bubble grows and buoyancy becomes large, the shape of the bubble deviates from a sphere. The buoyancy acts on the direction by which the bubble is vertically extended and necking is formed at the middle part of the bubble. Once the necking is formed, pressure P_2 at the necking becomes larger than pressure P_1 at the central part of the bubble. Then, the gas is squeezed from the necking part to

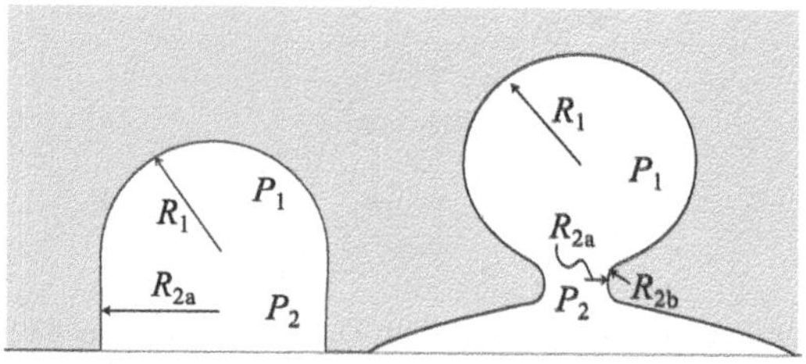

Fig. 6.12 (Left) The critical stable condition just before necking is formed. (Right) A bubble, pressure, and curvature radius when necking is formed. The shape of the main bubble in most part is a sphere with a radius of R_1. The interior radius of the necking part on the horizontal plane is R_{2a} and the curvature radius of the outside of the necking part on the vertical plane is $R_{2b} = \infty$ in the left bubble and is R_{2b} in the right bubble

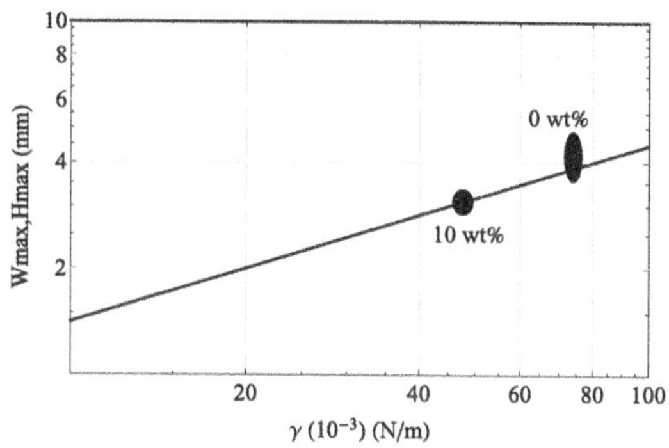

Fig. 6.13 Relationship between the maximum bubble size (width W_{max}, height H_{max}) adhering to the bedrock and the surface tension, obtained by the experiment of detachment of CO_2 bubbles in alcoholic aqueous solution. The straight line represents Eq. (6.84)

the bubble and the necking becomes increasingly narrow. In this manner, the necking part is torn off in the end and the bubble is detached. Let us examine at this phenomenon more quantitatively (Fig. 6.12).

Let us assume that gas pressure in the main bubble is determined by the Laplace equation as it is in mechanical equilibrium with the liquid pressure at the top of the bubble P_0: $P_1 = P_0 + 2\gamma/R_1$. Let us assume that gas pressure at the necking is determined as it is in mechanical equilibrium with the neighboring liquid pressure $P_0 + 2\rho_L g R_1$: $P_2 = P_0 + 2\rho_L g R_1 + \gamma(1/R_{2a} - 1/R_{2b})$. Then, difference in pressure $P_2 - P_1$, which drives the flow of gas from the necking part to the main bubble, can be calculated as $2\rho_L g R_1 - \gamma(2/R_1 - 1/R_{2a} + 1/R_{2b})$. In a state where necking is not formed, like a cylindrical pipe, because $R_{2b} = \infty$, $1/R_{2b} = 0$ is obtained and thus $P_2 - P_1 = 0$ is obtained. The instability condition is obtained by solving the quadratic expression with respect to R_1 as follows:

$$R_1^* = \left(\frac{\gamma}{2\Delta\rho g}\right)^{1/2} \tag{6.84}$$

This result shows that the size of a bubble detaching increases in proportion to the square root of the interfacial tension γ.

We conducted an experiment in which carbon dioxide is entrapped in pure water into which ethanol is dissolved and the gas-liquid interfacial tension is controlled. The surface tension is almost halved by addition of ethanol (National Research Laboratory of Metrology 1977). In the experiment conducted by changing the ethanol concentration, the bubble size just before the detachment or the maximum bubble size adhering to the bedrock was measured. Figure 6.13 shows a comparison between the experimental result and the model Eq. (6.84). As the model appears to explain the experimental result well, bubble detachment is considered to occur by the mechanism described above.

6.5.2 Bubble Ascent

Because of the balance between upward buoyancy and downward viscous resistance by the surrounding fluid, a single bubble in viscous fluid acquires a constant ascent velocity. This velocity is called a terminal velocity (U_t). For a spherical bubble, that is, when the bubble ascends with low Reynolds number (Re), steady-state ascent velocity (terminal velocity) can be expressed by the well-known Stokes velocity U_{St}:

$$U_{St} = \frac{\Delta \rho g R^2}{3\eta} \tag{6.85}$$

The coefficient differs from that of the Stokes velocity for particles because, unlike particles, friction is not produced on the surface of bubbles.

The terminal velocity greatly varies according to the shape of bubbles. When Re becomes large, the shape of bubbles deviates from a sphere and the terminal velocity varies. If the shape of bubbles is not a sphere, because the shape of bubbles and movement of the surrounding fluid are determined by mutually influencing each other, it is extremely difficult to handle them theoretically but they have been studied experimentally since long in the past.

The terminal velocity and the shape of bubbles are summarized by dimensionless numbers - the Reynolds number, the Bond number (Bo) (Eötvös number; Eo), and their combination, the Morton number (Mo)—and given as

$$\mathrm{Re} = \frac{U_t R \rho}{\eta} \tag{6.86}$$

$$\mathrm{Bo} = \frac{\rho g R^2}{\gamma} \tag{6.87}$$

$$\mathrm{Mo} = \frac{g\eta^4}{\rho\gamma^3} \approx \frac{\mathrm{Bo}^3}{\mathrm{Re}^2} \tag{6.88}$$

The relationship between the middle and the right-hand side of the last equation is obtained by assuming the Stokes velocity and ignoring coefficients. It should be noted that Mo is a dimensionless number consisting only of physical properties. Bo and Re are mutually related to each other through the physical properties. The last Mo represents the relationship and gives the coefficient: $\mathrm{Re} = \mathrm{Mo}^{-1/2}\mathrm{Bo}^{3/2}$. Figure 6.14 shows Mo as a function of viscosity. Mo for water is less than 10^{-10}. Mo for basaltic melt roughly ranges from 1 to 10^{10}, Mo for andesitic melt ranges from 10^{10} to 10^{20}, and Mo for rhyolitic melt exceeds 10^{20}.

Figure 6.15 shows the shape of bubbles as a function of Eo (Bo) and Re. The slanting contour lines represent Mo. In the region of spherical bubbles (denoted by s), the slope of contour lines for Mo is 3/2. This means that the terminal velocity is equal to the Stokes velocity and Eq. (6.88) holds. Bo

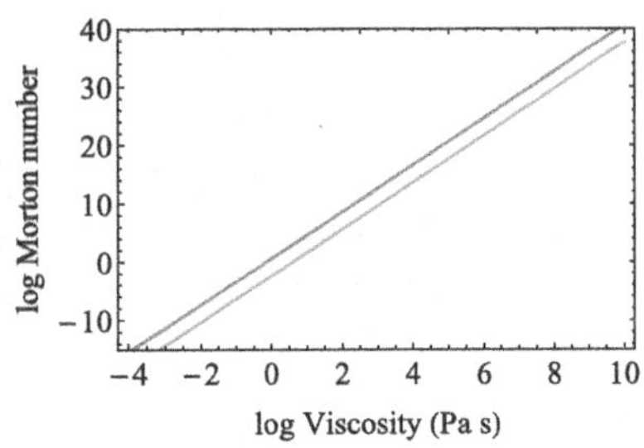

Fig. 6.14 Relationship between the viscosity and the Morton number. Difference in the density is assumed to be 2500 kg/m^3. The figure shows two examples where the surface tension is 0.1 N/m or 1 N/m

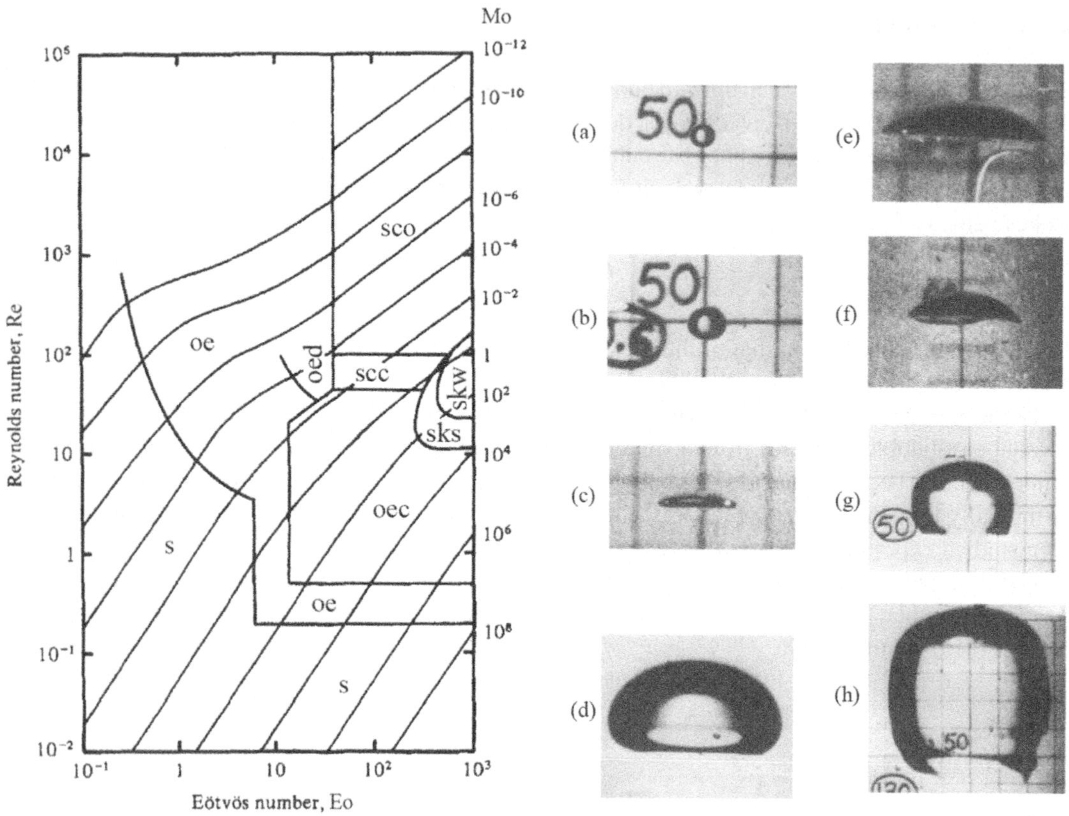

Fig. 6.15 The shape of bubbles ascending in viscous fluid with buoyancy. (Left) Phase diagram as a function of Bo (Eo) and Re. The lines represent contours of Mo (the left vertical axis). The symbols denote the shapes of the bubbles shown on the right. (Right) Typical shapes of bubbles. The squares in the background measure 2×2 cm. **a** s, spherical; **b** oe, oblate ellipsoid; **c** oed, oblate ellipsoidal (disk-like and wobbling); **d** oec, oblate ellipsoidal cap; **e** scc, spherical cap with closed, steady wake; **f** sco, spherical cap with open, unsteady wake; **g** sks, skirted with smooth, steady skirt; **h** skw, skirted with wavy, unsteady skirt. Bhaga and Weber (1981). Reprinted from Bhaga and Weber (1981), Copyright 1981, with permission from Cambridge University Press

(Eo) is estimated from Fig. 6.7 at 10^{-8} for a bubble radius of 1 μm, about 10^{-2} for 1 mm, and about 10^2 for 10 cm. In basaltic melt, the shape of bubbles changes from spherical to oblate ellipsoidal to skirted bubbles according to the bubble radius. In andesitic and rhyolitic melts, the possible range of Bo is 10^{-8}, extending to the smaller side (the left of the horizontal axis), and the range of Mo is 1020, extending to the larger side. Thus, the possible range of Re narrows to only about 10^{-10} (the lower part of the vertical axis). Therefore, bubble ascent in these melts can be considered as the ascent of almost spherical bubbles.

In the region where bubbles deviate from spheres ($\mathrm{Eo} \geq 40$, $\mathrm{Re} \geq 2$), the empirical equation below holds well between the ascent velocity U (cm/s) and the gas volume v (cm^3) (Bhaga and Weber 1981).

$$U = c_{\mathrm{Mo}} v^{\chi_{\mathrm{U}}} \tag{6.89}$$

where c_{Mo} and χ_{U} are given as a function of Mo as follows:

$$c_{\mathrm{Mo}} = 25/(1 + 0.33\mathrm{Mo}^{0.29}) \tag{6.90}$$

$$\chi_{\mathrm{U}} = 0.167(1 + 0.34\mathrm{Mo}^{0.24}) \tag{6.91}$$

6.5.3 Bubble Ascent and Advective Overpressure

Sahagian and Proussevitch (1992) demonstrated that in incompressible liquid contained in a system the volume of which cannot change, if a single bubble that was initially at the base of the system in mechanical equilibrium leaves the base and rises to the top of the system at the height of L, the pressure of the system increases by $\rho g L$. This means that the bubble that should expand cannot change its volume after moving to a position of a lower hydrostatic pressure because the liquid is incompressible and the volume of the system is constant, and to maintain the mechanical equilibrium, the pressure of the system has to be equal to the pressure of the bubble before the ascent. Although this can be experimentally confirmed, the fact is that, after the elapse of a certain period of time, the liquid or system changes its volume and accepts bubble expansion or the gas in the bubble redissolves into the liquid because of high pressure and the pressure relaxation (Bagdassarov 1994; Gro β and Peters, 2013; Pyle and Pyle 1995).

6.6 Bubble Shrinkage

There is a phenomenon called cavitation, which is related to the expansion and shrinkage of bubbles. This is a phenomenon in which cavities form in liquid under fluid-mechanical negative pressure (e.g. rotation of screw propellers) and has been considered to cause damage to the propellers by bubble collapse occurring at the moment of recovery of liquid pressure to the normal pressure. Moreover, in the case where the inertia term is effective in the expansion and shrinkage of bubbles, the oscillation phenomenon occurs. The oscillation phenomenon of bubbles considerably influences acoustic properties such as compressibility of bubble-dispersed liquid, the velocity of sound, and attenuation of pressure wave.

6.6.1 Rayleigh Collapse

Rayleigh (1917) was interested in the problem of cavitation and the sounds emitted by water boiling in a kettle. In the initial stage of boiling in a kettle, bubble generation at the bottom of the kettle in the overheating state and following bubble detachment, bubble ascent to undersaturated water, and bubble collapse there occur. Rayleigh considered that the bubble collapse at that time is the source of the sounds and calculated collapse time and increase in pressure at the time when cavities (bubbles the inside of which is a vacuum) produced by condensation in steam bubbles in water of the density ρ and liquid pressure P collapse. For Eq. (4.52) without the term of surface tension and the term of viscosity, the equation of motion with respect to the bubble radius R can be written as follows:

$$R(t)\ddot{R}(t) + \frac{3}{2}\dot{R}(t)^2 = -\frac{P}{\rho} \tag{6.92}$$

This corresponds to the case where the terms of viscosity and surface tension in the Rayleigh-Plesset equation (4.52) are ignored.

He did not start from this equation of motion but started from the following energy theory. He gave a velocity field around bubbles shrinking and collapsing from the mass conservation equation (4.53) in the same way as Eq. (4.63). Total kinetic energy for bubble shrinkage is as follows:

$$\frac{1}{2}\rho\int_R^{\infty} u^2 \cdot 4\pi r^2 dr = 2\pi\rho\dot{R}^2 R^3 \tag{6.93}$$

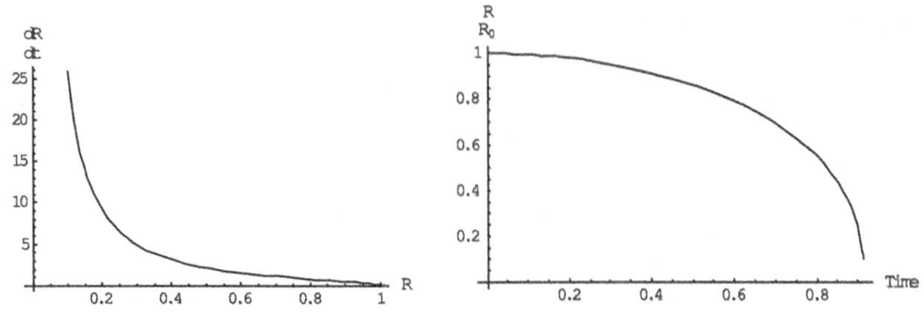

Fig. 6.16 Calculation example of the Rayleigh collapse. Calculated by assuming $R(0)(\rho/P)^{1/2} = 1$. (Left) The collapse rate $\dot{R}$ (absolute value) as a function of R. (Right) The bubble radius as a function of time. It is found that the time that elapsed before collapse becomes a coefficient of Eq. (6.96)

When a bubble with an initial radius of $R(0)$ shrinks to R, this can be seen as a result of the action of pressure P on fluid at infinity. Work done by the pressure is difference between work moved by $4\pi r^2 P$ from infinity to $R(0)$ and that from infinity to R, i.e.,

$$\int_{\infty}^{R} 4\pi r^2 \cdot P dr - \int_{\infty}^{R(0)} 4\pi r^2 \cdot P dr = \frac{4\pi P}{3}(R^3 - R(0)^3) \tag{6.94}$$

Because Eqs. (6.93) and (6.94) are equal, the change rate of the bubble radius $\dot{R}$ is obtained as follows.[8]

$$\dot{R} = -\sqrt{\frac{2P}{3\rho}\left(\frac{R(0)^3}{R^3} - 1\right)} \tag{6.95}$$

Although $\dot{R} = 0$ when $R = R(0)$, because $\ddot{R} \neq 0$, the change rate is accelerated by P_0/ρ. In the collapse of empty bubbles assumed here, because R can approach 0, the inside of the square root infinitely increases. That is, the collapse rate is found to explosively increase with decrease in the bubble radius (Fig. 6.16). The equation above is, of course, equal to the solution obtained by integrating the Rayleigh-Plesset equation (6.92) in this case. That is, it will be equal to the case where $\gamma = 0$ and $\Delta P = -P$ in Eq. (4.79). Rayleigh obtained the following result of the time that elapsed before collapse t_{col} by calculating $\int_{R(0)}^{0} \dot{R}^{-1} dR$.

$$t_{\text{col}} = R(0)\sqrt{\frac{\rho}{6P}}\frac{\Gamma(\frac{5}{6})\Gamma(\frac{1}{2})}{\Gamma(\frac{4}{3})} = 0.91468R(0)\sqrt{\frac{\rho}{P}} \tag{6.96}$$

The time for collapse of an empty bubble with a radius of 1 mm in water at 1 atm is 0.1 ms.

With increase in collapse rate, liquid pressure around bubbles explosively increases. This is because the surface area of bubbles that receives the momentum of liquid moving to the bubbles becomes infinitely small with the collapse of bubbles. Although change in pressure in liquid is described by the momentum conservation equation (4.54), because we ignore viscosity in this case, the equation will be

$$\rho\left(\frac{\partial u_r}{\partial t} + u_r\frac{\partial u_r}{\partial r}\right) = -\frac{\partial P}{\partial r} \tag{6.97}$$

Substituting the equation of the velocity field (4.63) into the equation above yields

[8] This equation corresponds to the case where we assume $\Delta P = -P$ and $\gamma = 0$ in Eq. (4.79).

$$\frac{\partial P}{\partial r} = -\rho\left\{\left(\frac{R}{r}\right)^2\left[2\frac{\dot{R}^2}{R} + \ddot{R}\right] - 2\left(\frac{R}{r}\right)^5\left[\frac{\dot{R}^2}{R}\right]\right\} \tag{6.98}$$

From this equation, it is found that the liquid pressure P formally takes on the local maximum value when $r = r_{\max}$ which satisfies

$$\left(\frac{r_{\max}}{R}\right)^3 = \frac{2\dot{R}^2}{2\dot{R}^2 + R\ddot{R}^2} = \frac{2\dot{R}^2}{\dot{R}^2/2 + \mathrm{RHRP}} \tag{6.99}$$

where RHRP is the right-hand side (= the left-hand side) of the Rayleigh-Plesset equation (4.52) and is equal to $-P/\rho$ from Eq. (6.92) in this case. Substituting the rate of the bubble radius, Eq. (6.95), into the middle part of the equation above and rewriting it gives

$$\left(\frac{r_{\max}}{R}\right)^3 = \frac{4\left[\left(\frac{R(0)}{R}\right)^3 - 1\right]}{\left(\frac{R(0)}{R}\right)^3 - 4} \tag{6.100}$$

We find from this equation that in the case where the bubble radius satisfies

$$1 < \left(\frac{R(0)}{R}\right)^3 \leq 4 \tag{6.101}$$

with respect to the initial bubble radius, the bubble radius takes on the maximum value at infinity and is equal to P_∞; when the bubble radius is smaller than $(R(0)/4)^{1/3}$, it takes on the local maximum value at the distance$r = r_{\max}$. As a result of the integral of Eq. (6.98), the distribution of liquid pressure $P(r, t)$ is calculated as follows:

$$P(r,t) = P_\infty\left[1 + \frac{1}{3}\left(\frac{R}{r}\right)\left(\left(\frac{R(0)}{R}\right)^3 - 1\right)\left(1 - \left(\frac{R}{r}\right)^3\right) - \frac{R}{r}\right] \tag{6.102}$$

(left chart of Fig. 6.17). When the bubble is larger than $(R(0)/4)^{1/3}$, the maximum pressure is equal to the initial liquid pressure P; when the bubble is smaller than $(R(0)/4)^{1/3}$, the maximum value $P_{\max}$ of liquid pressure is as follows:

$$P_{\max} = P(r_{\max}) = \left(1 + \frac{\left[\left(\frac{R(0)}{R}\right)^3 - 4\right]^{\frac{4}{3}}}{4^{\frac{4}{3}}\left[\left(\frac{R(0)}{R}\right)^3 - 1\right]^{\frac{1}{3}}}\right) P_\infty \tag{6.103}$$

(right chart of Fig. 6.17). Moreover, when R is sufficiently small, $P_{\max}$ can be approximated as follows:

$$P_{\max} = 4^{-\frac{4}{3}}\left(\frac{R(0)}{R}\right)^3 \tag{6.104}$$

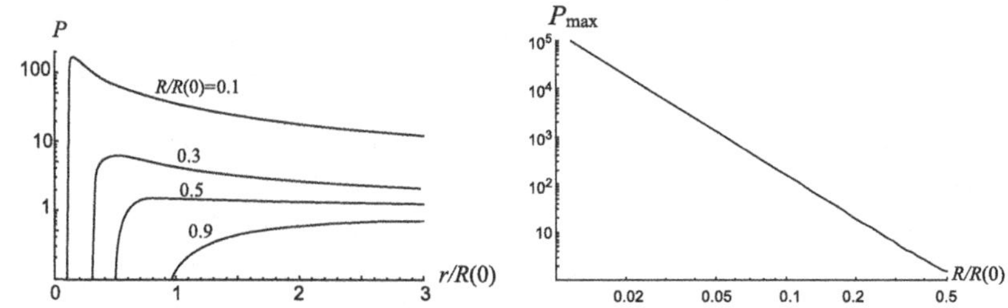

Fig. 6.17 (Left) Pressure distribution around a bubble at the Rayleigh collapse. Calculated by assuming $R(0)(\rho/P)^{1/2} = 1$. (Right) The maximum pressure as a function of the bubble radius

This indicates that the local pressure infinitely diverges as the bubble radius approaches zero. In the engineering field, this constitutes major grounds for the fact that bubble formation (cavitation) by rotation of screw propellers of ships is a cause of damage to propellers themselves.

6.6.2 The Influence of Gas in Bubbles on Bubble Shrinkage

The same discussion is applicable to the case where bubbles contain gas. When not considering the terms for the surface tension and viscosity, the Rayleigh-Plesset equation (4.52) will be as follows:

$$R(t)\ddot{R}(t) + \frac{3}{2}\dot{R}(t)^2 = \frac{1}{\rho}\left[P_G(R) - P_\infty\right] \tag{6.105}$$

Gas pressure $P_G(R)$ is a function of the bubble radius and, assuming that, the initial bubble radius is $R(0)$ and the initial gas pressure is P_{G0}, $P_G(R)$ can be written as follows:

$$P_G(R) = P_{G0}\left(\frac{R}{R(0)}\right)^{-3\gamma_h} \tag{6.106}$$

where γ_h is the ratio of specific heats and $\gamma_h = 1$ for the isothermal process and $\gamma_h = 1.33$ for the adiabatic process of steam. The rate of change in the bubble radius $\dot{R}$ as a function of the bubble radius R can be solved by the same method as the case for inviscid fluid. Consequently, change in the bubble radius in respective cases can be calculated as follows.

In the isothermal process,

$$\dot{R}^2 = \frac{2P_\infty}{3\rho}\left(\frac{R(0)}{R}\right)^3\left\{3\left(\frac{P_{G0}}{P_\infty}\right)\ln\left(\frac{R}{R(0)}\right) - \left[\left(\frac{R}{R(0)}\right)^3 - 1\right]\right\} \tag{6.107}$$

is obtained and, in the adiabatic process,

$$\dot{R}^2 = \frac{2P_\infty}{3\rho}\left(\frac{R(0)}{R}\right)^3\left\{\frac{1}{1-\gamma_h}\left(\frac{P_{G0}}{P_\infty}\right)\left[\left(\frac{R}{R(0)}\right)^{3(1-\gamma_h)} - 1\right] - \left[\left(\frac{R}{R(0)}\right)^3 - 1\right]\right\} \tag{6.108}$$

is obtained. Figure 6.18 shows a comparison of bubble shrinkage between these processes.

Although the distribution of pressure in liquid is expressed by Eqs. (6.98) and (6.99), the Rayleigh-Plesset term (RHRP) in Eq. (6.99) becomes complicated and is hard to obtain.

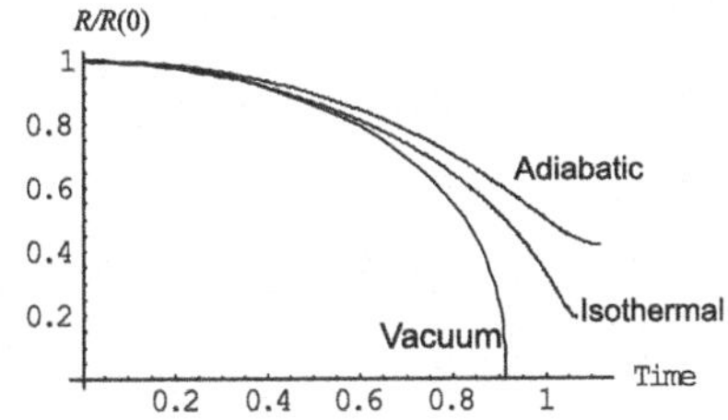

Fig. 6.18 Comparison of the change in the bubble radius at the Rayleigh collapse among the vacuum, isothermal, and adiabatic processes

6.7 Bubble Oscillation

6.7.1 The Case Where the Amount of Gas in a Bubble Is Constant

If the bubble is not empty but contains gas when a bubble shrinks, the bubble stops shrinking at a certain bubble size. Then, it rebounds, turns to expansion, and oscillates. Whether it oscillates or not depends on the magnitude of viscosity. When the liquid is water, oscillation occurs; when the liquid is magma, oscillation does not generally occur. What will happen can be obtained by numerically solving the Rayleigh-Plesset equation (4.52) (Fig. 6.19). Assuming that the initial bubble radius is $R(0)$, the characteristic timescale (t_c; obtained by changing a coefficient of Eq. 6.96, often called Rayleigh collapse time, to 1) is

$$t_c = R(0)\sqrt{\frac{\rho_l}{P_\infty}} \tag{6.109}$$

and normalizing variables as $X = R/R(0)$ and $\tilde{t} = t/t_c$, the Rayleigh-Plesset equation is expressed in the form of a simultaneous first-order ordinary differential equation as follows:

$$\frac{dX}{d\tilde{t}} = Y \tag{6.110}$$

$$\frac{dY}{d\tilde{t}} = \frac{1}{X}\left(\tilde{P}_{G0}X^{-3\gamma_h} - 1 - \frac{\Lambda}{X} - \Pi\frac{Y}{X} - \frac{3}{2}Y^2\right) \tag{6.111}$$

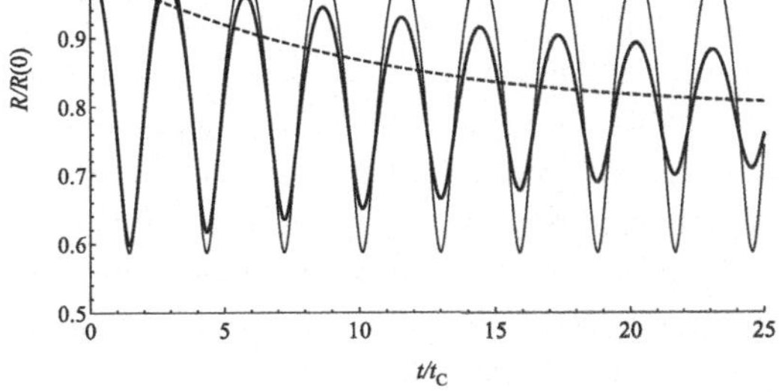

Fig. 6.19 The presence or absence of oscillation at bubble shrinkage and difference in characteristics. Thin solid line: Bubble shrinkage in water (Π (scale of the viscosity) = 0.00011, Λ (scale of the interfacial tension) = 0.0011). Thick solid line: Bubble shrinkage in picritic melt ($\Pi = 0.046$, $\Lambda = 0.00002$). Broken line: Bubble shrinkage in basaltic melt ($\Pi = 25$, $\Lambda = 0.00002$). The initial pressure in a bubble is set at 1/2 of the equilibrium pressure. Liquid pressure is 1, 100, and 100 atm, respectively. Difference in Λ between water and silicate is owing to difference in liquid pressure

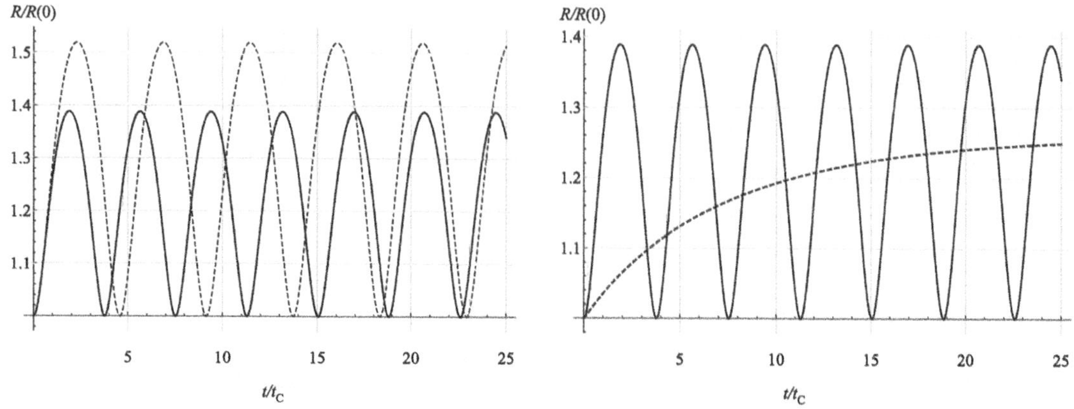

Fig. 6.20 Occurrence of oscillation at bubble expansion and difference in characteristics. (Left) Isothermal expansion (thin solid line) and adiabatic expansion (thick solid line) in the case of water. (Right) Comparison of isothermal expansion of bubbles between the case of water (solid line) and that of basaltic melt (broken line). The initial pressure in a bubble is set at twice as much as the equilibrium pressure. Other parameters are the same as Fig. 6.19

where $\tilde{P}_{G0}$, Λ, and Π are dimensionless numbers defined as follows:

$$\tilde{P}_{G0} = \frac{P_{G0}}{P_\infty} \tag{6.112}$$

$$\Lambda = \frac{2\gamma}{R(0)P_\infty} \tag{6.113}$$

$$\Pi = \frac{4\eta}{t_c P_\infty} \tag{6.114}$$

where $R(0)$ is the initial bubble radius and P_{G0} is the initial internal pressure of the bubble. These equations mean the initial mechanical disequilibrium as a driving force, contribution of the interfacial tension, and the effect of viscosity, respectively.

Moreover, bubble oscillation occurs when bubbles with overpressure expand. Assuming that the initial overpressure in bubbles is ΔP and the number of molecules of gas in bubbles is constant, an underlying equation of motion and equation of gas pressure are Eqs. (6.105) and (6.106), the same as the case of bubble shrinkage. Consequently, the rate of change in the bubble radius $\dot{R}$ for an isothermal process and an adiabatic process will be Eqs. (6.107) and (6.108), respectively.

Although temporal development of the bubble radius can be obtained by integrating these equations, when the amount of gas in the bubble is constant, the bubble stops expanding at the maximum bubble radius, rebounds and shrinks to the minimum radius, and then expands again to oscillate (Fig. 6.20). The oscillation period of the adiabatic process is slightly lesser than that of the isothermal process and the amplitude of the adiabatic process is smaller. Whether oscillation occurs or not depends on the magnitude of viscosity of liquid. When the liquid is water, oscillation occurs; when the liquid is magma, oscillation occurs in some cases but not in others according to viscosity and the bubble radius (left chart of Fig. 6.20). To study such characteristics of oscillation and the conditions for oscillation in more detail, we will conduct linear analysis of the Rayleigh-Plesset equation in the next subsection.

6.7.2 Linear Analysis of the Rayleigh-Plesset Equation

Given $R = R_{\rm eq}(1 + X')$ and $P_\infty = P_{\rm G0} - \frac{2\gamma}{R_{\rm eq}} + P'_\infty$, linearizing the Rayleigh-Plesset equation (4.52) yields

$$\ddot{X}' + 2b_{\rm v}\dot{X}' + \omega_{\rm G}^2 X' = -\frac{P'_\infty}{\rho_{\rm L} R_{\rm eq}^2} \tag{6.115}$$

Ichihara and Nishimura (2009),[9] where

$$b_{\rm v} = \frac{2\eta}{\rho_{\rm L} R_{\rm eq}^2} \tag{6.116}$$

$$\omega_{\rm G} = \frac{1}{R_{\rm eq}}\sqrt{\frac{3\gamma_{\rm h} P_{\rm G0} - 2\frac{\gamma}{R_{\rm eq}}}{\rho_{\rm L}}} \tag{6.117}$$

where $R_{\rm eq}$ is the equilibrium radius and $b_{\rm v}$ is a damping coefficient, which represents viscous dissipation with change in the bubble radius. $\omega_{\rm G}$ corresponds to the linear bulk modulus of the gas in the bubble. This is easy to understand when ignoring the term of the surface tension. $\omega_{\rm G}^2 = 3\gamma_{\rm h} P_{\rm G0}/(R_{\rm eq}^2 \rho_{\rm L})$ holds and $P_{\rm G} V^{\gamma_{\rm h}}$ =constant holds for change in volume of the bubble; thus, the modulus of elasticity of the bubble $K_{\rm G}$ is$K_{\rm G} = -(dV/(VdP_{\rm G})^{-1} = \gamma_{\rm h} P_{\rm G}$. The coefficient 3 is attributed to $V \approx R^3$. Here, assuming that a modulus of linear elasticity of gas is $k_{\rm G} = 3K_{\rm G}$, $\omega_{\rm G}^2 = k_{\rm G}/P_{\rm G0}/t_{\rm c}^2$ holds, which clarifies that this is oscillation of gas. It is found from the equation above that oscillation of a bubble forms a harmonic oscillator system in which the elasticity of pressure of the gas in the bubble serves as a spring and the surrounding liquid serves as a dashpot.

Because the oscillation frequency is $\omega_{\rm G}/2\pi$, the period of bubble oscillation becomes long (low frequency) as the bubble radius becomes large and becomes short (high frequency) as the internal pressure becomes large. Because the ratio of specific heats $\gamma_{\rm h}$ is 1.33 for the adiabatic process and can be considered as 1 for the isothermal process, the adiabatic process becomes higher frequency (shorter period) by the difference than the isothermal process. In the case where the bubble radius is larger than 1 cm and under very low pressure, the oscillation frequency is expected to be lower 100 Hz. It is found that the influence of the interfacial tension is exerted only when the bubble radius is very small, less than 1 μm. Moreover, the damping coefficient becomes large as the viscosity and the bubble radius become large. These relationships are shown in Fig. 6.21.

For harmonic oscillation with damping, the critical state in which damped oscillation shifts to overdamping occurs when $b_v = \omega_{\rm G}$. Figure 6.22 shows the boundary between damped oscillation and overdamping for water and silicate melt as a function of bubble radius and viscosity of liquid with the bubble pressure as a parameter. It is found from this figure that the surface tension is effective only at low pressure around 1 atm, when the bubble radius is 1 μm or smaller and that there is not much difference between adiabatic and isothermal processes. Moreover, in the case of water, oscillation occurs in almost all regions of bubble radius; in the case of silicate melt, oscillation occurs in basaltic melt and andesitic melt when the bubble radius is large but does not occur even in these melts if the bubble radius is small and does not occur in dacitic and rhyolitic melts. This supports the result of the numerical calculation described earlier.

[9] It should be noted that this equation is not nondimensionalized with respect to time.

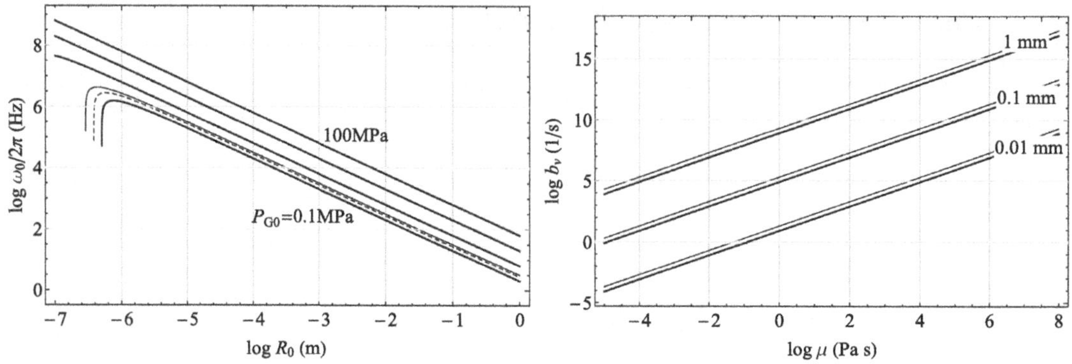

Fig. 6.21 (Left) Relationship between the oscillation frequency and the bubble radius in the case of the adiabatic process. The internal pressure of bubbles is used as a parameter. The thick lines and thin line correspond to silicate melt and water (only $P_{G0} = 0.1$ MPa is shown), respectively. The bubble pressure $P_{G0} = 0.1$, 1, 10, and 100 are shown. The broken line represents the isothermal process (only $P_{G0} = 0.1$ MPa in water is shown). (Right) Relationship between the damping coefficient and the viscosity of liquid. The bubble radius is used as a parameter. The line types indicate the same as those in the left chart

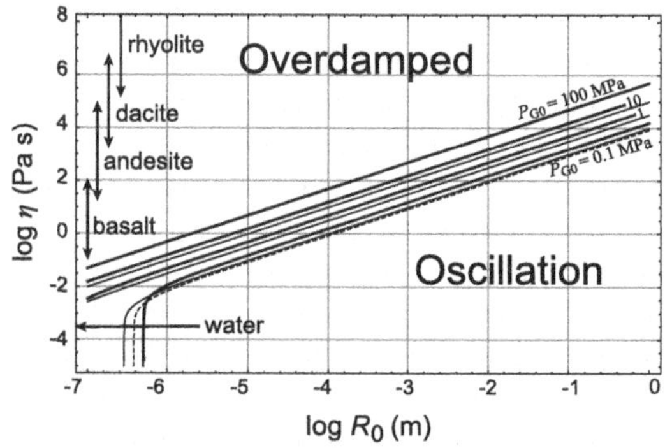

Fig. 6.22 The boundary between damped oscillation and overdamping of bubbles obtained by linear analysis of the Rayleigh-Plesset equation in the adiabatic process. The thick lines and thin lines correspond to silicate melt and water (only $P_{G0} = 0.1$ MPa is shown for water), respectively. The bubble pressure $P_{G0} = 0.1$, 1, 10, and 100 MPa are shown. The broken line represents the isothermal process. There is not much difference in the isothermal process

6.8 Influence of Viscoelasticity of Liquid on Collapse and Oscillation of Bubbles

Fogler and Goddard (1970) conducted a linear analysis of Old royd viscoelastic liquid on the basis of Eq. (4.135), an extended Rayleigh-Plesset equation. To investigate collapse of an empty bubble, assuming that $P_G = 0$, they rewrote Eq. (4.135) using $\tilde{R} = X_e + X'$ and linearized it, yielding

$$\ddot{X}' + \frac{4\eta t_c}{\rho_1 X_e^2}\dot{X}' - \frac{4K_L}{P_0}\left[\frac{\ln X_e^3}{X_e^3(1 - X_e^3)}\right]X' = 0 \tag{6.118}$$

where X_e is the equilibrium bubble radius and satisfies $X_e < 1$. Because the third term becomes positive, this equation takes the form of harmonic oscillator with the damping term. Comparison between Eqs. (6.118) and (6.115) revealed that—note that Eq. (6.115) is not nondimensionalized with respect to time—they are different only in that while the term of characteristic oscillation in Eq. (6.115) is basically the elasticity (bulk modulus) of the gas in the bubble K_G, the corresponding term in Eq. (6.118) is the elasticity of liquid K_L. This characteristic angular frequency ω_L is written as follows:

$$\omega_L^2 = -\frac{4K_L}{P_0 t_c^2}\left[\frac{\ln X_e^3}{X_e^3(1-X_e^3)}\right] \tag{6.119}$$

Using this equation,

$$\ddot{X}' + 2b_v\dot{X}' + \left(\omega_G^2 + \omega_L^2\right)X' = 0 \tag{6.120}$$

is given as an equation of linear oscillation for the case of gas bubbles in the form nondimensionalized with respect to time. Because K_L is about two orders of magnitude larger than K_G in general, the characteristic frequency of the case where the elasticity of liquid is more effective will be one order of magnitude larger if the bubble radius is the same. More detailed discussion was conducted by Ichihara et al. (2004) and Ichihara and Kameda (2004).

References

Acrivos A, Lo TS (1978) Deformation and breakup of a single slender drop in an extensional flow. J Fluid Mech 86(4):641–672

Bagdassarov N (1994) Pressure and volume changes in magmatic systems due to the vertical displacement of compressible materials. J Volcanol Geotherm Res 63(1–2):95–100

Batchelor GK (1967) An introduction to fluid dynamics. Cambridge University Press

Bhaga D, Weber ME (1981) Bubbles in viscous liquids: shapes, wakes and velocities. J Fluid Mech 105:61–85

Blower JD, Keating JP, Mader HM, Phillips JC (2002) The evolution of bubble size distributions in volcanic eruptions. J Volcanol Geotherm Res 20:1–23

Carey VP (2007) Liquid vapor phase change phenomena: an introduction to the thermophysics of vaporization and condensation processes in heat transfer equipment. CRC Press

Castro JM, Burgisser A, Schipper CI, Mancini S (2012) Mechanisms of bubble coalescence in silicic magmas. Bull Volcanol 74(10):2339–2352

Eberl DD, Drits VA, Srodon J (1998) Deducing growth mechanisms for minerals from the shapes of crystal size distributions. Am J Sci 298(6):499–533

Eichelberger JC, Carrigan CR, Westrich HR, Price RH (1986) Non-explosive silicic volcanism. Nature 323(6089):598

Eshelby JD (1957) The determination of the elastic field of an ellipsoidal inclusion, and related problems. Proc R Soc Lond A 241(1226):376–396

Fogler HS, Goddard JD (1970) Collapse of spherical cavities in viscoelastic fluids. Phys Fluid 13:1135–1141

Friedlander SK (1977) Smoke, dust and haze. Fundamentals of aerosol behaviour. John Wiley and Sons

Friedlander SK (2000) Smoke, dust, and haze, vol 198. Oxford University Press, New York

Friedlander SK, Wang CS (1966) The self-preserving particle size distribution for coagulation by Brownian motion. J Colloid Interface Sci 22(2):126–132

Gaonac'h H, Stix J, Lovejoy S (1996a) Scaling effects on vesicle shape, size and heterogeneity of lavas from Mount Etna. J Volcanol Geotherm Res 74:131–153

Gaonac'h H, Lovejoy S, Stix J, Schertzer D (1996b) A scaling growth model for bubbles in basaltic lava flows. Earth Planet Sci Lett 139:395–409

Gaonac'h H, Lovejoy S, Schertzer D (2005) Scaling vesicle distributions and volcanic eruptions. Bull Volcanol 67:350–357

Gardner JE (2007) Bubble coalescence in rhyolitic melts during decompression from high pressure. J Volcanol Geotherm Res 166:161–176

Giordano D, Russell JK, Dingwell D (2008) Viscosity of magmatic liquids: A model. Earth Planet Sci Lett 271:123–134

Groβ TF, Peters F (2013) The pressure in an enclosed volume of liquid in case of a rising bubble. Acta Mechanica 224(8):1685–1694

Gupta A, Kumar R (2008) Lattice Boltzmann simulation to study multiple bubble dynamics. Int J Heat Mass Transfer 51(21–22):5192–5203

Hinch EJ, Acrivos A (1980) Long slender drops in a simple shear flow. J Fluid Mech 98:305–328

Huber C, Su Y, Nguyen CT, Parmigiani A, Gonnermann HM, Dufek JA (2014) New bubble dynamics model to study bubble growth, deformation, and coalescence. J Geophys Res Solid Earth 119(1):216–239

Ichihara M, Kameda M (2004) Propagation of acoustic waves in a visco-elastic two-phase system: influences of the liquid viscosity and the internal diffusion. J Volcanol Geotherm Res 137:73–91

Ichihara M, Okunitani H, Ida Y, Kameda M (2004) Dynamics of bubble oscillation and wave propagation in viscoelastic liquids. J Volcanol Geotherm Res 129:37–60

Ichihara M, Nishimura T (2009) Pressure impulses generated by bubbles interacting with ambient perturbation. In: Lee WHK (ed) Complexity in earthquakes, tsunamis and volcanoes and forecasting and early warning of their hazards, encyclopedia of complexity and system science. Springer, pp 6955–6977

Jackson NE, Tucker CL III (2003) A model for large deformation of an ellipsoidal droplet with interfacial tension. J Rheol 47(3):659–682

Jaupart C, Allègre CJ (1991) Gas content, eruption rate and instabilities of eruption regime in silicic volcanoes. Earth Planet Sci Lett 102:413–429

Kingery WD, Bowen HK, Uhlmann DR (1976) Introduction to ceramics. Wiley, New York

Klug C, Cashman KV (1996) Permeability development in vesiculating magmas: implications for fragmentation. Bull Volcanol 58:87–100

Klug C, Cashman K, Bacon C (2002) Structure and physical characteristics of pumice from the climactic eruption of Mount Mazama (Crater Lake). Oregon Bull Volcanol 64:486–501

Larsen JF, Gardner JE (2000) Experimental constraints on bubble interactions in rhyolite melts: implications for vesicle size distributions. Earth Planet Sci Lett 180:201–214

Larsen JF, Denis M-H, Gardner JE (2004) Experimental study of bubble coalescence in rhyolitic and phonolitic melts. Geochim Cosmochim Acta 68:333–344

Lautze NC, Sisson TW, Mangan MT, Grove TL (2011) Segregating gas from melt: an experimental study of the Ostwald ripening of vapor bubbles in magmas. Contrib Mineral Petrol 161(2):331–347

Lifshitz IM, Slyozov VV (1961) The kinetics of precipitation from supersaturated solid solutions. J Phys Chem Solids 19:35–50

Linde AT, Sacks IS, Johnston MJS, Hill DP, Billharm RG (1994) Increased pressure from rising bubbles as a mechanism for remotely triggered seismicity. Nature 371:408–410

Mancini S, Forestier-Coste L, Burgisser A, James F, Castro J (2016) An expansion-coalescence model to track gas bubble populations in magmas. J Volcanol Geotherm Res 313:44–58

Manga M, Stone HA (1994) Interactions between bubbles in magmas and lavas: effects of bubble deformation. J Volcanol Geotherm Res 63(3–4):267–279

Masotta M, Ni H, Keppler H (2014) In situ observations of bubble growth in basaltic, andesitic and rhyodacitic melts. Contrib Mineral Petrol 167(2):976

Meakin P (1990) Coalescence of drifting droplets and bubbles in two- and three-dimensional space. Phys Rev A 42:4678–4687

Ohashi M, Ichihara M, Toramaru A (2018) Bubble deformation in magma under transient flow conditions. J Volcanol Geotherm Res 364:59–75

Okumura S, Nakamura M, Tsuchiyama A, Nakano T, Uesugi K (2008) Evolution of bubble microstructure in sheared rhyolite: formation of a channel-like bubble network. J Geophys Res Solid Earth 113(B7). https://doi.org/10.1029/2007JB005362

Proussevitch AA, Sahagian DL, Kutolin VA (1993) Stability of foams in silicate melts. J Volcanol Geotherm Res 59:161–178

Pyle DM, Pyle DL (1995) Bubble migration and the initiation of volcanic eruptions. J Volcanol Geotherm Res 67:227–232

Ramabhadran TE, Peterson TW, Seinfeld JH (1976) Dynamics of aerosol coagulation and condensation. AIChE J 22(5):840–851

Rayleigh, Lord (1917) On the pressure developed in a liquid during the collapse of a spherical cavity. Philos Mag J Sci 34:94–98

Rumpf H, Gupte AR (1971) Einflusse der porositat und korngrossenvertellung in widerstandsgesetz der perenstromung. Chemie-Ing Tehc 43:367–375

Rust AC, Manga M (2002) Bubble shapes and orientations in low Re simple shear flow. J Coll Int Sci 249:476–480

Saar MO, Manga M (2002) Continuum percolation for randomly oriented soft-core prisms. Phys Rev E 65:056131

Sahagian D (1985) Bubble migration and coalescence during the solidification of basaltic lava flows. J Geol 93(2):205–211

Sahagian DL, Proussevitch AA (1992) Bubbles in volcanic systems. Nature 359:485

Sahagian DL, Anderson AT, Ward B (1989) Bubble coalescence in basalt flows: comparison of a numerical model with natural examples. Bull Volcanol 52(1):49–56

Shante VKS, Kirkpatrick S (1971) An introduction to percolation theory. Adv Phys 20:325–357

Spickenbom K, Sierralta M, Nowak M (2010) Carbon dioxide and argon diffusion in silicate melts: insights into the CO_2 speciation in magmas. Geochim Cosmochim Acta 74:6541–6564

Stauffer D, Aharony A (2014) Introduction to percolation theory: revised, 2nd edn. CRC Press

Stewart CW (1995) Bubble interaction in low-viscosity liquids. Int J Multiph Flow 21(6):1037–1046

Takeuchi S, Nakashima S, Tomiya A (2008) Permeability measurements of natural and experimental volcanic materials with a simple permeameter: toward an understanding of magmatic degassing processes. J Volcanol Geotherm Res 177(2):329–339

Takeuchi S, Tomiya A, Shinohara H (2009) Degassing conditions for permeable silicic magmas: Implications from decompression experiments with constant rates. Earth Planet Sci Lett 283(1–4):101–110

Taylor GI (1932) The viscosity of a fluid containing small drops of another fluid. Proc R Soc A Lond 138:41–48

Taylor GI (1934) The formation of emulsions in definable fields of flow. Proc R Soc A Lond 146:501–523

Toramaru A (1988) Formation of propagation pattern in two-phase flow systems with application to volcanic eruptions. Geophys J 95(3):613–623

Traykov TT, Manev ED, Ivanov IB (1977) Hydrodynamics of thin liquid films. Experimental investigation of the effect of surfactant on the drainage of emulsion films. Int J Multiph Flow 3:485–494

Weaire D, Hutzler S (2000) The physics of foams, Oxford University Press

Wetzel ED, Tucker CL (2001) Droplet deformation in dispersions with unequal viscosities and zero interfacial tension. J Fluid Mech 426:199–228

Part III
Crystallization of Magma

Rock texture is an important indicator to determine the names of igneous rocks, which are classified according to the colors and grain size of rocks. The rock color is primarily determined by its chemical compositions—basic rocks (poor in SiO_2) are often blackish and acidic rocks (rich in SiO_2) are often Whitish—the color reflects magma formation and histories of change in compositions. On the other hand, crystal grain size, or more generally, rock texture is determined by crystallization kinetics in magma and disequilibrium processes such as motion as multiphase flow and a cooling and solidification process.

Both chemical composition and rock texture are attributed to the accumulation of multistep processes. To date, for chemical composition, because relatively solid foundations for equilibrium (including phase equilibrium relationships and the distribution of elements) have been established and because recent rapid advances in technology have improved the measurement limits of microconcentrations of trace elements and isotopic compositions in a microspace, many studies have been performed; moreover, many, not to say almost all, petrologists have been involved in these studies.

In comparison, despite its importance, there are very few researchers examining rock texture. One reason is that the disequilibrium process is the foundation of rock texture and the conclusions drawn from it are impractical. Moreover, a general theory of disequilibrium thermodynamics has not yet been established, and, when individual problems are considered, there are multiple other problems that need to be experimentally verified. Furthermore, another reason may be that hardly any of the relational expressions by which we can quantitatively determine the unknown quantities we want to determine from observables, i.e., observation tools, have been produced from disequilibrium process-related studies. For crystallization-related disequilibrium processes, studies of the classical theory, which are relatively firmly established, and its experimental verification have been conducted; therefore, creating tools for observing rock texture has become possible.

In Part III, regarding magma crystallization, we begin in Chap. 7 by quantitatively examining cooling crystallization based on equilibrium and the classical theory of kinetics and comparison with experiments. By developing our understanding of cooling crystallization, we explain decompression-induced crystallization in Chap. 8, which has been uniquely developing in volcanology and handles conditions involving considerable changes in pressure. In these two chapters, to understand the process of texture formation, we use crystal number density as a quantitative index for rock texture. In Chap. 9, we will explain the theoretical handling of crystal size distribution (CSD), which includes more comprehensive information on rock texture. In each chapter, comparisons between theories and experiments are conducted to provide materials for exploring the future direction of studies.

Cooling Crystallization of Magma

7

Abstract

In this chapter, we will discuss the kinetics of the cooling crystallization of magma, i.e., nucleation and growth, to use it as a foundation for making tools for observing unknown information from crystallization texture similar to the vesiculation processes (equilibrium conditions Chap. 2, nucleation process Chap. 3, growth process Chap. 4, temporal development of vesiculation Chap. 5, and complex processes related to vesiculation Chap. 6) dealt with in previous chapters. In this chapter, for crystallization-related problems, we will explain the similar contents in one go because both vesiculation and crystallization are the formation of a new phase from solution or a phase transition process and have similarity. First, using a binary eutectic system as an example of a multi-component system, we will explain crystallization-related conditions from a liquid and define a solid—liquid equilibrium as well as supersaturation in terms of thermodynamics. Next, we will conduct the formulation of crystal nucleation and crystal growth, evaluate experiments using these results, and examine the classical theory. As the crystal growth-related problems, the compositional zoning that is often observed in crystals is discussed for simplified cases, involving the movement of the interface, which has been extensively studied in metallurgy.

7.1 Thermodynamics of Cooling Crystallization

7.1.1 Melting Points of Crystals and Equilibrium Phase Diagrams

Equilibrium phase diagrams show phases that exist in equilibrium states, e.g., in the equilibrium phase diagram of the diopside—anorthite binary eutectic system shown in the right chart of Fig. 7.1, if a temperature and composition of a certain system is located in the L + Di region, the system comprises two phases: the liquid phase (L) and diopside phase (Di). In the binary system (liquidus), the boundary between the L + Di region and the L region on the higher temperature side shows melting points. The leftmost, Di = 100% point on the liquidus, represents the melting point of the pure substance; we can read it as 1390 °C from the figure. Because the anorthite (An) component increases (moving to the

A. Toramaru, *Vesiculation and Crystallization of Magma*, Advances in Volcanology,
https://doi.org/10.1007/978-981-16-4209-8_7

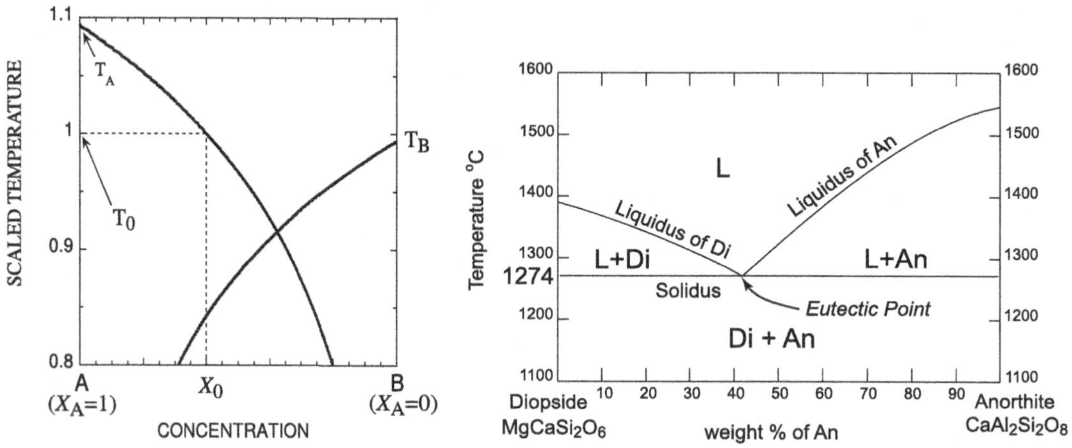

Fig. 7.1 (Left) An equilibrium phase diagram of liquidus of each solid phase in a binary eutectic system to investigate cooling crystallization. (Right) An equilibrium phase diagram of a diopside—anorthite binary eutectic system at 1 atm (Bowen 1915). L: Liquid, Di: Diopside, An: Anorthite

right on the horizontal axis), the liquidus moves to the lower temperature side; it corresponds to the freezing point depression. The eutectic point is the point at which the liquidus of Di and the liquidus of An intersect, whereas solidus is the line through the eutectic point and parallel to the horizontal axis (showing a constant temperature). A multi-component system is unlike a one-component system in which melting or solidification occurs in a temperature range between solidus and liquidus. Equilibrium phase diagrams inform us that both phase existing in equilibrium and the ratio of quantity of phases by the level rule. While equilibrium phase diagrams are drawn using theoretical considerations in terms of thermodynamics, they have been established only after the experimental confirmation of the presence of phases and determination of physical properties such as the latent heat of melting.

As shown in the left chart of Fig. 7.1, in the following theoretical consideration of crystallization, an equilibrium phase diagram of a eutectic system is assumed. In the right chart of Fig. 7.1, this can be considered to be representative of the diopside—anorthite system. Moreover, for relatively mafic magma from basaltic to andesitic, diopside (an end member of pyroxene solid solution) and anorthite (an end member of plagioclase solid solution) are always two major constituent minerals. The rock texture of intrusions is named from the geometric relation between plagioclase and pyroxene (Chap. 10). Therefore, the consideration of the diopside—anorthite system is generally applicable.

Equilibrium conditions through a flat interfacefor two states that are different in temperature, state I (temperature T_0 and molar fraction X_0) and state II (temperature T and molar fraction X_0), (Figs. 7.2 and 7.3) are expressed as follows:

$$\mu^{\mathrm{A}}_{\mathrm{Liquid}}(T_0, P, X_0) = \mu^{\mathrm{A}}_{\mathrm{Solid}}(T_0, P) \tag{7.1}$$

$$\mu^{\mathrm{A}}_{\mathrm{Liquid}}(T, P, X_{\mathrm{eq}}) = \mu^{\mathrm{A}}_{\mathrm{Solid}}(T, P) \tag{7.2}$$

using chemical potential. $\mu^{\mathrm{A}}_{\mathrm{Liquid}}$ and $\mu^{\mathrm{A}}_{\mathrm{Solid}}$ denote chemical potential per molecule of the component A in the liquid phase and the solid phase, respectively. As shown in Sect. 2.7, we assume the chemical potential of an ideal solution $\mu = \mu_0 + k_{\mathrm{B}}T \ln X$, subtract Eq. (7.2) from Eq. (7.1), and assume that $T = T_0 + dT$ and then integrate the obtained equation using $\partial\mu/\partial T = -s$ (s is molecular entropy) to obtain

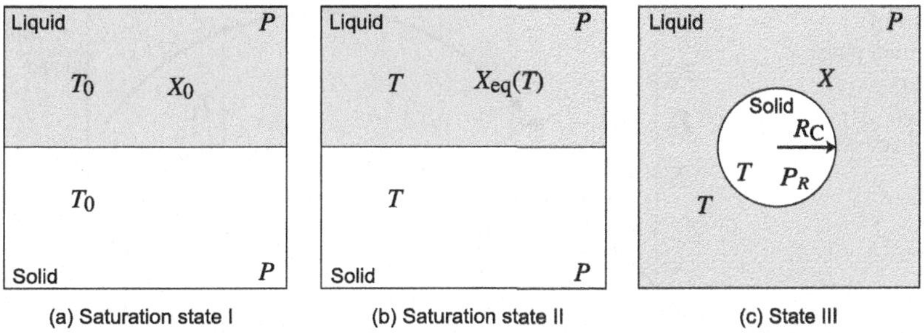

Fig. 7.2 **a** Equilibrium state through a flat interface at temperature T_0 (Saturated state I). **b** Equilibrium state through a flat interface at temperature T different from T_0 (Saturated state II). **c** Local interfacial equilibrium state through a curved interface (State III). Note that while pressure of the liquid phase and the solid phase in State I and State II is equal to P, solid pressure in State III is different, P_R, according to the curvature radius R. Moreover, while State I and State II are assumed to have a homogeneous system, in State III only melt at the interface is focused on and the entire melt has higher concentration than the composition in the saturated state at the pressure P and temperature T, i.e., State III is in a supersaturated state. Each state is indicated in the phase diagram in Fig. 7.3

$$\frac{X_{eq}}{X_0} = \frac{C_{eq}}{C_0} = \exp\left[-\frac{\Delta s_A}{k_B}\left(\frac{T_0}{T} - 1\right)\right] \tag{7.3}$$

which is similar to Eq. (2.66). This equation defines the liquidus, which is the phase boundary between the region in which liquid and solid phase coexist (L + A), as well as the region in which only liquid exists (L) in the temperature—composition space. For this equation, Δs_A is the entropy of dissolution of the pure solid phase A into the liquid of composition X_0 at temperature T_0 and is related to latent heat (per molecule) of pure A at its melting point T_A. Moreover, $\Delta h_A^0 = \Delta s_A^0 T_A$,[1] where Δs_A^0 is change in entropy associated with melting (solidification) of the pure substance A at temperature T_A. In the left chart of Fig. 7.1, $\Delta s_A^0/k_B = \Delta s_B^0/k_B = 6.0$, and $T_A/T_B = 1.1$ are assumed.

Writing $\Delta s_A T_0$, equivalent to the latent heat at temperature T_0, as simply Δh and assuming that it is constant, liquidus $T_L(C)$ can be rewritten from Eq. (7.3) as a relative relation from a reference point (C_0, T_0) on the liquidus as follows:

$$T_L(C) = \frac{T_0}{1 - \frac{k_B T_0}{\Delta h}\ln\frac{C}{C_0}} \tag{7.8}$$

Linearizing this equation with respect to concentration C will be convenient.

1

$$\begin{aligned}
\Delta s_A &= \Delta s_A(T_0, X_0) = s^A_{Liquid}(T_0, X_0) - s^A_{Solid}(T_0, 1) & (7.4)\\
&= s^A_{Liquid}(T_0, X_0) - s^A_{Liquid}(T_0, 1) + s^A_{Liquid}(T_0, 1) - s^A_{Solid}(T_0, 1) & (7.5)\\
&= \frac{\partial s^A_{Liquid}}{\partial X}(X_0 - 1) + \Delta s_A(T_0, 1) & (7.6)\\
&= \Delta s_A(T_A, 1) + \frac{\partial s^A_{Liquid}}{\partial X}(X_0 - 1) + \frac{\partial \Delta s_A}{\partial T}(T_0 - T_A) & (7.7)
\end{aligned}$$

Because $\Delta s_A(T_A, 1) = \Delta s_A^0$ holds, $\Delta s_A = \Delta s_A^0$ can be considered to hold when the correction term for composition in the second term of the right-hand side and the correction term for temperature in the third term are small.

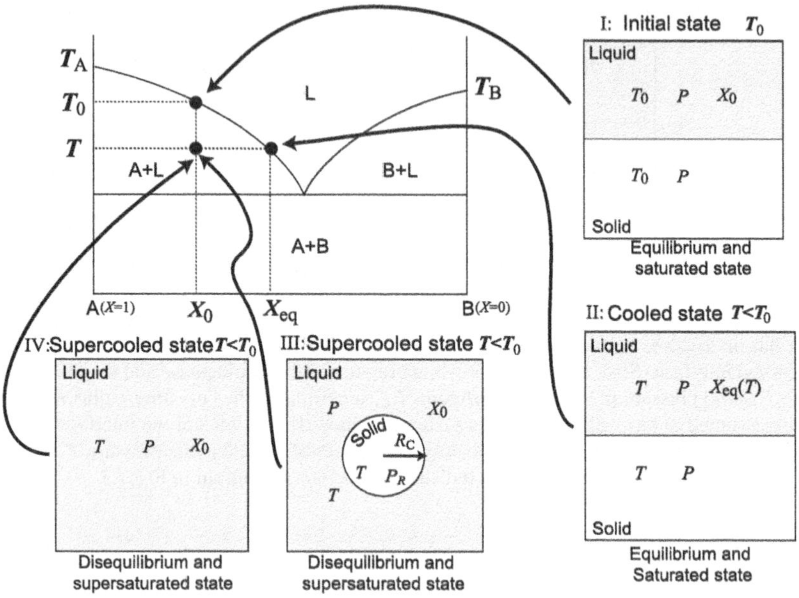

Fig. 7.3 Positions of respective states on a phase diagram of a binary eutectic system for thermodynamic discussion

$$T_L(C) = T_0\left[1 + \frac{k_B T_0}{\Delta h}\left(\frac{C}{C_0} - 1\right)\right] \tag{7.9}$$

Here, if a dimensionless number

$$\alpha_2 = \Delta\tilde{h} = \frac{\Delta h}{k_B T_0} = \frac{\Delta s}{k_B} \tag{7.10}$$

which defines the slope of liquidus, is defined, Eq. (7.9) can then be rewritten as follows:

$$T_L(C) = T_0\left[1 + \frac{1}{\alpha_2}\left(\frac{C}{C_0} - 1\right)\right] \tag{7.11}$$

7.1.2 Thermodynamic Discussion of Crystal Nuclei and Gibbs-Thomson Relation

As per Figs. 7.2 and 7.3, we perform the thermodynamic discussion of crystallization. The purpose is to understand the degree of supersaturation and to derive a critical radius. Because the critical solid phase with a critical radius is in chemical equilibrium with a supersaturated solution, the critical nucleus size can be determined from chemical equilibrium (see Box in Sect. 7.1.2 for relationship between the critical nucleus size and energy theory of nucleation). We examine a saturated state I of crystals and liquid through a plane interface with zero curvature at T_0 and an equilibrium state III through a curved interface in undercooled state at a temperature T that is lower than T_0. Because pressure is different through the curved interface, the equilibrium conditions are given as follows:

$$\mu^A_{\text{Liquid}}(T, P, X_0) = \mu^A_{\text{Solid}}(T, P_R) \tag{7.12}$$

To determine the critical nucleus size corresponding to the degree of undercooling $\Delta T = T_0 - T > 0$, we determine the difference in chemical potential between the saturated state I shown in Fig. 7.2 and the undercooled state III.

We assume that molar fraction corresponding to a reference concentration is X_0 and subtract Eq. (7.12) from Eq. (7.1), thus yielding

$$\mu^{A}_{\text{Liquid}}(T_0, P, X_0) - \mu^{A}_{\text{Liquid}}(T, P, X_0) = \mu^{A}_{\text{Solid}}(T_0, P) - \mu^{A}_{\text{Solid}}(T, P_R) \tag{7.13}$$

$\partial\mu/\partial P = v$ (v is a partial molecular volume) to Eq. (7.13), we obtain

$$-s^{A}_{\text{Liquid}}(X_0)(T_0 - T) = -s^{A}_{\text{Solid}}(T_0 - T) + v^{A}_{\text{Solid}}(P - P_R) \tag{7.14}$$

Rewriting this equation using an equation of mechanical equilibrium with respect to the curvature radius R of the solid phase (the Laplace equation)

$$P_R - P = \frac{2\gamma}{R}, \tag{7.15}$$

[2] we obtain

$$\Delta h\left(1 - \frac{T}{T_0}\right) = \frac{2\gamma v_S}{R} \tag{7.16}$$

where v_S is a molecular volume v^{A}_{Solid} of the crystallized solid phase, which is assumed to be a constant that is independent of pressure.

Saturation, supersaturation, undercooling, equilibrium, and disequilibrium

Saturation means that the concentration of liquid is on a saturation curve or equilibrium curve at the temperature or pressure of the liquid. Supersaturation means that concentration of liquid is richer in a component of interest than a corresponding value on a saturation curve (equilibrium curve). Undercooling means that a state is located at the lower temperature side (the region where crystals coexist) of the saturation curve (equilibrium curve: melting point) along with the temperature axis on the equilibrium phase diagram. Note that when factors other than temperature, such as pressure and water content, cause undercooling, effective undercooling (supersaturation) is used. Equilibrium in chemical equilibrium is a state in which the chemical potential of a component of interest in each phase is equal through the phase interface between the two phases. Equilibrium in mechanical equilibrium means a state where static mechanical equilibrium holds through the interfaceand the Laplace equation holds there. Thermal equilibrium means that temperatures in each phase are equal, whereas disequilibrium indicates that the state is not in any of these equilibrium states.

Equation (7.16) shows the relationship between T_0 at which liquid of a certain composition is in equilibrium with a solid phase through a flat interface (curvature radius $R = \infty$), as well as T at which the liquid of the composition is in equilibrium with the solid phase through a curved interface (convex to the liquid side; curvature radius R); thus, the equilibrium temperature T is given as follows:

$$\frac{T}{T_0} = 1 - \frac{2\gamma v_S}{\Delta h}\frac{1}{R} = 1 - \frac{R^*_T}{R} \tag{7.17}$$

[2] In the case of crystals in liquid, mechanical equilibrium can be considered to always hold, unlike in the case of bubbles.

where R_T^* is defined by

$$R_T^* = \frac{2\gamma v_S}{\Delta h} \tag{7.18}$$

and gives a scale for the critical radius used when a degree of undercooling is defined by temperature. If $R = \infty$, $T = T_0$, and if $R > 0$, $T < T_0$, the state has to be undercooled. Equation (7.17) is the Gibbs-Thomson formula written in terms of temperature. R in Eq. (7.17) defines the critical radius R_C for the undercooling of $T < T_0$:

$$R_C = \frac{R_T^*}{(1 - T/T_0)}, \quad \left(\text{or } \frac{1}{R_C} = \frac{1}{R_T^*}\left(1 - \frac{T}{T_0}\right)\right) \tag{7.19}$$

is obtained and the degree of undercooling $(T_0 - T)$ and the inverse of the critical radius are found to be synonymous. In terms of energy theory, this is the same as Eq. (7.36) (Box in Sect. 7.1.2) obtained from discussion of nucleation. Moreover, generalizing a reference point (C_0, T_0) on the liquidus as $(C, T_L(C))$ and replacing $T_0 - T$ with $T_L(C) - T$ in the above equation of the critical radius yields the critical radius (the inverse of the degree of undercooling) as a function of composition and temperature. $T_L(C)$ is obtained using Eqs. (7.8)–(7.11).

To derive a Gibbs-Thomson formula with respect to composition, we examine the change of equilibrium conditions with difference in curvature of the interface at a certain temperature T. Considering the difference between the state II (Eq. 7.2) and the state III (Eq. 7.12) at the same temperature T and using an expression for chemical potential of an ideal solution ($\mu = \mu_0 + k_B T \ln X$) yields the following Gibbs-Thomson equation:

$$\frac{X}{X_{eq}} = \frac{C_R}{C_{eq}} = \exp\left(\frac{2\gamma v_S}{k_B T}\frac{1}{R}\right) = \exp\left(\frac{R_X^*}{R}\right) \tag{7.20}$$

where R_X^* is defined by

$$R_X^* = \frac{2\gamma v_S}{k_B T} \tag{7.21}$$

which gives the scale of the critical radius used when the degree of supersaturation is defined by composition at a reference undercooling temperature T.[3] Assuming that equilibrium concentration C_R of the curved interface is equal to that of supersaturated solution C ($> C_{eq}$), the equilibrium radius R will be the critical radius R_C at nucleation in the supersaturated solution. That is,

[3]When temperature T_0 is assumed to be undercooling temperature, T_L higher than T_0 has to be assumed. From Eq. (7.8), $T_L - T_0 = \Delta\tilde{h}^{-1} T_L \ln(C_L/C_0)$ holds. On the other hand, from Eq. (7.17), $1 - T_0/T_L = R_T^*/R$ holds. From these two equations,

$$\frac{C_L}{C_0} = \exp\left[\Delta\tilde{h}\frac{R_T^*}{R}\right] = \exp\left(\frac{R_X^*}{R}\right) \tag{22}$$

where $\Delta\tilde{h} = \Delta h/(k_B T_0)$. R_X^* at this time is $2\gamma v_S/k_B T_0$ and T_0 is used as temperature.

$$R_C = \frac{R_X^*}{\ln\left(\frac{C}{C_{eq}}\right)}, \quad \left(\text{or } \frac{1}{R_C} = \frac{1}{R_X^*}\ln\left(\frac{C}{C_{eq}}\right)\right) \tag{7.23}$$

is obtained.[4]

In the actual crystallization process, because both concentration C and temperature T change and comprehensively contribute to the degree of supersaturation (the degree of undercooling), it will be convenient to clearly express the manner of contribution as change from the initial reference state (C_0, T_0). Although in the critical radius defined by temperature difference (Eq. 7.19), the reference temperature T_0 is assumed to be constant, the temperature changes with Eq. (7.8) as a function of concentration. That is, the critical radius when temperature T and concentration C are provided is written by replacing T_0 with $T_L(C)$ as follows:

$$\frac{1}{R_C} = \frac{1}{R_T^*}\left(1 - \frac{T}{T_L(C)}\right) \tag{7.24}$$

Moreover, using $T/T_L(C) = T/T_0 \times T_0/T_L(C)$ and substituting Eq. (7.8) into the above equation yields

$$\frac{1}{R_C} = \frac{1}{R_X^*}\ln\frac{C}{C_0} + \frac{1}{R_T^*}\left(1 - \frac{T}{T_0}\right) \tag{7.25}$$

$$= \underbrace{\frac{k_B T}{2\gamma v_S}\ln\frac{C}{C_0}}_{\text{Concentration Change}} + \underbrace{\frac{\Delta h}{2\gamma v_S}\left(1 - \frac{T}{T_0}\right)}_{\text{Temperature Change}} \tag{7.26}$$

The first term of the right-hand side is the contribution of change in concentration to the degree of supersaturation (the inverse of the critical radius), and then the critical radius increases when the concentration C decreases from the initial concentration C_0. The second term of the right-hand side is the contribution of change in temperature. The degree of supersaturation increases when the temperature decreases from the initial temperature T_0. Both R_X^* and R_T^* in Eq. (7.25) are defined by Eqs. (7.21) and (7.18).

$$R_X^* = \frac{2\gamma v_S}{k_B T} \tag{7.27}$$

$$R_T^* = \frac{2\gamma v_S}{\Delta h} \tag{7.28}$$

R_P^* and R_T^*, letters with subscripts X and T, in addition to R_P^* defined in Eq. (4.10), are scales of critical radius corresponding to supersaturated states caused by change in composition, temperature, and pressure. Moreover, the ratio of them

$$\Delta\tilde{h} = \frac{R_X^*}{R_T^*} = \frac{\Delta h}{k_B T} \approx \frac{\Delta s}{k_B} \tag{7.29}$$

---

[4]For this equation, when supersaturated concentration C is considered as a reference concentration C_0, the ratio of equilibrium concentration at the degree of undercooling $T_0 - T$ (or difference in concentration, i.e., the degree of supersaturation $1 - C_0/C_{eq}$) as first order approximation of logarithm) is defined using the critical radius. In other words, this equation becomes Eq. (7.19) when Eq. (7.3), which defines liquidus, is used.

is a dimensionless entropy of crystallization or the latent heat and is defined as $\Delta\tilde{h}$.[5]

The critical radius: Relationship with chemical equilibrium and with energy theory of nucleation

From the discussion of nucleation in terms of energy theory, the critical radius R_C is given as follows:

$$R_C = \frac{2\gamma}{\Delta\mathcal{G}_V^*} \tag{7.31}$$

where $\Delta\mathcal{G}_V^*$ is the difference in Gibbs free energy of a crystallizing component in the solid phase (a pure substance) and the liquid phase (molar fraction X_0) under temperature and pressure (T, P) common to solid and liquid phases (Sect. 3.1.2). That is,

$$\Delta\mathcal{G}_V^* = \frac{1}{v}\left(\mu_{\text{Melt}}^{\text{A}}(T, P, X_0) - \mu_{\text{Solid}}^{\text{A}}(T, P)\right) \tag{7.32}$$

where $\mu_{\text{Melt}}^{\text{A}}(T, P, X_0)$ is equivalent to the free energy in supersaturated (undercooled) melt and $\mu_{\text{Solid}}^{\text{A}}(T, P)$ is equivalent to the free energy in crystals (μ(J/molecule)). v is an effective molecular volume of the crystallizing component (v (m^3/molecule))(v in the liquid phase and that in the solid phase are assumed to be equal). Because the solid phase at that time is in equilibrium with the liquid of composition X_{eq}, $\mu_{\text{Melt}}^{\text{A}}(T, P, X_{\text{eq}}) = \mu_{\text{Solid}}^{\text{A}}(T, P)$ holds based on chemical equilibrium. Therefore,

$$\Delta\mathcal{G}_V^* = \frac{1}{v}\left(\mu_{\text{Melt}}^{\text{A}}(T, P, X_0) - \mu_{\text{Melt}}^{\text{A}}(T, P, X_{\text{eq}})\right) \tag{7.33}$$

holds and difference in free energy is reduced only to difference in the liquid compositions. Using an expression for chemical potential of an ideal solution ($\mu = \mu_0 + k_B T \ln X$) and $C_0 = C_{\text{eq}} + (C_0 - C_{\text{eq}})$ for the above equation yields the following:

$$\Delta\mathcal{G}_V^* = \frac{k_B T}{v}\ln\frac{X_0}{X_{\text{eq}}} = \frac{k_B T}{v}\ln\frac{C_0}{C_{\text{eq}}} \approx \frac{k_B T(C_0 - C_{\text{eq}})}{v C_{\text{eq}}} \tag{7.34}$$

This is energy driving crystallization expressed by the degree of supersaturation defined by difference in composition. The critical radius is defined as follows:

$$R_C = \frac{2\gamma v}{k_B T \ln\frac{C_0}{C_{\text{eq}}}} = \frac{R_X^*}{\ln C_0/C_{\text{eq}}} \tag{7.35}$$

This is the critical radius defined in terms of difference in concentration. On the other hand, using Eq. (7.3) of liquidus obtained from chemical equilibrium to Eq. (7.34) yields

$$R_C = \frac{2\gamma v}{\Delta h(1 - T/T_0)}, \quad \left(\text{or } \frac{1}{R_C} = \frac{\Delta s(T_0 - T)}{2\gamma v}\right) \tag{7.36}$$

This is energy driving crystallization and the critical radius defined in terms of the degree of undercooling $T_0 - T$. These equations agree with the equations derived from chemical equilibrium (Eqs. 7.23 and 7.19).

---

[5]The ratio R_X^*/R_T^* is equal to the dimensionless number α_2 (Eq. 7.10), which defines the slope of liquidus:

$$\frac{R_X^*}{R_T^*} = \frac{\Delta s}{k_B} = \alpha_2. \tag{7.30}$$

Although equilibrium concentration at the curved interface C_R in the Gibbs-Thomson equation of Eq. (7.20) is described using equilibrium concentration through the flat interface C_{eq} as a reference, the concentration of supersaturated melt C can be used as a reference. Eliminating C_{eq} from Eqs. (7.20) and (7.23) yields

$$\frac{C_R}{C} = \exp\left[\frac{2\gamma v_S}{k_B T}\left(\frac{1}{R} - \frac{1}{R_C}\right)\right] \tag{7.37}$$

$$\approx 1 - R_X^*\left(\frac{1}{R_C} - \frac{1}{R}\right) \tag{7.38}$$

where R_X^* is defined in Eq. (7.21).[6] Under an isothermal condition, Eq. (7.38) is then useful for considering the growth of crystals close to the critical nucleus size and Ostwald ripening. Note that crystals larger than the critical radius in supersaturated melt have local equilibrium concentration that is lower than the concentration of liquid C, whereas crystals smaller than the critical radius have a higher concentration, i.e., concentration in liquid heterogeneously distributes and its boundary condition is local interfacial equilibrium dependent on curvature. Note that heterogeneous concentration drives mass transfer by diffusion and the solid phase will dissolve or exsolve (grow) to maintain the interfacial equilibrium.

Let us summarize the above from the point of view of difference in free energy as a driving force for either nucleation or growth. The difference in free energy for phase change, including nucleation and growth (per molecule) Δg^*, can be determined as follows:

$$\Delta g^* = k_B T \ln \frac{C}{C_{eq}} = k_B T \frac{R_X^*}{R_C} \ (= \Delta h(1 - T/T_0)) \tag{7.39}$$

(Δ g* per unit volume is $\Delta \mathcal{G}_V^* = \Delta g^*/v$). This is the driving energy for growth at a flat interface $\Delta g^*(R = \infty)$. The driving energy for growth at a curved interface (curvature radius R) $\Delta g^*(R)$ can be determined as follows:

$$\Delta g^*(R) = k_B T \ln \frac{C}{C_R} = k_B T \left(\frac{R_X^*}{R_C} - \frac{R_X^*}{R}\right) \tag{7.40}$$

Furthermore, difference in free energy for nucleation is larger than the difference in free energy for growth by R_X^*/R and is equivalent to the case where $R \to \infty$ in $\Delta g^*(R)$.

7.2 Classical Understanding of Igneous Rock Texture Using Nucleation Rate and Growth Rate

In addition to the abovementioned equilibrium, understanding the crystallization-related kinetics, including the nucleation and growth processes to be explained in the following sections, allows us to provide a interepretation to the rock texture. To examine the significance of the nucleation and growth rates, the approach in the classical understanding of rock texture is explained.

Information on rock texture such as the number density and crystal size formed by cooling crystallization has been classically understood qualitatively using the dependence of nucleation rate and

[6]Note that it differs from R_X^* in the case of bubbles only by a factor y^2.

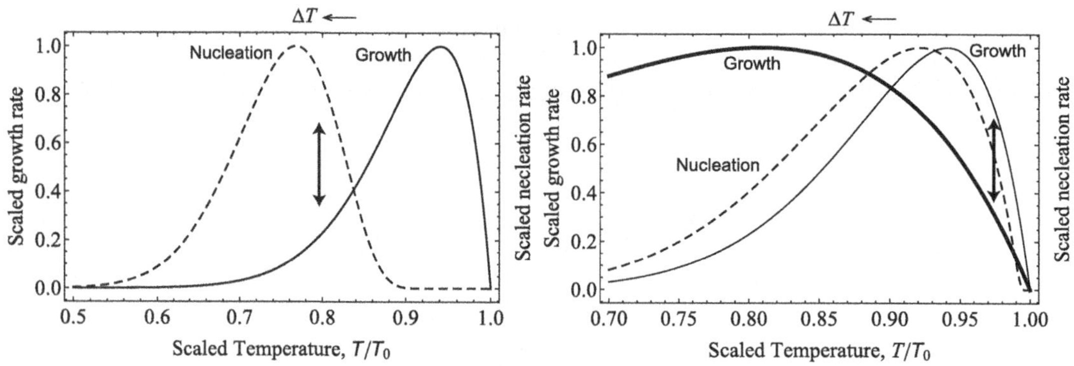

Fig. 7.4 Comparison between a master curve of the nucleation rate (broken line) and that of the growth rate (solid line). Both of them are scaled by peak values effective to the problem under consideration. Note that the horizontal axis is not the degree of undercooling but temperature. The activation energy of diffusion (discussed later) $\tilde{q} = 11.7$ is shared. The hypothetical nucleation point is indicated by an arrow. (Left) The interfacial tension equivalent to homogeneous nucleation (discussed later) $\Gamma = 0.084$. (Right) $\Gamma = 0.0001$, which is equivalent to heterogeneous nucleation. The thick solid line of the growth rate corresponds to the case where the activation energy of diffusion is reduced to an extreme (1/10) and the degree of undercooling giving the maximum growth rate becomes larger than the degree of undercooling of nucleation

growth rate on the degree of undercooling (Winkler 1948; Wager 1961). Let us examine it concretely using the nucleation and growth rates to be derived in the following sections. The dependence of nucleation rate on the degree of undercooling is determined by Eq. (7.48), whereas that of growth rate is obtained by Eq. (7.103). Both these parameters are shown in Fig. 7.4. Master curves are the curves of nucleation and growth rates as a function of the degree of undercooling because they control rock texture, which records the crystallization history as shown below; it is extremely important to experimentally determine master curves. As described in the following section, the shape of master curves depends on physical properties such as the activation energy of diffusion and surface tension.

For cooling crystallization, both master curves are bell-shaped curves against temperature because the effect of increase in the degree of undercooling and the effect of decrease of diffusion by lowering of temperature compete. Generally, for homogeneous nucleation, because the occurrence of nucleation requires considerable undercooling, temperature giving the maximum nucleation rate will be lesser. Because growth rate is high even at relatively high temperatures, the relationship between the nucleation and growth rates shown in the left chart of Fig. 7.4 holds. In this case, when the melt is cooled, it is meaningless even if crystal growth rate is large because nucleation does not effectively occur at high temperatures (nucleation rate is extremely small), leading to lack of crystals. If melt is undercooled to a sufficiently low temperature to cause nucleation, growth rate at the temperature becomes so small, and nucleation predominantly occurs and small crystals are formed at a high crystal number density.

For heterogeneous nucleation, however, because nucleation easily occurs even with a low degree of undercooling, the curves will be similar to that shown in the right chart of Fig. 7.4. In this case, even at a low degree of undercooling, although nucleation easily occurs with decrease in temperature, the growth rate is so large that, as soon as nucleation occurs, crystals grow and then consume the crystallization component, resulting in the prevention of additional nucleation. Consequently, a reduced number of large crystals are formed. However, for some reason (denoted by the thick line in the figure), if the growth rate is suppressed, then growth does not effectively work to decrease the degree of supersaturation and the nucleation continues, resulting in the formation of numerous small crystals. Thus, both size and number density of crystals are determined by the relative relationship between nucleation and growth (see Box).

Cooling rate is a factor that influences relative relationship between nucleation and growth rates. Difference in rock texture, characterized by the number density of crystals, directly reflects differences in cooling rate. Based on the relative relationship between nucleation and growth rates, the influence of the cooling rate can be understood.

Scaling of crystallization using the characteristic magnitude of nucleation and growth

The scale of nucleation rate $J_{\rm m}$ is the number of crystals per unit time per unit volume and has a dimension of [$\rm m^{-3}s^{-1}$]. The scale of crystal growth rate $G_{\rm m}$ is length per unit time and has a dimension of [$\rm m\ s^{-1}$]. From a dimensional analysis of them, a quantity having a dimension of length is determined as follows:

$$L = \left(\frac{G_{\rm m}}{J_{\rm m}}\right)^{\frac{1}{4}} \tag{7.41}$$

and a quantity having a dimension of time is determined as follows:

$$\tau = \left(J_{\rm m} G_{\rm m}^3\right)^{-\frac{1}{4}} \tag{7.42}$$

Because the crystal number density has a dimension of [$\rm m^{-3}$] and is equal to L^{-3},

$$N = \left(\frac{J_{\rm m}}{G_{\rm m}}\right)^{\frac{3}{4}} \tag{7.43}$$

is obtained. Such discussion on scaling holds only when the scales of the nucleation rate and the growth rate $J_{\rm m}$ and $G_{\rm m}$, respectively, are appropriately selected. Using the maximum values of the nucleation rate and the growth rate just selected for parameterization (the maximum values of the master curves) for it is not appropriate.

Assuming that the volume fraction of the crystallization component ϕ is determined using the chemical composition of the liquid and equilibrium phase diagram and is independent of the cooling rate, if the crystal number density N is fixed, the size will be automatically determined. For example, if a sphere is assumed, L is then determined from both N and ϕ

$$\phi = \frac{4\pi}{3} L^3 N \tag{7.44}$$

Thus, the crystal number density is the most important quantity, for which $J_{\rm m}$ and $G_{\rm m}$ appearing in Eq. (7.43) have to be representative values at the stage of nucleation, i.e., at the moment the number of crystals is determined.

7.3 Nucleation of Crystals

7.3.1 Basic Characteristics of Homogeneous Nucleation

For understanding the crystallization process, to understand the rate of formation of crystal nuclei that instantaneously form in response to a certain degree of undercooling, i.e., to know the nucleation rate, is the most fundamental point. In other words, in the crystallization process, the nucleation rate as a function of the degree of undercooling is a master curve (governing curve).

For homogeneous nucleation, the nucleation rate of crystals can be derived using the same procedure as the nucleation of bubbles. An important different from the case of bubbles is the growth rate reported in the next section. For bubbles, because of the dissociation reaction, the activity of water in liquid becomes similar to that of a nonideal solution; this behavior is essential for determining vesiculation conditions and the degree of supersaturation. Consequently, nonideality appears in both diffusion growth rate and nucleation rate. However, for crystals, the solution can be considered as an ideal solution as the first approximation; therefore, derivation of the growth and nucleation rates is less complex.

Using exactly the same procedure as that in Sect. 3.3.4 and the diffusion growth rate (Eq. 7.68) to be obtained in the next section and describing the degree of supersaturation using the critical nucleus radius R_C (Eq. 7.26), we obtain the following equation to determine the nucleation rate of crystals.

$$J = Z\frac{v_S DC^2}{4\pi R_C}\exp\left(-\frac{4\pi\gamma R_C^2}{3k_B T}\right) = \frac{DC\phi_S}{2\pi R_C}\sqrt{\frac{\gamma}{k_B T}}\exp\left(-\frac{4\pi\gamma R_C^2}{3k_B T}\right) \tag{7.45}$$

where $\phi_S = v_S C$ is a volume concentration (fraction) of the crystallization component in the melt.

To more concretely examine the dependence of the degree of undercooling $1 - T/T_0$, concentration is assumed to be constant. Assuming that dimensionless temperature is $\tilde{T} = T/T_0$, the critical nucleus radius R_C (Eq. 7.26) can be written as follows:

$$R_C = \frac{R_T^*}{1-\tilde{T}} \tag{7.46}$$

We then define the degree of undercooling of a crystallization component $\Delta\tilde{T}$ as $1 - \tilde{T}$. Note that diffusivity D has a thermally activated temperature dependence and plays an important role along with temperature reduction.

$$D = D_0 \exp\left(-\frac{q}{k_B T}\right) \tag{7.47}$$

where q is the activation energy of diffusion (per molecule). Substituting this equation into Eq. (7.45) yields

$$J = J_0 \cdot \tilde{T}^{-\frac{1}{2}} \cdot \Delta\tilde{T} \exp\left(-\frac{\tilde{q}}{\tilde{T}} - \frac{\Gamma}{\tilde{T}(1-\tilde{T})^2}\right) \tag{7.48}$$

where J_0, called a pre-exponential factor, is

$$J_0 = \frac{D_0 C\phi_S}{2\pi R_T^*}\sqrt{\frac{\gamma}{k_B T_0}} \tag{7.49}$$

and R_T^* is a scale of critical nucleus radius when temperature is used as a reference, which is defined by Eq. (7.18) or Eq. (7.28). A dimensionless number Γ is then defined by

$$\Gamma = \frac{4\pi\gamma(R_T^*)^2}{3k_B T_0} \tag{7.50}$$

which gives an approximate magnitude of a nucleation barrier when temperature is used as a reference; it considerably depends on interfacial energy (Right chart of Fig. 7.5). Although the pre-exponential factor J_0 depends on physical properties such as diffusivity and interfacial energy, it generally considers

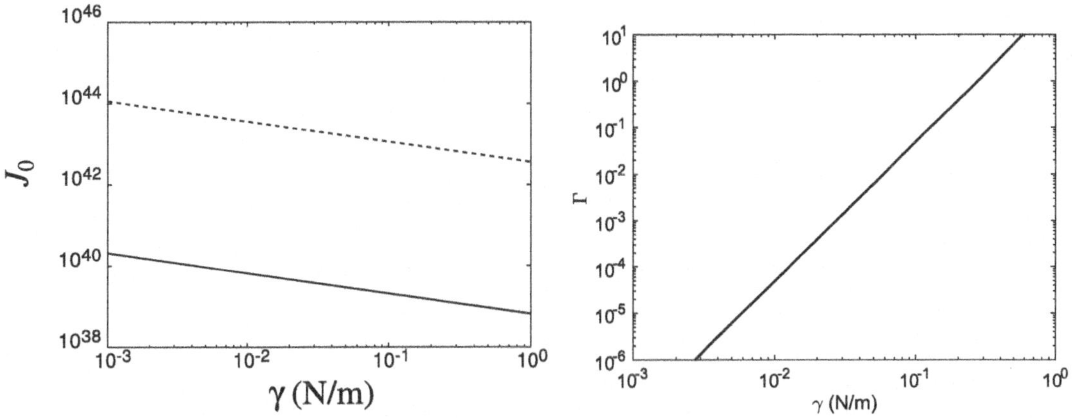

Fig. 7.5 (Left) Relationship between the constant part of the pre-exponential factor obtained from theoretical equation (7.49) and interfacial energy. The case where temperature $T_0 = 1400$ (K), diffusivity $D = 10^{-11}$ (m^2/s), the molecular number density of crystallization component $C = 10^{28}$ (m^{-3}), $v_S = 5 \times 10^{-29}$ (m^3) ($\phi_0 = 0.5$), and $\tilde{q} = 8.6$. The solid line represents the case including $\exp(-\tilde{q})$. The broken line represents the case using D_0 in $D = D_0 \exp(-\tilde{q})$. (Right) Relationship between the parameter Γ and the interfacial energy γ

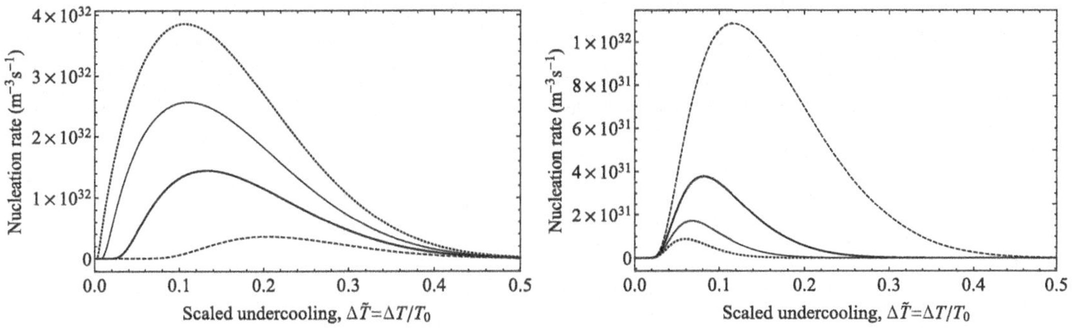

Fig. 7.6 (Left) The nucleation rate as a function of undercooling ΔT with the interfacial energy as a parameter. For $\tilde{q} = 8.0$, values of parameter α_1 are as follows: broken line ($\alpha_1 = 1$, $\Gamma = 0.028$, $\gamma = 0.056$ (J/m^2)), thick solid line ($\alpha_1 = 0.1$, $\Gamma = 0.0028$, $\gamma = 0.026$ (J/m^2)), thin solid line ($\alpha_1 = 0.01$, $\Gamma = 0.00028$, $\gamma = 0.012$ (J/m^2)), and dotted line ($\alpha_1 = 0.001$, $\Gamma = 0.000028$, $\gamma = 0.0056$ (J/m^2)). Γ and α_1 are then defined by Eqs. (7.50) and (7.60). (Right) Against parameter $\alpha_1 = 0.1$, values of parameter $\tilde{q}$ are as follows: broken line ($\tilde{q} = 10$), thick solid line ($\tilde{q} = 20$), thin solid line ($\tilde{q} = 30$), and dotted line ($\tilde{q} = 40$)

large values (Left chart of Fig. 7.5). $\tilde{q}$ is a dimensionless value of activation energy of diffusion defined below:

$$\tilde{q} = \frac{q}{k_B T_0} \tag{7.51}$$

Plotting nucleation rate (Eq. 7.48) as a function of degree of undercooling $\Delta\tilde{T}$ gives a characteristic bell-shaped curve (Fig. 7.6), known as a master curve of the nucleation rate. This is attributed to the fact that although nucleation rate initially increases with increase in the degree of undercooling, diffusivity decreases with additional decrease in temperature, i.e., it becomes difficult for molecules to move. Figure 7.6 shows the influence of interfacial energy and activation energy of diffusion on the master curve. With increase in interfacial energy, because it becomes hard for nucleation to occur, the peak nucleation rate reduces and the degree of undercooling giving it becomes large. With increase in activation energy, the degree of undercooling giving the peak nucleation rate reduces and the peak

Table 7.1 Typical physical properties related to crystallization and values of related dimensionless quantities. $T_0 = 1300$ K is assumed. q and Δh are activation energy and latent heat per molecule, given as $q = Q/N_{Av}$ and $\Delta h = \Delta\mathcal{H}/N_{Av}$, respectively, when Avogadro's number is assumed to be $N_{Av} = 6 \times 10^{23}$

Quantity	Notatio	Value	Dimensionless number	Value
Activation energy of diffusivity	Q	80–250 (kJ/mol)	$\tilde{q} = \frac{q}{k_B T_0}$	7.4–23
Enthalpy of fusion	$\Delta\mathcal{H}$	100 (kJ/mol)	$\Delta\tilde{h} = \frac{\Delta h}{k_B T_0}$	9.2
Molecular volume of crystallization component	v_S	5×10^{-29} (m^3)		
Scale of the critical nucleus size R_T^*	$\frac{2\gamma v}{\Delta h}$	6×10^{-11}		
Solid-melt interfacial energy	γ	0.02–0.2 (J/m^2)		
Scale of the interfacial tension R_X^*	$\frac{2\gamma v}{k_B T_0}$	5.6×10^{-10}		
Number of molecules of	C_0	10^{28} (m^{-3})		

nucleation rate itself reduces. This is because the larger activation energy leads to the steaper decrease in diffusivity with cooling even at a high temperature.

As a function of the degree of undercooling, the behavior of the nucleation rate is greatly influenced by interfacial energy and activation energy of diffusion. This behavior is characterized by peak nucleation rate J_{PEAK} and undercooling temperature $\tilde{T}_{JPEAK}$, giving J_{PEAK}. The peak nucleation rate is approximately determined from conditions for the extremum of the exponential part of Eq. (7.48)

$$-\frac{\tilde{q}}{\tilde{T}} - \frac{\Gamma}{\tilde{T}(1-\tilde{T})^2} \tag{7.52}$$

That is,

$$\frac{\tilde{q}}{\tilde{T}^2} + \frac{\Gamma(1-3\tilde{T})}{\tilde{T}^2(1-\tilde{T})^3} = 0 \tag{7.53}$$

This is a cubic equation with respect to $\Delta\tilde{T}_{JPEAK} = 1 - \tilde{T}_{JPEAK}$ and its solution is as follows:

$$\Delta\tilde{T}_{JPEAK} = -\frac{\Gamma_q}{(\Gamma_q + (\Gamma_q^2 + \Gamma_q^3)^{1/2})^{1/3}} + (\Gamma_q + (\Gamma_q^2 + \Gamma_q^3)^{1/2})^{1/3} \tag{7.54}$$

$$\Gamma_q = \frac{\Gamma}{\tilde{q}} \tag{7.55}$$

As found from the abovementioned equations, the degree of undercooling giving the peak nucleation rate is determined only by the ratio of Γ and $\tilde{q}$. This relationship is shown in the left chart of Fig. 7.7. As shown in Fig. 7.5, in response to increase in interfacial energy γ, Γ varies from 10^{-6} to 1. Because $\tilde{q}$ takes values from 8 to about 50 (Table 7.1), the degree of undercooling giving the peak nucleation rate is also found to vary in a considerably wide range. Moreover, the peak nucleation rate J_{PEAK} changes in response to it and, when nucleation easily occurs, $\Delta\tilde{T}_{JPEAK}$ becomes small (left chart of Fig. 7.7) and J_{PEAK} becomes large (right chart of Fig. 7.7).

The response of nucleation rate to the degree of undercooling corresponds to the case in which the degree of undercooling is given by instantaneously lowering temperature; moreover, it can be compared with experiments, as shown in the next section. Note that when temperature is continuously reduced from a temperature higher than the melting point for solidification, the maximum nucleation

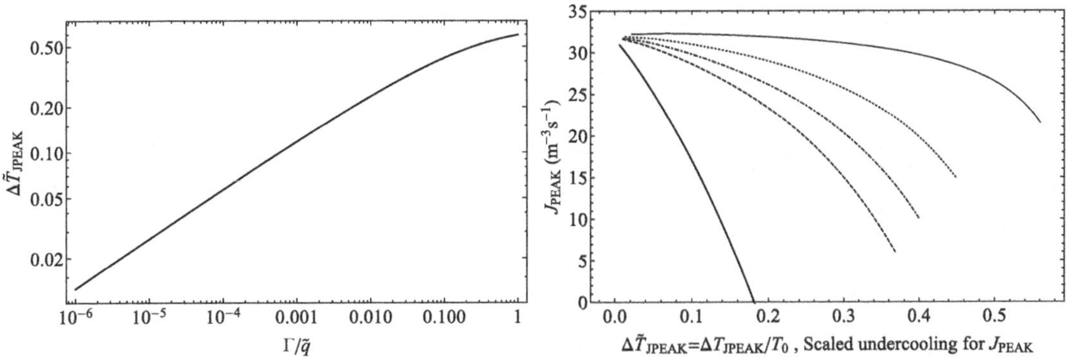

Fig. 7.7 (Left) Relationship between the dimensionless degree of undercooling $\Delta\tilde{T}_{\mathrm{JPEAK}}$ giving the peak nucleation rate and the parameter $\Gamma/\tilde{q}$. (Right) Relationship between the peak nucleation rate and the dimensionless degree of undercooling giving the peak nucleation rate. Line types reflect values of $\tilde{q}$: thin solid line ($\tilde{q} = 5$), dotted line ($\tilde{q} = 20$), chain line ($\tilde{q} = 35$), broken line ($\tilde{q} = 50$), and thick solid line ($\tilde{q} = 200$)

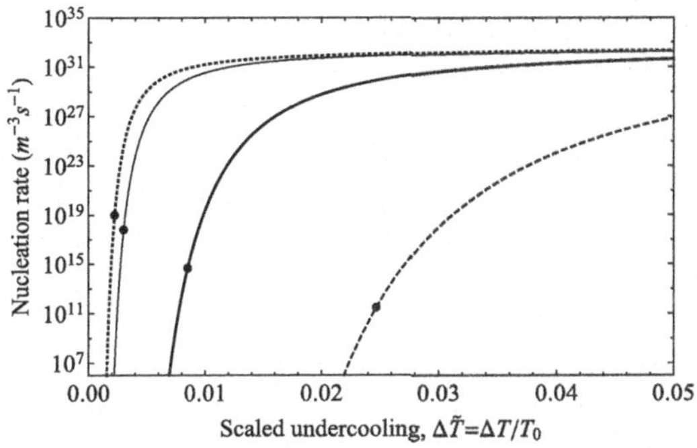

Fig. 7.8 Relationship between the instantaneous nucleation rate and the degree of undercooling. $\tilde{q} = 8$ is assumed. Line types reflect values of α_{I}: broken line ($\alpha_{\mathrm{I}} = 1$), thick solid line ($\alpha_{\mathrm{I}} = 0.1$), thin solid line ($\alpha_{\mathrm{I}} = 0.01$), and dotted line ($\alpha_{\mathrm{I}} = 0.005$). The black circles represent the maximum values experienced in continuous crystallization at a constant cooling rate

rate and the corresponding degree of undercooling in the complete history of the crystallization process differs from such instantaneous maximum nucleation rate. As shown in Fig. 7.8, the maximum values of the nucleation rate actually experienced are obtained with reference to, for example, the result of numerical calculation. At this time, similar to a pulse, nucleation occurs in a certain brief duration. From this figure, it is found that, compared with the peak value J_{PEAK} of the bell-shaped master curve of the nucleation rate, the maximum nucleation rate J_{MAX} reached in the actual cooling history is considerably smaller. Moreover, this maximum nucleation rate J_{MAX} is the maximum value reached for a moment in the cooling history. However, when the actual pre-exponential factor is small and J_{PEAK} is smaller than the maximum nucleation rate J_{MAX}, the nucleation rate remains constant, thus maintaining its maximum value in the cooling history. As described above, for understanding the crystallization process, the pre-exponential factor of the nucleation rate plays an important role. This will be explained, as required, in sections on comparison with experiments, outline of crystallization process, and crystal size distribution.

The above expressions are applicable to the case in which the degree of supersaturation or the degree of undercooling is a function of only temperature. Because, for actual crystallization of a multi-component system magma, the degree of supersaturation depends on both temperature and composition, it is convenient to obtain an expression for the case. Assuming that dimensionless concentration is

$\tilde{C} = C/C_0$ and using

$$(\tilde{R}_{CT})^{-1} = \left(\frac{R_C}{R_T^*}\right)^{-1} = \frac{R_T^*}{R_X^*}\ln\tilde{C} + \left(1 - \tilde{T}\right) \tag{7.56}$$

which is obtained by nondimensionalizing the critical radius (Eq. 7.26) using the radius scale of the degree of undercooling R_T^*, as an approximate estimation of the degree of supersaturation, the nucleation rate is given as follows:

$$J = J_0\tilde{T}^{-\frac{1}{2}}\Delta\tilde{T}\exp\left(-\frac{\tilde{q}}{\tilde{T}} - \Gamma\frac{(\tilde{R}_{CT}^*)^2}{\tilde{T}}\right) \tag{7.57}$$

When $\tilde{C} = 1$, this equation results in Eq. (7.48).

If using

$$(\tilde{R}_{CX})^{-1} = \left(\frac{R_C}{R_X^*}\right)^{-1} = \ln\tilde{C} + \alpha_2\left(1 - \tilde{T}\right) \tag{7.58}$$

which is obtained by normalizing the critical radius using the radius scale of the degree of supersaturation R_X^* (α_2 is defined by Eq. 7.30), the nucleation rate will be

$$J = J_0\tilde{T}^{-\frac{1}{2}}\Delta\tilde{T}\exp\left(-\frac{\tilde{q}}{\tilde{T}} - \alpha_1\frac{(\tilde{R}_{CX}^*)^2}{\tilde{T}}\right) \tag{7.59}$$

α_1 is then defined by,

$$\alpha_1 = \frac{4\pi\gamma(R_X^*)^2}{3k_B T_0} \tag{7.60}$$

and represents the magnitude of nucleation barrier in the same way as the problem of nucleation and growth of bubbles. The relationship between α_1 and Γ is then determined as follows:

$$\Gamma = \frac{4\pi\gamma(R_X^*)^2}{3k_B T_0}\left(\frac{R_T^*}{R_X^*}\right)^2 = \alpha_1\left(\frac{R_T^*}{R_X^*}\right)^2 = \alpha_1\cdot\alpha_2^{-2} \tag{7.61}$$

7.3.1.1 Heterogeneous Nucleation

For heterogeneous nucleation, the same concept of nucleation of bubbles is applicable to crystals. It exerts two influences on the nucleation rate. One is nucleation barrier, i.e., the surface energy γ in Eq. (7.45) or the parameter Γ in the exponential part of Eq. (7.48). The other is the reduced pre-exponential factor under situations where nucleation occurs on the surface of crystals or crystal-constituting molecules form clusters in liquid at a temperature under the liquidus. The influences on the pre-exponential factor result from the variable number of heterogeneous nucleation sites controlled by disorder on the crystal surface, and considerable decrease in effective number of molecules and in the molecular mobility or diffusivity. The influences appear when the peak of the master curve of the nucleation rate is lowered to the same level as the maximum nucleation rate that shall be actually reached. Consequently, the huge reduction in the pre-exponential factor leads to completely different characteristics in the temporal development of nucleation rate, the resultant crystal size distribution (CSD), and dependence on the cooling rate.

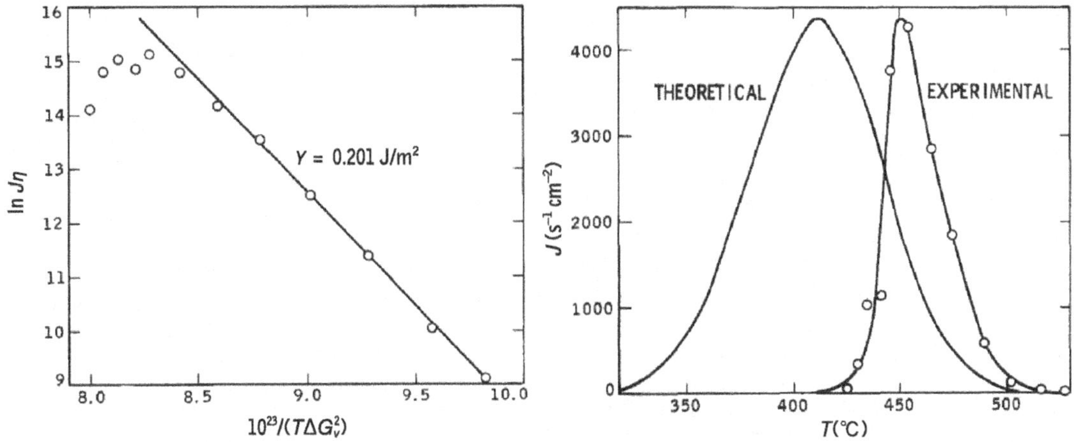

Fig. 7.9 (Left) The experimental result is summarized using $\ln J\eta$ and $(T\Delta G_v^2)^{-1}$ is the degree of undercooling used in their paper. γ is estimated at 0.201 (J/m^2) from the slope. (Right) Comparison between a theoretical curve calculated using γ estimated from the left figure and the experimental result (adapted from Neilson and Weinberg 1979). Copyright 1979, with permission from Elsevier. The value of J_0 was selected as the maxima become equal. The authors argue that the degree of undercooling giving the maximum value and the broadening of nucleation predicted by the theory are different from that observed in the experiments

7.3.2 Comparison with Nucleation Experiments

Many experiments with silicate melt used as industrial glass, have been conducted using the following procedure: (1) Completely melt silicates at a temperature higher than the melting point, (2) instantaneously cool the melt to a temperature lower than the melting point and maintain the melt at the temperature for a certain period of time, (3) rapidly cool the sample to a low temperature for a short time period, and (4) gradually increase the temperature of the sample to allow crystals that have nucleated at Step (2) to grow to observable size (nucleation does not newly occur at this step). As shown in the previous section, such experiments, called instantaneous undercooling experiments, are considered to show a nucleation rate as a function of temperature and allow us to compare the theory with experiments (Kirkpatrick 1981; Neilson and Weinberg 1979; James 1985). Beginning with the conclusion, although experimental results of the peak value of nucleation and the degree of undercooling giving the peak value can be theoretically explained if latent heat, activation energy of diffusion, and interfacial energy are appropriately selected, the form of the nucleation rate as a function of temperature, i.e., width and skewness of the master curve cannot be reproduced. Moreover, optimum values for the latent heat, the activation energy, and the interfacial energy deviate from the values determined by the experiments.

For example, Neilson and Weinberg (1979) attempted to explain the data of an experiment with $Li_2O{\cdot}2SiO_2$ using $\tilde{q} \approx 50$ and $\gamma = 0.201$ (Fig. 7.9). However, the experimental result is not explained using these values. The peak nucleation rate J_{PEAK} and the degree of undercooling $\Delta\tilde{T}_{JPEAK}$ giving J_{PEAK} are uniquely determined by Γ and $\tilde{q}$. Their experiment resulted in $J_{PEAK} = 4.3 \times 10^9$ (m^{-3}s^{-1}) and $\Delta\tilde{T}_{JPEAK} = 1 - 723/1300 = 0.4438$, corresponding to $\Gamma = 3.5$ and $\tilde{q} = 26.8$. $\Gamma = 3.5$ corresponds to the interfacial energy $\gamma = 0.28$ (J/m^2). The left chart of Fig. 7.10 shows the relationship between the nucleation rate and temperature drawn using these values and their experimental data. As evident from this figure, even if the degree of undercooling corresponding to the maximum value of the nucleation rate is given, broadening of the nucleation rate cannot be explained. In the classical nucleation theory, this point has been recognized as a defect.

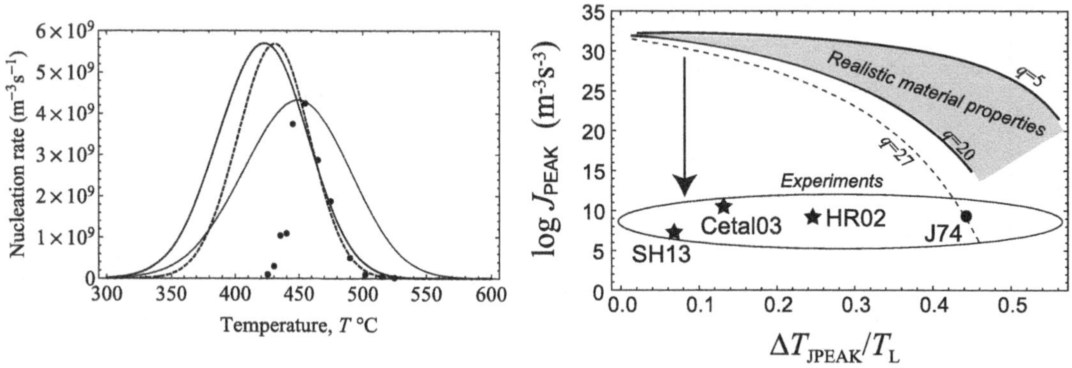

Fig. 7.10 Comparison between the experimental result (black circles) by James (1974) and the theories described so far. The melting point is 1033 °C. The degree of undercooling giving the peak nucleation rate is 583 (K), $\Delta\tilde{T}_{\mathrm{JPEAK}} = 0.45$. Thin solid line: $\gamma = 0.28$ (J/m^2), $\tilde{q} = 26.9$. J_0 is not adjusted. The maximum value of J and the degree of undercooling giving it agree with the experimental data. Thick solid line: $\gamma = 0.33$ (J/m^2), $\tilde{q} = 26.9$. J_0 is amplified by multiplying 10^{12}. Broken line: $\gamma = 0.55$ (J/m^2), $\tilde{q} = 55$. J_0 is amplified by multiplying 10^{30}. Trends to the maxima are fitted to the experimental data. Comparison between the thick solid line and the broken line revealed that if γ and $\tilde{q}$ are increased, the variance becomes small and the maximum also becomes small (multiplying 10^{30} is best fit). Similarly, to match the theoretical data with the experimental data requires extraordinarily large amplification of J_0. (Right) Relationship between the instantaneous peak nucleation rate J_{PEAK} predicted from the homogeneous nucleation theory and the degree of undercooling $\Delta\tilde{T}_{\mathrm{JPEAK}}$ giving the J_{PEAK}, and the experimental results by instantaneous undercooling. J74 (James 1974): A cooling experiment with a $Li_2O{\cdot}2SiO_2$ system. HR02 (Hammer and Rutherford 2002): Feldspar in a decompression-induced crystallization with Pinatubo dacite. Cetal03 (Couch et al. 2003b): Feldspar in dacite found in Soufriere Hills. SH13 (Shea and Hammer 2013): Clinopyroxene in basaltic melt

From various aspects, clarification of the cause of this defect and correction of this defect, such as reconsideration of experimental conditions, reconsideration of classical theories, and reconsideration of physical properties, have been conducted. The following is one of the attempts (James 1985). (1) Using the inverse of viscosity rather than diffusivity: although relationship between viscosity and diffusion in silicate melt does not follow a simple Stokes-Einstein relation (Liang et al. 1997), viscosity obtained by measurement are related to mobility of molecules and is used. (2) Taking dependence of the surface tension on temperature into consideration.

Sometimes, the peak nucleation rate in experimental data can be explained by making the pre-exponential factor J_0 of the nucleation rate larger or smaller than the theoretical value. The necessity to make the J_0 value larger is that if parameters for the degree of undercooling corresponding to the peak position in the experimental data are given according to the classical nucleation theory (Fig. 7.7), the value of the peak nucleation rate itself becomes too small to explain the value obtained by the experiment. In this manner, for example, James's data of $\Delta\tilde{T}_{\mathrm{JPEAK}}$ and J_{PEAK} can be explained; however, even if J_0 is varied using $\tilde{q} = 27$, a relatively realistic value of the activation energy of diffusion, i.e., variance of the master curve of the nucleation rate, cannot be explained (the left chart of Fig. 7.10). Furthermore, a physical meaning of making J_0 larger is unclear.

On the contrary, the necessity to make the value of J_0 smaller is that in many experiments using silicate melt (Hammer and Rutherford 2002; Couch et al. 2003b; Shea and Hammer 2013), small peak nucleation rates have been obtained by relatively small degrees of undercooling (the right chart of Fig. 7.10). If theoretical relationship between $\Delta\tilde{T}_{\mathrm{JPEAK}}$ and J_{PEAK} truly holds, the peak nucleation rate should become large by more than 10^{20} times against a small degree of undercooling. However, the experiments did not yield such results. To make the J_0 value smaller than the theoretical value corresponds to considering the number of heterogeneous nucleus sites formed by heterogeneous nucleation, which is realistically probable.

Considering heterogeneous nucleation, if we make the value of J_0 small to the degree that can explain the experimental result, another issue is observed. It will appear in the case of continuous cooling. When the J_0 value is large to some extent, nucleation occurs as a single event in a certain narrow temperature range. In this case, the dependence of nucleation on the degree of supersaturation is exponential and the nucleus number density will drastically increase with only a small increase in the degree of supersaturation. Thus, the degree of supersaturation dramatically decreases and the nucleation event ends in a short time period. However, if the value of J_0 is sufficiently small, the nucleation rate reaches a limit at the J_0 value even if the degree of undercooling is increased. This is because even if the degree of supersaturation is increased, the value of the part of the exponential function of the nucleation rate does not exceed 1. Therefore, nucleation continues at the constant maximum and peak value for a certain extent of time until sufficient nucleus number density is formed by an inactive but maximum and peak nucleation rate. This phenomenon will exert considerable influence on the interpretation of the resultant crystal size distribution and rock texture through their dependence on the cooling rate (described later).

In cases where the nucleation rate is small in experiments, sample size may affect the measurement of nucleation rate (the sample size effect) because nucleation events cannot be completely found in a small sample. In such cases, the probability of identifying crystal nuclei in the sample have to be considered. For this purpose, to determine the probability $\mathcal{P}_x$ of identifying crystal nuclei as a function of time, multiple experiments with the same degree of undercooling ΔT should be conducted. Assuming that the probability of identifying crystal nuclei is determined using the Poisson process, the probability is influenced by sample size and experiment time and, because the size is large or the time is extended, the probability increases. Therefore, probability is expressed as a function of the nucleation rate J, which is as a function of the experiment time t, and the sample size V, as follows (Toschev et al. 1972):

$$\mathcal{P}_x = 1 - \exp\left(-\int_0^t J(t')V dt'\right) \tag{7.62}$$

Using the melt of Di_{80} composition (a wire-loop method using several millimeter-sized samples), Tsuchiyama (1983) conducted an experiment on an An-Di system. For $\Delta T = 70\,°C$ where the undercooling temperature is small, he fitted the time change of the probability determined by experiments repeated 31–34 times to the time change curve of the equation above in the case of $J(t) =$ constant and then the nucleation rate was determined at 1.3×10^{-2} $(cm^{-3}s^{-1}) \approx 10^4$ $(m^{-3}s^{-1})$. Using only one set of data of the degree of undercooling and the nucleation rate, the nucleation barrier Γ and the value of the pre-exponential factor J_0 cannot be uniquely determined even if the value of $\tilde{q}$ is determined.

7.4 Diffusion-Limited Growth

7.4.1 Steady-State Diffusion Growth of Spherical Crystals

In a spherically symmetric field, a steady-state solution to a diffusion equation is given using Eq. (4.20), i.e.,

$$C(r) = \frac{c_2}{r} + C_\infty \tag{7.63}$$

where the constant c_2 is determined by the boundary condition, which is specified by local equilibrium at the interface, i.e., equilibrium concentration C_R at the curvature radius R and assuming C_∞ when $r \to \infty$ yields

$$c_2 = R(C_R - C_\infty) \tag{7.64}$$

Note that whether the local equilibrium holds or not is not obvious. As seen later, crystallization is a process in which the structure comprising molecules rearranges from a liquid structure to a crystal structure and its reaction rate is not necessarily sufficiently faster than that of local diffusion; this will be discussed later.

Assuming the solution is dilute, the growth rate of spherical crystals is determined by Eq. (4.23) (note that v_S = molecular volume of the crystallization component is used rather than v_G) and Eq. (4.25). Using Eq. (7.20) in Eq. (4.25), the diffusion-limited crystal growth rate is obtained as follows:

$$\frac{dR}{dt} = \frac{v_S \cdot D \cdot C_{eq}}{R}\left(\frac{C_\infty}{C_{eq}} - \frac{C_R}{C_{eq}}\right) \tag{7.65}$$

$$= \frac{v_S \cdot D \cdot C_{eq}}{R}\left[\frac{C_\infty}{C_{eq}} - \exp\left(\frac{2\gamma v_S}{k_B T_0}\frac{1}{R}\right)\right] \tag{7.66}$$

Note that C_∞ / C_{eq} is the scaled supersaturation concentration. In cases where the critical nucleus radius R_C is used in the ratio of supersaturation concentration, assuming $C = C_\infty$ and substituting Eq. (7.37) into Eq. (7.66) yields

$$\frac{dR}{dt} = \frac{v_S \cdot D \cdot C_\infty}{R}\left\{1 - \exp\left[-R_X^*\left(\frac{1}{R_C} - \frac{1}{R}\right)\right]\right\} \tag{7.67}$$

$$= \frac{v_S \cdot D \cdot C_\infty \cdot R_X^*}{R}\left(\frac{1}{R_C} - \frac{1}{R}\right) \tag{7.68}$$

.[7] Relationship between the scaled degree of supersaturation Δ and the critical nucleus radius R_C is written as follows:

$$\Delta = \frac{R_X^*}{R_C} = \frac{C_\infty - C_{eq}}{C_{eq}} \tag{7.70}$$

Note that this degree of supersaturation is not a value defined by the interface but defined by the difference between the concentration C_∞ at a distance (infinity) unaffected by the interface and the equilibrium concentration at a flat interface (liquidus concentration at the temperature C_{eq}).

To see basic properties of the diffusion-limited growth law, R and t in Eq. (7.66) is scaled as follows.[8]

$$R = \frac{R_X^*}{\Delta}\tilde{R} = R_C\tilde{R} \tag{7.71}$$

$$t = \frac{R_0^2}{\Delta^3 \cdot v_S \cdot D \cdot C_\infty}\tilde{t} = \frac{R_C^2}{\Delta \cdot v_S \cdot D \cdot C_\infty}\tilde{t} \tag{7.72}$$

---

[7]When $\Delta g^* = (C_\infty - C_R)/C_\infty = R_X^*/R_C - R_X^*/R$ is considered as the driving energy for growth, Eq. (7.67) then becomes

$$\frac{dR}{dt} = \frac{v_S \cdot D \cdot C_\infty}{R}\left\{1 - \exp\left(-\frac{\Delta g^*}{k_B T}\right)\right\} \tag{69}$$

which is equivalent in the form to the expression of the reaction-limited growth explained in the next section, although the coefficient differs. However, note that the difference in concentration used for the driving energy for growth is the difference in concentration due to the difference in distance between the interface and a distant point. .

[8]The case in which R_C (or Δ) is assumed to change with time and used as time in transforming the equation will be the Lifshitz-Slyozov theory, which is discussed in Sect. 6.1.2.

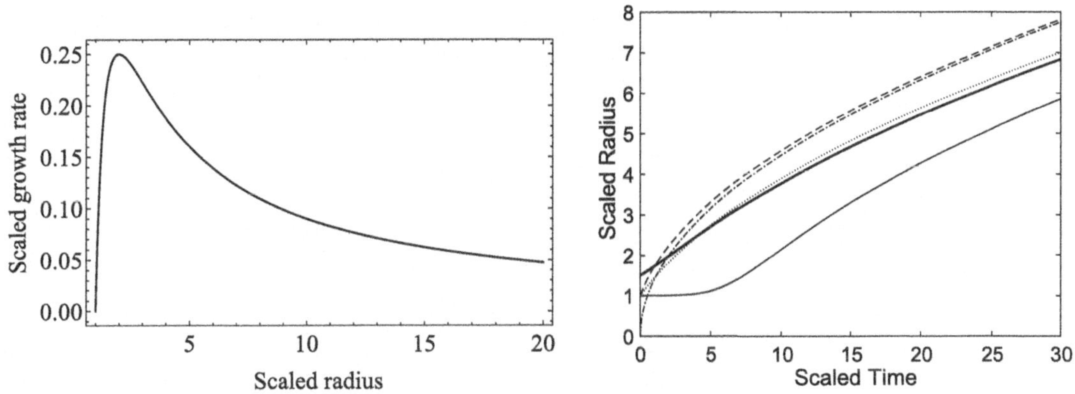

Fig. 7.11 Time change of the growth rate and the crystal radius in diffusion growth. (Left) Relationship between the scaled growth rate and the scaled crystal radius (Eq. 7.73). (Right) Comparison of the analytic solution (Eq. 7.76) of the diffusion-limited growth under different assumption of terms. Thin solid line: all terms are included. Broken line: $(\tilde{R}^2 - \tilde{R}^2(0)) = 2\tilde{t}$. Chain line: $\tilde{R} = (2\tilde{t})^{1/2}$. Dotted line: the term of ln is excluded. $\tilde{R}(0) = 1.001$ was used for the above. Thick solid line: $\tilde{R}(0) = 1.5$ and all terms are included

Hereafter, to understand the simplest temporal development of the crystal size under constant degree of supersaturation, equations are transformed assuming that R_C and Δ are constant.[9]

Using dimensionless crystal radius $\tilde{R}$ and time $\tilde{t}$, Eq. (7.66) is written as follows:

$$\frac{d\tilde{R}}{d\tilde{t}} = \frac{1}{\tilde{R}}\left(1 - \frac{1}{\tilde{R}}\right) \tag{7.73}$$

This equation shows that the growth rate reaches the maximum $d\tilde{R}/d\tilde{t} = 1/4$, i.e., $v_\mathrm{S} D C \Delta^2/(4R_X^*)$ at $\tilde{R} = 2$, i.e., $R = 2R_\mathrm{C}$ (the left chart of Fig. 7.11).

The solution to Eq. (7.73) is obtained by separating variables and, under the following initial condition,

$$\tilde{R} = \tilde{R}(0) \tag{7.74}$$

$$\tilde{t} = 0 \tag{7.75}$$

the solution is

$$\frac{1}{2}\left(\tilde{R}^2 - \tilde{R}^2(0)\right) + \tilde{R} - \tilde{R}(0) + \ln\left(\frac{\tilde{R} - 1}{\tilde{R}(0) - 1}\right) = \tilde{t} \tag{7.76}$$

The solution for $\tilde{R}(0) = 1.001$ is shown in the right chart of Fig. 7.11 using a thin solid line. A notable point is that delay in growth is observed in the initial stage. This initial delay is attributed to the natural logarithm (ln) term in Eq. (7.76) and occurs when $\tilde{R}(0)$ is close to 1. As $\tilde{R}(0)$ is close to 1, the delay becomes long. $\tilde{R}(0) = 1$ means that the initial size is the critical size, and the fact that growth is gradually accelerated from unstable equilibrium is represented in this chart. Calculation on the assumption that $\tilde{R}(0) = 1.5$ (the initial size is 1.5 times the critical size) gives a thick solid line in the figure in which the initial delay is not observed. That is, when the crystal radius is more than 1.5 times the critical radius, a so-called square root law of the diffusion-limited growth holds. Moreover,

[9]Diffusion growth (bubbles) in the case where the degree of supersaturation $1/R_\mathrm{C} - 1/R$, on the basis of the inverse of the critical radius is constant in Eq. (7.66), was dealt with in Sect. 4.3.

in both cases, when the growth advances and $\tilde{R} \gg \tilde{R}(0)$ holds, terms other than the term of $\tilde{R}^2$ can be ignored; therefore, $\tilde{R}^2 - \tilde{R}(0)^2 \approx 2\tilde{t}$ holds and the equation approaches the square root law

$$\tilde{R} \propto \tilde{t}^{1/2} \tag{7.77}$$

In this case, Eq. (7.73) approaches $d\tilde{R}/d\tilde{t} = \text{const}/\tilde{R}$.

Laplace transform and analytic solution of partial

Laplace transform is defined by

$$\hat{C}(x,s) = \int_0^\infty e^{-st} C(x,t)dt = \mathcal{L}(C(x,t)) \tag{7.78}$$

where s is a dummy variable. For simplicity, let us consider a simple 1D diffusion equation that does not involve the motion of the interface:

$$\frac{\partial C}{\partial t} = D\frac{\partial^2 C}{\partial x^2} \tag{7.79}$$

The initial condition is $C(x,0) = C_0$ when $t = 0$ and the boundary conditions are $C(0,t) = C_e$ when $x = 0$ and $C(\infty, t) = C_0$ when $x \to \infty$. By multiplying both sides of Eq. (7.79) by e^{-st} and integrating it, the left-hand side becomes

$$\text{LHS} = \int_0^\infty e^{-st}\frac{\partial C(x,t)}{\partial t}dt = \left[C(x,t)e^{-st}\right]_0^\infty - \int_0^\infty (-s)e^{-st}C(x,t)dt \tag{7.80}$$

$$= -C_0 + s\hat{C}(x,s) \tag{7.81}$$

using integration by parts. Because the right-hand side does not include differentiation of t, integration can be moved into differentiation and C can be replaced with $\hat{C}$. Therefore, the diffusion equation becomes the following ordinary differential equation with respect to $\hat{C}$.

$$D\frac{\partial^2 \hat{C}(x,s)}{\partial x^2} + C_0 - s\hat{C}(x,s) = 0 \tag{7.82}$$

Considering the Laplace transform of the boundary conditions

$$\hat{C}(0,s) = \int_0^\infty C_e e^{-st}dt = \frac{C_e}{s}, \quad \hat{C}(\infty,s) = \int_0^\infty C_0 e^{-st}dt = \frac{C_0}{s} \tag{7.83}$$

the solution to Eq. (7.82) will be

$$\hat{C}(x,s) = \frac{C_e - C_0}{s}e^{-x\sqrt{s/D}} + \frac{C_0}{s} \tag{7.84}$$

$C(x,t)$ can be obtained using the inverse Laplace transform of $\hat{C}(x,s)$.

$$C(x,t) = (C_e - C_0)\mathcal{L}^{-1}\left(\frac{e^{-x\sqrt{s/D}}}{s}\right) + C_0\mathcal{L}^{-1}\left(\frac{1}{s}\right) \tag{7.85}$$

$$= (C_e - C_0)\text{erfc}\left(\frac{x}{2\sqrt{Dt}}\right) + C_0 \tag{7.86}$$

where $\mathcal{L}^{-1}(y(x,s))$ is the inverse Laplace transform of y and defined as follows:

$$\mathcal{L}^{-1}(y(x,s)) = \frac{1}{2\pi i}\int_{\gamma-i\infty}^{\gamma+i\infty} y(x,s)e^{-st}ds \tag{7.87}$$

Although this equation can be calculated by the complex path integral, the results are given in tables because the calculation is often complicated.

7.4.2 Non-steady State Diffusion Growth on the Plane Crystal Face

This section discusses the case where the moving velocity of the interface is so sufficiently slow that it does not influence the concentration profile due to diffusion. A fundamental equation is the diffusion equation (7.79) and the following conditional equation represents a mass balance between the diffusion flux at the interface and crystal growth:

$$D\left.\frac{\partial C}{\partial x}\right|_{x=0} = (C_S - C_{\text{int}})G = (K_D - 1)C_{\text{int}}G \tag{7.88}$$

where K_D is the distribution coefficient C_S/C_{int} and G is the moving velocity (growth rate) of the interface. For cases where a solid of a pure substance crystallizes in a dilute solution, $C_S = 1$ and $C_S \gg C_{\text{int}}$ holds and the equation will be the conditional equation for the diffusion growth of spherical crystals, which has been discussed earlier. If the interfacial concentration C_{int} in the right-hand side and the composition of the solid phase (distribution coefficient) are constant, the concentration profile in liquid is determined independent of Eq. (7.88) and the moving velocity is determined based on this equation. However, on the right-hand side, the interfacial concentration C_{int} and the moving velocity G are generally a function of the interfacial concentration; if the interfacial concentration changes with time, we have to solve the diffusion equation with this conditional equation as the boundary condition. That is difficult in terms of mathematics. The problem of the diffusion field after considering the interface movement will be dealt with later regarding zonal structure of the chemical composition of crystals.

(1) The Case in Which the Interfacial Concentration Is Constant

For cases in which the interfacial concentration C_{int} is equal to the constant value of equilibrium concentration C_e, the moving velocity of interface G is determined from boundary condition (Eq. 7.88) by giving the concentration profile. A satisfying solution to diffusion equation can be obtained by the method of the Laplace transform explained in the Box in Sect. 7.4.1 as well as a solution to the concentration profile will be Eq. (7.86) (the left chart of Fig. 7.12). In this case, diffusion flux is given by the concentration gradient at the interface as follows:

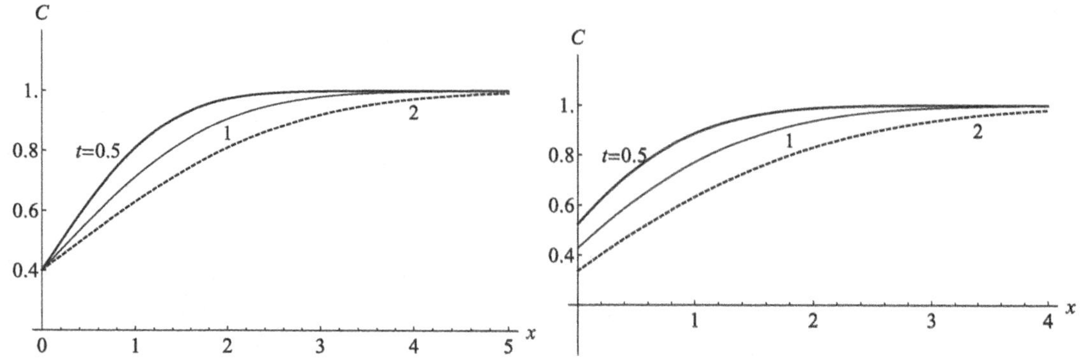

Fig. 7.12 Temporal change of the concentration profile in liquid when the interface does not move. Line types correspond to the time elapsed. (Left) The case where the interfacial concentration is constant (Eq. 7.86). (Right) The case where the diffusion growth rate is constant (Eq. 7.91)

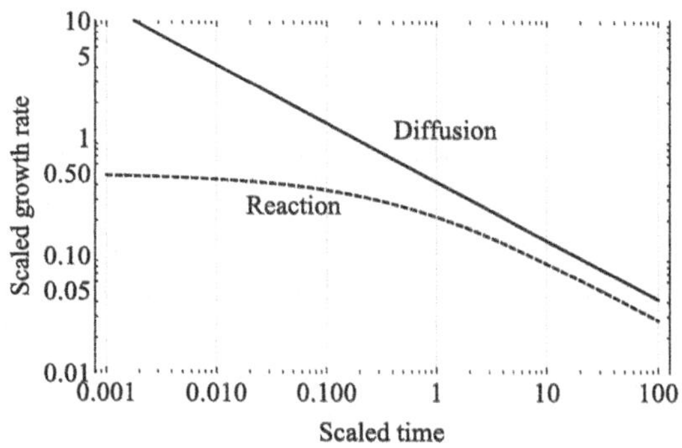

Fig. 7.13 Comparison of the time change of growth rate of a flat interface in the case where the interface does not move. The solid line (Diffusion) represents the case where interfacial concentration is constant (Eq. 7.90); the growth rate is determined only by the concentration gradient of interface. The growth rate is proportional to $t^{-1/2}$. The broken line (Reaction) represents the case where the interfacial concentration is considered to be constant but the growth rate is proportional to the difference between the interfacial and the equilibrium concentrations (the degree of supersaturation) (Eq. 7.96)

$$\left|\frac{\partial C}{\partial x}\right|_{x=0} = \frac{(C_0 - C_{eq})}{(\pi D t)^{1/2}} \tag{7.89}$$

From Eq. (7.88), the moving velocity of the interface G is as follows:

$$G = \frac{(C_0 - C_{eq})}{(C_S - C_{eq})(\pi D t)^{1/2}} \tag{7.90}$$

which is found to be inversely proportional to the square root of time (Fig. 7.13).

(2) The Case Where the Interfacial Concentration Changes

(2-1)The case where the growth rate is constant

When the moving velocity of interface G is constant, the interfacial concentration $C_{int} = C(0, t)$ is determined as a part of the solution to the diffusion equation, thus satisfying the boundary condition (Eq. 7.88). Moreover, in this case, the solution of the diffusion equation can be obtained using the method of the Laplace transform explained in Box in Sect. 7.4.1; consequently, the concentration profile is given as follows:

$$C(x,t) = C_0\{1 - \psi(x,t)\} \tag{7.91}$$

(the right chart of Fig. 7.12), where $\psi(x,t)$ is given by

$$\psi(x,t) = \left\{\mathrm{erfc}\left[\frac{x}{2(Dt)^{1/2}}\right] - \exp\left[\frac{G'}{D}x + \left(\frac{G'}{D}\right)^2 Dt\right] \cdot \mathrm{erfc}\left[\frac{x}{2(Dt)^{1/2}} + \left(\frac{G'}{D}\right)(Dt)^{1/2}\right]\right\} \tag{7.92}$$

In this case, $G' = (K_\mathrm{D} - 1)G$. The moving velocity of the interface G is given by $G \approx dC/dx|_{x=0}/C(0,t)$, which is, of course, a constant value, as assumed in this case. A solution to the diffusion equation considering the interface movement is given by Smith et al. (1955) (Eq. 7.242).

(2-2)The case where the growth rate is proportional to the degree of supersaturation at the interface (explained in detail in the next section).

In this case, the interfacial concentration $C_\mathrm{int} = C(0,t)$ included in the boundary condition (Eq. 7.88) is an unknown variable determined as a part of the solution to the diffusion equation. If, similar to the case of reaction-limited growth, the moving velocity of the interface G is proportional to the difference $(C_\mathrm{int} - C_\mathrm{eq})$ between the interfacial concentration C_int and the equilibrium concentration C_eq,

$$D\left.\frac{\partial C}{\partial x}\right|_{x=0} = (K_\mathrm{D} - 1)C_\mathrm{int}G_0(C_\mathrm{int} - C_\mathrm{eq}) \tag{7.93}$$

is obtained. Note that equilibrium concentration is nonlinear with respect to C_int and difficult to be analyzed. When the influence of change in C_int itself on the boundary condition and the influence of change in $C_\mathrm{int} - C_\mathrm{eq}$ on the boundary condition is compared, the latter is expected to be large. In this case, $(K_\mathrm{D} - 1)C_\mathrm{int}$ can be approximately considered as a constant $(K_\mathrm{D} - 1)C_\mathrm{eq}$. Therefore,

$$D\left.\frac{\partial C}{\partial x}\right|_{x=0} = G'(C_\mathrm{int} - C_\mathrm{eq}) \tag{7.94}$$

where $G' = (K_\mathrm{D} - 1)C_\mathrm{eq}G_0$ is a constant. A solution of the diffusion equation, which satisfies the boundary condition, is given by Carslaw and Jaeger (1959) as follows:

$$C(x,t) = C_0\left\{1 - \left(1 - \frac{C_\mathrm{eq}}{C_0}\right)\psi(x,t)\right\} \tag{7.95}$$

The concentration gradient at the interface is

$$\left.\frac{\partial C}{\partial x}\right|_{x=0} = (C_0 - C_\mathrm{eq})\exp\left[\left(\frac{G'}{D}\right)^2 Dt\right]\mathrm{erfc}\left[\left(\frac{G'}{D}\right)(Dt)^{1/2}\right] \tag{7.96}$$

As t becomes large, the moving velocity of the interface approaches that in the case where the interfacial concentration is constant (Fig. 7.13). $(Dt)^{1/2}$ gives an effective diffusion distance x_D. The content of the error function (erfc) will be the Péclet number (Pe) (Pe $= Gx_D/D$). Moreover, Pe appears in a problem of diffusion considering the movement of the interface, dealt with later.

Thus, in cases where diffusion transport of the crystallization component driven by the concentration gradient controls the crystal growth rate, the growth rate decreases in inverse proportion to the square root of time. Consequently, the growth length, or the crystal size, L increases in proportion to the square root of time $L \approx (Dt)^{1/2}$.

7.5 Reaction-Limited Growth

7.5.1 Theoretical Consideration

Reaction-limited growth is the case in which a reaction at the liquid-crystal interface, i.e., a reaction process in which molecules in melt are taken in the crystal lattice controls the growth rate. Actual reaction progresses using steps and dislocation on the crystal surface. Substances are fed to the vicinity of the crystal surface sufficiently fast by diffusion. Here, we do not mention details of the reaction and phenomenologically consider the reaction as a thermally activated process to model it.

We define the liquid-crystal interface as a region with thickness on the crystal surface. Assuming that the number of molecules incorporated in the crystal in the region is $\mathcal{M}$, crystallization reaction can be written as follows:

$$\mathcal{M} \longrightarrow \mathcal{M} + 1 \tag{7.97}$$

and dissolution reaction can be written as follows:

$$\mathcal{M} \longleftarrow \mathcal{M} + 1 \tag{7.98}$$

Assuming that the moving velocities (m/s) of the interface due to the reaction are G_+ and G_-, respectively,

$$G_+ = k_+ \exp\left(-\frac{\Delta g_+}{k_B T}\right) \tag{7.99}$$

$$G_- = k_- \exp\left(-\frac{\Delta g_-}{k_B T}\right) \tag{7.100}$$

are obtained from the Arrhenius relationship. In these equations, k_+ and k_- are reaction constants and Δg_+ and Δg_- are activation energy of the reaction (Fig. 7.14). Because true growth rate G is the balance between crystallization and dissolution,

$$G_{\text{react}} = G_+ - G_- = k_+ \exp\left(-\frac{\Delta g_+}{k_B T}\right)\left[1 - \frac{k_-}{k_+}\exp\left(-\frac{\Delta g^*}{k_B T}\right)\right] \tag{7.101}$$

is obtained. In this equation, $\Delta g^* = \Delta g_- - \Delta g_+$, which is the difference between free energy of a molecule in melt and free energy of a molecule on a crystal. From Sect. 7.1.2, this can be considered as the free energy of crystallization at a flat interface, $\Delta g^* = k_B T \ln C/C_{eq}$, where C is the concentration in liquid at the interface. Moreover, for crystallization at a curved interface, C_R have to be used as equilibrium concentration instead to C_{eq}. For a saturated state $\Delta g^* = 0$ (at the melting point), to satisfy $G_+ - G_- = 0$, the reaction constant should be $k_+ = k_- = G_0$= constant.[10]. Ultimately, Eq. (7.101) becomes

[10] Refer to the Footnote 7.

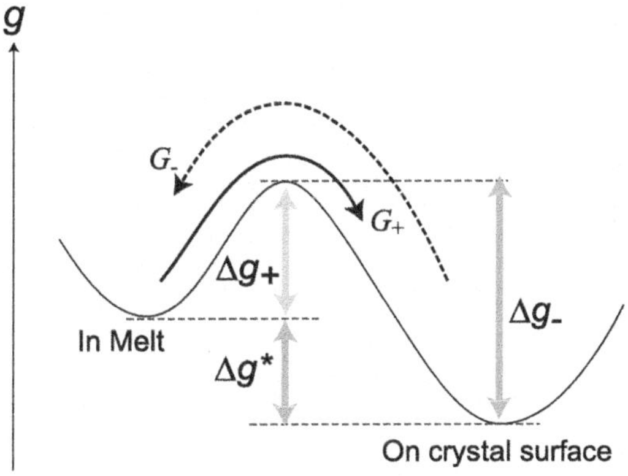

Fig. 7.14 A conceptual diagram of reaction-limited growth

$$G_{\text{react}} = \underbrace{G_{\text{react0}} \exp\left(-\frac{q}{k_B T}\right)}_{\text{Diffusion}} \cdot \underbrace{\left[1 - \exp\left(-\frac{\Delta g^*}{k_B T}\right)\right]}_{\text{Driving force of growth}} \tag{7.102}$$

In this equation, the magnitude of energy barrier when molecules move from liquid to the crystal surface Δ g+ is considered as an activation energy at molecular fluctuation in liquid, that is, the activation energy of diffusion q.[11] Moreover, using $\Delta g^*(R)$ instead to Δg^* gives the growth rate at a curved interface.

Because the free energy Δg^*, a driving force of growth, is zero at the melting point, $\exp(-\Delta g^*/k_B T)$ is 1 and the driving force of growth is 0. The driving force of growth drastically increases with decrease in temperature or increase in the degree of undercooling. However, diffusion becomes inactive with decrease in temperature. Therefore, the growth rate will take the maximum at a certain degree of undercooling because of a competition between them.

Given a flat interface ($R \to \infty$), $\Delta g^* = \Delta s \cdot (T_0 - T) = \Delta h(1 - T/T_0)$ is obtained from Eq. (7.39), and given $\tilde{T} = T/T_0$,

$$G_{\text{react}} = G_{\text{react0}} \exp\left(-\frac{\tilde{q}}{\tilde{T}}\right) \cdot \left[1 - \exp\left(-\Delta\tilde{h}\frac{(1-\tilde{T})}{\tilde{T}}\right)\right] \tag{7.103}$$

is thus obtained. $\tilde{q}$ in this equation is the scaled activation energy of diffusion in liquid and is defined using Eq. (7.51). $\Delta\tilde{h}$ is dimensionless latent heat and defined by Eq. (7.29). Drawing a graph of the growth law on a logarithmic scale yields the left chart of Fig. 7.15. As the temperature decreases from the melting point ($\tilde{T} = 1$), the growth rate drastically increases; however, as the temperature decreases further, the thermally activated process becomes inactive and the growth rate decreases.

To ensure generality, an interface having curvature is assumed and the driving energy of growth is expressed in terms of concentration (Eq. 7.40). Moreover, because $\Delta g^*(R) = k_B T \ln C/C_R$ (C is the concentration in liquid at the interface C_{int}) holds,

$$G_{\text{react}} = G_{\text{react0}} \exp\left(-\frac{\tilde{q}}{\tilde{T}}\right) \cdot \left[\frac{C - C_R}{C}\right] \tag{7.104}$$

[11] Because $\Delta g^* + \Delta g_+ = \Delta g_-$ is assumed, if Δg_- is assumed to be considered as the activation energy of diffusion in the solid phase here, the free energy of solid-liquid phase transition would be equal to the difference in the activation energy of diffusion between solid and liquid. However, it is unclear whether this relation holds.

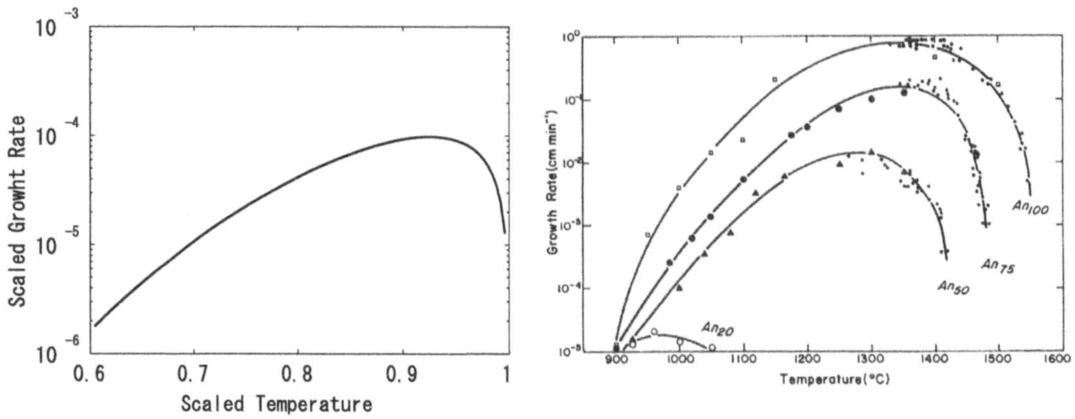

Fig. 7.15 (Left) Temperature dependence of G when $G_0 = 1$, $\tilde{q} = 8$, and $\Delta h/k_B T_0 = 10$ are assumed. (Right) Experimental results of the growth rate of plagioclase, conducted by Kirkpatrick et al. (1979). Curves are fitted curves

holds. In reaction-limited growth, the growth rate increases in proportion to $C - C_R$.

From the Gibbs-Thomson equation (Eq. 7.20), equilibrium interfacial concentration of spherical crystals C_R is $C_R = C_{\text{eq}} \exp(R_X^*/R)$, which rapidly approaches C_{eq} as R becomes larger than the scale of microscopic crystal nucleus R_X^*. Hence, the growth rate is proportional to the degree of supersaturation, and the moving velocity of the interface does not depend on the crystal radius. Consequently, the crystal radius is proportional to time.

$$\tilde{R} \propto \tilde{t} \tag{7.105}$$

For $C_R = C_{\text{eq}}$, this is called a classical Frenkel-Wilson law. $G_0 \exp(-\tilde{q}/\tilde{T})$ is a kinetic factor of crystal growth. The kinetic factor is dependent on the reaction mechanism on the crystal surface; the mechanism is separated into a process of continuous formation of roughness (steps) on the crystal surface and a process in which molecules attach to the steps. As the step formation process, surface nucleation and dislocation (the Frank mechanism) are considered, and studies on them have been conducted in the field of crystal growth.

7.5.2 Comparison with Experiments

The right chart of Fig. 7.15 shows the result of the growth rate of plagioclase, obtained by an in situ observation experiment by Kirkpatrick et al. (1979). The general trends of the growth rate are consistent. The degree of undercooling $\tilde{T}_{G_{\text{PEAK}}}$ giving the maximum growth rate is approximately determined from the extrema in the exponential part of Eq. (7.103) and given as

$$\tilde{T}_{G_{\text{PEAK}}} = \frac{\Delta\tilde{h}}{\Delta\tilde{h} + \ln\left(1 + \frac{\Delta\tilde{h}}{\tilde{q}}\right)} \tag{7.106}$$

.[12] $\tilde{T}_{G_{\text{PEAK}}}$ is determined only by $\tilde{q}$ and $\Delta\tilde{h}$. Because both $\tilde{q}$ and $\Delta\tilde{h}$ have values from 5 to about 20, their ratio $\Delta\tilde{h}/\tilde{q}$ will be of the order of 1. Their relationship is shown in the left chart of Fig. 7.16.

[12] Although the right-hand side is an approximate equation when the Frenkel-Wilson law is used, because $\Delta\tilde{h}/\tilde{q}$ cannot become much smaller than 1, it cannot be considered as an approximation from the middle part of the equation.

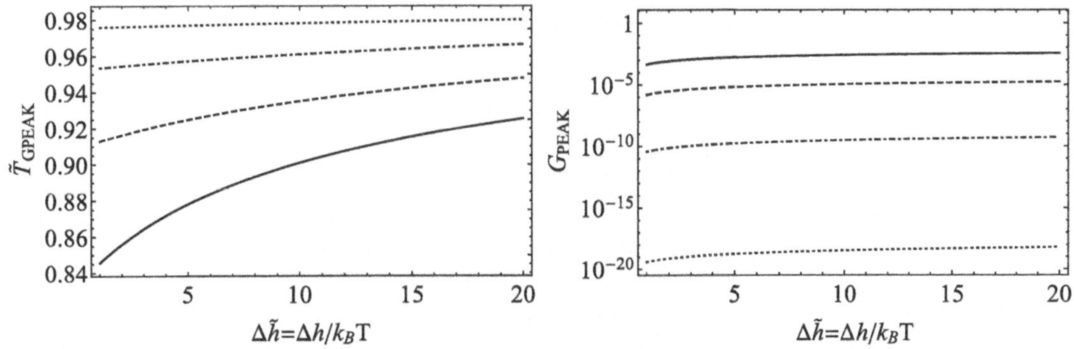

Fig. 7.16 (Left) Scaled temperature giving the maximum value of growth rate as a function of driving force of crystallization (latent heat $\Delta h/k_B T$), calculated by Eq. (7.106). Scaled activation energy of diffusion $\tilde{q} = q/k_B T$ is used as a parameter. Line types represent difference in $\tilde{q}$ as follows: dotted line ($\tilde{q} = 5$), chain line ($\tilde{q} = 10$), broken line ($\tilde{q} = 20$), and solid line ($\tilde{q} = 40$). (Right) Corresponding maximum growth rate. Line types represent the same as the left figure. $G_0 = 1$ is then assumed

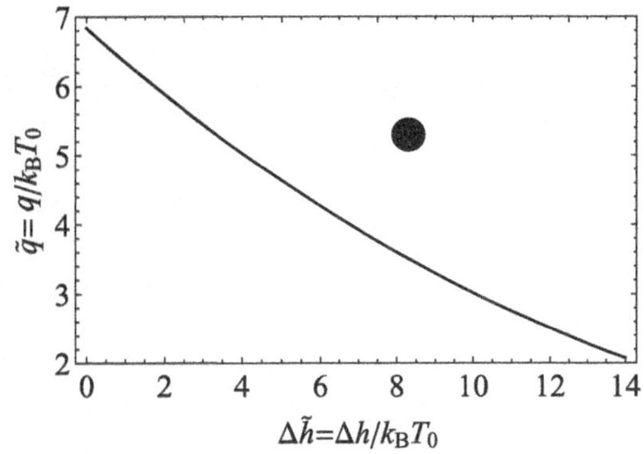

Fig. 7.17 The curve represents the relationship between latent heat ($\Delta\tilde{h}$) in the case of $\tilde{T}_{\text{G}_{\text{PEAK}}} = 0.87$, expected from a simple reaction-limited growth model, and the activation energy of diffusion ($\tilde{q}$). The black circle denotes the value of the activation energy and the latent heat that is experimentally measured

Rewriting Eq. (7.106) to $\exp(\Delta\tilde{h}(1 - \tilde{T}_{\text{G}_{\text{PEAK}}})/\tilde{T}_{\text{G}_{\text{PEAK}}}) = 1 + \Delta\tilde{h}/\tilde{q}$ and substituting it into Eq. (7.103) yields a corresponding peak value of the growth rate G_{PEAK},

$$G_{\text{PEAK}} = G_0 \frac{\Delta\tilde{h}}{\tilde{q} + \Delta\tilde{h}} \exp\left(-\frac{\tilde{q}}{\tilde{T}_{\text{G}_{\text{PEAK}}}}\right) \tag{7.107}$$

The values of $\Delta\tilde{h}$, $\tilde{q}$, and G_0 are restricted from this peak growth rate. This is shown in the right chart of Fig. 7.16.

Because this experiment with Anorthite = 100% by Kirkpatrick results in $\tilde{T}_{\text{G}_{\text{PEAK}}} \approx 0.87$, the values of $\Delta\tilde{h}$ and $\tilde{q}$ are restricted from Eq. (7.106) (Fig. 7.17). On the other hand, from the melting point $T_{\text{An}} = 1826$ K and $\Delta\mathcal{H} = 126$ kJ/mol, the experimental conditions and physical properties (Table 7.1) of Anorthite100, $\Delta h/k_B T_{\text{An}} = 8.3$ is obtained. Furthermore, even if the activation energy of diffusivity is the minimum value, i.e., 80 kJ/mol, $\tilde{q} = q/k_B T_0 = 5.3$ is obtained and there is a difference from the curve in the figure, which is a restriction from the model. The model requires smaller $\tilde{q}$ and $\Delta\tilde{h}$. However, as reported by Lasaga (1982), if composition dependence of diffusivity[13] is considered, agreement with the experimental data becomes better. The composition dependence of diffusivity

[13] Lasaga used the inverse of viscosity as the index.

($D = D_0 \exp(q/k_B T)$) appears in the pre-exponential factor D_0 and the activation energy q; moreover, there is a fixed relation (the compensation law) between them.

Expressions of concentration: C, C_{eq}, C_R, C_{int}, and $\tilde{C}$

In this book, concentration (or composition) in a unit of [Number of molecules/m^3] is expressed with C. $\tilde{C}$ is a dimensionless concentration in wt% (100 times). C_{eq} is the concentration of liquid that is in equilibrium with bubbles or crystals through a flat interface (the curvature radius is infinity) and corresponds to the concentration on the phase boundaries such as liquidus on an equilibrium phase diagram and solubility curves. C_R is the concentration of liquid that is in equilibrium with bubbles or crystals through an interface with a curvature radius R (being positive when the liquid side is convex) and $C_R(R \to \infty) = C_{eq}$ holds. C_{int} is the concentration in liquid at the interface, and if the interfacial equilibrium holds, $C_{int} = C_R$ holds.

7.5.3 Balanced Growth Between Diffusion and Reaction

Assuming the case of spherical crystals, the influence of the interface movement on the diffusion field is ignored. For diffusion-limited growth, it is assumed that concentration at the solid-liquid interface with a radius of R is equilibrium concentration C_R and the growth rate is determined by molecular flux driven by the difference between C_R and concentration at a distant point C_∞, i.e., the concentration gradient. For reaction-limited growth, however, the concentration in liquid at the solid-liquid interface C_{int} has a larger supersaturation concentration than equilibrium concentration in liquid C_R and the degree of supersaturation $C_{int} - C_R$ at the point becomes the driving energy for crystal growth. Because an infinitely fast reaction is assumed in the diffusion-limited growth, $C_{int} = C_R$ always holds. For reaction-limited growth, on the other hand, infinitely fast diffusion is assumed and it is implicitly assumed that the concentration in liquid at the solid-liquid interface represents that in the entire liquid. Actually, the growth is considered to occur in the intermediate state. In this section, let us examine the parameters that control the balanced state and how much C_{int} and C_R differ.

First, assuming that the concentration of a crystal component in a liquid near the interface to be general concentration C_{int} does not follow the Gibbs-Thomson law, an equation for steady-state diffusion-limited growth is derived.[14] In an equation for diffusion, assuming $r = R$ as the boundary condition and $C = C_{int}$, the concentration profile can be written as follows:

$$C(r) = \frac{R(C_{int} - C_\infty)}{r} + C_\infty \tag{7.108}$$

The gradient at the solid-liquid interface is determined as follows:

$$\left[\frac{dC}{dr}\right]_{r=R} = \frac{C_\infty - C_{int}}{R} \tag{7.109}$$

A solid phase comprising a single component is then assumed. In this case, the concentration of the solid phase C_S (number of molecules/m^3) is $1/v_S$. Moreover, the molar volume of the crystallization

[14]The case following the G-T law in dilute solution was dealt with in Sect. 7.4.

component in the liquid is assumed to be equal to that of the solid phase. Conservation of mass at the solid-liquid interface is then written as follows:

$$D\frac{C_\infty - C_{\mathrm{int}}}{R} \times 4\pi R^2 = (C_{\mathrm{S}} - C_{\mathrm{int}})G_{\mathrm{dif}} \times 4\pi R^2 \tag{7.110}$$

$$[\text{Inflowbydiffusion}] = [\text{Consumptionbycrystalgrowth}] \tag{7.111}$$

Note that G_{dif} is the rate of diffusion-limited growth:

$$G_{\mathrm{dif}} = \frac{D}{R}\frac{\tilde{C}_\infty - \tilde{C}}{\tilde{C}_{\mathrm{S}} - \tilde{C}_{\mathrm{int}}} \tag{7.112}$$

where $\tilde{C} = C/C_{\mathrm{eq}}$, $\tilde{C}_\infty = C_\infty/C_{\mathrm{eq}}$, and $\tilde{C}_{\mathrm{S}} = C_{\mathrm{S}}/C_{\mathrm{eq}}$. Assuming that $\tilde{C}_{\mathrm{int}} = \tilde{C}_R$ and $\tilde{C}_{\mathrm{S}} - \tilde{C}_{\mathrm{int}} \approx C_{\mathrm{S}}/C_{\mathrm{eq}}$ hold, the equation becomes the equation of diffusion-limited growth in dilute solution (Eq. 7.66).

When this becomes equal to reaction-limited growth, the reaction and diffusion are balanced. The reaction-limited growth rate in this case is proportional to the difference between the supersaturated concentration at the solid-liquid interface C_{int} and the equilibrium concentration C_R. Assuming that the coefficient of proportionality is $G_0 \exp(-\tilde{q}/\tilde{T}) = \hat{G}_0$, the rate of the reaction-limited growth G_{react} becomes

$$G_{\mathrm{react}} = \hat{G}_0(\tilde{C}_{\mathrm{int}} - \tilde{C}_R) \tag{7.113}$$

where $\tilde{C}_R = C_R/C_{\mathrm{eq}}$, in which C_R is equilibrium concentration at the interface with a curvature radius of R and is determined from the Gibbs-Thomson relation.

Because $G_{\mathrm{dif}} = G_{\mathrm{react}}$, solving a quadratic equation for $\tilde{C}_{\mathrm{int}}$ yields the balanced interfacial concentration $\tilde{C}^*_{\mathrm{int}} = C^*_{\mathrm{int}}/C_{\mathrm{eq}}$ as follows:

$$\tilde{C}^*_{\mathrm{int}} = \frac{C^*_{\mathrm{int}}}{C_{\mathrm{eq}}} = \frac{1}{2}(\tilde{C}_R + (\mathrm{Pe}\tilde{R})^{-1} + \tilde{C}_{\mathrm{S}})\left[1 - \sqrt{1 - \frac{4((\mathrm{Pe}\tilde{R})^{-1}\tilde{C}_\infty + \tilde{C}_{\mathrm{S}}\tilde{C}_R)}{(\tilde{C}_R + (\mathrm{Pe}\tilde{R})^{-1} + \tilde{C}_{\mathrm{S}})^2}}\right] \tag{7.114}$$

In this equation,

$$\mathrm{Pe} = \frac{\hat{G}_0 R^*_X}{D} \tag{7.115}$$

is the Péclet number defined by the moving velocity of the interface $\hat{G}_0$ and is a constant. The following equations are given as a function of $\tilde{R} = R/R^*_X$, the crystal radius is scaled by the radius R^*_X.

$$\tilde{C}_R(\tilde{R}) = \frac{C_R}{C_{\mathrm{eq}}} = \exp\left(\frac{1}{\tilde{R}}\right) \tag{7.116}$$

and,

$$\tilde{R}_{\mathrm{C}} = \frac{R_C}{R^*_X} = \left(\ln \frac{C}{C_{\mathrm{eq}}}\right)^{-1} \tag{7.117}$$

Naturally, based on the Gibbs-Thomson relation, this balanced interfacial concentration becomes larger than the saturated concentration CR. If this concentration is used as interfacial concentration, the growth rate will become smaller than the diffusion growth rate in the case of the interfacial equilibrium. As

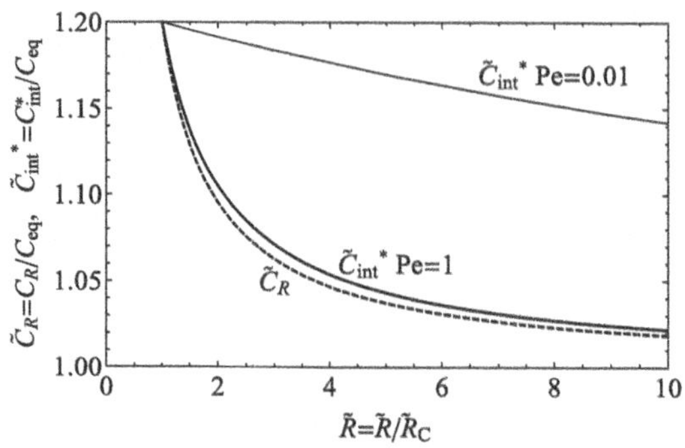

Fig. 7.18 Difference between interfacial equilibrium concentration (broken line) following the Gibbs-Thomson relation and interfacial concentration (solid line) in which growth reaction and diffusion is balanced. To avoid the uncertainty of the relationship between R^*_X and R, $\tilde{R}/\tilde{R}_C$ is considered as the crystal radius. As the Péclet number becomes large (as the kinetic factor $\hat{G}_0$ becomes large), the interfacial concentration approaches the interfacial equilibrium concentration determined by the Gibbs-Thomson law and the growth rate approaches the diffusion-limited growth

the kinetic factor $\hat{G}_0$ becomes large, the Péclet number becomes large, and the growth approaches the diffusion-limited growth. In this case, the approximation $\tilde{C}_{int} \approx \tilde{C}_R$ can be used and Eq. (7.67) will be obtained. Moreover, if D is sufficiently large, $\tilde{C}_{int} \approx (\mathrm{Pe}^{-1}\tilde{C}_\infty \tilde{R}^{-1} + \tilde{C}_S\tilde{C}_R)/(\tilde{C}_R + \tilde{C}_S) \to \tilde{C}_\infty$ holds whenPe $\to 0$, thus representing the reaction-limited growth. This is shown in Fig. 7.18.

In cases where the interface is flat, the width of the boundary layer on the side of the liquid phase is adjusted as it is determined by D/G against any values of the growth rate G. This is a point greatly different from the spherically symmetric diffusion field.

7.6 Temporal Development of the Crystallization Process: Crystallization of a Binary Eutectic System

7.6.1 Scaling and Controlling Parameters

Similar to the case of bubble size distribution, crystal size distribution caused by cooling crystallization can be understood by a scheme considering nucleation/growth processes. As explained in the chapter on temporal development of bubble size distribution, there are two approaches of understanding temporal development of size distribution: the Eulerian formulation and the Lagrangian formulation.

Brandeis and Jaupart (1987) solved the problem of cooling crystallization, including size distribution, and they assumed a one-component system and solved the problem using the Lagrangian formulation. They assumed the phenomenological nucleation rate and reaction-limited growth rate and attempted to understand the obtained numerical result through scaling using the scale of the nucleation rate J_m and the scale of the growth rate G_m. However, they did not investigate these values at the point of time when the nucleation rate reaches the maximum in the whole history of crystallization, i.e., when the number density is determined. Moreover, they did not systematically investigate the dependence of crystal number density on cooling rate.

Subsequently, Hort and colleagues (Spohn et al. 1988; Hort and Spohn 1991a) assumed a binary eutectic system and solved the problem using the Lagrangian formulation to investigate change in the grain size in both dikes and sills. Moreover, their studies assumed that the growth is reaction-limited growth. They nondimensionalized the nucleation rate and crystal growth rate using the peak values of master curves, J_{PEAK} and G_{PEAK}, and defined the following Avrami number (Av), which naturally appears in a dimensionless governing equation system, as a parameter that characterizes crystallization.

$$\mathrm{Av} = \frac{(J_{\mathrm{PEAK}} G_{\mathrm{PEAK}}^3)^{-1}}{t_{\mathrm{th}}^4} \tag{7.118}$$

where t_{th} is the timescale of cooling. Because the kinetic time constant t_{ki} can be defined as

$$t_{\mathrm{ki}} = \left(J_{\mathrm{PEAK}} G_{\mathrm{PEAK}}^3\right)^{-\frac{1}{4}} \tag{7.119}$$

(see Box in Sect. 7.2), the Av number is a dimensionless number of the ratio of the kinetic timescale to the timescale of cooling to the power of four.

Although there is no doubt that the Av number controls the crystallization behavior, note that the J_{PEAK} and G_{PEAK} Hort et al. used for the Av number are not the maximum values through the whole crystallization history, i.e., the values at the time when the nucleation rate takes the local maximum value. In other words, because the Av number is not a parameter that directly controls crystallization, it becomes difficult to understand an empirical rule for the result of numerical experiments by simple analysis of scaling using the Av number (see Box in Sect. 7.2).

This Av number can be thought as a rough indication of the physical properties and cooling rate (the inverse of t_{th}) contained in both JJ_{PEAK} and G_{PEAK}. As seen in the previous section, however, J_{PEAK} and G_{PEAK} are determined by the balance of the values of Γ, $\tilde{q}$, and $\Delta\tilde{h}$ in the exponential terms of master equations of nucleation and growth rates; however, they do not include explicitly these values. Considering these points, this approach of considering the Av number is mathematically elegant but not useful for the physical interpretation of results.

Moreover, something similar to this parameter appears in Eulerian formulation (Toramaru 1991, 2001).

Toramaru (1991, 2001):

$$\alpha_3 = \frac{\text{Cooling time}}{\text{Growth kinetic time}} = \frac{t_{\mathrm{th}}}{t_{\mathrm{g}}} = \frac{\phi_0 D t_{\mathrm{th}}}{(R_X^*)^2} \propto \left|\frac{dT}{dt}\right|^{-1} \tag{7.120}$$

The parameter α_3 is the ratio of the timescale of the cooling time to the timescale of the crystal growth. t_{g} is determined by the kinetics of crystal growth and given as follows:

$$t_{\mathrm{g}} = \frac{R_X^*}{G_0} \tag{7.121}$$

where R_X^* is the size scale of the crystal nucleus defined by Eq. (7.21) and is considerably influenced by surface tension (energy). Therefore, it exerts considerable influence when the cooling rate and α_3 is related. G_0 is a coefficient of the growth law and determined by a physical property representing the growth rate. For example, for diffusion growth, it will be $(\phi_{\mathrm{S}} D)/R_X^*$ and is related to diffusivity. Figure 7.19 shows the relationship between the interfacial energy and cooling rate.[15]

[15]The Av number and α_3 are mutually related. Assuming that the coefficient of the nucleation rate is J_0,

$$\alpha_3 = \frac{(G_0/J_0)^{1/4}}{R_X^*} \mathrm{Av}^{-1/4} \tag{7.122}$$

holds.

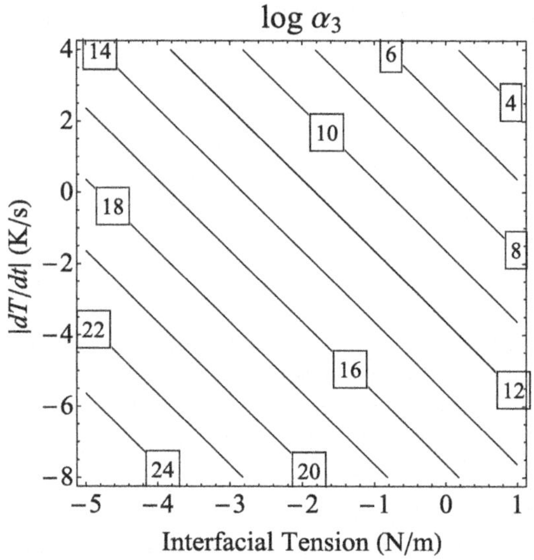

Fig. 7.19 α_3 as a function of the relationship between cooling rate and surface tension

Crystallization involves the release of latent heat and then thermal feedback occurs. The influence of this feedback is characterized by the Stefan number (St) defined as follows:

$$\mathrm{St} = \frac{\Delta\mathcal{H}}{\rho C_P T_0} \tag{7.123}$$

where $\Delta\mathcal{H}$ is enthalpy of melting per unit volume and C_P is specific heat at constant pressure (J kg^{-1} K^{-1}). The Stefan number considers values from about 0.1 to 0.4 dependent of the type of minerals to be crystallized.

In the following, to explain the crystallization process in a binary eutectic system, we use parameters appearing in Eulerian description for ease of numerical experiments. Basically, the parameters are the same as α_1 and α_3 used in the modeling of the vesiculation process and defined by Eqs. (7.60) and (7.120), respectively. The procedure of numerical calculation is almost the same as that described in the chapter on vesiculation process. In the discussion on crystallization process above, although only a phase that first crystallizes from melt below the liquidus (called the liquidus phase) (phase A) has been discussed, the discussion is applicable to phase B or other phases in a multi-component system. Therefore, the calculation of a binary eutectic system differs from calculating vesiculation in that there are two or more crystalizing phases and the supersaturated state, which is sequentially calculated for each phase.

7.6.2 Basic Behavior of the Crystallization Process Under Constant Heat Loss

Note that constant heat loss means the extraction of heat H_{L} (J/s) from the system per unit time, i.e., temperature change can be described as follows:

$$\rho C_P \frac{dT}{dt} = -H_{\mathrm{L}} + \Delta\mathcal{H}\frac{d\phi}{dt} \tag{7.124}$$

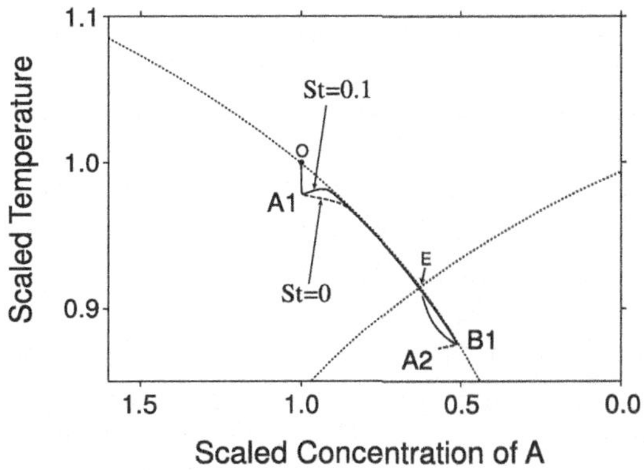

Fig. 7.20 Trajectory of crystallization in an equilibrium phase diagram of a binary eutectic system (Fig. 7.1). The dotted lines indicate the liquidus of each phase. Composition and temperature are normalized by the initial values (point O). St is the value of the Stefan number. The broken line indicates the case of constant cooling rate without release of the latent heat. The solid line indicates the case of constant heat loss with release of the latent heat. A1, A2, and B1 indicate the first and second nucleation events of phase A and the first nucleation event of phase B

where ϕ is the volume fraction of crystals and the sum of the volume fraction of phase A and that of phase B, $\phi_A + \phi_B$. Nondimensionalizing temperature and time by $\tilde{T} = T/T_0$, $\tilde{t} = t/t_{th}$, and characteristic time of heat loss $t_{th} = \rho C_P T_0/H_L$ yields

$$\frac{d\tilde{T}}{d\tilde{t}} = -1 + \mathrm{St}\frac{d\phi}{d\tilde{t}} \tag{7.125}$$

and one of the controlling parameters explained earlier, the Stefan number, appears.

Although we explain the case of the diffusion-limited growth as an example, behavior in the reaction-limited growth is qualitatively similar. As initial state, melt having a certain composition is assumed to be just at a temperature on the liquidus (point O in Fig. 7.20). Assuming that heat is released at a constant rate, we consider a case in which the latent heat is released and a case where the latent heat is not released. The case in which the latent heat is released corresponds to the cooling of magma in nature; the temperature change of the system (the cooling rate) is determined by competition between the rate of heat loss and release of the latent heat. However, in the case where the latent heat is not released, the rate of heat loss is equal to the constant cooling rate and corresponds to programmed cooling in laboratory experiments.[16]

First, the behavior in the equilibrium phase diagram (Fig. 7.20) is shown. We consider the cooling of melt of the composition represented by point O on the liquidus. In this composition, solid phase A crystalizes first. Even if cooling starts and temperature falls below the liquidus of the solid phase A, nucleation does not occur down to a certain temperature because of the nucleation barrier associated with surface energy. This is a point of difference from the equilibrium crystallization process that we studied while learning how to read an equilibrium phase diagram. Therefore, the chemical composition of the liquid does not change. The fact that the degree of supersaturation or the degree of undercooling is expressed by the distance from the liquidus was learned from Eq. (7.26).

In Fig. 7.20, the nucleation of solid phase A occurs at a temperature corresponding to point A1. Once nucleation occurs, the chemical composition of the liquid changes to the direction in which the crystallization component decreases due to crystal growth and approaches the liquidus of A. Whether

[16]A programmed cooling system in which temperature change occurs because of release of the latent heat is regulated by electrical circuits or computer programs to ensure cooling at a constant rate.

temperature increase or not on this occasion depends on the efficiency of release of the latent heat. When the cooling proceeds further, the melt evolves along the liquidus A with a small supersaturation (quasi-equilibrium stage), becomes supersaturated with solid phase B below the eutectic point E, i.e., the undercooled state for both solid phases while sometimes showing complex behavior. In cases where latent heat will be released, represented in the figure, the melt is undercooled once to a certain temperature B1 below the eutectic point, then nucleation and growth of solid phase B occurs, and, because of consequent latent heat release and mass conservation, the trajectory on the phase diagram ultimately converges to the eutectic point.

On the other hand, in cases where the latent heat will not be released, nucleation and growth of solid phase B occurs at a certain temperature below the eutectic point and, because of consequent increase in the concentration of component A, the second nucleation of solid phase A sometimes occurs. The example represented in Fig. 7.20 shows the case. Occurrence of second nucleation of solid phase A primarily depends on the interfacial energy of solid phase A; second nucleation occurs if the interfacial energy is small. On the basis of reaction-limited growth rate, Hort and Spohn (1991b) performed calculations using a crystal model by the Lagrangian description and showed oscillatory or chaotic behavior near the eutectic point. However, such behavior has not occurred in the calculation shown in Fig. 7.20, in which diffusion growth is assumed.

Figure 7.21 shows the time change of the nucleation rate, the crystal number density, and the crystal size. Although the behavior is fairly similar to that in the case of vesiculation, there are considerable differences in that the case of crystals involves two crystalizing phases and thermal feedback due to release of latent heat.[17] The graphs in the figure are of temperature, amount of crystals and liquid composition, crystal number density, and nucleation rate, from the top to the bottom; the left graphs are of the case without release of the latent heat and the right graphs are of the case with release of the latent heat. From the behavior in these graphs, the following aspects are noted: (1) the presence or absence of latent heat release results in whether temperature increases or not due to sudden increase in the amount of crystals at the time of nucleation; (2) liquid composition is determined by the amount of crystals; (3) depending on the interfacial energy, nucleation events of a phase that first crosses the liquid phase and crystallizes (liquidus phase) occur twice; and (4) the crystal number density is determined by nucleation events.

In the following section, we will understand parameters that characterize crystallization, which were extracted from numerical experiments, using controlling parameters.

7.6.3 Crystallization Parameters Characterizing Crystallization Process

Although the complete image of the crystallization process is understood by the crystal size distribution (CSD), which is explained in the following chapter, we use the crystal number density, a representative quantity of CSD, to understand it here. Basically, the number density (or the mean grain size), a parameter that characterizes crystallization texture, is symbolically related to the nucleation rate or growth rate via relational expressions of scaling. The problem is the manner in which the representative nucleation rate or growth rate determining crystallization texture is related to cooling rate (or, decompression rate or water exsolution rate, for decompression-induced crystallization). Putting aside symbolic scaling relations, we examine it via practical nucleation rate and nucleation duration.

Here, we will describe it by classifying it into the following three cases: (1) the case of diffusion-limited growth, (2) the case of reaction-limited growth, and (3) the case where the pre-exponential

[17] In the case of vesiculation, if a system is finite and mechanically interacts with its surroundings, the system receives pressure-related feedback.

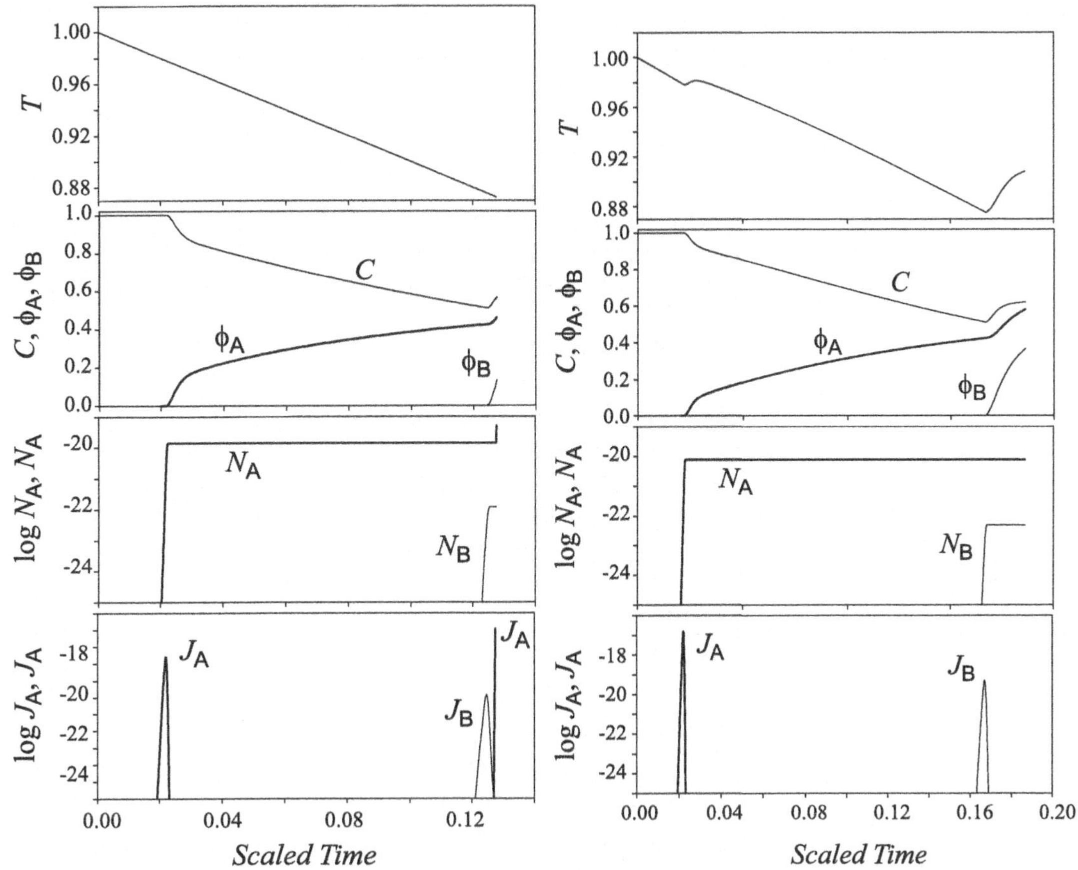

Fig. 7.21 Calculation results of crystallization process of a binary eutectic system for constant heat loss rate. The graphs correspond to the trajectory in Fig. 7.20. The graphs are of temperature, concentration, and volume fraction of each phase, crystal number density of each phase, and nucleation rate of each phase from the top to bottom. (Left) Without release of the latent heat. Note that phase A experienced the second nucleation below the eutectic point. (Right) With release of latent heat, St $= 0.1$. Note that because crystallization suddenly progresses in the event of nucleation, the influence of release of the latent heat appears and the temperature suddenly increases

factor of the nucleation rate is extremely small. For all cases, the crystal number density is determined by the product of the maximum nucleation rate J_{MAX} and the nucleation duration δt_{n}, i.e.,[18]

$$N \approx J_{\mathrm{MAX}} \times \delta t_{\mathrm{n}} \tag{7.126}$$

(1) The Case of Diffusion-Limited Growth

From a parametric study of a numerical experiment in which a binary eutectic system and diffusion-limited growth were assumed (Toramaru 2001),

$$J_{\mathrm{MAX}} = \mathrm{const} \times \alpha_1^{-2.2} \alpha_3^{-1.4} \tag{7.127}$$

$$\delta t_{\mathrm{n}} = \mathrm{const} \times \alpha_1^{0.5} \alpha_3^{-0.06} \tag{7.128}$$

[18]Although this equation approximately holds, it does not hold with respect to the exponent denoting the parameter dependence, to be accurate. This is because sudden change in the nucleation rate is represented only by J_{MAX} and δt_{n}.

are obtained. The cooling rate is contained in α_3 (Eq. 7.120); the dependence of the nucleation duration on cooling rate is found to be small. Moreover, the crystal number density is obtained as follows:

$$N = 10^{3.5} C_0 \alpha_1^{-1.7} \alpha_3^{-1.5} \propto \left| \frac{dT}{dt} \right|^{3/2} \tag{7.129}$$

Note that the exponent (cooling rate exponent) representing the dependence of the crystal number density on the cooling rate is about 3/2; these relations hold for both solid phase A and solid phase B.

The degree of undercooling by which nucleation becomes maximum, $\Delta T_n = T_0 - T_n$, is given by

$$\Delta T_n = 10^{-1.3} T_0 \alpha_1^{0.5} \alpha_3^{-0.02} \tag{7.130}$$

The degree of undercooling at the climax of nucleation is almost totally controlled by α_1 or the solid-liquid interfacial tension γ; moreover, it is proportional to the 3/2 power of γ. On the other hand, the cooling rate and diffusivity are found to have almost no influence from the fact that the exponent of α_3 is - 0.02, i.e., the dependence on α_3 is low. The dependence on parameters is quantitatively almost the same as the case of vesiculation in the diffusion regime (Eqs. 5.68–5.71), which has common physical processes.

(2) The Case of Reaction-Limited Growth

$$J_{MAX} = \text{const} \times \alpha_1^{-3} \alpha_3^{-2.6} \tag{7.131}$$

$$\delta t_n = \text{const} \times \alpha_1^{0.5} \alpha_3^{-0.1} \tag{7.132}$$

Although the dependence of nucleation duration on cooling rate ($\alpha_3^{-0.1}$) is larger than that for diffusion-limited growth ($\alpha_3^{-0.06}$), both are sufficiently small and the dependence of J_{MAX} on the cooling rate is dominant. For the crystal number density,

$$N \approx 10^{8.3} C_0 \alpha_1^{-2.7} \alpha_3^{-2.8} \propto \left| \frac{dT}{dt} \right|^{2.8} \tag{7.133}$$

is obtained. The cooling rate exponent of the crystal number density is about 3, which is found to be larger than that for diffusion-limited growth. The fact that these relations hold for both solid phase A and solid phase B is the same as in the case of the diffusion-limited growth.

The degree of undercooling by which nucleation becomes maximum is given by

$$\Delta T_n = 10^{-1.3} T_0 \alpha_1^{0.5} \alpha_3^{-0.03} \tag{7.134}$$

almost the same as in the case of diffusion growth.

(3) The Case Where the Peak Nucleation Rate Is Extremely Small

The above cases are those in which the peak nucleation rate on nucleation master curves takes on theoretical values and the actual maximum nucleation rate becomes sufficiently smaller than theoretical values. In these cases, as described in the section on heterogeneous nucleation of crystals (Sect. 7.3.1.1), because the peak nucleation rate takes on a sufficiently small value and maintains the peak value, the nucleation rate can be considered to be constant. In this case, to be dealt with in detail in the chapter on CSD, the nucleation duration becomes extremely long and the crystal number density becomes proportional to the cooling time. Therefore, the crystal number density is inversely proportional to the cooling rate and the cooling rate exponent will be-1 (Toramaru and Kichise, manuscript in preparation).

$$N \propto \left| \frac{dT}{dt} \right|^{-1} \tag{7.135}$$

This equation simply expresses the fact that with a higher cooling rate, the nucleation duration is reduced. Moreover, because the nucleation rate is effectively small, a certain constant supersaturated state is maintained. Under the condition where the nucleation rate is constant and the supersaturated state is constant, the rate of increase of the total volume of crystals is constant. The total surface area of crystals increases with increase in the number of crystals; consequently, the crystal growth rate of a single crystal is inversely proportional to time (Sect. 9.3.5).

As the pre-exponential factor of nucleation rate, or the peak nucleation rate becomes small, the state gradually enters the kinetic regime in which the nucleation rate is constant and the increasing rate of the total volume of crystals is constant. Moreover, the cooling rate exponent, which represents the dependence of the crystal number density on the cooling rate, changes from 3/2 to 0 to -1 and takes on a complicated aspect. In laboratory experiments, although cases in which cooling is started from temperatures higher than the liquidus yield about 3/2 of the dependence on the cooling rate for many silicates such as olivine, pyroxene, and plagioclase, as shown in the section on comparison with experiments, cases in which cooling is started from temperatures lower than the liquidus yield variable cooling rate exponents. This fact suggests that the peak nucleation rate is governed by the existing heterogeneous nuclei and that these experimental systems are in the vicinity of this type of a kinetic regime.

7.6.4 Relationship Between the Cooling Rate Dependence of the Crystal Number Density and the Crystal Growth Law

For the case of cooling at a constant rate, let us examine the role of nucleation and growth in the crystal number density and the dependence of the crystal number density on the cooling rate through a discussion of simple scaling.[19] The crystal number density is determined by the product of nucleation rate and nucleation duration. The nucleation rate at this moment is the maximum value in the history of nucleation and is considerably influenced by the corresponding effective degree of undercooling. The effect of the cooling rate works on the effective degree of undercooling through time dependence of crystal growth and, as the cooling rate becomes large, the effective crystal growth is suppressed. Consequently, the degree of undercooling required for the maximum nucleation becomes large and the nucleation rate becomes large.

A characteristic of the behavior of nucleation is that it exponentially increases with a slight change in the effective degree of undercooling. The certain crystal number density that determines the climax of nucleation is determined by a certain single point where it drastically changes with the degree of undercooling. We assume that the crystal number density is determined by the balance among the crystal growth, nucleation and cooling, moreover, and that it has a certain value during the crystallization time (active nucleation stage) that determines the end of nucleation. Although the nucleation rate at this moment depends on the cooling rate, it is ultimately determined by the balance with growth and is constant regardless of time in this time regime where it competes with growth.

[19]The discussion here assumes the case where the pre-exponential factor of the nucleation rate is sufficiently large and nucleation does not continue for a long time.

On the other hand, crystal growth progresses according to a time law that expresses the crystal size R as a power function of time:

$$R \propto t^p \tag{7.136}$$

where the exponent p characterizes the growth of crystals, and $p = 1/2$ in the case of the diffusion-limited growth (Eq. 7.77) and $p = 1$ in the case of the reaction-limited growth (Eq. 7.105). Because the reaction-limited growth shows stronger time dependence, the crystal growth progresses faster.

First, a system with a certain cooling rate is set as a reference system. The number of crystallizing molecules δC is

$$\delta C = \frac{\frac{4}{3}\pi N R^3}{v_S} \tag{7.137}$$

The decrement of the crystallization component required to finish nucleation of the crystals, δC_{thres}, corresponds to the decrement of the degree of supersaturation and is independent of the cooling rate.

When we consider a system with a cooling rate k times faster than that of the reference system, time required for crystal growth is reduced to $1/k$ with reference to the timescale of cooling. That is,

$$R_k \propto \left(\frac{t}{k}\right)^p \tag{7.138}$$

Similarly, the number of crystallizing molecules of the crystallization component is

$$\delta C_k = \frac{\frac{4}{3}\pi N_k R_k^3}{v_S} \tag{7.139}$$

Because δC_{thres} does not depend on the cooling rate, $\delta C_{\text{thres}} = \delta C = \delta C_k$ holds and δC equals δC_k. Thus,

$$N t^{3p} = N_k \left(\frac{t}{k}\right)^{3p} \tag{7.140}$$

is obtained. From this equation,

$$\frac{N_k}{N} = k^{3p} \tag{7.141}$$

is derived. This fact shows that the crystal number density of a system the cooling rate of which is k times faster becomes k^{3p} times larger. That is, the cooling rate exponent becomes $3p$. Note that an exponent p that defines the growth law appears here. Although crystals grow while nucleation continues, the time used for the growth will be effectively reduced with increase in cooling rate. Because crystals continue nucleation during that time, the number of crystals thereby formed becomes large. In other words, because crystals grow only by the amount corresponding to $1/k$ of time, the time required to decrease the degree of supersaturation as it makes nucleation end becomes long; consequently, a large number of crystals is required.

7.6.5 Comparison with Laboratory Experiments

In the 1970s, efforts to clarify the origins of lunar rocks brought back by the Apollo program led to the start of laboratory experiments on crystallization. Many experiments have been conducted using liquid of the same composition as basalt on the moon (Lofgren et al. 1974; Walker et al. 1976, 1978; Grove and

Walker 1977; Grove, 1978, 1990). These experiments were conducted to replicate the rock texture of the moon rocks, which may give a key to the timescale of the cooling of the Moon. These experiments were primarily conducted using a wire-loop method under 1 atm, and then technical know-how for later crystallization experiments was accumulated.

As a result of such experiments, our understanding of the crystallization process in basaltic melt has been progressed in many ways. One point is the complexity of crystal growth. Compared with naturally occurring rock texture, much more complicated crystal forms, such as feathery crystals, needle crystals, and regular dendrite crystals, have been reported as experimental products. Another aspect is the quantification of rock texture. Quantitative evaluation of the influence of the cooling rate on rock texture has become possible owing to acquisition of data on the crystal size, the crystal number density, and the dendrite arm spacing of experimental products. Although crystallization experiments had been conducted to understand rock texture on the Earth (Swanson 1977; Sato 1995), quantitative examination based on theory had not been extensively conducted.

In recent years, experiments on kinetics of crystallization, focusing on decompression-induced crystallization (Hammer and Rutherford 2002; Couch et al. 2003b; Shea and Hammer 2013; Befus et al. 2015) and crystal size distribution (Zieg and Lofgren 2006; Pupier 2008; Schiavi et al. 2009), have been conducted. These topics will be dealt with in chapters "Crystallization Induced by Vesiculation" and "CSD (Crystal Size Distribution)". Moreover, by an in situ high-temperature observation experiment using a moissanite (SiC) cell, the dynamic images of the interaction of crystals, which is not dealt with in this book, were investigated (Ni et al. 2014).

Before examining experimental results, we will summarize theoretical estimates for the dependence of crystal number density on cooling rate. As shown above, the cooling rate dependence of the crystal number density N in the case of homogeneous nucleation is related to the exponent p in the time law of the crystal growth rate

$$R \propto t^p \tag{7.142}$$

Note that $p = 1/2$ in the case of diffusion-limited growth and $p = 1$ in the case of reaction-limited growth. Using this exponent, the dependence of the crystal number density on the cooling rate $((|\dot{T}| = |dT/dt|))$ is written as follows:

$$N \propto |\dot{T}|^{3p} \tag{7.143}$$

This is the crystal number density in a 3D space, and the number density in a 2D observation plane N_A (Number/m^2) was used in measurement in the experimental results. Between both N and N_A, there is the relationship $N_A \approx N^{2/3}$; thus, the cooling rate dependence is written as follows:

$$N_A \propto |\dot{T}|^{2p} \tag{7.144}$$

The relation $N_A \propto |\dot{T}|^1$ holds for diffusion-limited growth and the relation $N_A \propto |\dot{T}|^2$ holds for reaction-limited growth. Moreover, because the 3D crystal grain size is proportional to the $-1/3$ power of the 3D number density, $R \propto |\dot{T}|^{-1/2}$ holds for diffusion-limited growth and $R \propto |\dot{T}|^{-1}$ holds for reaction-limited growth. Similarly, because the 2D crystal grain size is proportional to the $-1/2$ power of the 2D number density, $R_A \propto |\dot{T}|^{-1/2}$ holds for diffusion-limited growth and $R_A \propto |\dot{T}|^{-1}$ holds for reaction-limited growth. This comes from the requirement that the volume fraction of each crystal phase does not depend on the cooling rate, i.e., $R^3 N \approx R_A^2 N_A \approx$ const.

Table 7.2 summarizes the classical experimental results in the early years. Comparison between these data and abovementioned theoretical estimates indicates the following. Data on the 2D crystal number density of olivine is consistent with the data in the case of diffusion growth. Moreover, data on the 2D crystal number density of pyroxene does not agree with that on diffusion growth; however,

Table 7.2 Cooling rate exponent ξ_2 indicating cooling rate dependence $\propto |\dot{T}|^{\xi_2}$ of the crystal number density per unit area N_A and the size per unit area R_A. (References are a Walker et al. (1978), b Lofgren et al. (1979), c Sato (1995), d Grove and Walker (1977); Walker et al. (1978); Grove, 1978, 1990, and e Walker et al. (1976)). Arm spacing of the dendrite pyroxene

	Theory	Olivine	Pyroxene	Plagioclase
N_A (m^{-2})	1 (Diffusion)	1^a	$\frac{1}{2}^b$	$\frac{1}{3}^c$
	2 (Reaction)	($|\dot{T}| < 10°$C/hr)		($T_{init} < T_{liq}$)
R_A (m)	$-\frac{1}{2}$ (Diffusion)	no data	$-\frac{1}{2}^b$	$-\frac{1}{2}^d$
				($T_{init} > T_{liq}$)
	-1 (Reaction)		$-\frac{1}{3}^e$	$-\frac{1}{5}^d$
				($T_{init} < T_{liq}$)

data on the grain size in the case where the initial temperature is higher than the liquidus agrees. For plagioclase, in the case where the initial temperature is higher than the liquidus, data on the grain size agrees with the model of diffusion growth; in the case where the initial temperature is lower than the liquidus, neither the number density nor the grain size agrees.

7.6.6 Experiments in Nature

(1) Cooling of Dikes and Data on Number Density

The cooling crystallization of dikes is a natural laboratory. Dikes, the occurrence of which is well known, can be considered to solidify by cooling by heat conduction.[20] As described in the Appendix (Chap. 11), for cooling by heat conduction, the cooling rate decreases in proportion to the- 2 power of the distance y from the cooling surface.

$$|\dot{T}| \propto y^{-2} \tag{7.145}$$

Using this equation for the above relation Eq. (7.143), the dependence of the crystal number density on distance is expected to be

$$N \propto y^{-6p} \tag{7.146}$$

That is, $N \propto y^{-3}$ holds for diffusion-limited growth and $N \propto y^{-6}$ holds for reaction-limited growth. The number density of diffusion-limited growth slowly (that said, at the -3 power of the distance) decreases compared with that of reaction-limited growth. Because the crystal grain size is proportional to the $-1/3$ power of number density, the distance dependence is $R \propto y^1$ for diffusion-limited growth and $R \propto y^2$ for reaction-limited growth.

Gray (1970, 1978) measured the number density of dikes in Canada as a function of distance from the contact; the results are shown in Fig. 7.22. Straight lines denote $p = 1/2$, i.e., the distance dependence in the case of diffusion-limited growth. All data cannot be explained by dependence for reaction-limited growth, $N \propto y^{-6}$. For plagioclase, although almost all data on dikes cannot be explained by either the diffusion-limited growth or the reaction-limited growth, the data enclosed in the broken line may follow diffusion-limited growth. For pyroxene, certain data seem to follow the diffusion-limited growth and

[20]For narrow dikes, magma movement due to forced convection caused by cooling from the wall of the dike is negligible compared with the cooling time. Similarly, for thin sills, convection does not occur because the Rayleigh number is small and forced convection caused by cooling from the side boundary is negligible.

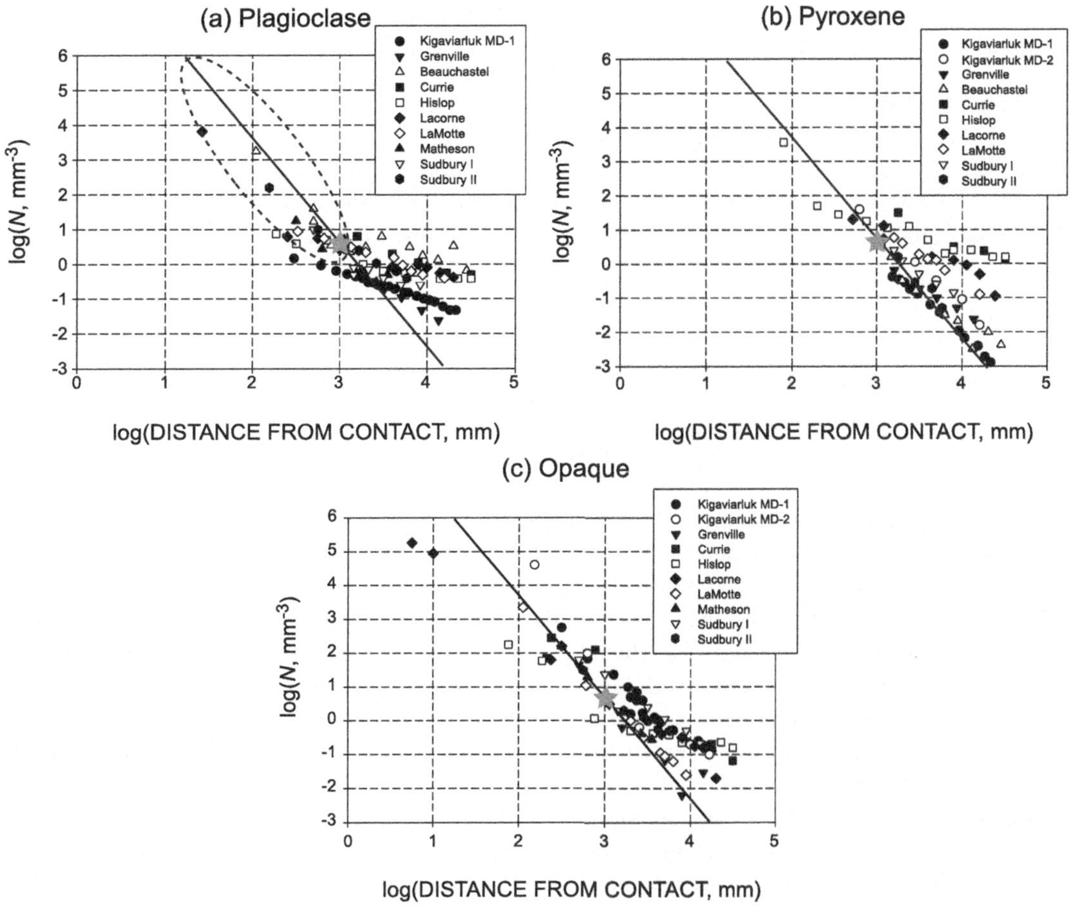

Fig. 7.22 Measurement results of crystal number density by Gray (1970, 1978). (**a**) Plagioclase, (**b**) pyroxene, and (**c**) opaque minerals (oxides such as magnetite and hematite). The horizontal axis represents a common logarithm of the distance (mm) from the cooling surface (contact); the vertical axis represents a common logarithm of the number density (number/mm^3) of groundmass per unit volume. The stars are reference points to determine kinetic factors (Toramaru et al. 2008). Reprinted from (Toramaru et al. (2008)), Copyright 2008, with permission from Elsevier

some do not. For opaque minerals, although many of them appear to follow diffusion-limited growth, some dikes showed questionable values at distances of around 10 m or more.

As a cause of deviation from trends expected from homogeneous nucleation and diffusion-limited growth (in the sense that the distance dependence weakens), heterogeneous nucleation is considered. In fact, when the CIPW norm[21] is calculated from whole-rock chemical composition and plotted in an equilibrium phase diagram of plagioclase solid solution and diopside, all liquidus phases are in the region of plagioclase. This fact suggests that the temperature of magma will have already been lower than the liquidus of plagioclase when the dike intruded; moreover, it can be considered that molecular clusters, which become heterogeneous nucleus sites, had already been formed. This result is consistent with the result that in the laboratory experiments. Note that the cooling rate dependence of the crystal

[21] A method in which a realistic crystallization sequence is assumed and by which the ratio of each mineral contained in rocks after complete crystallization of magma is calculated. Using this method, chemical composition of magma can be expressed by the ratios of each mineral component and the chemical composition of magma can be plotted on an equilibrium phase diagram.

number density in the case where the initial temperature is higher than the liquidus is large; however, that in the case where the initial temperature is lower than the liquidus is almost none or weak.

(2) Kinetic Factors

In cases where the pre-exponential factor is sufficiently large and the growth is diffusion-limited, the number density of crystals formed by cooling crystallization, N, is determined by Eq. (7.129) from numerical calculation. Bearing in mind that T_0/t_{th} is the cooling rate $|\dot{T}| = |dT/dt|$, substituting definitional equations of α_1 and α_3, Eqs. (7.60) and (7.120), into Eq. (7.129) yields

$$N = a\left|\dot{T}\right|^{\frac{3}{2}} \tag{7.147}$$

The coefficient a is

$$a = 10^{3.5} \times C_0 \left(\frac{4\pi\gamma(R_X^*)^2}{3k_B T_0}\right)^{-1.7} \times \left(\frac{\phi_0 D}{T_0(R_X^*)^2}\right)^{-1.5} \tag{7.148}$$

Because the coefficient a is determined by the kinetics of crystallization, i.e., the mechanism of nucleation and growth, it is called a kinetic factor. It is difficult to determine the kinetic factor from physical properties such as the diffusivity D, the molecular volume v_S, and the interfacial energy γ (contained in R_X^*). This is because the molecular volume is the effective molecular volume of the crystallization component. Moreover, the effective interfacial energy, including the presence or absence of heterogeneous nucleation, cannot be given with high accuracy. Avoiding the uncertainty, we consider dikes as the results of natural experiments and determine the kinetic factor from the number density and a cooling model by heat conduction. Nevertheless, ambiguity of about one order of magnitude remains as seen in the following.

Sing an equation for semi-infinite cooling (11.106) (not considering the change in temperature of country rock) or (11.107) (considering change in the temperature of country rock) in the Appendix (Chapter"Appendix"), we determine the kinetic factor as it will give about $5 \times 10^9\,\mathrm{m}^{-3}$ of the number density at a distance of 1 m. Note that diffusivity, which has large composition dependence, for basaltic melt is assumed to be $10^{-11}\,\mathrm{m}^2/\mathrm{s}$. The effect of composition can be considered in diffusivity later.

In cases where temperature change of country rock is not considered, assuming that the initial temperature of magma is 1400 K, the temperature of country rock is 500 K, and the crystallization temperature is $0.95 \times 1400 = 1330$ K, the cooling rate at a distance of 1 m is calculated to be 8.32×10^{-4} K/s. A kinetic factor that gives about $5 \times 10^9\,\mathrm{m}^{-3}$ of the crystal number density for this cooling rate is calculated as 2×10^{14}. In cases where temperature change of country rock is considered, the cooling rate is calculated to be 3.75×10^{-4} K/s and the kinetic factor to be 6.89×10^{14}. Incidentally, the estimates of interfacial energy corresponding to these values are 2×10^{-3} and $3 \times 10^{-2}\,\mathrm{J/m}^2$, respectively.

Assuming that the crystallization temperature is $0.99 \times 1400 = 1386$ K, the cooling rate at a distance of 1 m is calculated at 5.5×10^{-4} K/s for cases where temperature change of country rock is not considered and 1.96×10^{-4} K/s for the case where temperature change of country rock is considered; furthermore, the kinetic factor is calculated at 3.93×10^{14} and 1.82×10^{15}, respectively. Thus, estimates for kinetic factors have a certain range. Here, we assume that the kinetic factor for anhydrous basaltic melt ($SiO_2 = 50$ wt%, initial temperature = 1400 K) is $a = 10^{15\pm1}$.

(3) Crystal Number Density Cooling Rate Meter

In the above calculation, the effective diffusivity of the crystallization components in basaltic melt was assumed to be D$D = 10^{-11}$ m^2/s. To estimate kinetic factors in a wide range other than basaltic melt, the dependence of diffusivity on chemical composition has to be considered. However, to experimentally determine the effective diffusivity of the crystallization components is difficult because of interdiffusion in the multi-component system; moreover, crystal nucleation and growth processes have to be reconsidered and radical measures are required.

Therefore, as a rough estimate, we will evaluate the effect of composition using the relationship between viscosity and diffusion. Compared with diffusivity, viscosity has been experimentally obtained against various compositions. Furthermore, a viscosity coefficient is considered to represent an average mobility of molecules as a bulk and to show a mobility of atoms of major components such as Si, Al, Mg, Ca, H_2O, Na, and K. Although the relationship between diffusivity D and viscosity η of simple spherical molecules is a well-known Stokes-Einstein relation ($D\eta$ = constant), this relation does not hold for liquid with complex molecular structure, including silicate melt. The element that diffuses most slowly among crystallization components is Si; Liang et al. (1996a, b) $D\eta$ = const. experimentally showed that a relation

$$D_{\mathrm{Si}} \propto \eta^{-0.8} \tag{7.149}$$

holds.

The viscosity of silicate melt is approximately given as $\eta = \exp[0.46(\tilde{C}_{\mathrm{Si}} - 72) - 1.25\tilde{C}_{\mathrm{W}} + 2 \times 10^4/T]$ using the SiO_2 content $\tilde{C}_{\mathrm{Si}}$ (wt%), the water content $\tilde{C}_{\mathrm{W}}$ (wt%), and temperature T (K) (Toramaru 1995). Moreover, the temperature of magma has an apparent relationship in which the temperature decreases as the SiO_2 content increases and can be approximately expressed as $T = 10^3/(0.16 + 0.01\tilde{C}_{\mathrm{Si}})$. Substituting these equations into Eq. (7.149) gives a relational equation

$$D_{\mathrm{Si}} \propto \exp(-0.53\tilde{C}_{\mathrm{Si}} + \tilde{C}_{\mathrm{W}}) \tag{7.150}$$

Because the kinetic factor has the dependence on the diffusivity at $D^{-1.5}$ as found from Eq. (7.148), eliminating $D_{\mathrm{Si}}^{-1.5}$ from Eq. (7.150) and transforming the power of e to the power of 10 yield the kinetic factor

$$a = 10^{15\pm1+0.345\Delta\tilde{C}_{\mathrm{Si}}-0.65\tilde{C}_{\mathrm{W}}} \tag{7.151}$$

where $\Delta\tilde{C}_{\mathrm{Si}} = \tilde{C}_{\mathrm{Si}} - 50$. When this kinetic factor is used, Eq. (7.147) has a meaning as a cooling rate meter for every SiO_2 melt.

$$|\dot{T}| = \left(\frac{N}{a}\right)^{\frac{2}{3}} \tag{7.152}$$

Substituting the crystal number density N, the chemical composition $\tilde{C}_{\mathrm{Si}}$ wt%, and the estimated water content $\tilde{C}_{\mathrm{W}}$ wt% into this equation as observables yields the cooling rate $|\dot{T}|$.

Figure 7.23 shows contours of the cooling rate as a function of SiO_2 content and crystal number density. Note that, as seen in the above, this is based on the cooling rate model by heat conduction, which considers the temperature change of country rock, and is in the case where the degree of undercooling of crystal nucleation is 1% (14 K) of liquidus temperature. If the degree of undercooling of nucleation

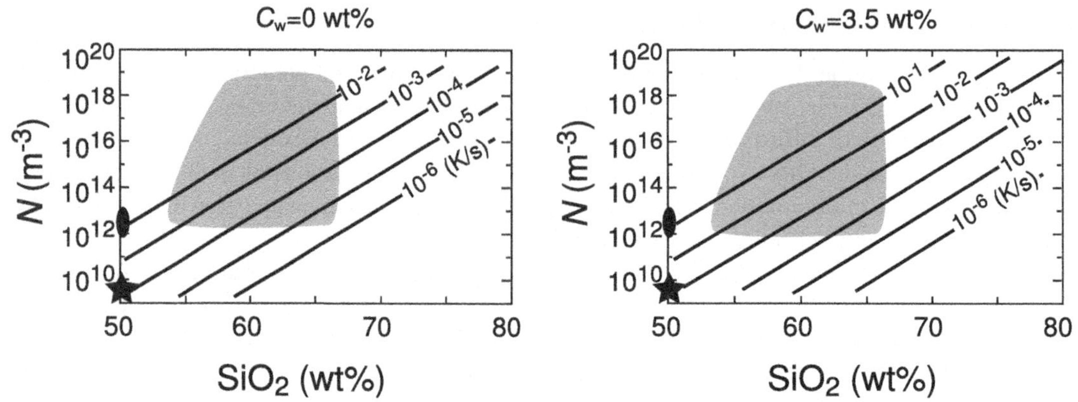

Fig. 7.23 Cooling rate meter as a function of silica content in magma. (Left) and (Right) correspond to differences in water content. Rough cooling rate is estimated by plotting silica content and crystal number density in this contour plot. Data on dikes by Gray are represented by black ellipses (at the contact) and stars (at 1-m distance from the contact). The shaded regions are ranges of observed values of microlites in natural eruptive rocks (Adapted from Toramaru et al. 2008). Reprinted from (Toramaru et al. 2008), Copyright 2008, with permission from Elsevier

increases to 5% (70 K), the cooling rate will be estimated at about twice for the same crystal number density. Moreover, because diffusivity becomes large when water is added, a higher cooling rate is required to obtain the same crystal number density. For example, if 3.5 wt% of water is contained, the kinetic factor will be increased by $10^{-0.65\times3.5}$; therefore, the cooling rate has to be increased $10^{0.65\times3.5\times2/3} \approx 10^{1.5}$ times.

7.6.7 Summary of Factors Controlling the Crystal Number Density

By integrating theoretical understanding demonstrated to date and experimental results, the dependence of crystal number density and grain size on cooling rate in crystallization under a constant cooling rate can be summarized as follows:

1. In cooling crystallization from the initial temperature higher than the liquidus, the dependence of the crystal number density N on the cooling rate $|\dot{T}|^{3/2}$ by the classical nucleation theory and the diffusion-limited growth holds for many silicates and oxides, i.e., $N \propto |\dot{T}|^{3/2}$.
2. In cooling crystallization from the initial temperature lower than the liquidus, the dependence of the crystal number density on the cooling rate using the classical nucleation theory and the diffusion-limited growth does not hold for many silicates; in fact, the dependence is weak. This behavior is attributed to the decrease in the effective number of molecules because of the presence of heterogeneous nucleus clusters and lowering of the solid-liquid interfacial energy.

7.7 Chemical Composition of Crystals

For crystallizing multi-component melt, the way in which chemical composition of a crystal is determined in natural conditions is intriguing because data obtained by analyzing actual samples may allow us to estimate conditions of formation. However, in many cases, chemical composition of the inside of a natural crystal is not uniform and often systematically changes from the core to the rim of the crystal. Such a heterogeneous structure of chemical composition of the inside of crystals is called a compositional zoning structure.

The manner in which we view the compositional zoning structure can be roughly divided into two. One idea is that the structure maintains the chemical composition of the solid-liquid interface at the time of crystal growth. Another idea is that the structure is modified by subsequent change in temperature, pressure, and liquid composition. In this book, only the former idea is addressed. Moreover, there are two ideas for the former: chemical equilibrium holds at the solid-liquid interface of growing crystals, and disequilibrium state determines certain factors.

In cases where the system maintains chemical composition at the interface at the time of growth and intracrystalline diffusion can be ignored because it is sufficiently slow, the solid-liquid interfacial concentration at the time of crystal growth is recorded on the inside of crystal grains. Such a zoning structure (or zoning) is called growth zoning; the following explanation is provided with a focus on growth zoning. Even for cases in which chemical equilibrium at the interface does not strictly hold, if some distribution relation directly related to temperature and pressure exists, growth zoning can be observed. However, a factor determining a certain distribution relation under disequilibrium state has not been elucidated to date; it will be discussed in the corresponding subsection in Sect. 7.7.1.

First, in Sect. 7.7.1, we organize the determinants of chemical composition of the solid phase by considering a binary solid solution system that serves as a basis for considering compositional zoning structure. Here, we either assume that composition of the inside of the solid and liquid is uniform or that spatial extent is ignored.[22] The simplest multi-component system is a binary solid solution system in which composition of the solid phase changes; if interfacial chemical equilibrium is assumed, the chemical composition of the liquid and the solid phase that coexist according to temperature and pressure and the ratio of the endmember compositions in the solid solution are uniquely determined. First, we understand this for the case in which chemical equilibrium holds. However, using experimental results, we show that this solid-liquid chemical equilibrium does not necessary hold for crystals formed from a disequilibrium state and explain the manner of thinking to understand it. In this case, whether zoning structure formation in crystals uniquely corresponds to temperature and pressure or not has to be examined in more detail.

Next, in Sect. 7.7.2, an equation for growth zoning is derived, which describes the spatial change of composition in crystals in the case where the interfacial concentration is determined only by either temperature or pressure. Here, we examine the fact that the growth law of crystals determines a composition profile of growth zoning.

[22]In typical binary solid solution such as plagioclase and olivine, chemical composition is sequentially changed by reaction between the liquid phase and solid phase on crystallization. If chemical equilibrium in the whole system (chemical potential of any chemical component in the system is equal everywhere in the space) holds perfectly, crystal composition should be homogeneous in grains at every step of crystallization. Moreover, chemical equilibrium in the whole system holds in the case where reaction in the solid-liquid interface and diffusion in the solid phase and the liquid phase are sufficiently fast or the case in which physical conditions such as temperature and pressure change sufficiently slowly. As per the situation, because diffusion in solid is slow, the composition at the vicinity of the interface often differs from the composition at the interior of grains. This is observed as growth zoning.

Zoning structure of binary solid solution

For an equilibrium state of a binary solid solution system, the chemical composition of crystals is a function only of temperature under a given pressure; it is expressed by Eq. (7.163). We assume a state in which intracrystalline diffusion does not work, the inside of the crystals and liquid do not react, chemical equilibrium is maintained only at the interface, and cooling crystallization occurs (a type of fractional crystallization). In this case, zoning structure is formed in the crystals in which composition of the solidus is recorded from the core to the rim. Although the influence of diffusion in solid and that of the crystal growth rate are included in actuality, the crystals qualitatively show a concentration profile in which high-temperature components are abundant in the core and gradually decrease toward the rim. This is called normal zoning. However, reverse zoning is the zoning structure showing a concentration profile in which high-temperature components increase from the core to the rim. It is considered to be caused by increase in water content in the melt and change in composition of the melt due to magma mixing.

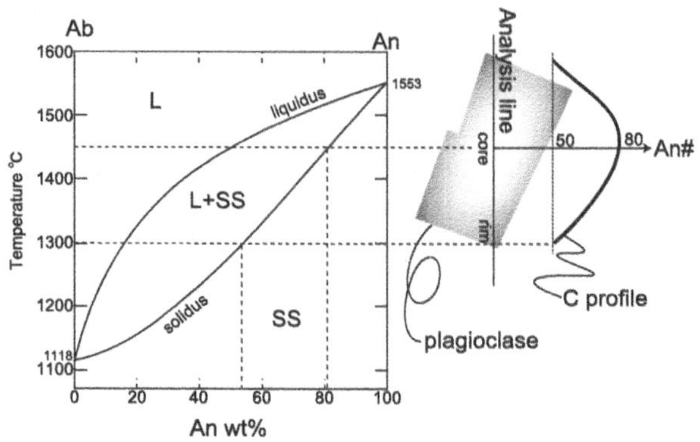

Figure An equilibrium phase diagram of plagioclase at 1 atm. Schematic of normal zoning for melt that starts crystallizing at around 1450 °C and is rapidly cooled at around 1300 °C. The high-temperature component in this case is Anorthite (An: $CaAl_2Si_2O_8$)). L denotes the liquid phase, SS denotes plagioclase solid solution, "C profile" is the concentration profile (plot of the distance in the crystal (the vertical axis) against concentration (the horizontal axis)), and An# is the An value (%).

Section 7.4.2 discusses the diffusion profile in liquid in the case where the crystal growth rate can be ignored, in which only time change in the crystal growth rate was discussed. However, in cases where the growth rate cannot be negligible against diffusion in liquid, movement of the interface between crystals and liquid influences the diffusion profile in liquid as well as the concentration profile of growth zoning. In Sect. 7.7.3, we will explain a general formulation for a chemical composition profile of crystals under such circumstances and understand the fact that the ratio of diffusivity in liquid to the moving velocity of the interface (the Péclet number) serves as a controlling parameter. In this case, zoning structure is formed in crystals due to mass transfer even under constant temperature and pressure.

7.7.1 The Solid-Liquid Equilibrium and Disequilibrium in Binary Solid Solution and Chemical Composition of Crystals

(1) Chemical Composition in Thermodynamic Equilibrium

The top chart of Fig. 7.24 shows an equilibrium phase diagram of a binary solid solution system. Figure 7.24 shows the relationship between free energy and chemical potential at a temperature T_0. In a solid solution, chemical equilibrium conditions are as follows:

$$\mu_{\rm L}^{\rm A} = \mu_{\rm S}^{\rm A} \tag{7.153}$$

$$\mu_{\rm L}^{\rm B} = \mu_{\rm S}^{\rm B} \tag{7.154}$$

Subscripts L and S denote the liquid phase and the solid phase, respectively; moreover,

$$\mu_{\rm L}^{\rm A} = \mathcal{G}_{\rm L}(X_{\rm L}^{\rm A}) + (1 - X_{\rm L}^{\rm A})\frac{\partial \mathcal{G}_{\rm L}}{\partial X_{\rm L}^{\rm A}} \tag{7.155}$$

$$\mu_{\rm S}^{\rm A} = \mathcal{G}_{\rm S}(X_{\rm S}^{\rm A}) + (1 - X_{\rm S}^{\rm A})\frac{\partial \mathcal{G}_{\rm S}}{\partial X_{\rm S}^{\rm A}} \tag{7.156}$$

can be given (see Box in Sect. 7.7.1). Chemical potential of component A in the liquid $\mu_{\rm L}^{\rm A}$ is given by a point ($\rm M_A$) in which a tangent from the point of the liquid composition $(X_{\rm L}^{\rm A}, \mathcal{G}_{\rm L}(X_{\rm L}^{\rm A}))$ on the free energy curve of the liquid $\mathcal{G}_{\rm L}(X_{\rm L}^{\rm A})$ intersects with the A-axis. Similarly, chemical potential of component A in the solid phase $\mu_{\rm S}^{\rm A}$ is given by a point where a tangent from the point $(X_{\rm S}^{\rm A}, \mathcal{G}_{\rm S}(X_{\rm S}^{\rm A}))$, where the composition of the solid phase is $X_{\rm S}^{\rm A}$ on the free energy curve of the solid phase $\mathcal{G}_{\rm S}(X_{\rm S}^{\rm A})$, intersects with the A-axis. Because the equilibrium condition for component A is that these two values

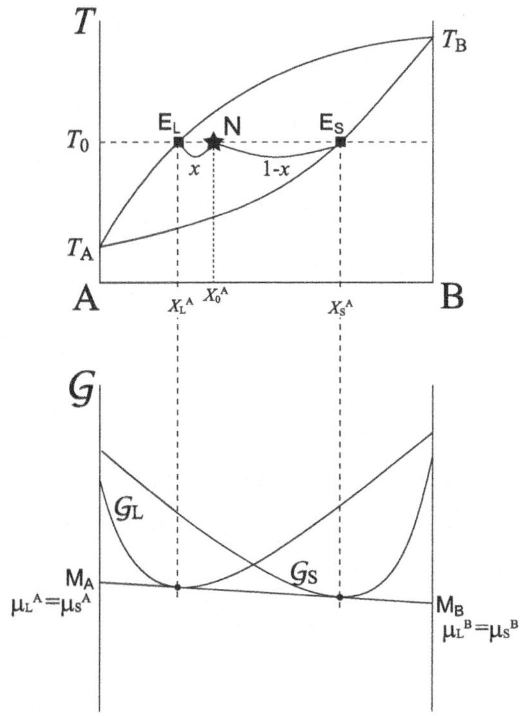

Fig. 7.24 (Top chart) Equilibrium phase diagram in a binary solid solution. (Lower chart) The relationship of chemical potential and Gibbs free energy of liquid and solid phases, which are in the equilibrium state

agree, the equilibrium condition in a graphical form is that a common tangent to the two curves $\mathcal{G}_L$ and $\mathcal{G}_S$ can be drawn, i.e., the two tangents share the point (M_A)) on the A-axis. The equilibrium composition is determined by the points of contact of the curves with the common tangent. Discussion on component B results in the same conclusion; the intercept of the common tangent on the B-axis gives the chemical composition of component B in the liquid and solid phases. Consequently, composition at which the liquid phase and the solid phase can coexist in equilibrium at a certain temperature T_0 is uniquely determined.

Because chemical potential of an ideal solution (see Box in Sect. 2.5.1) with respect to component i is $\mu_i = \mu(\text{pure}) + k_B T \ln X_i$,

$$k_B T \ln \frac{X_L^A}{X_S^A} = \mu_S(\text{pureA}) - \mu_L(\text{pureA}) = -(\Delta h_A - T\Delta s_A) \tag{7.157}$$

is obtained from Eq. (7.153). For pure A, because $X_L^A = X_S^A = 1$ and $\Delta h_A = T_A \Delta s_A$ hold at the melting point T_A, Eq. (7.157) is rewritten as follows:

$$\ln \frac{X_L^A}{X_S^A} = \frac{\Delta h_A}{k_B}\left(\frac{1}{T_A} - \frac{1}{T}\right) \tag{7.158}$$

This is an equation that determines the distribution of component A between the solid phase and the liquid phase. In the same way,

$$\ln \frac{X_L^B}{X_S^B} = \frac{\Delta h_B}{k_B}\left(\frac{1}{T_B} - \frac{1}{T}\right) \tag{7.159}$$

holds for component B. Using these equations and a self-evident mass conservation law

$$X_L^A + X_L^B = 1 \tag{7.160}$$

$$X_S^A + X_S^B = 1 \tag{7.161}$$

yields an equation that defines the solidus,[23]

$$X_S^A \exp\left[\frac{\Delta h_A}{k_B}\left(\frac{1}{T_A} - \frac{1}{T}\right)\right] + (1 - X_S^A)\exp\left[\frac{\Delta h_B}{k_B}\left(\frac{1}{T_B} - \frac{1}{T}\right)\right] = 1 \tag{7.162}$$

From this equation, the ratio of molar fractions of end members A and B in the solid phase, X_S^B / X_S^A, is determined as follows:

$$\frac{X_S^B}{X_S^A} = \frac{\exp\left[\frac{\Delta h_A}{k_B}\left(\frac{1}{T_A} - \frac{1}{T}\right)\right] - 1}{1 - \exp\left[\frac{\Delta h_B}{k_B}\left(\frac{1}{T_B} - \frac{1}{T}\right)\right]} \tag{7.163}$$

Thus, under the equilibrium conditions, the end-member ratio of the crystals in the binary solid solution system is determined only using temperature.

Similarly, an equation that defines the liquidus is given as follows:

$$X_L^A \exp\left[-\frac{\Delta h_A}{k_B}\left(\frac{1}{T_A} - \frac{1}{T}\right)\right] + (1 - X_L^A)\exp\left[-\frac{\Delta h_B}{k_B}\left(\frac{1}{T_B} - \frac{1}{T}\right)\right] = 1 \tag{7.164}$$

---

[23]Rewrite Eqs. (7.158) and (7.159) to the form of $X_L^A =$ and $X_L^B =$ and substitute them into Eq. (7.160).

Derivation of Eq. (7.155) or Eq. (7.156)

Free energy of liquid $\mathcal{G}_L$ is expressed using chemical potentials of the components μ_L^A and μ_L^B and molar fractions X_L^A and X_L^B as

$$\mathcal{G}_L = X_L^A \mu_L^A + X_L^B \mu_L^B \tag{7.165}$$

Focusing attention on one phase (the liquid phase in the following), the following Gibbs-Duhem equation holds in the inside of the phase under constant P and T.[a]

$$X_L^A d\mu_L^A + X_L^B d\mu_L^B = 0 \tag{7.167}$$

Differentiating Eq. (7.165) with respect to X_L^A and using this equation gives

$$\left(\frac{\partial \mathcal{G}_L}{\partial X_L^A}\right)_{T,P} = \mu_L^A - \mu_L^B \tag{7.168}$$

Rewriting this equation to the form of $\mu_L^B =$ and substituting it into Eq. (7.165) yields

$$\mathcal{G}_L = \mu_L^A - X_L^B \left(\frac{\partial \mathcal{G}_L}{\partial X_L^A}\right)_{T,P} \tag{7.169}$$

From this equation, Eq. (7.155) can be derived.

[a] Because chemical potential

$$\mu_L^A = \left(\frac{\partial \mathcal{G}_L}{\partial N_A}\right)_{T,P} \tag{7.166}$$

has a characteristic of partial molar quantity, chemical potentials of individual components are mutually related under Eq. (7.167). In this case, because temperature and pressure are constant, the same constraint as the Gibbs-Duhem equation holds.

Let us assume that the liquid of a certain composition X_0^A is in an undercooled state (represented by point N on a phase diagram). In that case, the liquid is separated into the liquid phase of composition X_L^A (point E_L) and the solid phase of composition X_S^A (point E_S). At that time, the mass ratio of the phases (solid: liquid = x: $1-x$) is determined by the mass conservation of the system

$$x_0^A = xX_S^A + (1-x)X_L^A \tag{7.170}$$

The way in which the separation reaction temporally develops, including the kinetics of crystallization and diffusion in a liquid, will be a dynamic problem.

(2) Disequilibrium Chemical Composition

For natural crystals, heterogeneous chemical composition is observed in the diverse type of zoning structures. Such heterogeneous composition in the crystals is often attributed to change in temperature, pressure, or chemical composition with growth. That is, the idea is that undercooled solution is separated into the liquid and solid phase while maintaining interfacial equilibrium to crystallize. This means that in a reaction in which undercooled melt → melt + the solid phase, a local system that satisfies the two conditions, i.e., (1) mass conservation before and after the reaction and (2) the interfacial equilibrium between reaction products, exists in the vicinity of the solid-liquid interface. In this case, the compositions of the liquid phase and the solid phase are uniquely determined by temperature. As mentioned earlier, the way in which the separation reaction temporally develops is a dynamic problem, including the kinetics of crystallization and diffusion in liquid. However, another idea is that a solid phase of a certain composition determined by some causes can be formed under disequilibrium conditions from undercooled solution composition with which the solid compositions are not in equilibrium.

In the case where a local system satisfying the equilibrium conditions and mass conservation is not assumed, a certain fixed distribution relation or exchange equilibrium

$$K_D = \frac{\frac{X_S^A}{X_S^B}}{\frac{X_L^A}{X_L^B}} \tag{7.171}$$

is assumed to hold. In the dynamic problem reported later, a model in which composition of the solid phase formed from the liquid phase of a certain undercooled (supersaturated) composition is determined by an effective distribution coefficient is used. Although this model may empirically hold, it is unclear whether it holds in terms of physics and what are the conditions under which it is physically justified. The free energy of a local system comprising the liquid phase + the solid phase after crystallization has to be smaller than that of the local system before crystallization (i.e., undercooled solution). Then, a question arises: if a local equilibrium system cannot be realized, can a solid phase coexisting in equilibrium with undercooled melt without the liquidus compositions be possible? However, in the following, we will show that the chemical composition of such a solid phase is not uniquely determined, by experimental results and thermodynamic consideration.

In crystallization experiments using silicate melt, various zoning structures are observed corresponding to various cooling rates and temperature histories (e.g. Lofgren 1974, 1980) (the right chart of Figure). For minerals in igneous rocks, normal zoning is defined as a concentration profile in which a high-temperature component decreases from the core to the rim, whereas reverse zoning is defined as a concentration profile in which a high-temperature component increases from the core to the rim. Using a melt of a Di-An system, Tsuchiyama (1985) conducted crystallization experiments at constant temperature (the left chart of Fig. 7.25 and the left chart of Fig. 7.26) and experiments in which temperature continuously decreased (constant cooling rate) (the right chart of Fig. 7.26) to observe the zoning structure of the crystals that were formed. For example, for the melt of composition Di80, the initial phase is diopside and Al dissolves in it. As a result of investigation of Al concentration in a diopside, in the crystallization experiments at a constant temperature, the crystal had a core composition that was higher than the expected equilibrium concentration and showed reverse zoning in which Al concentration decreased toward the rim. Although the Al amount in diopside increased with decrease in temperature when a certain distribution coefficient close to chemical equilibrium is assumed, the zoning structure showed the opposite tendency. On the other hand, in the experiments at a constant cooling rate, the crystal showed normal zoning in which Al concentration increased from the core to the rim. Thus, these experimental results revealed that the solid phase crystallized from the undercooled solution has a composition different from that in equilibrium at the temperature and pressure (Fig. 7.26).

With respect to components of interest, whether the crystal composition is rich in components or poor determines whether the zoning structure formed in the following crystal growth will be either reverse zoning or normal zoning. For cases where a melt is first given a discontinuous degree of

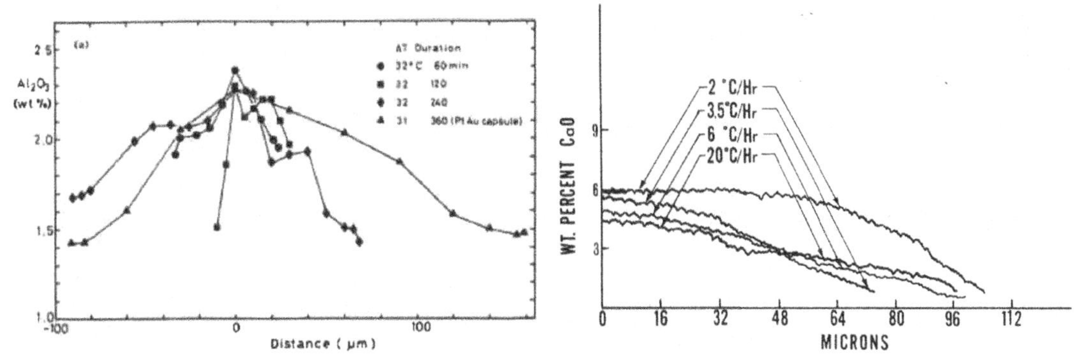

Fig. 7.25 (Left) Al concentration in Di formed by the cooling crystallization experiment at a constant temperature using melt of Di80 composition of the Di-An system. When the distribution coefficient is constant (at a similar level to the value in equilibrium state), Al concentration decreases (from the core to the rim of a crystal) with time, resulting in reverse zoning (Tsuchiyama 1985, with permission from the Mineralogical Society of America). (Right) Concentration change of CaO in plagioclase formed by the crystallization experiment at a constant cooling rate using An_{15}-H_2O. When the distribution coefficient is constant (at a similar level to the value in equilibrium state), CaO concentration decreases (from the core to the rim of a crystal) with time, resulting in normal zoning (Lofgren 1980). From PHYSICS OF MAGMATIC PROCESSES by Robert Bero Hargraves. Copyright © 1980 by Princeton University Press. Reprinted by permission

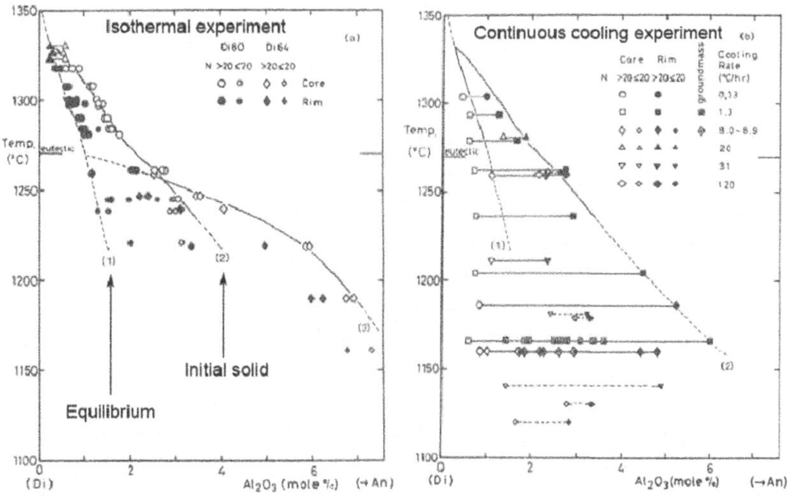

Fig. 7.26 Al concentration in Di formed by the cooling crystallization experiment using melt of Di80 composition of the Di-An system. (Left) The crystallization experiments under a constant temperature. (Right) The crystallization experiments under a constant cooling rate. Lines (1) and (2) represent the expected solid-phase equilibrium composition (Equilibrium) and the initial solid-phase (core) composition (Initial solid) in crystallization at a constant temperature. The white symbols denote the core composition and the black symbols denote the rim composition (note that the tendency is reversed in right and left figures). (Adapted from Tsuchiyama 1985, with permission from the Mineralogical Society of America)

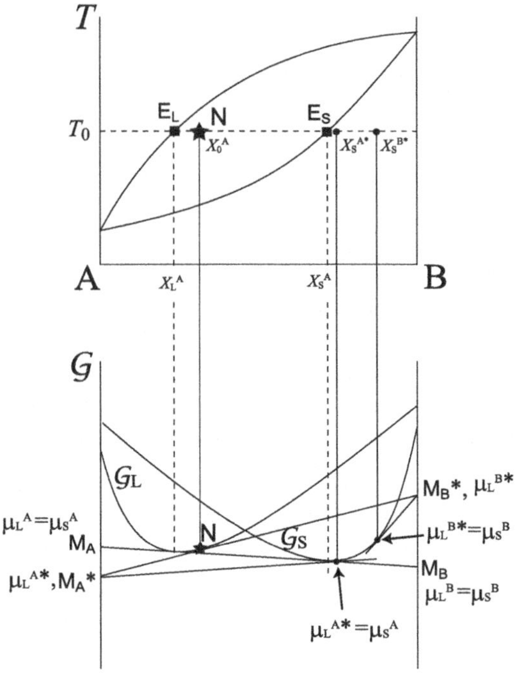

Fig. 7.27 Equilibrium phase diagram of a binary solid solution and relationship of chemical potential between disequilibrium melt and solid phase

undercooling and then crystallized at a constant temperature, the crystal composition initially formed is in many cases rich in low-temperature components (e.g. Ab in the case of plagioclase) compared with the equilibrium composition at the temperature. With progress in crystal growth, the crystal becomes rich in An and the remaining melt becomes rich in Ab. Consequently, reverse zoning showing a gradation from Ab-rich core to An-rich rim is generated. However, the physical reason why the first forming crystal becomes richer in Ab than the equilibrium composition is unclear. If a distribution coefficient does not considerably differ from the value in equilibrium state, then as influenced by the initial composition of undercooled solution, a crystal composition that is richer in An than the equilibrium composition should be formed and the crystal should become rich in Ab as approaching equilibrium composition by the subsequent crystal growth, resulting in the formation of normal zoning. In reality, normal zoning has been formed in the experiment at a constant cooling rate.

The problem that chemical composition of the solid phase crystallized from undercooled solution is not uniquely determined can be graphically understood as follows using a free energy-composition (molar fraction) chart, as shown in Fig. 7.27. From the previous discussion, the chemical potential of component A in melt at temperature T_0 and chemical composition denoted by point N is represented by point M_A^* where the tangent from the free energy curve intersects with the A-axis and is different from M_A of equilibrium conditions. If the component A in the solid phase satisfies the chemical equilibrium from Eq. (7.153), a point of contact (μ LA* = μ SA) of the tangent from point M_A^* to curve $\mathcal{G}_S$ should determine the chemical composition of the solid phase X_S^{A*}. On the other hand, the chemical potential of component B is represented by point M_B^* where the tangent from the free energy curve intersects with the B-axis. In the same manner, if the component B satisfies the chemical equilibrium (Eq. 7.154), a point of contact (($\mu_L^{B*} = \mu_S^B$) of the tangent from point M_B^* to curve $\mathcal{G}_S$ should determine the chemical composition of the solid phase X_S^{B*}. As can be easily observed from the figure, the compositions of the two points, X_S^{A*} and X_S^{B*}, do not agree. Thus, the chemical composition of the solid phase being in equilibrium with the undercooled solution of different composition from the equilibrium composition at the temperature is not determined from the equilibrium theory.

In cases where $\mu_{\mathrm{L}}^{\mathrm{A}*} = \mu_{\mathrm{S}}^{\mathrm{A}}$ in the example shown in the figure and described above, because the experiment by Tsuchiyama (1985), in which cooling rate is constant, seems to be explainable, it seems best for matters to be determined by conditions of A. However, the result of the experiment of cooling crystallization at a constant temperature cannot be explained. Based on thermodynamic examination by Baker and Cahn (1971) and Hopper and Uhlmann (1974), Tsuchiyama (1985) inferred as follows:

Assuming that the composition of undercooled solution[24] is X_0 and composition of a solid phase formed from the solution by crystallization is X_{S}, from mass conservation,

$$X_0 = xX_{\mathrm{S}} + (1-x)X_{\mathrm{L}} \tag{7.172}$$

is obtained. x is the mass ratio (fraction) of the solid and the liquid and, by rewriting this equation, x is related to compositions of the solid and the liquid, X_{S} and X_{L}, as

$$x = \frac{X_{\mathrm{L}} - X_0}{X_{\mathrm{L}} - X_{\mathrm{S}}} \tag{7.173}$$

That is, x is a function of X_{S} and X_{L}: $x(X_{\mathrm{S}}, X_{\mathrm{L}})$. Difference in the free energy before and after crystallization, $\Delta\mathcal{G}$, is determined as follows:

$$\Delta\mathcal{G} = x\mu_{\mathrm{S}}(X_{\mathrm{S}}) + (1-x)\mu_{\mathrm{L}}(X_{\mathrm{L}}) - \mu_{\mathrm{L}}(X_0) \tag{7.174}$$

At this point in time, considering Eq. (7.173), the variables of $\Delta\mathcal{G}$ are two among x, X_{L}, and X_{S}.

When the composition of the liquid is constant, the condition under which the difference in the free energy becomes largest is

$$\left(\frac{\partial\Delta\mathcal{G}}{\partial X_{\mathrm{S}}}\right)_{X_{\mathrm{L}}} = 0 \tag{7.175}$$

and substituting x in Eq. (7.173) into this equation and calculating it yields

$$\mu_{\mathrm{L}}(X_{\mathrm{L}}) = \mu_{\mathrm{S}}(X_{\mathrm{S}}) + (X_{\mathrm{L}} - X_{\mathrm{S}})\left(\frac{\mu_{\mathrm{S}}(X_{\mathrm{S}})}{\partial X_{\mathrm{S}}}\right)_{X_{\mathrm{L}}} \tag{7.176}$$

(the line L_1 in Fig. 7.28). This graphically means that the point of contact of the tangent from the liquid of composition X_{L} to μ_{S} is X_{S} (Fig. 7.28). We will write it as X_{S}^*. Because this composition X_{S}^* is richer in low-temperature components than equilibrium composition $X_{\mathrm{S}}^{\mathrm{E}}$, the result of the experiment under the condition of constant temperature by Tsuchiyama (1985) seems to be explainable.

What will constitute another constraint condition? That is, what are compositions of the solid and the liquid that make the difference in the free energy largest when the mass ratio x after crystallization is constant? The condition can be given by

$$\left(\frac{\partial\Delta\mathcal{G}}{\partial X_{\mathrm{S}}}\right)_{x} = 0 \tag{7.177}$$

and, calculating this equation yields

[24]Hereafter, the composition (molar fraction) X indicates that of component B unless otherwise stated.

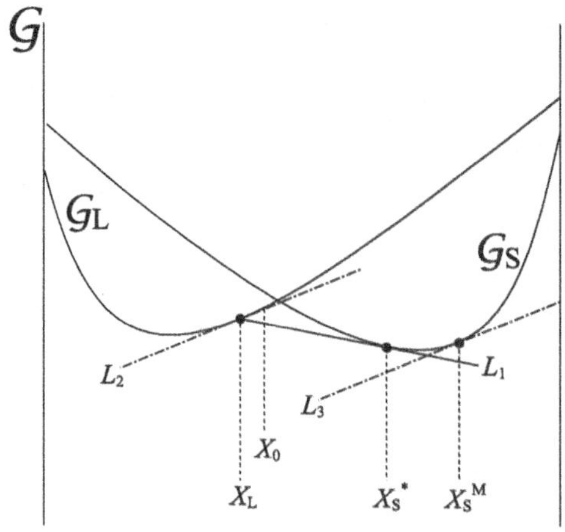

Fig. 7.28 Relationship between free energy and composition in the binary solid solution. In crystallization from undercooled melt with initial composition X_0 in disequilibrium, the composition of solid phase X_S^* at which difference in the free energy becomes largest when the composition of the liquid X_L is constant and the composition of the solid phase X_S^M at which difference in the free energy becomes largest when the solid-liquid ratio is constant (For brevity in the figure, the mass ratio selected as the composition of the liquid is X_L and the composition of solid phase is X_S against the initial composition X_0.)

$$\left(\frac{\partial \mu_S}{\partial X_S}\right)_x = \left(\frac{\partial \mu_L}{\partial X_L}\right)_x \tag{7.178}$$

This is a set of compositions of solid and liquid phases (X_L and X_S^M in Fig. 7.28), the slope of the tangent of the free energy of which becomes equal (that is, parallel tangents can be drawn; lines L_2 and L_3 in Fig. 7.28). However, as shown from this figure, because this solid phase composition is poorer in low-temperature components than equilibrium composition X_S^E, the result of the experiment by Tsuchiyama (1985) in which cooling rate is constant seems to be explainable; however, the result of the experiment in which temperature is constant cannot be explained.

The experimental results and the abovementioned thermodynamic examination result in the following conclusions:

1. Effective distribution coefficient is a function of the degree of disequilibrium and is a function of time.
2. The physical basis that determines the crystal composition at the time of crystallization in the disequilibrium state has not yet been established.

7.7.2 Relationship Between the Growth Law and Zoning Structure

In case where the chemical composition of a crystal is uniquely determined by the temperature, pressure, and liquid composition of the system, if the liquid composition is determined and one of two intensive variables, temperature and pressure, is made constant, the chemical composition of the formed crystal will be a function of the remaining intensive variable. For this case, the chemical composition profile is uniquely determined from the crystal growth behavior. We show it in the following.

First, we will obtain the relationship between chemical composition and temperature or pressure. For cooling crystallization, we assume that concentration of a component in a crystal $\tilde{C}$ (wt%)[25] is a linear function of only temperature.

$$\tilde{C} = \tilde{C}_0 - \frac{d\tilde{C}}{dT}(T_0 - T) = \tilde{C}_0 - \frac{d\tilde{C}}{dT}\left|\frac{dT}{dt}\right| t = \tilde{C}_0 - \tilde{C}_0 t_z^{-1} t \tag{7.179}$$

where $\tilde{C}_0$ is concentration of the component in the crystal at the reference temperature T_0 and the coefficient of time t_z^{-1} has a dimension of 1/time (s), which is inversely proportional to the cooling time t_{th} and represents the influence of temperature change in cooling crystallization on the zoning of the chemical composition of the crystal.

$$t_z^{-1} = \frac{1}{\tilde{C}_0}\frac{d\tilde{C}}{dT}\left|\frac{dT}{dt}\right| = \frac{1}{\tilde{C}_0}\tilde{C}'_T|\dot{T}| \tag{7.180}$$

As $d\tilde{C}/dT = \tilde{C}'_T$ in this equation is determined from the equilibrium phase diagram or distribution of elements, it can be called as a thermodynamic factor. Subscript T indicates a differential coefficient concerning temperature.

For decompression-induced crystallization to be explained in the next chapter, we assume that the concentration of a component in the crystal $\tilde{C}$ is a linear function of only pressure. Assuming that the process is an isothermal process and the total pressure = partial pressure of water P,

$$\tilde{C} = \tilde{C}_0 - \frac{d\tilde{C}}{dP}(P_0 - P) = \tilde{C}_0 - \frac{d\tilde{C}}{dP}\left|\frac{dP}{dt}\right| t = \tilde{C}_0 - \tilde{C}_0 t_z^{-1} t \tag{7.181}$$

where $\tilde{C}_0$ is concentration in the crystal at the reference pressure P_0; the coefficient t_z^{-1} in this case is

$$t_z^{-1} = \frac{1}{\tilde{C}_0}\frac{d\tilde{C}}{dP}\left|\frac{dP}{dt}\right| = \frac{1}{\tilde{C}_0}\tilde{C}'_P|\dot{P}| \tag{7.182}$$

In the same manner as above, $d\tilde{C}/dP = \tilde{C}'_P$ is a thermodynamic factor. For decompression-induced crystallization, t_z is proportional to the decompression time t_{dec} and represents the influence of pressure change in decompression-induced crystallization on the zoning of the crystal chemical composition.

An equation of time change in the composition (Eq. 7.179) connects the time to the core composition or the rim composition. To avoid confusion, note that the nucleation time and common time are separately expressed by t' and t, respectively, in the following. Using these expressions, the crystal size can be written as a function of two types of time, the nucleation time and the common time, as $R(t', t)$.

The nucleation time t' is then connected to the core composition of crystals that nucleate at the time $\tilde{C}_{core}(R(t', t))$.

$$t' = \frac{\tilde{C}_0 - \tilde{C}_{core}(R(t', t))}{\tilde{C}_0} t_z \tag{7.183}$$

Moreover, the common time t is connected to the rim composition of crystals that nucleated at the earlier time t', $\tilde{C}_{rim}(R)$, as follows:

[25] Any units such as the molar fraction X and the number of molecules per unit volume C can be used for concentration but wt% is used here to maintain consistency with the discussion on decompression-induced crystallization (Chap. 8).

$$t = \frac{\tilde{C}_0 - \tilde{C}_{\rm rim}(R(t', t))}{\tilde{C}_0} t_{\rm z} \tag{7.184}$$

Let us consider the following three cases in which the growth rate $G(t)$ is a function of time: (1) $G(t) \propto (t - t')^{p-1}$ $(p > 0)$, (2) $G(t) \propto t^{-1}$, and (3) $G(t) = G_0 \exp(t/t_{\rm g})$.

Case (1) $G(t) \propto t^{p-1}$ $(p \neq 0)$

In this case, the integration can be performed as follows and the size $R(t', t)$ of the crystals at time t which nucleated at time t' can be obtained.

$$\frac{dR}{dt} = G_0 \left(\frac{t - t'}{t_{\rm g}}\right)^{p-1} \tag{7.185}$$

$$R = R_0 \left(\frac{t - t'}{t_{\rm g}}\right)^{p} \tag{7.186}$$

where R_0 is a proportionality constant that characterizes the crystal size, and $t_{\rm g}$ is the timescale of crystal growth and has the relationship

$$R_0 = \frac{G_0 t_{\rm g}}{p} \tag{7.187}$$

with G_0.

Based on the relationship between time and crystal composition (Eqs. 7.184 and 7.183), substituting t' and t into Eq. (7.186) and rewriting the equation yields

$$\tilde{C}_{\rm rim}(R) = \tilde{C}_{\rm core}(R) - \tilde{C}_0 \left(\frac{t_{\rm g}}{t_{\rm z}}\right) \left(\frac{R}{R_0}\right)^{1/p} \tag{7.188}$$

Although this equation gives the composition of the rim of the size R, reading the size R as the distance from the core $r = R$ yields the crystal composition at the distance $r \leq R$ in the crystal with a size of R. That is, this will be the composition profile $\tilde{C}(r)$ of the zoning structure,

$$\tilde{C}(r) = \tilde{C}_{\rm core}(R) - \tilde{C}_0 \left(\frac{t_{\rm g}}{t_{\rm z}}\right) \left(\frac{r}{R_0}\right)^{1/p} \tag{7.189}$$

The size R of the crystal observed, the core composition $\tilde{C}_{\rm core}(R)$, and the rim composition $\tilde{C}_{\rm rim}(R)$ are mutually related through $t_{\rm g}/t_{\rm z}$, and from Eq. (7.188)

$$\tilde{C}_0 t_{\rm g}/t_{\rm z} = \left(\tilde{C}_{\rm core}(R) - \tilde{C}_{\rm rim}(R)\right) \left(\frac{R}{R_0}\right)^{-1/p} \tag{7.190}$$

is obtained. Substituting this equation into Eq. (7.189) yields the composition profile, the parameters of which are only the core composition $\tilde{C}_{\rm core}$ and the rim composition $\tilde{C}_{\rm rim}$

$$\tilde{C}(r) = \tilde{C}_{\rm core}(R) - (\tilde{C}_{\rm core}(R) - \tilde{C}_{\rm rim}(R)) \left(\frac{r}{R}\right)^{1/p} \tag{7.191}$$

for the crystal with a size of R (the left chart of Fig. 7.29). Note that by considering the distance from the core normalized by the crystal size, r/R, and normalizing the composition by the core composition, $\tilde{C}(r)/\tilde{C}_{\rm core}$, the composition profile is determined only by the exponent of the growth law p and the ratio between the core composition and the rim composition $\tilde{C}_{\rm rim}/\tilde{C}_{\rm core}$

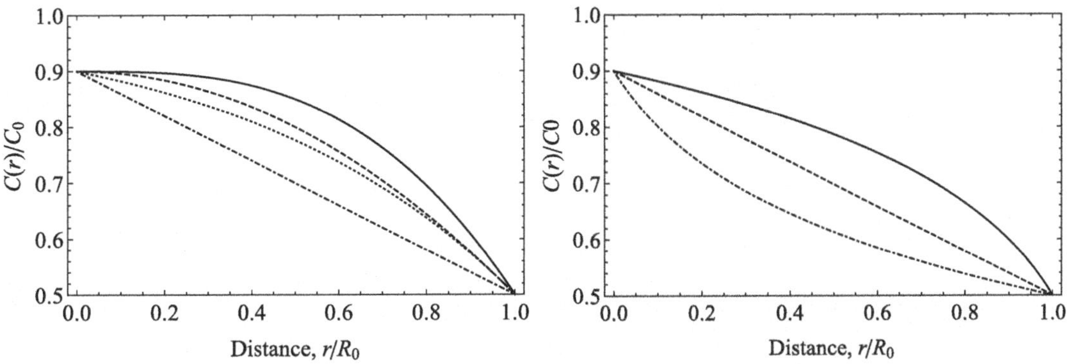

Fig. 7.29 (Left) Zoning structure profiles of case (1) $G(t) \propto t^{p-1}$ $(p \neq 0)$ (solid line $p = 1/3$, broken line $p = 0.5$, chain line $p = 1$) and case (2) G(t) ∝ t-1 (dotted line) (the core composition = 0.9 and the rim composition = 0.5 are assumed). (Right) Zoning structure profiles of case (3) $G(t) = G_0 \exp(t/t_g)$ (solid line $t_z/t_g = -5$, broken line $t_z/t_g = 0.1$, chain line $t_z/t_g = 5$)

$$\frac{\tilde{C}(r)}{\tilde{C}_{\text{core}}(R)} = 1 - \left(1 - \frac{\tilde{C}_{\text{rim}}(R)}{\tilde{C}_{\text{core}}(R)}\right)\left(\frac{r}{R}\right)^{1/p} \tag{7.192}$$

Moreover, when the maximum grain size among observed crystals is assumed to be $R_{\max}$,

$$\tilde{C}_{\text{rim}}(R_{\max}) = \tilde{C}_{\text{core}}(R_{\max}) - \tilde{C}_0\left(\frac{t_g}{t_z}\right)\left(\frac{R_{\max}}{R_0}\right)^{1/p} \tag{7.193}$$

is obtained, and eliminating tg/tz from Eqs. (7.193) and (7.188) yields the relationship between the core composition with an arbitrary grain size R, $\tilde{C}_{\text{core}}(R)$, and the size R in the form in which those at the maximum grain size are used as a reference (the left chart of Fig. 7.30).

$$\tilde{C}_{\text{core}}(R) = \tilde{C}_{\text{rim}}(R) + (\tilde{C}_{\text{core}}(R_{\max}) - \tilde{C}_{\text{rim}}(R_{\max}))\left(\frac{R}{R_{\max}}\right)^{1/p} \tag{7.194}$$

Case (2) $G(t) \propto t^{-1}$

This case corresponds to the case in which the nucleation rate is constant (J_1) and the rate of increase in the total volume of crystals is constant ($\dot{\phi}$), and has an interesting nature in that CSD shows exponential distribution (Sect. 9.3.5). When the crystal growth law is written as follows:

$$\frac{dR}{dt} = \frac{R_4}{t} \tag{7.195}$$

R_4 is a characteristic size in this kinetic model and has the following relation with the rate of increase in the amount of crystals $\dot{\phi} = 4\pi\dot{M}_3/3$ and the nucleation rate J_1 (Sect. 9.3.5, Eq. 9.162). t in the last equation is global time connected with T and P.

$$R_4 = \left(\frac{\dot{\phi}}{6\alpha_s J_1}\right)^{1/3} \tag{7.196}$$

The growth law can then be integrated as follows:

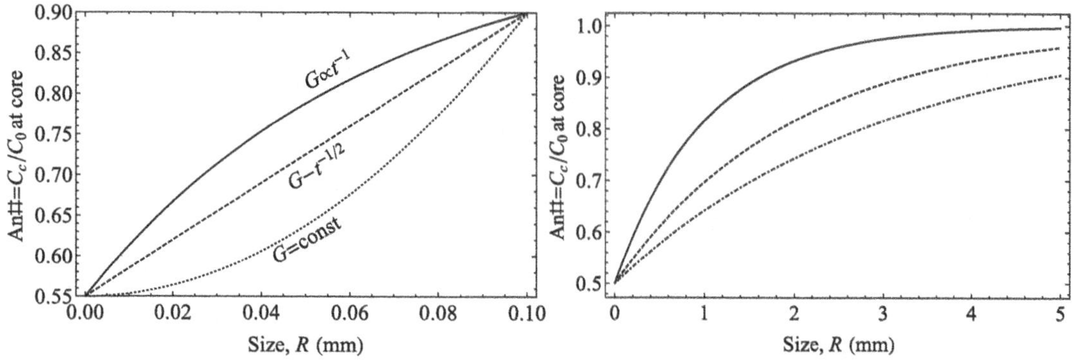

Fig. 7.30 (Left) Difference in relationship between core composition and size for different growth law. (Right) Relationship between core composition and size when $G(t) \propto t^{-1}$. Relationship between core composition and size when the value of $R_4 = (\dot{\phi}/6\alpha_s J_1)^{1/3}$ is varied. The core composition is scaled; however, the size is on the real scale. Solid line: $R_4 = 1$ (mm), broken line: 2 (mm), chain line: 3 (mm). Because R_4 becomes large as the crystallization rate is large and the nucleation rate is small, the gentler the curve, and the smaller the nucleation rate. This indicates that as the nucleation rate is small, individual crystals grow larger. The rim composition is assumed to be $\tilde{C}_{\text{rim}}(R)/C_0 = 0.5$

$$R = R_4 \ln\left(\frac{t}{t'}\right) \tag{7.197}$$

and substituting t' and t from Eqs. (7.184) and (7.183) into this equation gives

$$R = R_4 \ln\left(\frac{\tilde{C}_0 - \tilde{C}_{\text{rim}}(R)}{C_0 - \tilde{C}_{\text{core}}(R)}\right) \tag{7.198}$$

Rewriting this gives

$$\tilde{C}_{\text{rim}}(R) = \tilde{C}_0 - (\tilde{C}_0 - \tilde{C}_{\text{core}}(R))e^{R/R_4} \tag{7.199}$$

From this equation, the core composition $C_{\text{core}}(R)$ of the crystal with a size of R is given as follows:

$$\tilde{C}_{\text{core}}(R) = \tilde{C}_0 - (\tilde{C}_0 - \tilde{C}_{\text{rim}}(R))e^{-R/R_4} \tag{7.200}$$

(Fig. 7.30).

In the same manner as case (1), by reading this size R as the distance from the core (substituting r for R), the rim composition $\tilde{C}_{\text{rim}}(R)$ (Eq. 7.199) yields a composition profile $\tilde{C}(r)$ in crystals with a size of $R > r$.

$$\tilde{C}(r) = \tilde{C}_0 - (\tilde{C}_0 - \tilde{C}_{\text{core}}(R))e^{r/R_4} \tag{7.201}$$

Eliminating R_4 using Eq. (7.199) gives

$$\tilde{C}(r) = \tilde{C}_0 - (\tilde{C}_0 - \tilde{C}_{\text{core}}(R))\left(\frac{\tilde{C}_0 - \tilde{C}_{\text{rim}}(R)}{\tilde{C}_0 - \tilde{C}_{\text{core}}(R)}\right)^{r/R} \tag{7.202}$$

(the left chart of Fig. 7.29).

To normalize the grain size by the observed maximum grain size $R_{\max}$, a relational equation

$$\tilde{C}_{\text{core}}(R_{\max}) = \tilde{C}_0 - (\tilde{C}_0 - \tilde{C}_{\text{rim}}(R_{\max}))e^{-R_{\max}/R_4} \tag{7.203}$$

is used to eliminate R_4, and the core composition $\tilde{C}_{\text{core}}(R)$ as a function of the grain size is given as follows:

$$\tilde{C}_{\text{core}}(R) = \tilde{C}_0 - (\tilde{C}_0 - \tilde{C}_{\text{rim}}(R)) \left(\frac{\tilde{C}_0 - \tilde{C}_{\text{core}}(R_{\max})}{\tilde{C}_0 - \tilde{C}_{\text{rim}}(R_{\max})} \right)^{R/R_{\max}} \tag{7.204}$$

(the right chart of Fig. 7.30). Note that the equilibrium crystal composition $\tilde{C}_0$ corresponding to the initial pressure P_0 does not agree with the core composition at the maximum grain size $\tilde{C}_{\text{core}}(R_{\max})$. In this crystal growth model, the size of crystals that nucleated at the time $t' = 0$ becomes infinity. Such crystals are not observed in reality.

Case (3) $G(t) = G_0 \exp(t/t_g)$

In this case, the time constant characterizing the rate of increase in the crystal growth rate is t_g. As a result of similar calculations, an expression of zoning profile, although this is somewhat complicated, becomes

$$\tilde{C}(r) = \tilde{C}_0 - \tilde{C}_0 \left(\frac{t_g}{t_z} \right) \ln \left[(1 - \frac{r}{R}) e^{(t_z/t_g)(\tilde{C}_0 - \tilde{C}_C)/\tilde{C}_0} + \frac{r}{R} e^{(t_z/t_g)(\tilde{C}_0 - \tilde{C}_R)/\tilde{C}_0} \right] \tag{7.205}$$

t_z/t_g, appearing in the cases above, gives the ratio of the cooling time or the decompression time to the characteristic time of the crystal growth ($t_z/t_g = t_{\text{cool}}/t_g$ or $= t_{\text{dec}}/t_g$). It is very interesting that this is equal to the controlling parameter of crystallization α_3.

7.7.3 Diffusion Profile Taking the Moving Interface into Consideration and Chemical Composition of Crystals

7.7.3.1 An Equation for Conservation of the Liquid Composition from the Point of View of a Coordinate System Moving Together with the Solid-Liquid Interface

Let us consider chemical composition in the case where crystals nucleate in the undercooled state and grow. Here, we consider that, as discussed above, the chemical composition of crystals that are in equilibrium with a certain undercooled melt is not accurately defined in terms of thermodynamics but is determined because it satisfies a certain distribution relation. That is, assuming that the distribution coefficient is K_D and the molar fraction in the liquid phase is X_L, a molar fraction in the solid phase X_S is determined by

$$K_D = \frac{X_S}{X_L} \tag{7.206}$$

as $X_S = K_D \cdot X_L$. Moreover, the liquid composition close to the interface is continuously recorded in the crystals with the crystal growth.

On the other hand, even though the liquid composition is homogeneous at first, some components in the liquid are consumed by growth of the solid phase with a composition different from that of the liquid phase; moreover, a concentration gradient is generated toward the crystals. That is, the composition near the solid-liquid interface and that at a distance from there are different, and mass transfer by diffusion occurs between them. Furthermore, mass conservation responding to the growth rate G and distribution holds at the solid-liquid interface and provides the boundary condition for mass transfer in the liquid.

Assuming that $x' \leq 0$ in the solid phase and $x' > 0$ in the liquid phase, when the solid-liquid interface moves at a constant growth rate G, as seen from the point of view of a coordinate x' moving together with the solid-liquid interface, the infinitesimal change $dC \cdot dx'$ due to coming in and out of flux $I(x')$ in the infinitesimal spatial width dx' in the liquid is written as follows:

$$dC \cdot dx' = \left(I(x') - I(x' + dx')\right) \cdot dt = -\frac{\partial I(x')}{\partial x'} dx' \cdot dt \tag{7.207}$$

Therefore, the conservation of mass of the crystallization components in the liquid is

$$\frac{\partial C}{\partial t} = -\frac{\partial I(x')}{\partial x'} \tag{7.208}$$

On the other hand, considering the fact that the width dx' moves with the movement of the coordinate at a rate of G in the positive direction, flux I is written as follows:

$$I(x') = -D \cdot \frac{\partial C}{\partial x'} - G \cdot C \tag{7.209}$$

The last term is the effect of relative movement. Substituting this equation into Eq. (7.208) yields the diffusion equation seen from the point of view of the coordinate system moving together with the advancing interface.

$$\frac{\partial C}{\partial t} = D \frac{\partial^2 C}{\partial x'^2} + G \frac{\partial C}{\partial x'} \tag{7.210}$$

Note that this equation holds if G depends on time (see Box).

Derivation of a diffusion equation in the case where the interface moves by coordinate transformation

We start from a diffusion equation in a coordinate system that does not move

$$\frac{\partial C}{\partial t} = D \frac{\partial^2 C}{\partial x^2} \tag{7.211}$$

This equation is an equation followed by a dependent variable $C(t, x)$, the independent variables of which are t and x. We then transform this equation into an equation that is followed by $C'(t, x')$, the independent variable of which is a coordinate system x' that moves at a rate of $G(t)$ to the positive direction of x with respect to a space coordinate. The relationship of the variables in the respective coordinate systems is as follows:

$$C(t, x) = C(t, x(x', t)) = C'(t, x'(x, t)) \tag{7.212}$$

$$x(x', t) = x' + \int_0^t G(t') dt' \tag{7.213}$$

Therefore,

$$\frac{\partial x}{\partial t} = G = -\frac{\partial x'}{\partial t} \tag{7.214}$$

The right equation has a minus sign because, with the movement of the interface (the direction of $x > 0$), the value of the position x' seen from the coordinate x becomes small (because the origin of the coordinate system x moves in the direction of $x' < 0$)

$$\frac{\partial x}{\partial x'} = 1 = \frac{\partial x'}{\partial x} \tag{7.215}$$

$$\frac{\partial C(t,x)}{\partial t} = \frac{\partial C'(t,x')}{\partial t} + \frac{\partial C'(t,x')}{\partial x'}\frac{\partial x'}{\partial t} \tag{7.216}$$

$$= \frac{\partial C'(t,x')}{\partial t} - G\frac{\partial C'(t,x')}{\partial x'} \tag{7.217}$$

$$\frac{\partial C(t,x)}{\partial x} = \frac{\partial C'(t,x')}{\partial x'}\frac{\partial x'}{\partial x} = \frac{\partial C'(t,x')}{\partial x'} \tag{7.218}$$

Similarly,

$$\frac{\partial^2 C(t,x)}{\partial x^2} = \frac{\partial^2 C'(t,x')}{\partial x'^2} \tag{7.219}$$

From the above, substituting Eqs. (7.217) and (7.219) into Eq. (7.211) and removing $'$ in C' yields Eq. (7.210).

7.7.3.2 Steady-State Solution When the Growth Rate Is Constant

Tiller et al. (1953) dealt with the problem of crystallization of metal and investigated the concentration distribution of crystallization components in a liquid at a steady state when G is constant. For the steady state, a fundamental equation is

$$D\frac{\partial^2 C}{\partial x'^2} + G\frac{\partial C}{\partial x'} = 0 \tag{7.220}$$

by ignoring the term of time derivative. The boundary conditions are then determined as follows:

$$(1)\quad C(\infty) = C_0 \tag{7.221}$$

$$(2)\quad C(0) = \frac{C_S}{K_D} \tag{7.222}$$

$$(3)\quad \left.\frac{\partial C}{\partial x'}\right|_{x'=\infty} = 0 \tag{7.223}$$

$$(4)\quad D\left.\frac{\partial C}{\partial x'}\right|_{x'=0} = G(C_S - C(0)) \tag{7.224}$$

where C_0 is the initial melt composition and C_S is the composition of the solid phase. For the steady state to hold,

$$(5)\qquad C_S = C_0 \tag{7.225}$$

holds as mass conservation. Substituting this equation into (2) and (4) of the boundary conditions gives

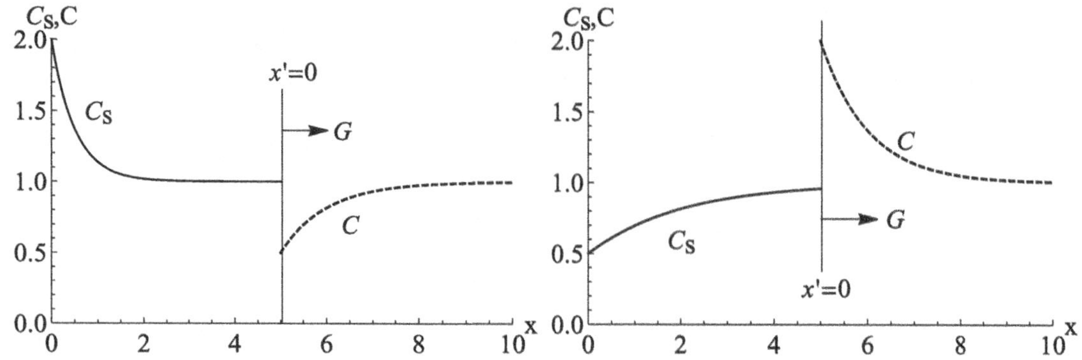

Fig. 7.31 The concentration profile in the liquid and in the solid on the basis of the analytic solution in the steady state by Tiller et al. (1953). (Left) $K = 2$. (Right) $K = 0.5$. $x' = 0$ represents the position of the interface. $C_0 = 1$ and $x_0 = 1$ are assumed

$$(2')\quad C(0) = \frac{C_0}{K_D} \tag{7.226}$$

$$(4')\quad D\left.\frac{\partial C}{\partial x'}\right|_{x'=0} = \frac{K_D - 1}{K_D} G C_0 \tag{7.227}$$

Integrating Eq. 7.220 once more and using the boundary conditions gives

$$\frac{\partial C}{\partial x'} = -\frac{G}{D}(C - C_0) \tag{7.228}$$

Integrating this once more and using the boundary conditions gives

$$C(x') = C_0 \left[1 + \frac{1 - K_D}{K_D} \exp\left(-\frac{Gx'}{D} \right) \right] \tag{7.229}$$

From this equation, the concentration profile is determined only using K_D and G/D (Fig. 7.31). The concentration profile in the steady state is found to have a boundary layer with a characteristic length of $x_0 = D/G$ and to be attenuated to the concentration in the liquid C_0 from the interface. Moreover, the time required to grow using the characteristic length, $t_0 = x_0/G = D/G^2$ can be called the characteristic time for crystal growth.

Under the assumption of the steady state, change in composition of the solid phase side $C_S(x)$, or zoning, cannot be found out.[26] One estimation method is a model in which the rate of change in deviation from the steady state $C_0 - C_S$ relaxes in proportion to itself. That is,

$$\frac{d}{dx}(C_0 - C_S) = -\frac{C_0 - C_S}{x_S} \tag{7.230}$$

where x_S is a characteristic relaxation scale of the solid phase side and determined by consistency in the concentration profile (mass conservation between the solid and the liquid). As a result of the integration of this equation,

[26]Note that the space coordinate x independent of the movement of the interface is used here.

$$C_S(x) = C_0 \left[1 - (1 - K_D) \exp\left(-\frac{x}{x_S} \right) \right] \tag{7.231}$$

From mass conservation, the integral of deviation of the solid phase concentration from the steady-state value, $C_0 - C_S$, from $x = 0$ to $x = \infty$ and the integral of deviation of the liquid phase concentration from the steady-state value, $C_0 - C$, from $x' = 0$ to $x' = \infty$ should be equivalent. Therefore, x_S is given as follows:

$$x_S = \frac{D}{K_D G} = \frac{x_0}{K_D} \tag{7.232}$$

(Fig. 7.31).

7.7.3.3 Non-steady State Solution in the Case of Constant Growth Rate: Zoning Pattern of Crystals

By a method using the Laplace transform explained in Sect. 7.4.1, Smith et al. (1955) obtained a solution of a time-changing concentration profile in the liquid that satisfies Eq. (7.210) when G is constant. By multiplying both the sides of Eq. (7.210) by e^{-st} (s is a dummy variable of the Laplace transform) and integrating it, the left-hand side is calculated using partial integration as follows:

$$\text{LHS} = \int_0^\infty e^{-st} \frac{\partial C(x', t)}{\partial t} dt = \left[C(x', t) e^{-st} \right]_0^\infty - \int_0^\infty (-s) e^{-st} C(x', t) dt \tag{7.233}$$

$$= -C_0 + s\hat{C}(x', s) \tag{7.234}$$

Here, the Laplace transform corresponding to the initial condition

$$C(x', 0) = C_0 \tag{7.235}$$

and that to the boundary condition

$$\left. \frac{\partial C}{\partial x'} \right|_{x'=0} = -\frac{G}{D}(1 - K_D) C(0, t) \tag{7.236}$$

are used. Moreover, because the right-hand side does not include differentiation of t, integration can be moved into differentiation and C can be replaced with the Laplace transform $\hat{C}$. Therefore, an ordinary differential equation that $\hat{C}$ (the Laplace transform of C) satisfies is

$$D \frac{\partial^2 \hat{C}(x', s)}{\partial x'^2} + G \frac{\partial \hat{C}(x', s)}{\partial x'} + C_0 - s\hat{C}(x', s) = 0 \tag{7.237}$$

Substituting $\hat{C} - C_0/s \propto e^{-\lambda x'}$ into this equation and obtaining an eigenvalue λ yields

$$\lambda_\pm = \frac{G}{2D} \pm \sqrt{\left(\frac{G}{2D} \right)^2 + \frac{s}{D}} \tag{7.238}$$

Considering the finiteness of the solution at $x' \to \infty$, + is adopted and a general solution

$$\hat{C}(x', s) = ce^{-\lambda_+ x'} + \frac{C_0}{s} \tag{7.239}$$

is obtained. The coefficient c can be obtained from the boundary condition as

$$c = \frac{1 - K_D}{\frac{D}{G}\lambda_+ - (1 - K_D)} \frac{C_0}{s} \tag{7.240}$$

Using this to rewrite $\hat{C}$ yields

$$\frac{\hat{C}(x', s)}{C_0} = \frac{(1 - K_D)\frac{G}{2\sqrt{D}} e^{-\frac{G}{2D}x'} \exp\left[-\frac{1}{\sqrt{D}}\left(\frac{G^2}{4D} + s\right)^{1/2} x'\right]}{s\left[\frac{G}{\sqrt{2D}}(2K_D - 1) + \left(\frac{G^2}{4D} + s\right)^{1/2}\right]} + \frac{1}{s} \tag{7.241}$$

The solution can be obtained by the inverse Laplace transform of this equation with respect to s. The first term of the right-hand side takes the form of const$\times \exp(-a(b+s)^{1/2})/(s(c+(b+s)^{1/2}))$ as a function of the variable s. Equation 827.1 in the table of transform pairs by Campbell and Foster (1931) corresponds to this.[27] The inverse Laplace transform of the second term on the right-hand side, $1/s$, is 1. Consequently,

$$\begin{aligned}\frac{C(x', t)}{C_0} &= 1 + \frac{1 - K_D}{2K_D} \exp\left(-\frac{Gx'}{D}\right) \mathrm{Erfc}\left[\frac{x' - Gt}{2\sqrt{Dt}}\right] - \frac{1}{2}\mathrm{Erfc}\left[\frac{x' + Gt}{2\sqrt{Dt}}\right] \\ &+ \frac{1}{2}\left(\frac{2K_D - 1}{K_D}\right) \exp\left(-(1 - K_D)\frac{G(x' + K_D G t)}{D}\right) \\ &\cdot \mathrm{Erfc}\left[\frac{x' - (2K_D - 1)Gt}{2\sqrt{Dt}}\right]\end{aligned} \tag{7.242}$$

This equation becomes Eq. (7.229) when $t \to \infty$. On the other hand, because the chemical composition of a crystal at the moment it grows is given by $K_D C(0, t)$, the zoning pattern recorded in the crystal is calculated by $C_S(x) = K_D C(0, x/G)$. This can be compared with zoning patterns found in nature or in experiments. This solution is shown in Fig. 7.32.

In Eq. (7.242) what influences time change is $(1 - K_D)G(x' + K_D G t)/D$ in the exponential part ahead of the last Erfc and the time it changes from 1 to 2 can be seen as relaxation time to the steady-state solution, t_{relax}. When $x' = 0$, $(1 - K_D)K_D G^2 t_{\mathrm{relax}}/D = 1$ and $t_{\mathrm{relax}} = D/((1 - K_D)K_D G^2) = t_0/(1 - K_D)K_D$ are obtained. When $K_D = 0.2$, $t_{\mathrm{relax}} = 6t_0$ to $12t_0$ and time that the concentration profile in the liquid relaxes to the steady-state solution is about an order of magnitude larger than the characteristic time of the crystal growth t_0. This can be found out from the figure.

7.7.3.4 Time-Dependent Solution in the Case Where the Growth Rate Depends on Time or Liquid Composition and Oscillatory Zoning

To derive an expression of liquid concentration at the interface, Lasaga (1982) did not perform the Laplace transform on time, as Smith et al. (1955) conducted, but performed the Laplace transform on space, which is also a solution to the problem of balance between diffusion and crystal growth. Using the expression, he obtained formal expressions of analytical solutions to the composition of solid

[27] This is not listed in the tables often referred to for integral transforms by Erdèlyi (1954). Campbell and Foster (1931) is provided as a reference instead.

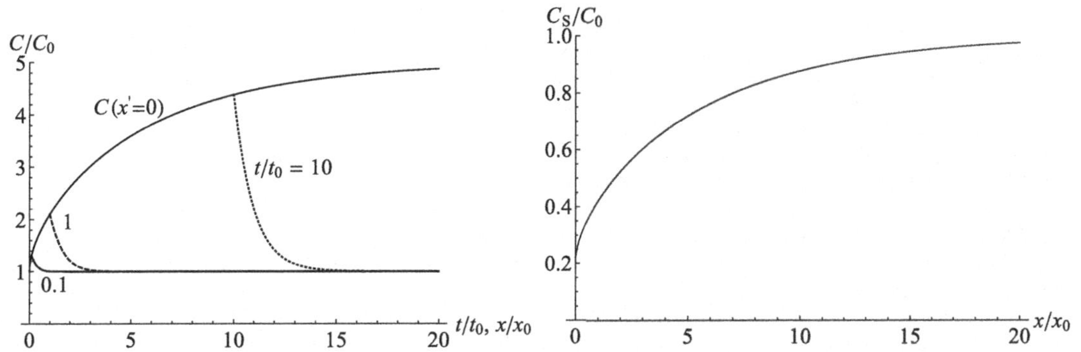

Fig. 7.32 (Left) Time change of concentration profile in the liquid based on the analytic solution by Smith et al. (1955). $K_D = 0.2$. $C(x' = 0)$ represents the concentration at the interface. The case of $t \approx 10t_0$ is close to the steady-state solution ($C(x' = 0, t) = 1/K_D = 5$) (Right) Zoning profile from the center of the crystal. $x_0 = G/D$

phase in cases where G has dependence on the liquid composition at the interface and where G has dependence on time. On the basis of a model of the reaction-limited crystal growth, he parameterized the experimental result by Kirkpatrick et al. (1979), discussed the stability of balanced state between the diffusion growth and the interface reaction-limited growth, and showed it to be stable. Consequently, he reached the conclusion that, for plagioclase, oscillatory zoning does not spontaneously arise using the published crystal growth model.

References

Baker JC, Cahn JW (1971) Thermodynamics of solidification. Paper from Solidification, ASM, Metals Park, pp 23–58

Befus KS, Manga M, Gardner JE, Williams M (2015) Ascent and emplacement dynamics of obsidian lavas inferred from microlite textures. Bull Volcanol 77–88. https://doi.org/10.1007/s00445-015-0971-6

Bowen NL (1915) The crystallization of haplobasaltic, haplodioritic, and related magmas. Am J Sci 236:161–185

Brandeis G, Jaupart C (1987) The kinetics of nucleation and crystal growth and scaling laws for magmatic crystallization. Contrib Mineral Petrol 96:24–34

Campbell GA, Foster RM (1931) Fourier integrals for practical applications, Bell Telephone System, Technical publications, Monograph B-584

Carslaw HS, Jaeger JC (1959) Conduction of heat in solids, 2nd edn. Oxford University Publications

Couch S, Sparks RSJ, Carroll MR (2003) The kinetics of degassing-induced crystallization at the Soufriere Hills Volcano. Montserrat J Petrol 44:1477–1502

Erdèl Yi A (ed) (1954) Tables of integral transforms, vol I. McGraw-Hill, New York

Gray NH (1970) Crystal growth and nucleation in two large diabase dikes. Can J Earth Sci 7:366–375

Gray NH (1978) Crystal growth and nucleation in flash-injected diabase dikes. Can J Earth Sci 15:1904–1923

Grove TL (1978) Cooling histories of Luna 24 very low Ti (VLT) ferrobasalts: an experimental study. In: Proceedings of the 9th lunar planetary science conferences, pp 565–584

Grove TL (1990) Cooling histories of lavas from Seroki volcano. In: Proceedings of ODP, science research, vol 106/109, pp 3–8

Grove TL, Walker D (1977) Cooling histories of Apollo 15 quartz-normative basalts. In: Proceedings of the 8th lunar planetary science conferences, pp 1501–1520

Hammer JE, Rutherford MJ (2002) An experimental study of the kinetics of decompression-induced crystallization in silicic melt. J Geophys Res 107. https://doi.org/10.1029/2001JB000281

Hopper RW, Uhlmann DR (1974) Solute redistribution during crystallization at constant velocity and constant temperature. J Crystal Growth 21(2):203–213

Hort M, Spohn T (1991) Numerical simulation of the crystallization of multicomponent melt in thin dikes or sills. 2: effects of heterocatalytic nucleation and composition. J Geophys Res 96:485–499

Hort M, Spohn T (1991) Crystallization calculations for a binary melt cooling at constant rates of heat removal: implications for the crystallization of magma bodies. Earth Plant Sci Lett 107:463–474

James PF (1974) Kinetics of crystal nucleation in lithium-silicate glasses. Phys Chem Glasses 15:95–105

James PF (1985) Kinetics of crystal nucleation in silicate glasses. J Non-Crystal Solids 73:517–540

Kirkpatrick RJ (1981) Kinetics of crystallization of igneous rocks. Rev Mineral 8:321–398

Kirkpatrick RJ, Klein L, Uhlmann DR, Hays JF (1979) Rates and processes of crystal growth in the system anorthite-albite. J Geophys Res 84:3671–3676

Lasaga AC (1982) Toward a master equation in crystal growth. Am J Sci 282:1264–1288

Liang Y, Richter FM, Davis AM, Watson EB (1996a) Diffusion in silicate melts: I. Self diffusion in Ca-Al_2O_3-SiO_2 at 1500°C and 1 GPa. Geochim Cosmochim Acta 60:4353–4367

Liang Y, Richter FM, Watson EB (1996b) Diffusion in silicate melts: II. Multicomponent diffusion in Ca-Al_2O_3-SiO_2 at 1500°C and 1 GPa. Geochim Cosmochim Acta 60:5021–5035

Liang Y, Richter FM, Chmberlin L (1997) Diffusion in silicate melts: III. Emprical models for multicomponent diffusion. Geochim Cosmochim Acta 61:5295–5312

Lofgren GE, Donaldson CH, Williams RJ, Mullins O, Usselman TM (1974) Experimentally reproduced textures and mineral chemistry of Appolo 15 quartz normative basalts. In: Proceedings of the 5th lunar planetary science, pp 549–567

Lofgren GE (1980) Experimental studies on the dynamic crystallization of silicate melts. In: Hargrabes RB (ed) Physics of magmatic process. Princeton University Press, Princeton, pp 488–551

Lofgren GE, Grove TL, Brown RW, Smith DP (1979) Composition of dynamic crystallization techniques on Apollo 15 quartz normative basalts. In: Proceedings of the 10th lunar planetary science conference, pp 423–438

Lofgren G (1974) Temperature induced zoning in synthetic plagioclase feldspar. The Feldspars, NATO Feldspar Institute, Manchester University Press, Manchester, pp 362–375

Neilson GF, Weinberg MC (1979) A test of classical nucleation theory: crystal nucleation of lithium disilicate glass. J Non-Crystal Solids 34:137–147

Ni H, Keppler H, Walte N, Schiavi F, Chen Y, Masotta M, Li Z (2014) In situ observation of crystal growth in a basalt melt and the development of crystal size distribution in igneous rocks. Contrib Mineral Petrol 167:1003. https://doi.org/10.1007/s00410-014-1003-9

Pupier E, Duchene S, Toplis MJ (2008) Experimental quantification of plagioclase crystal size distribution during cooling of a basaltic liquid. Contrib Mineral Petrol 155(5):555–570

Sato H (1995) Textural difference of pahoehoe and aa lava of Izu-Oshima volcano, Japan–an experimental study on population density of plagioclase. J Volcanol Geotherm Res 66:101–113

Schiavi F, Walte N, Keppler H (2009) First in situ observation of crystallization processes in a basaltic-andesitic melt with the moissanite cell. Geology 37(11):963–966

Shea T, Hammer JE (2013) Kinetics of cooling- and decompression-induced crystallization in hydrous mafic-intermediate magmas. J Volcanol Geotherm Res 260:127–145

Smith VG, Tiller WA, Rutter JW (1955) A mathematical analysis of solute redistribution during solidification. Canad J Phys 33:723–745

Spohn T, Hort M, Fischer H (1988) Numerical simulation of the crystallization of multicomponent melts in thin dikes or sills. 1: the liquidus phase. J Geophys Res 93:4880–4894

Swanson SE (1977) Relation of nucleation and crystal-growth rate to the development of granitic texture. Am Mineral 62:966–978

Tiller WA, Jackson KA, Rutter JW, Chalmers B (1953) The redistribution of solute atoms during the solidification of metals. Acta Metal 1:428–437

Toramaru A (1991) Model of nucleation and growth of crystals in cooling magmas. Contrib Mineral Petrol 108:106–117

Toramaru A (1995) Numerical study of nucleation and growth of bubbles in viscous magmas. J Geophys Res 100:1913–1931

Toramaru A (2001) A numerical experiment of crystallization for a binary eutectic system with application to igneous textures. J Geophys Res 106:4037–4060

Toramaru A, Noguchi S, Oyoshihara S, Tsune A (2008) MND (microlite number density) water exsolution rate meter. J Volcanol Geotherm Res 175:156–167

Toschev S, Milchev A, Stoyanov S (1972) On some probabilistic aspects of the nucleation process. J Crystal Growth 13:123–127

Tsuchiyama A (1983) Crystallization kinetics in the system $CaMgSi_2O_6$-$CaAl_2Si_2O_8$; the delay in nucleation of diopside and anorthite. Am Mineral 68(7–8):687–698

Tsuchiyama A (1985) Crystallization kinetics in the system $CaMgSi_2O_6$-$CaAl_2Si_2O_8$; development of zoning and kinetics effects on element partitioning. Am Mineral 70(5–6):474–486

Wager LR (1961) A note on the origin of ophitic texture in the chilled olivine gabbro of the Skaergaard intrusion. Geol Mag 1008:353–369

Walker D, Kirkpatrick RJ, Longhi J, Hays JF (1976) Crystallization history of lunar picritic basalt sample 12002: phase-equilibria and cooling rate studies. Geol Soc Am Bull 87:646–656

Walker D, Powell MA, Lofgren GE, Hays JF (1978) Dynamic crystallization of a eucrite basalt. In: Proceedings of the 9th lunar planetary science conference, pp 1369–1391

Winkler HFG (1948) Crystallization of basaltic magma as recorded by variation of crystal size in dikes. Mineral Mag 28:557–574

Zieg MJ, Lofgren GE (2006) An experimental investigation of texture evolution during continuous cooling. J Volcanol Geotherm Res 154(1):74–88

8 Crystallization Induced by Vesiculation

Abstract

In Sect. 2.8, we learned that magma vesiculation, associated with decompression, induces decrease in water content in the silicate melt, leading to rise in the melting point of silicates. Therefore, for natural volcanic eruptions, when magma (which is saturated with water at depths) at a constant temperature spouts to the surface along with vesiculation, it becomes a supercooled state in most cases. In this chapter, we will summarize a point of similarity and a point of difference between cooling crystallization and decompression-vesiculation-induced crystallization (hereafter referred to as decompression-induced crystallization) from a thermodynamic perspective. Next, in cases where vesiculation progresses in equilibrium with decompression, we will explain a crystallization process. In this case, water content in melt is equivalent to solubility with respect to pressure. Then, because of the influence of bubble nucleation and growth processes, we will examine the case in which water content can change in disequilibrium with pressure. In the latter case, the rate of change in water content is greatly different between the disequilibrium and quasi-equilibrium regimes. Because there is no difference between both cases in the facts that crystallization undergoes crystal nucleation and growth, we use knowledge on cooling crystallization.

8.1 Similarity and Difference Between Decompression-Induced Crystallization and Cooling Crystallization

8.1.1 Phase Equilibrium Relation

As shown in Sect. 2.8, magma vesiculation is the exsolution of water from melt and decrease in water content leads to increase in the melting point of crystal phases. Therefore, most groundmass crystals in volcanic rocks are formed by decompression-induced crystallization. Crystal nucleation and growth occur also in the decompression-induced crystallization because melt is supersaturated with a certain crystallization components. Therefore, similar to the case of bubble formation by vesiculation and crystal formation in cooling crystallization, the discussion flow is as follows: a degree of supersaturation

A. Toramaru, *Vesiculation and Crystallization of Magma*, Advances in Volcanology,
https://doi.org/10.1007/978-981-16-4209-8_8

or an effective degree of supercooling is defined based on phase equilibrium relation and nucleation and growth are dealt with as kinetics. Note there are some differences in the way of change of the degree of supersaturation as a function of temperature and compositions.

In Sect. 2.8, silicate was treated as a single component and discussed as a binary system of water and silicate. In this section, considering a slightly more practical case, we will examine the influence of water on the binary eutectic system (Fig. 7.1) examined in the previous chapter. The left chart of Fig. 8.1 shows how an equilibrium phase diagram of the An (Anorthite)—Di (Diopside) eutectic system depends on the partial pressure of water.

First, we consider a cooling crystallization process of a system having temperature and composition represented by point Q saturated with water. This system, comprising a single melt phase, is present at a point higher than the liquidus of diopside at a pressure of 5 kbar. In this melt, water equivalent to the solubility (≈9 wt.%) at a pressure of 5 kbar is dissolved. When this system cools, it moves from point Q toward the direction denoted by the arrow, crosses the diopside liquidus at about 1250 °C, and then crystallizes diopside. This process was examined in the previous chapter.

Next, let us examine the case in which vesiculation occurs with decrease in the pressure and subsequently the partial pressure of water decreases. Because of decrease in the partial pressure of water, the diopside liquidus itself moves to the higher-temperature side (in the direction of up-pointing arrows), the system then crosses the liquidus at a temperature of point Q and enters a supersaturated state.

The right chart of Fig. 8.1 shows pressure change in water-saturated states of eutectic points and liquidus of various systems. This chart indicates that the water content dependence of liquidus differs among minerals. Plagioclase shows larger dependence than diopside. This means that change in the same amount of water content results in different degree of supersaturation. That is, with the same decompression rate, plagioclase and diopside have different effective cooling rates in magma. A thermodynamic factor that will be introduced in the following section defines the effective degree of supercooling determined by such an equilibrium relationship.

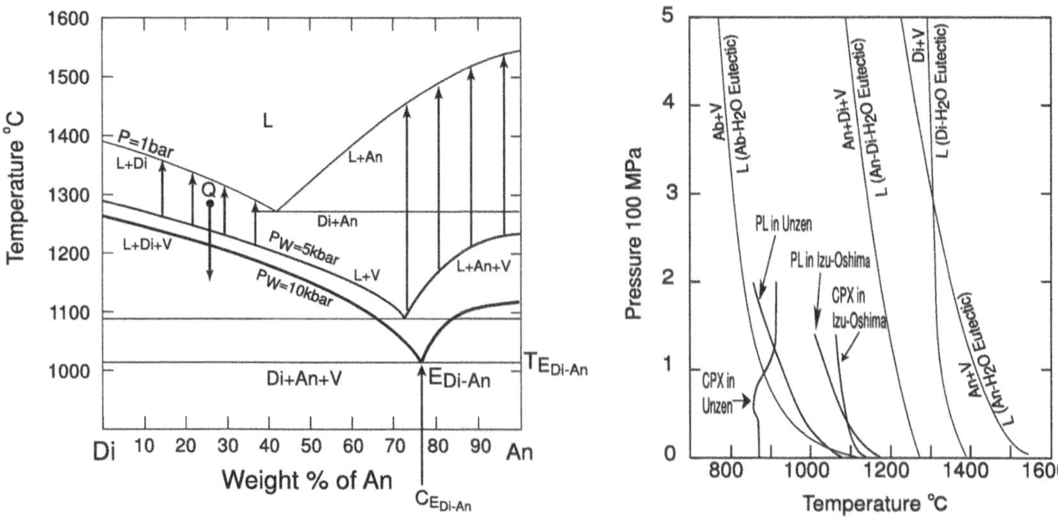

Fig. 8.1 (Left) The influence of the partial pressure of water in the anorthite—diopside system (Yoder 1965; Kushiro 1979). (Right) Dependence of eutectic points and liquidus of crystals in various systems on the pressure (total pressure = saturation pressure) (Burnham and Davis 1974; Yoder 1965; Hodges 1974). V means a vapor phase. Reprinted from Toramaru et al. (2008), Copyright 2008, with permission from Elsevier

Moreover, the gradients of melting points against pressure vary depending on whether the liquid is saturated with water or unsaturated or is supersaturated with water because of incomplete vesiculation (the right chart of Fig. 8.1). This is determined by the kinetics of vesiculation.

8.1.2 The Degree of Supercooling in Decompression-Induced Crystallization

Although cooling crystallization and decompression-induced crystallization are equivalent when seen as relative motion of the system to the liquidus in the equilibrium phase diagram, i.e., a degree of supercooling, there is a difference between them in the behavior of the nucleation rate or the growth rate as a function of intensive variables (temperature, pressure, and compositions). In cooling crystallization, temperature decreases with increase in degree of supercooling and the consequent inactivation of molecular motion leads to bell-shaped master curves of nucleation and growth rates, as shown in Fig. 7.4. In decompression-induced crystallization, however, instantaneous nucleation and instantaneous growth rates associated with decrease in pressure or decrease in water content only increase because the temperature does not decrease. Note that viscosity increases with decrease in water content and may influence molecular motion (Hammer 2004).

In any case, for a multi-component system, the maximum value of nucleation in the actual crystallization history is determined by the competition between the rate of decrease in the degree of supersaturation (due to increase in the amount of crystals caused by progress of nucleation and crystal growth) and the rate of increase in the degree of supersaturation (due to decrease in water content and cooling). Understanding obtained in cooling crystallization (in the case where the peak nucleation rate (the local maximum value of the master curve) is sufficiently large) can be used as it is by replacing the cooling rate with the effective cooling rate determined by change in the water content (the water exsolution rate). That is, the kinetic factor introduced in Sect. 7.6.6 is applicable to the relationship between the effective cooling rate and crystal number density.

Figure 8.2 shows the change in water content in melt caused by vesiculation induced by decompression due to magma ascent and water solubility (the solid line and the dotted line in the left chart) as well as the corresponding change of liquidus (the broken line in the right chart). After this figure, we will explain a kinetic path of microlite crystallization associated with magma vesiculation (a decompression-induced crystallization path). First, before and after nucleation of bubbles, the slope of the melting point, or liquidus, of microlite changes. Because the water content in the liquid is constant before bubble nucleation (higher pressure than point T_{BN}), the melting point tends to increase with depth by the influence of pressure. However, at depths shallower than the point of bubble nucleation, the liquidus rises as the pressure decreases or the depth becomes shallow because of decrease in water content. Depending on difference in the initial temperature of the liquid, a path of the degree of supercooling of microlites (the distance between liquid temperature and liquidus ΔT) differs.

For path A, the initial temperature of which is low, the system is in the supercooling state from the initial state and contains crystals (probably phenocrysts). The degree of supercooling decreases to point T_{BN} with magma ascent but increases after the vesiculation onset. Because the degree of supercooling decreases once, crystals may melt around point T_{BN}. Similar to the case of second nucleation of bubbles, whether the nucleation of microlites (second nucleation of crystals) occurs in magma containing phenocrysts with increase in the degree of supercooling probably depends on number density and size of initial phenocrysts.

For path C, the initial temperature of which is high, the system has a higher temperature than the liquidus in the initial state, is undersaturated with crystals, and contains no crystals. The degree of supercooling of the system increases with the onset of vesiculation caused by magma ascent and the system becomes saturated with crystals at a certain depth S_3, and nucleation of microlites occurs at

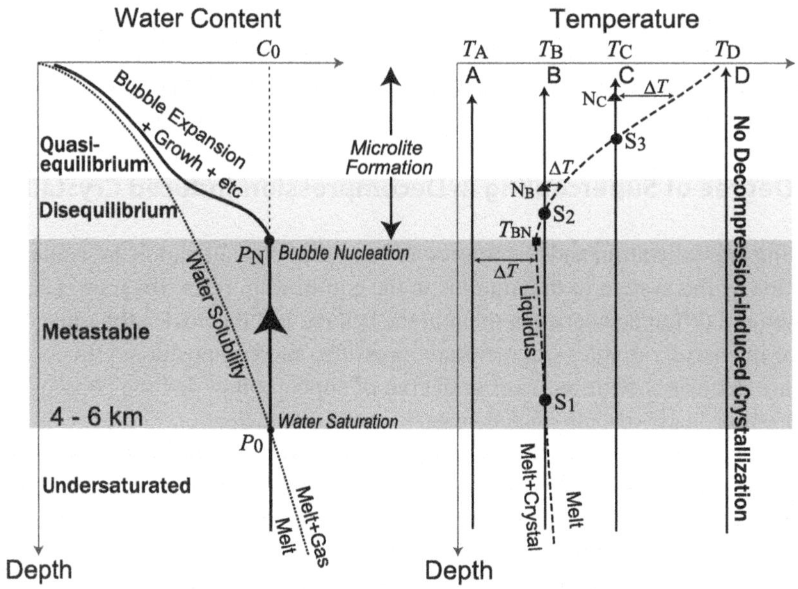

Fig. 8.2 The change in the water content in melt caused by vesiculation induced by decompression due to magma ascent and the water solubility (the solid line and the dotted line in the left chart) and corresponding change of liquidus (the broken line in the right chart). P_0 is the water saturation point, S_1, S_2, and S_3 are the saturation point of microlites, P_N is the bubble nucleation point, and N_B and N_C are the microlite nucleation points. The way in which the water content decreases is governed by kinetics of bubble nucleation and growth in the disequilibrium regime and follows the solubility in the quasi-equilibrium regime. Microlite crystallization shows several patterns according to relationship between the regime in which the water content decreases and the initial temperatures of liquid (A, B, C, D)

N_C. Homogeneous nucleation probably occurs at a relatively large degree of supercooling because the initial temperature is higher than the liquidus.

For path B, the initial temperature of which is moderate, the system is almost totally saturated with crystals at the initial stage (point S_1 and deeper) and is divided into two cases depending on whether the system contains phenocrysts or not. The system containing phenocrysts follows a path similar to path A. At a depth around a point of bubble nucleation T_{BN} where the system will be undersaturated, phenocrysts tend to melt (between S_1 and S_2). For path A in which the initial temperature is low and for path B in which the system is almost completely saturated, there is a possibility that clusters of crystals have already been formed in the liquid. For such cases, the degree of supercooling at nucleation of microlites is expected to be reduced (ΔT at point N_B along path B in the figure).

When the nucleation of microlites occurs in paths A and B, the nucleation of microlites is considered to occur in the regime where the water content is in disequilibrium immediately after bubble nucleation (there is a large gap with the water solubility curve) because the degree of supercooling is small. For path C, the initial temperature of which is higher than the liquidus, the microlite nucleation depth is at approximately the equilibrium regime because the degree of supercooling required for microlite nucleation is large. In the case where the initial temperature is further higher (path D), the magma has a higher temperature than the liquidus even after it reached the surface of the Earth because water completely exsolves from magma and decompression-induced crystallization does not occur and only cooling crystallization causes microlite formation.

In the following sections, for cases where melt is saturated with water (in equilibrium) and then the disequilibrium case, the dependence of the number density of crystals obtained as a result of the crystallization process on controlling parameters will be discussed. In any case, an important thing for kinetics of crystallization is the degree of supercooling; moreover, the degree of supercooling is the distance from the melting point, i.e., the liquidus. As seen in the above, the melting point or the liquidus depends on the water content dissolved in melt. For decompression-induced crystallization, the degree of supercooling is a function of water content. Moreover, the effective cooling rate is the ascent rate of melting point associated with decrease in water content. To clarify these facts, we first explain the relationship between the effective supercooling rate and change in the water content, i.e., the water exsolution rate in the following explanation of the equilibrium vesiculation process.

8.2 Crystallization in the Equilibrium Vesiculation Regime

8.2.1 Thermodynamic Factors in the Equilibrium Vesiculation Regime

In the equilibrium vesiculation process, water content changes with the water solubility curve. When the water solubility curve is approximated using the simplified Eq. (2.26), the melting point of the crystal (liquidus) T_L is described by Eq. (2.81) as a function of water content $\tilde{C}_{W0}$, as discussed in Sect. 2.8. For practical application, we will consider the concentration $\tilde{C}_{W0}$ corresponding to a virtual molar fraction of 1 as a reference point when we consider this equation. For dacitic melt, because a molar fraction of 0.5 is found to correspond to about 6 wt.% (a saturation pressure of 200 MPa) from Fig. 2.2, we virtually set the reference concentration at a molar fraction of 1 at $\tilde{C}_{W0}$ = 12 wt.% $\tilde{C}_{W0} = 12$ wt.%.[1] The solubility curve at this time can be written as $P = K_W(\tilde{C}_W/\tilde{C}_{W0})^2$, where the coefficient K_W is 8×10^8. Accordingly, Eq. (2.81) can be written as follows:

$$T_L(\tilde{C}_W) = \frac{T_L(0)\Delta s + K_W\left(\frac{\tilde{C}_W}{\tilde{C}_{W0}}\right)^2 \Delta v}{\Delta s - k_B \ln\left(1 - \frac{\tilde{C}_W}{\tilde{C}_{W0}}\right)^2} \tag{8.1}$$

Because the effective cooling rate $|dT/dt|$ is equivalent to the ascent rate of liquidus with decrease in water content, $|dT_L/dt|$, it is given as first order approximation as follows:

$$\left|\frac{dT}{dt}\right| = \frac{dT_L}{dt} = \left|\frac{dT_L}{d\tilde{C}_W}\right|\left|\frac{d\tilde{C}_W}{dt}\right| \tag{8.2}$$

This equation connects the water exsolution rate $d\tilde{C}_W/dt$ with the cooling rate $|dT/dt|$. The coefficient $dT_L/d\tilde{C}_W$ is the slope of the liquidus, which is determined by thermodynamics. Therefore, this coefficient is called a thermodynamic factor and written as T'_L. Because differentiating $T_L(\tilde{C}_W)$ with respect to $\tilde{C}_W/\tilde{C}_{W0}$ (the reference concentration $\tilde{C}_{W0}$ is 12 wt.% and its equilibrium pressure is sufficiently high (higher than about 1 GPa)) becomes somewhat complicated, considering the Taylor expansion up to the second-order terms of the equation yields

$$\begin{aligned} T_L(\tilde{C}_W/\tilde{C}_{W0}) &= T_L(0) - \frac{2k_B T_L(0)}{\Delta s}\left(\frac{\tilde{C}_W}{\tilde{C}_{W0}}\right) \\ &\quad + \left(\frac{\Delta v \cdot K}{k_B T_0} - 1 + \frac{4k_B}{\Delta s}\right)\frac{k_B T_L(0)}{\Delta s}\left(\frac{\tilde{C}_W}{\tilde{C}_{W0}}\right)^2 \end{aligned} \tag{8.3}$$

[1]This is only an expedient for calculation.

Because the coefficient of the first-order term ($\tilde{C}_W$) takes on values from 10 to 50 but the coefficient of the second-order term ($\tilde{C}_W^2$) is less than 1, the second-order and higher terms can be ignored (Toramaru et al. 2008). This is also found from the fact that the equation is represented by an almost straight line in the left chart of Fig. 2.12. Therefore, the thermodynamic factor concerning the liquidus and the water content is determined as follows:

$$T'_L = \left|\frac{dT_L}{d\tilde{C}_W}\right| = \frac{2k_B T_L(0)}{\Delta s \cdot \tilde{C}_{W0}} \tag{8.4}$$

where the thermodynamic factor T'_L is defined as a positive quantity.

From the abovementioned results, rewriting Eq. (8.2) again yields

$$\left|\frac{dT}{dt}\right| = T'_L \left|\frac{d\tilde{C}_W}{dt}\right| \tag{8.5}$$

From the equilibrium phase diagram and calculation by MELTS,[2] values of T'_L for plagioclase (pl) and pyroxene (cpx) are obtained as follows:

$$T'_{Lpl} = 40 \tag{8.6}$$

$$T'_{Lcpx} = 17 \tag{8.7}$$

To use the slope of the liquidus experimentally obtained using the liquid of the actual compositions under study for these values is extremely reliable, and the value of plagioclase above is for the dacite of the Heisei eruption of Unzen and basaltic andesite compositions of 1986B eruption of Izu-Oshima and the value of pyroxene above is for the compositions of Izu-Oshima.

The gas—liquid equilibrium is assumed, and the water exsolution rate $d\tilde{C}_W/dt$ is connected with change in pressure using the solubility curve.

$$\left|\frac{d\tilde{C}_W}{dt}\right| = \frac{d\tilde{C}_W}{dP}\left|\frac{dP}{dt}\right| \tag{8.8}$$

Therefore, the water exsolution rate can be connected with the decompression rate, where $d\tilde{C}_W/dP$ is the slope of the solubility curve and is a thermodynamic factor written as $\tilde{C}'_W$. Assuming that the relationship between the solubility and the pressure is $P = P_0(\tilde{C}_W/\tilde{C}_{W0})^2$,

$$\tilde{C}'_W = \frac{d\tilde{C}_W}{dP} = \frac{\tilde{C}^2_{W0}}{2P_0\tilde{C}_W} \tag{8.9}$$

holds. This thermodynamic factor concerning the water content and the pressure is a function of water content $\tilde{C}_W$. Substituting Eq. (8.9) into Eq. (8.5) yields the relationship between the effective cooling rate and decompression rate using thermodynamic factors as follows:

$$\left|\frac{dT}{dt}\right| = T'_L \tilde{C}'_W \left|\frac{dP}{dt}\right| \tag{8.10}$$

[2]MELTS is a program, developed by Ghiorso et al., to compute equilibrium phase diagrams. It estimates parameters of Gibbs free energy for every conceivable case as they can explain past experimental results and allows us to draw equilibrium phase diagrams of liquid of an arbitrary composition experimental results of which do not exist by Gibbs free energy minimization using estimated parameters.

8.2.2 Crystal Number Density in Equilibrium Vesiculation Regime

As discussed in Chap. 7 on cooling crystallization process, the result of crystal nucleation and growth will appear in crystal size distribution, crystal number density, or crystal size. Although crystal size distribution (CSD) contains all information related to the number density and size, it will be treated in the next chapter. In this chapter, we advance discussion using crystal number density. As described in Chap. 7, the crystal number density is determined by competition between increase in the degree of supercooling due to cooling, i.e., increase in the nucleation rate and decrease in the degree of supercooling (the degree of supersaturation) due to growth.

At that time, the pre-exponential factor of nucleation rate and a manner of crystal growth at nucleation stage become important. As convincing nucleation and growth models that can explain experimental results of dependence of crystal number density on cooling rate, the following two cases are considered: (1) the pre-exponential factor is equal to theoretical value or is sufficiently large and the growth style is diffusion-limited growth and (2) the pre-exponential factor is sufficiently small and the growth style is not important. The case (2) will be explained in other section because the crystal number density in that case is determined not by cooling rate but by crystallization time. Here, we advance discussion on the assumption of case (1).

In Sect. 7.6.6, the relationship between the crystal number density and cooling rate $|\dot{T}|$ is given by Eq. (7.147).

$$N = a\left|\dot{T}\right|^{\frac{3}{2}} \tag{8.11}$$

where a is the kinetic factor representing nucleation and growth of crystals and can be expressed using Eq. (7.148) in terms of theoretical numerical analysis; however, as seen in Chap. 7, the results of natural experiments can be used for it. Substituting the effective cooling rate (Eq. 8.5) expressed with the water exsolution rate $|\dot{\tilde{C}}'_{\mathrm{W}}| = |d\tilde{C}_{\mathrm{W}}/dt|$ into Eq. (8.11) yields

$$N = a\left(T'_{\mathrm{L}}|\dot{\tilde{C}}_{\mathrm{W}}|\right)^{\frac{3}{2}} \tag{8.12}$$

Because the relationship between pressure and water content is not concerned, this equation holds in the disequilibrium vesiculation regime.

Moreover, substituting the above-described effective cooling rate (Eq. 8.10) expressed with decompression rate $|\dot{P}| = |dP/dt|$ into Eq. (8.11) yields

$$N = a\left(T'_{\mathrm{L}}\tilde{C}'_{\mathrm{W}}|\dot{P}|\right)^{\frac{3}{2}}, \tag{8.13}$$

which shows the relationship between decompression rate and crystal number density.

8.2.3 MND Water Exsolution Rate Meter and Decompression Rate Meter

Relationship between crystal number density and decompression rate or water exsolution rate forms the basis of tools for estimating unknown information such as water exsolution rate and decompression rate from observable crystal number density. Solving Eq. (8.12) with respect to the water exsolution rate, we obtain

$$|\dot{\tilde{C}}_{\mathrm{W}}| = \frac{1}{T'_{\mathrm{L}}}\left(\frac{N}{a}\right)^{\frac{2}{3}} \tag{8.14}$$

Using the thermodynamic factor in Eq. (8.4) (more specifically, Eqs. (8.6) or (8.7)) and a value obtained by substituting SiO_2 in the chemical composition of the liquid and the estimated water content at crystal nucleation pressure $\tilde{C}_W$ into the kinetic factor in Eq. (7.148)) for Eq. (8.14), we can estimate the water exsolution rate $|\dot{\tilde{C}}_W|$ from observable crystal number density N.

Solving Eq. (8.13) with respect to the decompression rate, we obtain

$$|\dot{P}| = \frac{1}{T_L' \tilde{C}_W'} \left(\frac{N}{a}\right)^{\frac{2}{3}} \tag{8.15}$$

Using the thermodynamic factor in Eq. (8.4), the thermodynamic factor obtained by substituting the estimated water content at crystal nucleation pressure $\tilde{C}_W$ (Eq. 8.9), and a value obtained by substituting SiO_2 in the chemical composition of the liquid and estimated water content at crystal nucleation pressure $\tilde{C}_W$ into the kinetic factor a in Eq. (7.148) for Eq. (8.15), we can estimate the decompression rate $|\dot{P}|$ from the observable crystal number density N. As discussed in the chapter on bubbles, the decompression rate is related to the ascent velocity of magma.

8.3 Crystallization in the Disequilibrium Vesiculation Regime

8.3.1 Thermodynamic Factor in the Disequilibrium Vesiculation Regime

Because decompression-induced crystallization in the disequilibrium vesiculation regime occurs in a short relaxation process from bubble nucleation to a quasi-equilibrium state, pressure can be approximated as a constant value of bubble nucleation pressure. As discussed in Sect. 2.8, the melting point of crystals (liquidus) T_L is described as a function of the water content $\tilde{C}_W$ and the pressure P using Eq. (2.81). Although the same discussion as the thermodynamic factors in the equilibrium vesiculation process will be conducted, the pressure term of Eq. (2.81) will be left unchanged.

Because the effective cooling rate $|dT/dt|$ is equivalent to ascent rate of liquidus with decrease in water content, $|dT_L/dt|$, it is given as first order approximation as follows:

$$\left|\frac{dT}{dt}\right| = \frac{dT_L(0)}{dt} = \left|\frac{dT_L}{d\tilde{C}_W}\right| \left|\frac{d\tilde{C}_W}{dt}\right| \tag{8.16}$$

he discussion up to this point is the same as that for equilibrium vesiculation.

Assuming that the pressure is constant, the thermodynamic factor is given as follows:

$$T_L' = \frac{2k_B(T_0 + \Delta v P/\Delta s)}{\Delta s \cdot \tilde{C}_{W0}} \tag{8.17}$$

which contains a correction term $\Delta v P/\Delta s$. Because the contribution of this correction term is 10% at most, Eq. (8.6) for plagioclase and Eq. (8.7) for pyroxene can be used as approximate values. However, they are upper limits and the thermodynamic factor will be overestimated to some extent. If the accuracy of experiments and observations improves, this approximation will require more strict discussion.

8.3.2 Water Exsolution Rate in the Disequilibrium Vesiculation Regime

In the disequilibrium vesiculation regime, crystallization is considered to occur in the single-step decompression experiment (SDE) in which pressure is reduced stepwise to induce crystallization. On the basis of Toramaru et al. (2008), let us consider the relationship between the water exsolution process and crystallization process in the diffusion-controlled regime. In this case, if the amount of decompression ΔP is sufficiently large, bubble nucleation immediately begins after instantaneous decompression and bubble nucleation duration is sufficiently short compared to the timescale of decrease in water content. As stated in Eq. (5.90) and surrounding discussion, the bubble nucleation duration is related to maximum nucleation rate, which depends on the amount of decompression ΔP; i.e., the larger the amount of decompression, the shorter the duration. Figure 8.3 shows the time change in water content where pressure is reduced by various amounts of decompression from the initial saturation pressure of water and then kept constant.

The water content in melt monotonously decreases and becomes almost equal to equilibrium concentration at the final pressure at a certain time. The water exsolution rate is given by Eq. (5.95) and the timescale of it is $t_{\mathrm{degasDif}} = (48\pi^2\phi_\mathrm{e})^{-1/3} D_\mathrm{W}^{-1} (\Delta C_\mathrm{e}/C)^{-1} N_{\mathrm{bub}}^{-2/3}$, where N_{bub} is the number density of bubbles generated by decompression, ΔC_e is the effective concentration difference that drives diffusion, and ϕ_e is the volume fraction of gases at microlite nucleation ($4\pi R^3 N_{\mathrm{bub}}/3$ can be approximated to be small). From this equation, water exsolution is found to proceed more effectively as the bubble number density is larger. Moreover, D_W is the diffusivity of water and is distinguished from the diffusivity of crystallization components, which is contained in the kinetic factor a.

The viscosity-controlled regime of bubble nucleation and growth, where the gas—liquid interfacial tension becomes effectively small, corresponds to the case where heterogeneous nucleation occurs. The water exsolution rate is given by Eq. (5.96) and the timescale is $t_{\mathrm{degasVis}} = 4\eta\phi_{\mathrm{g0}}\Delta P(3\phi_\mathrm{e})^{-1}$. In this case, although the water exsolution rate does not explicitly depend on bubble number density but depends on the amount of decompression, either may influence ϕ_e. As the amount of decompression becomes large, the characteristic time of water exsolution becomes short and the water exsolution rate increases.

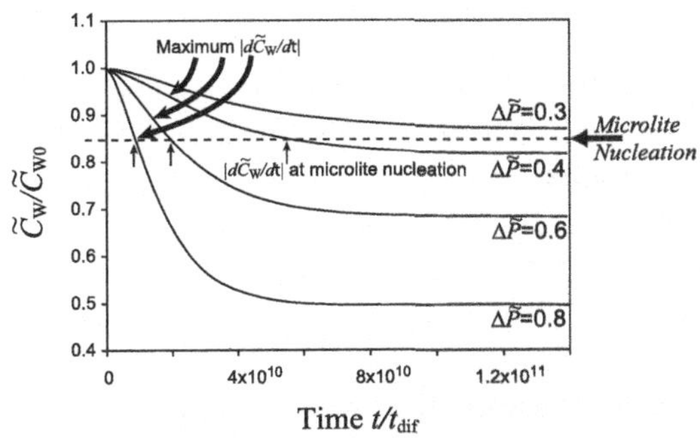

Fig. 8.3 Time change in water content in numerical experiments under a constant amount of decompression. The water content and time are normalized by initial value and characteristic time of diffusion growth, respectively. In this figure, it is assumed that microlite nucleation occurs actively when the water content decreases by 15%. Bubble nucleation almost ends at points close to the origin point of the time axis (Toramaru et al. 2008). Reprinted from Toramaru et al. (2008), Copyright 2008, with permission from Elsevier

8.3.3 The Crystal Number Density in the Disequilibrium Vesiculation Regime

Because the relationship between the crystal number density and the cooling rate (Eq. 8.11), the relationship between the effective cooling rate and the water exsolution rate (Eq. 8.5), and the relationship between the water exsolution rate and the crystal number density (Eq. 8.12) formally hold also in the disequilibrium vesiculation regime, substituting the abovementioned thermodynamic factor and the water exsolution rate (Eq. 5.95 or 5.96) into these equations yields an equation for the crystal number density N_{mic}.

In the diffusion-controlled regime where homogeneous nucleation of bubbles is expected, N_{mic} is determined as follows:

$$N_{\mathrm{mic}} = a(T_{\mathrm{L}}')^{3/2}(48\pi^2\phi_{\mathrm{e}})^{1/2}(D_{\mathrm{W}}\Delta C_{\mathrm{e}})^{3/2}N_{\mathrm{bub}} \tag{8.18}$$

Note that Eq. (8.17) underlies T_{L}'. In the equation, a is a kinetic factor characterizing crystal nucleation (Eqs. 7.148 and 7.151). The bubble number density is written as N_{bub} to distinguish it from the microlite number density N_{mic}. As confirmed from this equation, the number density of microlites in the disequilibrium vesiculation regime is proportional to the number density of bubbles. The coefficient depends on the diffusivity of water D_{W} and the effective vesicularity at the time of microlite nucleation ϕ_{e} and takes on values from 10^5 to 10^2. Because the bubble number density exponentially depends on the amount of decompression ΔP, the crystal number density of microlites is expected to exponentially increase with increase in the amount of decompression.

In the viscosity-controlled regime where heterogeneous nucleation is expected, the microlite number density is given as follows:

$$N_{\mathrm{mic}} = a(T_{\mathrm{L}}')^{3/2}\left(\frac{3\phi_{\mathrm{e}}}{4\eta v_{\mathrm{G}}}\Delta P\right)^{3/2} \tag{8.19}$$

which is expected to not explicitly depend on the bubble number density but depend on the amount of decompression and viscosity. The dependence on the amount of decompression is expected to not be exponential, similar to the case of diffusion-controlled regime, but to be gentler. Although behavior of the microlite number density is expected to be proportional to the 3/2 power of the amount of decompression from the above equation, this simplified model has not been able to take factors to be described below into consideration.

If crystallization occurs in the disequilibrium vesiculation regime in the case of continuous decompression, because the bubble number density is proportional to the 3/2 power of the decompression rate in the diffusion-controlled regime,

$$N_{\mathrm{mic}} \propto \left|\frac{dP}{dt}\right|_{\mathrm{BubbleNucleation}}^{\frac{3}{2}} \tag{8.20}$$

is obtained and the microlite number density becomes related to the decompression rate at the time of bubble nucleation.

For continuous decompression, whether a correspondence between heterogeneous nucleation and viscosity regime, as seen in the section on the constant amount of decompression, holds or not has not yet been clarified.

8.4 Calculation Combining the Vesiculation Process and the Crystallization Process

8.4.1 Problems of the Simplified Model

Up to this point, the vesiculation process and crystallization processes have been separately evaluated. This is, so to speak, the simplified analytic model, i.e., the influence on the crystal texture, such as the crystal number density, was investigated after going through a procedure in which the water exsolution rate obtained from the examination of the vesiculation process was assumed and subsequently crystallization process was evaluated. This simplified model allowed us to clarify the roles of individual factors in crystal texture. That is, in the equilibrium vesiculation regime, the water exsolution rate is determined by the solubility curve; moreover, in the disequilibrium vesiculation regime, the different water exsolution rates are obtained corresponding to diffusion-controlled vesiculation process and viscosity-controlled vesiculation process.

However, in actuality, because both processes proceed nonlinearly and simultaneously, some parameters, such as ΔC_{e} and ϕ_{e}, cannot be accurately evaluated. Furthermore, as described in Chap. 7 on cooling crystallization, there are many uncertainties in the classical nucleation theory on the homogeneous nucleation of crystals, e.g., the master curve of the nucleation rate against the water content has been experimentally reported but has not yet been theoretically justified and the causes of heterogeneous nucleation have not yet been elucidated. Moreover, this will become critical in experiments under the constant amount of decompression for bubble nucleation that determines the bubble number density. Furthermore, there may be causes not included in the abovementioned analytic examination such as the influence of heterogeneous nucleation of bubbles under the constant amount of decompression and viscosity-controlled regime.

Complex coupling between crystallization and vesiculation processes obviously manifests in the disequilibrium vesiculation regime where the bubble number density governs the water exsolution rate. In Part II and Chap. 7 in Part III, we learned procedures by which we calculated the entire picture of vesiculation and crystallization processes. Therefore, we will be able to calculate the crystallization process by combining the procedures and will arrive at a practical understanding of the coupling of vesiculation and crystallization. The combined calculation is simply performed by substituting the water content obtained by calculating vesiculation into the liquidus as a function of water content and then calculating the decompression-induced crystallization process. If the latent heat will be released, temperature is influenced by energy conservation, and the liquidus will become a function of temperature. Let us see the above using the decompression-induced crystallization process in the disequilibrium vesiculation regime at a constant amount of decompression. The below examination assumes the adiabatic process and release of latent heat is not considered.

8.4.2 The Case Where Homogeneous Nucleation of Bubbles and Crystals Occurs

First, we will describe a case that can be understood using the simplified model. Figure 8.4 is an example of calculation. Bubble nucleation finishes at the initial stage (figure (d)) and water content decreases (figure (c)) due to subsequent diffusion-limited bubble growth (figure (e)). With decrease in water content, the temperature of the liquidus increases (TL in figure (a)) and isothermal melt shifts to the supercooled state in which crystal nucleation occurs (J_{mic} in figure (b)). Then, decrease in the crystallization component in melt leads to lowering of liquidus and crystals, and then the melt maintains the quasi-equilibrium state. The dependence of bubble number density $\tilde{N}_{\mathrm{bub}}$ and the crystal number density $\tilde{N}_{\mathrm{mic}}$ on the amount of decompression $\Delta\tilde{P}$ obtained by a parametric study in this case

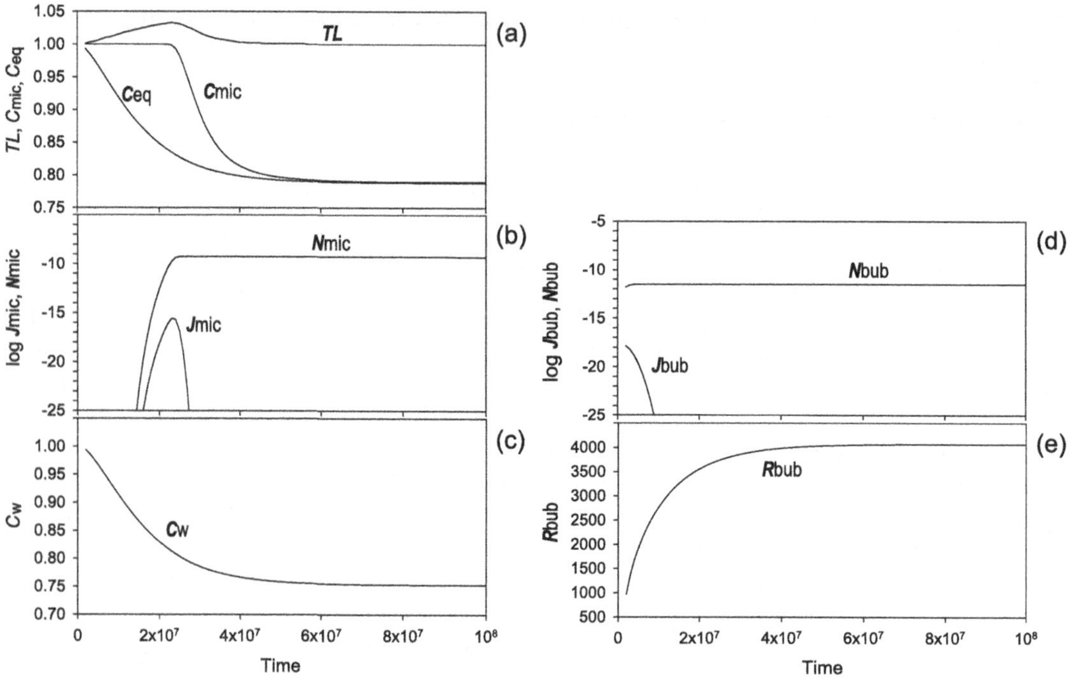

Fig. 8.4 The result of combined calculation of the vesiculation process and the crystallization process. Example of the behavior of the liquid phase of an assumed binary eutectic system handled in Chap. 7. The initial water content is assumed to be 4 wt.% at 150 MPa. α_1 of bubbles corresponding to the gas—liquid interfacial tension is 2.5, which is equivalent to about 0.5 N/m interfacial tension, as seen from Fig. 5.6. α_1 of crystals corresponding to the crystal—liquid interfacial tension is assumed to be 0.1. **a** The liquidus of crystals TL, the crystallization component concentration C_{mic}, and the equilibrium concentration of the crystallization component C_{eq}, **b** the nucleation rate of crystals J_{mic} and the crystal number density N_{mic}, **c** the water content in liquid C_{w}, **d** the bubble nucleation rate J_{bub} and the bubble number density N_{bub}, **e** the bubble radius. Variables are dimensionless and the time is normalized by the timescale of diffusion growth t_{dif}

is shown in the left chart of Fig. 8.5. Only change in the amount of decompression from 0.5 times the initial pressure to 0.7 times leads to change in the respective values of number density by nearly ten orders of magnitude; the crystal number density shows exponential dependence on the amount of decompression. This behavior of bubble number density has already been mentioned in Sect. 5.6.

The relationship between crystal number density and bubble number density (the right chart of Fig. 8.5) is in good agreement with proportional relation expected from the simplified model. Deviation from proportional relation is attributed to the influence of other parameters and deviation in this case is found to be small.

8.4.3 The Case Where Heterogeneous Nucleation of Bubbles and Crystals Occurs

In calculation in this case, the values of α_1 corresponding to gas—liquid interfacial tension and crystal—liquid interfacial tension are reduced by about an order of magnitude compared with the case of homogeneous nucleation to 0.2 and 0.1. These reductions correspond to reductions in the interfacial tension by 1/2 to 1/3. Figure 8.6 shows the result of a parametric study for this case. As can be seen from this figure, the heterogeneous nucleation of bubbles makes the dependence of the bubble number density on the amount of decompression extremely small (the trend labeled Hetero-Bubble in the left

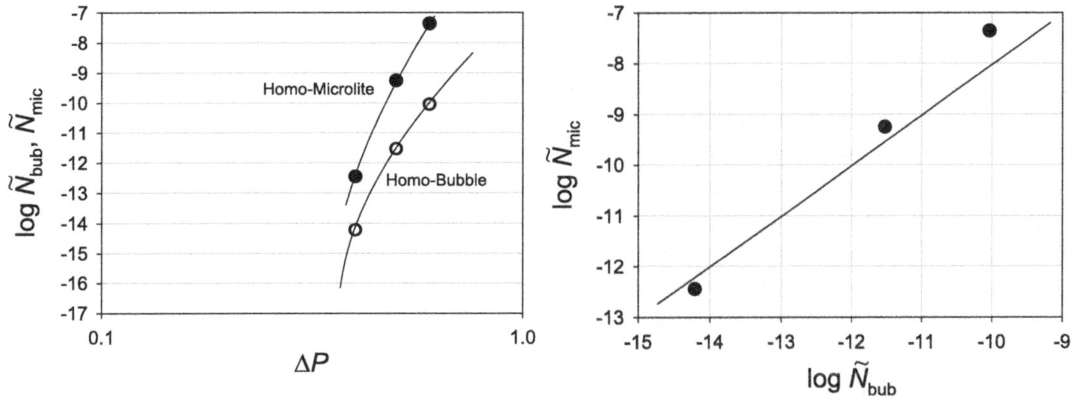

Fig. 8.5 The result of the parametric study by combined calculation of vesiculation and crystallization under a constant amount of decompression in the regime of homogeneous nucleation of bubbles and crystals (The circles in the figures: the hollow circles represent the bubble number density and the black circles represent the crystal number density). (Left) Dependence of the bubble number density $\tilde{N}_{bub}$ and the crystal number density $\tilde{N}_{mic}$ on the amount of decompression $\Delta \tilde{P} = \Delta P / P_0$. Homo-Bubble and Homo-Microlite correspond to homogeneous nucleation of bubbles and crystals, respectively. (Right) The bubble number density and the crystal number density are almost in proportion to each other

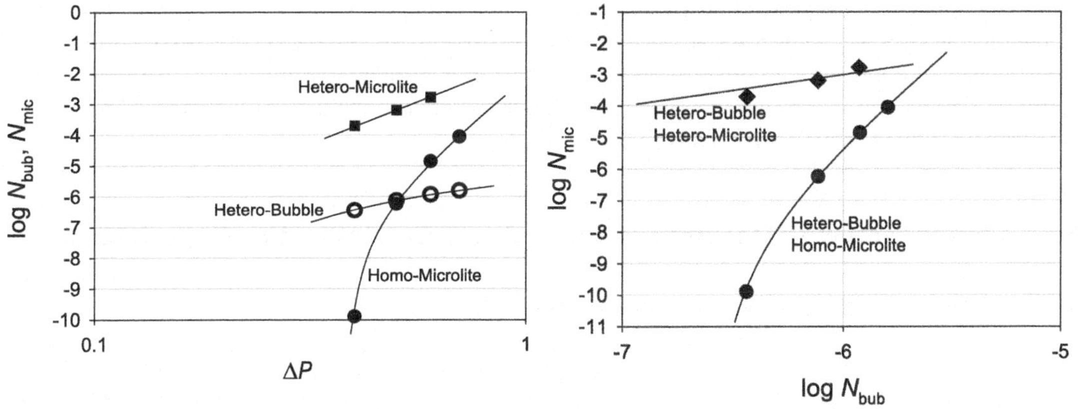

Fig. 8.6 The result of the parametric study by combined calculation of vesiculation and crystallization under a constant amount of decompression in the regime of heterogeneous nucleation of bubbles and crystals (The circles and lines in the figure represent the same as in Fig. 8.5). (Left) Dependence of the bubble number density $\tilde{N}_{bub}$ and the crystal number density $\tilde{N}_{mic}$ on the amount of decompression $\Delta \tilde{P} = \Delta P / P_0$. (Right) The relationship between the bubble number density and the crystal number density. Hetero-Bubble and Hetero-Microlite correspond to heterogeneous nucleation of bubbles and crystals, respectively

chart of Fig. 8.6). Moreover, when crystals heterogeneously nucleate, the dependence of crystal number density on the amount of decompression also becomes small (the trend labeled Hetero-Microlite in the left chart of Fig. 8.6). In this case, a proportional relation between the bubble number density and crystal number density roughly holds (the trend labeled Hetero-Bubble and Hetero-Microlite in the right chart of Fig. 8.6).

On the other hand, when crystals homogeneously nucleate (the trend labeled Homo-Microlite in the left chart of Fig. 8.6), the crystal number density exhibits a large exponential dependence on the amount of decompression and does not have a proportional relation with bubble number density (the trend labeled Hetero-Bubble and Homo-Microlite in the right chart of Fig. 8.6)When crystals homogeneously nucleate, the large dependence of crystal number density on the amount of decompression despite the relatively small dependence of the bubble number density on the amount of decompression can be understood from the relationship between the degree of supercooling at crystal nucleation and the amount of decompression, which was explained earlier (Fig. 8.3). That is, for homogeneous nucleation, when the degree of supercooling causing crystal nucleation becomes large and the amount of decompression is small, nucleation does not reach the maximum nucleation of crystals even if water exsolution sufficiently proceeds, and then nucleation continues at a low nucleation rate. Ultimately, because the nucleation rate is reduced, the number of crystals cannot be gained, regardless of the extended duration. In this case, the crystal number density is influenced by duration under decompression rather than the extent of decompression. However, when the amount of decompression is large, the sufficient degree of supercooling required for crystal nucleation is obtained by water exsolution, nucleation momentarily reaches the maximum nucleation rate, like a pulse, and the crystal number density becomes large.

An interesting case is where heterogeneous nucleation of both bubbles and crystals occurs. In this case, the mild dependence of crystal number density on the amount of decompression, which appears in the result of laboratory experiments described in the following section, appears by this calculation. Furthermore, when viewed from the viewpoint of parameters (Pe, $\alpha_1/\Delta\tilde{P}$), although there is a regime where the water exsolution rate is controlled by the viscosity and amount of decompression, and does not depend on the bubble number density, a proportional relation between the bubble number density and the crystal number density, which is expected to be observed in the case of homogeneous nucleation of bubbles and crystals and in the diffusion-controlled regime, is observed. This may be related to the existence of a transition regime between the diffusion-controlled regime and viscosity-controlled regimes in terms of water exsolution rate and to a relation between the degree of supercooling and the amount of decompression, which leads to crystal nucleation.

8.5 Comparison with Decompression-Induced Crystallization Experiments

8.5.1 Brief Summary of Experimental Studies

A number of decompression-induced crystallization experiments have been conducted after Geschwind and Rutherford (1995) conducted an experiment using dacite of the 1980 eruptions of Mount St. Helens. Most of the experiments have focused on the importance on determining the phase equilibrium relationship and reproducing crystal morphology. Experiments focusing on the crystallization kinetics have been conducted for the dacite of Mount St. Helens volcano by Hammer and Rutherford (2002), the dacite of Soufrière Hills volcano by Couch et al. (2003b), and a synthetic product having a differentiated composition (SiO_2 75 wt.%) that imitates a composition of glass inclusions of Soufrière Hills volcano by Martel and Schmid (2003). Subsequently, experiments using various samples have been conducted by many researchers: dacite of Unzen by Cichy et al. (2010), rhyodacite of Aniakchak volcano by Brugger and Hammer (2010), a synthetic product imitating plinian rhyolite compositions of Mt. Pelée by Martel (2012), trachytic obsidian by Arzilli and Carroll (2013), basaltic andesite by Shea and Hammer (2013), and obsidian of Yellowstone and the Mono-Inyo Craters by Befus et al. (2015). These experiments have been performed to reproduce the groundmass texture of eruptive materials and have clarified how the nucleation rate, the growth rate, the crystal size distribution, the crystallinity, the crystal morphology, and the composition are related to decompression conditions, which is the foundation of groundmass texture. Although many experiments have been conducted for microlites in the eruptive materials of nonexplosive eruptions, such as dome-forming eruptions involving composition from dacite to rhyolite, some experiments have been conducted for explosive eruptions (microlites are generally not contained) such as Mt. Pelée.

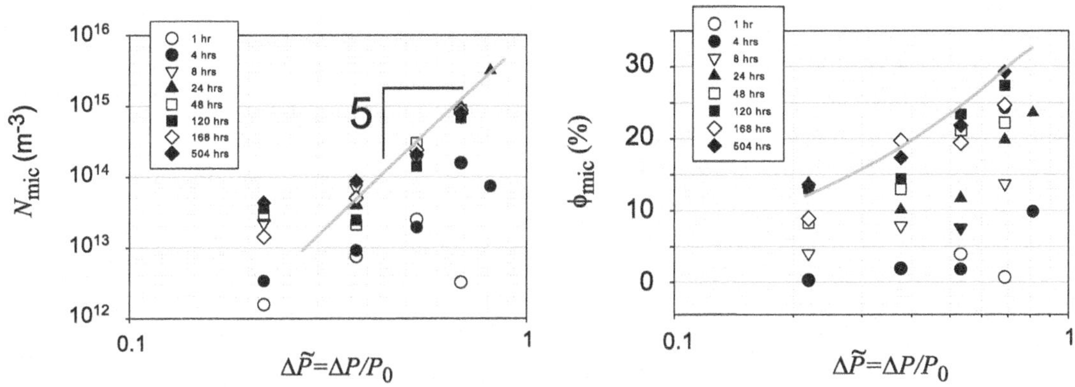

Fig. 8.7 The result of SDE (Couch et al. 2003b). (Left) Relationship between the crystallinity and the amount of decompression. (Right) Relationship between the crystal number density and the amount of decompression. The symbols represent the pressure retention time. The lines represent trends at which the relationships converge after sufficient time has elapsed. The value 5 in the figure indicates the slope in the logarithmic scale. There appears to be a relationship $N_{mic} \propto \Delta\tilde{P}^5$ in the left chart

There are three types of pressure change used for the experiments. In the single-step decompression experiment (SDE or SSD), pressure is instantaneously decreased by ΔP and is maintained at the pressure for a certain time period and samples are rapidly cooled. The parameters in this experiment are the decrease in pressure and retention time. This experiment simulates decompression-induced crystallization associated with discontinuous sudden ascent of magma. In the multi-step decompression experiment (MDE or MSD), pressure is reduced in multiple steps. This experiment can define an average decompression rate and simulates continuous ascent of magma. The continuous decompression experiment (CDE or CD) corresponds to continuous magma ascent itself. Although this experiment can be conducted for a relatively rapid decompression rate, it is technically difficult to correspond to a relatively small decompression rate. These studies are summarized in detail in Suzuki (2016).

8.5.2 SDE

In SDE, crystallization is considered to occur in the disequilibrium vesiculation regime. Hammer and Rutherford (2002) provided kinetic data such as the nucleation rate and the crystal growth rate but did not provide data on number density. Because data on number density is provided by Couch et al. (2003b), let us primarily interpret their experimental result theoretically. Figure 8.7 shows plots of crystallinity and number density of microlites against the amount of decompression for each experiment. These figures show that they converge to certain trends as the retention time increases. The fact that the amount of crystals increases with the amount of decompression suggests that the amount of crystals is determined by thermodynamic equilibrium. This is because, assuming that water exsolution from melt occurs in accordance with solubility as a function of pressure, it is expected that the initial degree of supercooling (the distance from the liquidus) becomes large and the crystallinity becomes large as the amount of decompression (the amount of water exsolution) is large.

It is not easy to understand the fact that the crystal number density increases with the amount of decompression. The crystal number density is governed by the effective supercooling rate, which is determined by the rate of water exsolution. It is found from numerical experiments and theoretical examination mentioned in the previous section that the rate of water exsolution is related to bubble number density and amount of decompression. Moreover, the bubble number density ultimately divides

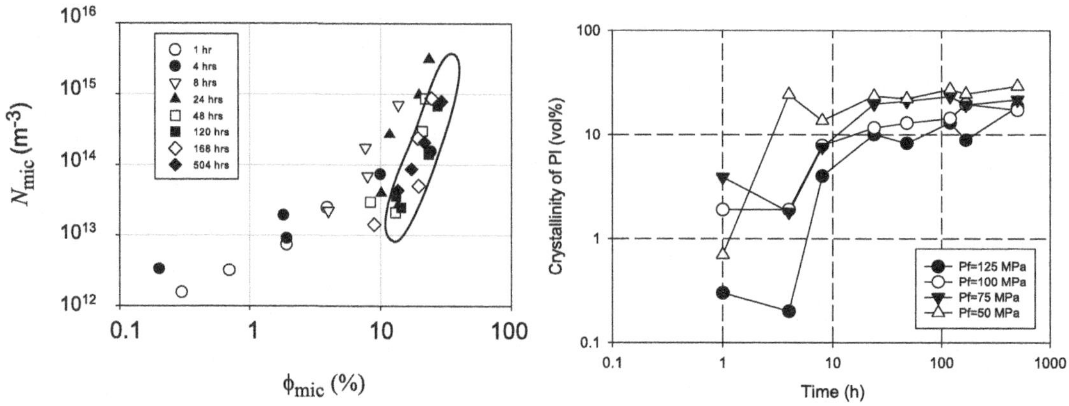

Fig. 8.8 Relationship between the crystallinity and the microlite number density (left) and time change in the crystallinity (right) observed in the SDE (Couch et al. 2003b). As the final pressure P_f is large, the amount of decompression is small

homogeneous nucleation and heterogeneous nucleation by the effective gas—liquid interfacial tension and influences diffusion control and viscosity control at the step of determination of bubble number density. We have to identify the actual bubble number density in SDE. Unfortunately, this experiment does not contain qualitative and quantitative description of vesicular texture.

The discussion in the previous section and the relationship (the left chart of Fig. 8.7) between the crystal number density and the amount of decompression after sufficient time has elapsed are similar to the case where heterogeneous nucleation of bubbles and that of crystals occur, as shown in Fig. 8.6. This leads us to surmise that in their experiment, heterogeneous nucleation of bubbles occurs and heterogeneous nucleation of crystals occurs. Kinetic data in the experiment can probably be estimated from theoretical parameters.

The left chart of Fig. 8.8 is a plot of relationship between the crystallinity and microlite number density from these data and shows that they have a positive correlation. A similar relationship is seen in MDE (Martel 2012). Interestingly, it is this relationship that is analogous to microlite systematics observed in nature, described in the next chapter (Sect. 10.3.4). Moreover, the amount of crystals reaches a constant value in several to dozens of hours and, as the amount of decompression is large, the final crystallinity becomes large (the right chart of Fig. 8.8). A similar result has been obtained by Hammer and Rutherford (2002).

Figure 8.9 shows a plot of the nucleation rate and the growth rate against the effective degree of supercooling reported in these experiments. For comparison, data obtained in cooling crystallization experiments conducted by Hammer and Rutherford (2002) and Shea and Hammer (2013) are plotted in the figure. For decompression-induced crystallization, the nucleation rate monotonously increases with increase in the effective degree of supercooling. On the other hand, for cooling crystallization, the nucleation rate increases once and then decreases with increase in the degree of supercooling, thus drawing a bell-shaped curve. The crystal growth rate has a smaller dependence on effective degree of supercooling compared to the nucleation rate. In the experiment by Hammer and Rutherford (2002), the effective degree of supercooling giving a peak nucleation rate is larger than that of crystal growth.

Because for decompression-induced crystallization, the effective degree of supercooling is an apparent one and the water exsolution process actually works to control the degree of supercooling, the interpretation of these experiments and comparison with theoretical examination requires attention. That is, because nucleation and growth of bubbles occur in response to sudden decompression, water exsolution from liquid proceeds, the liquidus rises, and consequently the effective supercooling occurs, time delay, and the effective cooling rate (the degree of supercooling is not constant) are contained.

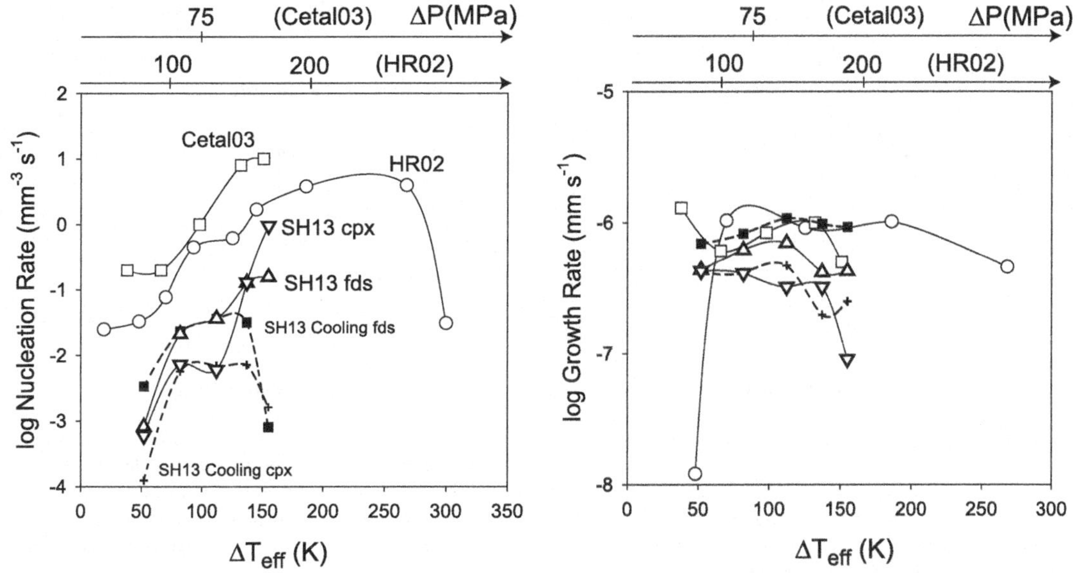

Fig. 8.9 Change in the nucleation rate (left) and the growth rate (right) of plagioclase against the effective degree of supercooling, obtained by decompression SDE. Cetal03 represent data from Couch et al. (2003b), HR02 from Hammer and Rutherford (2002), and SH13 from Shea and Hammer (2013). SH13 cooling cpx is the result of the cooling crystallization experiment under a constant pressure. The upper scales show rough indications of the amount of decompression

Furthermore, even if the pressure is instantaneously released, the system is decompressed at a certain finite decompression rate (1–10 MPa/s, according to Hammer and Rutherford (2002)).

8.5.3 MDE and CDE

In the multi-step decompression experiment (MDE), pressure is reduced by multi steps. This experiment can define an average decompression rate. In the experimental result by Couch et al. (2003b), there is a correlation between microlite number density and decompression rate (the left chart of Fig. 8.10). Plotting the crystal number density against crystallinity yields the right chart of Fig. 8.10, in which there seems to be no clear correlation at first glance. However, besides the highest decompression rate (Retention time of P_f of Highest $|dP/dt|$ is 8 h, whereas that of others is 86 and 192 h.), there appears to be a certain relationship. Regarding data on higher decompression rates, although the crystal number density is on an extension of the trend, the crystallinity is extremely small. This fact suggests that crystallization does not sufficiently proceed and the crystallinity does not reach equilibrium.

Cichy et al. (2010) conducted MDE for the decompression rate slower than 0.1 MPa/s and CDE for the decompression rate faster than it. Moreover, they analyzed both microlite texture and bubble texture. Consequently, for continuous decompression, while the bubble number density has a positive correlation with the decompression rate, the microlite number density does not have a clear correlation (Fig. 8.11). In the case where the average decompression rate is small in MDE, there is a negative correlation. This applies to the case where J_0 of crystal nucleation is small (Sect. 7.6.3). The result that the microlite number density does not depend on the decompression rate but depends on the amount of decompression and the elapsed time under decompression is obtained in MDE by Machida and Arai (2003) and CDE by Befus et al. (2015).

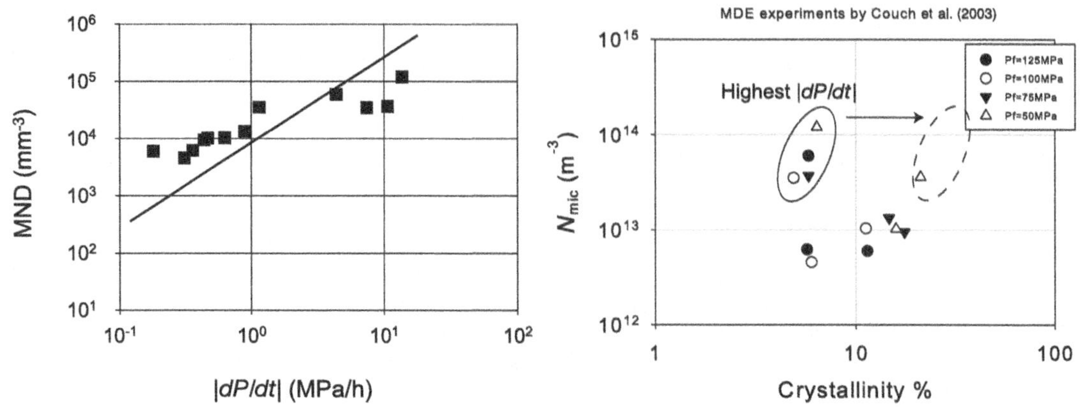

Fig. 8.10 Relationship between the number density and the crystallinity or the decompression rate of plagioclase microlites, observed in MDE (Couch et al. 2003b) (Left) Relationship between the microlite number density and the decompression rate. The data points are data of MND converted to 3D by Nakamura (2006). A straight line represents the relationship between the decompression rate and the number density expected from the model equation (8.13) on the assumption that SiO_2 = 71 wt.% and H_2O = 5 wt.%. (Right) Relationship between the crystallinity and the microlite number density. The different symbols represent different final pressures

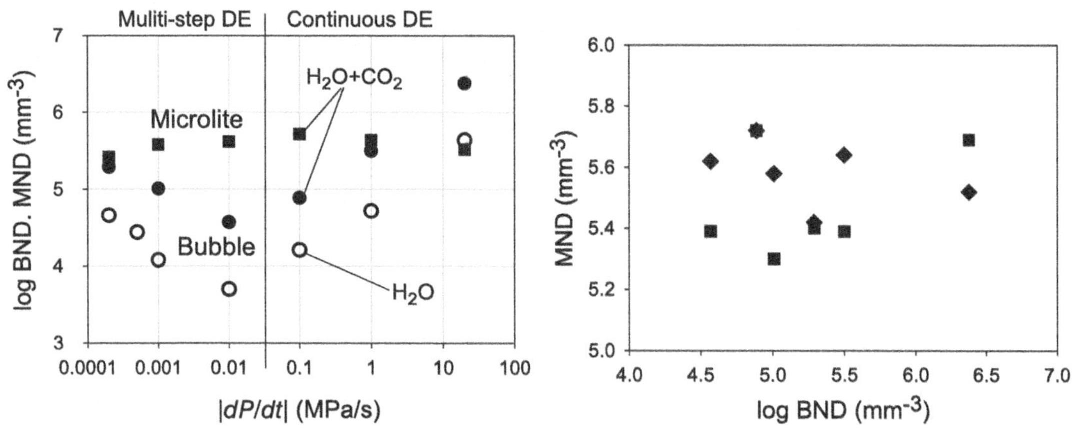

Fig. 8.11 Relationship between the decompression rate and the bubble number density or the microlite number density, observed in MDE (Cichy et al. 2010). The black symbols represent systems containing H_2O+CO_2 and the white symbols represent systems containing only H_2. The circles represent data on bubbles and the squares represent data on microlites. (Right) Relationship between the bubble number density and the microlite number density

Couch et al. (2003b) and Mollard et al. (2012) showed in MDE using a synthetic product of rhyodacite (SiO_2 = 71 wt.%) that the nucleation rate of crystals rapidly increases with the decompression amount, i.e., the effective supercooling. Moreover, as the decompression rate becomes large, the nucleation rate becomes slightly small. On the other hand, Brugger and Hammer (2010) made a starting material from natural pyroclasts and investigated by CDE the way in which the decompression rate influences the nucleation and growth rates (Fig. 8.12). Samples of rhyodacite (SiO_2 = 71 wt.%) saturated with water at 130 MPa are decompressed to a target pressure at a decompression rate of 10, 2, 1, and 0.5 MPa/s and rapidly cooled to observe them and to obtain the nucleation rate by dividing the crystal number by the decompression time. The difference between the target pressure and initial pressure is the amount of decompression and gives the effective degree of supercooling. Their experimental results demonstrate that there is the decompression rate dependence in the position of effective degree of su-

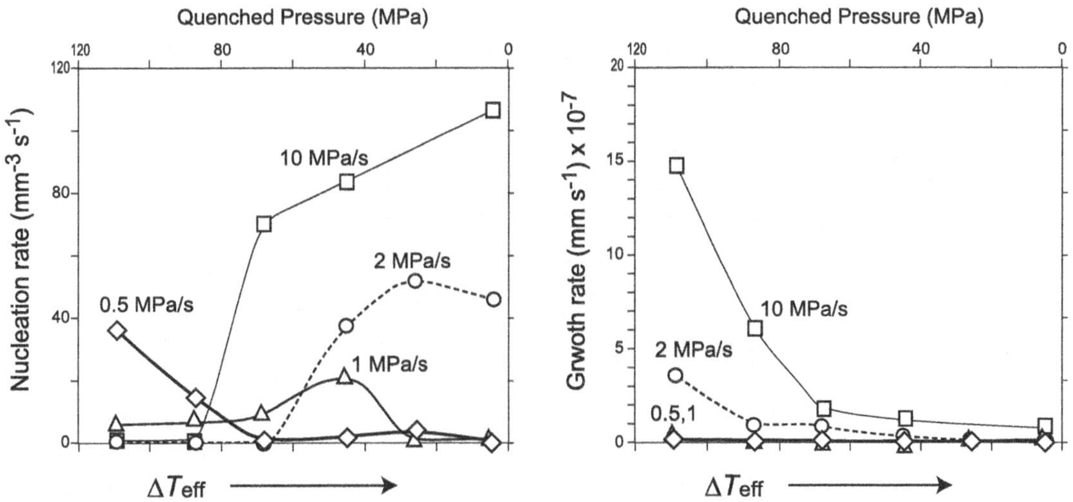

Fig. 8.12 The influence of the decompression rate on the relationship between the nucleation rate and the amount of decompression (left) and the influence of the decompression rate on the relationship between the crystal growth rate and the amount of decompression (right), observed in CDE (Brugger and Hammer 2010)

percooling or the pressure, which yields the nucleation peak. In other words, as the decompression rate becomes small, the nucleation pressure becomes small. According to theoretical examination, because the nucleation pressure is almost completely determined by the crystal—liquid interfacial energy, it should not greatly depend on the decompression rate. This experimental result may be related to the time delay taken for nucleated crystals to grow to observable size (1 μmm) or the delay in nucleation itself. In fact, Mollard et al. (2012) experimentally demonstrated that nucleation obviously appears at high pressure or with low effective supercooling. However, a clear correlation between time delay and decompression rate has not been identified.

8.6 Complexity of Crystal Growth

Various types of crystal forms represent the complexity of crystal growth. Crystals form in various shapes such as dendrite, hopper (having a hollow interior), swallowtail, tabular, and thready. These crystal shapes are observed in natural microlites (Fig. 1.6) and reproduction of these shapes by experiments enables us to presume the physical conditions for crystal formation.

The complexity of crystal growth is related to the temperature—pressure and the decompression rate. Figure 8.13 shows a variety of the crystal morphology of plagioclase found in SDE of decompression-induced crystallization at a constant pressure. When the degree of supercooling is less, cuboidal crystals with relatively small aspect ratios form and their size increases with time. When the degree of supercooling becomes about 70 K, the aspect ratios become large and crystals with hollow interiors, called hopper, start to form with time. When the degree of supercooling becomes about 100 K, the aspect ratios become small and selective growth occurs from the edges, leading to swallowtail crystals, sometimes hollow swallowtail crystals. When the degree of supercooling is large and the nucleation rate is dominant over the growth rate, needle crystals and tabular crystals of small size with large aspect ratios predominantly form and thready crystals form in the initial stage. In this experiment, although dendrite crystals did not form, it is known that in experiments using basaltic melt, dendrite, swallowtail,

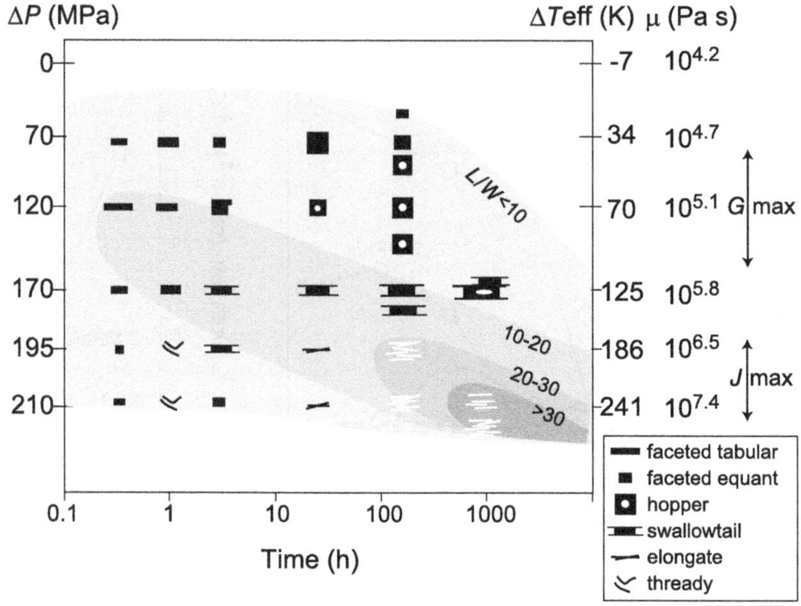

Fig. 8.13 Relationship between the variety of crystal forms and the degree of supercooling and the viscosity, observed in SDE. Shaded contours represent the aspect ratios L/W of crystals (adapted from Hammer and Rutherford (2002)

skeletal, and needle crystals form in the initial stage and they are transformed to euhedral crystals with time (Lofgren 1974b, 1980; Walker et al. 1978; Shea and Hammer 2013).

The results of these experiments suggest that various crystal shapes depends on both the physical conditions of crystallization and time. In particular, it is considered that for basaltic melt and basaltic andesite melt, unstable interface disappears with time and crystal forms shift to those with stable interfacial energy. Because current crystallization models approximate a crystal by a sphere, considering such various crystal forms and temporal development is a future issue.

References

Arzilli F, Carroll MR (2013) Crystallization kinetics of alkali feldspars in cooling and decompression-induced crystallization experiments in trachytic melt. Contrib Mineral Petrol 166(4):1011–1027

Befus KS, Manga M, Gardner JE, Williams M (2015) Ascent and emplacement dynamics of obsidian lavas inferred from microlite textures. Bull Volcanol 77–88. https://doi.org/10.1007/s00445-015-0971-6

Brugger CR, Hammer JE (2010) Crystallization kinetics in continuous decompression experiments: implications for interpreting natural magma ascent processes. J Petrol 51:1941–1965

Burnham CW, Davis NF (1974) The role of H_2O in silicate melts: II. Thermodynamic and phase relations in the system $NaAl\text{-}Si_3O_8\text{-}H_2O$ to 10 kilobars, 700°C to 1100°C. Am J Sci 274:902–940

Cichy SB, Botcharnikov RE, Holtz F, Behrens H (2010) Vesiculation and microlite crystallization induced by decompression: A case study of the 1991–1995 Mt Unzen eruption (Japan). J Petrol 52(7–8):1469–1492

Couch S, Sparks RSJ, Carroll MR (2003) The kinetics of degassing-induced crystallization at the Soufriere Hills Volcano. Montserrat J Petrol 44:1477–1502

Geschwind CH, Rutherford MJ (1995) Crystallization of microlites during magma ascent: the fluid mechanics of 1980–1986 eruptions at Mount St Helens. Bull Volcanol 57(5):356–370

Hammer JE, Rutherford MJ (2002) An experimental study of the kinetics of decompression-induced crystallization in silicic melt. J Geophys Res 107. https://doi.org/10.1029/2001JB000281

Hammer JE (2004) Crystal nucleation in hydrous rhyolite: experimental data applied to classical theory. Am Mineral 89(11–12):1673–1679

Hodges FN (1974) The solubility of H_2O in silicate melts. Carnegie Inst Wash Yearb 73:251–254

Kushiro I (1979) Fractional crystallization of basaltic magma. In: Yoder HS Jr (ed) The evolution of the igneous rocks. Princeton University Press, Princeton

Lofgren G (1974) An experimental study of plagioclase crystal morphology: isothermal crystallization. Am J Sci 274(3):243–273
Lofgren GE (1980) Experimental studies on the dynamic crystallization of silicate melts. In: Hargrabes RB (ed) Physics of magmatic process. Princeton University Press, Princeton, pp 488–551
Machida H, Arai F (2003) Atlas of tephra in and around Japan. University of Tokyo Press, p 360. (in Japanese)
Martel C (2012) Eruption dynamics inferred from microlite crystallization experiments: application to Plinian and dome-forming eruptions of Mt. Pelée (Martinique, Lesser Antilles). J Petrol 53(4):699–725
Martel C, Schmidt BC (2003) Decompression experiments as an insight into ascent rates of silicic magmas. Contrib Mineral Petrol 144:397–425
Mollard E, Martel C, Bourdier JL (2012) Decompression-induced crystallization in hydrated silica-rich melts: empirical models of experimental plagioclase nucleation and growth kinetics. J Petrol 53(8):1743–1766
Nakamura K (2006) Textures of plagioclase microlite and vesicles within volcanic products of the 1914–1915 eruption of Sakurajima Volcano, Kyushu. Japan J Mineral Petrol Sci 101:178–198
Shea T, Hammer JE (2013) Kinetics of cooling- and decompression-induced crystallization in hydrous mafic-intermediate magmas. J Volcanol Geotherm Res 260:127–145
Suzuki Y (2016) Methods to investigate syneruptive magma processes using textures in ejecta-a review of progress in last 10 years Kazan 61:367–384
Toramaru A, Noguchi S, Oyoshihara S, Tsune A (2008) MND (microlite number density) water exsolution rate meter. J Volcanol Geotherm Res 175:156–167
Walker D, Powell MA, Lofgren GE, Hays JF (1978) Dynamic crystallization of a eucrite basalt. In: Proceedings of the Ninth Lunar and Planetary Science Conference, pp 1369–1391
Yoder HS Jr (1965) Diopside-anorthite-water at five and ten kilobars and its bearing on explosive volcanism. Yearb Carnegie Inst Wash 64:82–89

9 CSD (Crystal Size Distribution)

Abstract

In this chapter, the basis for the tool to interpret crystal size distribution (CSD) will be explained. First, we will address the background of CSD introduction, the physical interpretation of the exponential CSD often observed in nature, and the examples of studies on CSD in nature. Then, we will introduce two theoretical methods of describing CSD's temporal development; the Eulerian and Lagrangian descriptions, which were introduced in the chapters on vesiculation and crystallization. In this chapter on CSD, we will obtain general solutions of CSD under various conditions in closed and open systems. The closed system is a system in which CSD is only determined by the dynamics of nucleation and growth of crystals, and liquid and crystals do not move. An open system is that in which exchange with the external system of crystals essentially influences CSD. An advantage of the mathematical disscussion of CSD is that when CSD obtained by models is compared with that obtained by experiments, the relationship between observable CSD and model parameters can be clearly undrestood than when we use an empirical rule obatained using numerical calculations. Finally, we will derive the CSD corresponding to the Avrami plot, which is often conducted in phade change experiments.

9.1 The Background of CSD Introduction in Rock Texture and the Current State of Research

Analysis using CSD (CSD analysis) has been extensively used after the research results of the size distribution of crystals in chemical engineering (e.g., Randolph and Larson 1988) were first applied to the crystal size distribution of volcanic rocks by Marsh (1988) and Cashman (1988). For the quantitative interpretation of rock texture, although attaching physical significance to CSD parameters is fundamentally important, there were only a few studies, including Marsh (1988, 1998) and Marsh (1988), in the early years. In recent years, some certain researchers have started to explore the relationship between CSD and the kinetics of nucleation growth by experiments to ensure the validity of CSD. Considering these results, we will describe the background of CSD introduction and introduce petrogenetic studies

A. Toramaru, *Vesiculation and Crystallization of Magma*, Advances in Volcanology,
https://doi.org/10.1007/978-981-16-4209-8_9

of log-linear (exponential) CSD as a typical CSD. Then, we will explain petrographic and experimental studies, define problems of previous studies, and describe a direction for future research.

9.1.1 Background of CSD Introduction

Requirements Arising from Rock Texture Observation

Marsh (1988) and Cashman and Marsh (1988) reported that the CSD of phenocryst and groundmass observed in igneous rocks can often be approximated using exponential distribution as defined below (Fig. 9.1).

$$F(R) = F_0 e^{-\frac{R}{R_S}} \tag{9.1}$$

where $F(R)$ is called the population density and is of the same as the size distribution function $F(R, t)$ used in this book, i.e., $F(R)dR$ gives the number of crystals with a size between R and $R + dR$ per unit volume. If the crystal size distribution follows this size distribution, plotting the measurement result of the crystal size distribution on a semilogarithmic coordinate paper (natural logarithm) yields a straight line. The intercept of the line is $\ln F_0$ and the negative slope is $1/R_S$. Note that F_0 in the intercept is called nucleation density and R_S yields a size in which $F(R)$ decreases to $1/e \approx 0.37$. An advantage of exponential distribution is that it can organize observation results using these two parameters. Surely, these observation parameters contain information such as petrogenetic controlling parameters, including the cooling rate.

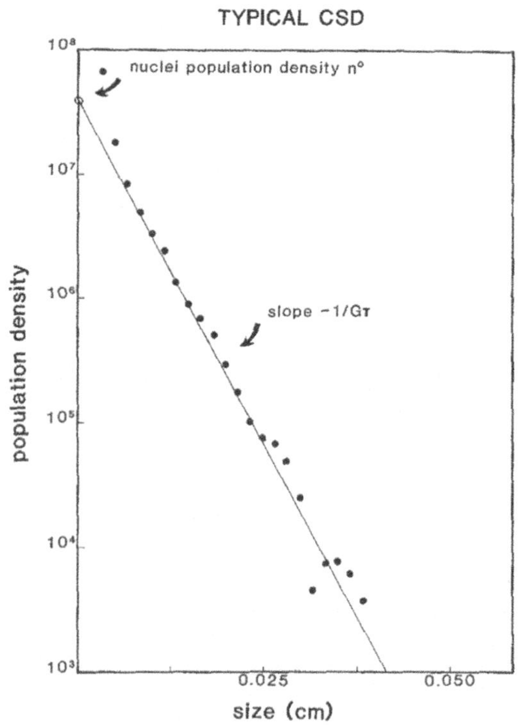

Fig. 9.1 CSD of groundmass plagioclase contained in drill samples from the Makaopuhi lava lake in Hawaii (Cashman and Marsh 1988). Reprinted by permission from Springer Nature (Cashman and Marsh 1988), COPYRIGHT (1988)

Research on the CSD in an Open System in Chemical Engineering

Discussion in the field of engineering has focused on a reactor tank in which the solution is cooled for crystallization and formed crystals are removed at a constant rate. Note that the CSD in the tank has been observed to be exponential distribution (Randolph and Larson 1988), which was considered as a steady-state solution of an equation of population balance (9.11); moreover, the intercept and the slope of the exponential CSD was interpreted. At the time, the characteristic size R_S (the inverse of the slope) was interpreted as the product of the growth rate G and the residence time $t_{residence}$, $R_S = Gt_{residence}$; moreover, the intercept F_0 was interpreted to be related with the nucleation rate J using $J = F_0 G$. This relationship will be explained below.

CSD in a Closed System

Marsh (1988) assumed that when the crystal growth rate G is constant, similar exponential CSD has been derived for closed-system crystallization

$$F(R,t) = \check{F}_0 \exp\left[-\frac{R}{Gt_0} + \frac{t}{t_0}\right] \tag{9.2}$$

where t_0 is a time constant that serves as an estimate of the crystal growth time; however, its relation to the nucleation duration t is not certain. Subsequently, Marsh (1988) revealed that t_0 characterizes an increase in the rate of the nucleation rate, which exponentially increases with time; however, t_0 has not been clearly distinguished from the crystallization time t_x. In this case, the slope $(1/R_S)$ is determined as $1/(Gt_x)$ by the product of the crystal growth rate and some type of timescale characterizing crystal growth $t_x (= t_0)$.

The number of crystals (the number density) of the size R or smaller, $N(R)$ (number/m^3), is obtained by integrating $F(R,t)$ of Eq. (9.1) from 0 to R:

$$N(R) = \int_0^R F(R')dR' = F_0 R_S \left(1 - \exp(-\frac{R}{R_S})\right) \tag{9.3}$$

Because the nucleation rate is the increasing rate of the number of crystals dN/dt, it can be defined as follows:

$$J = \left.\frac{dN}{dt}\right|_{R=0} \tag{9.4}$$

where the reason why a time derivative of N can be defined is that R is considered to be $R(t)$, which grows with time. Then, the nucleation rate can be written as follows:

$$J = \left.\frac{dN}{dR}\right|_{R=0} \cdot \left.\frac{dR}{dt}\right|_{R=0} \tag{9.5}$$

where $(dN/dR)_{R=0}$ is the slope at the origin of the integrated size distribution, and differentiating Eq. (9.3) with respect to R yields F_0, i.e., it is equal to the nucleation density of the exponential size distribution F_0. Moreover, because $dR/dt|_{R=0}$ is the crystal growth rate at the origin, it is $G(R(t) = 0))$. Consequently, a relational expression

$$J = F_0 G(R(t) = 0) \tag{9.6}$$

can be obtained. Moreover, the nucleation density F_0 is expressed using the ratio between the crystal growth rate and nucleation rate J as follows:

$$F_0 = \frac{J}{G(R = 0)} \tag{9.7}$$

Because Eq. (9.2) above is Eq. (9.1) when the nucleation rate is a function of the nucleation duration t, $J = G(R = 0)\check{F}_0 \exp(t/h)$. Moreover, it is found that derivation of Eq. (9.2) is on the assumption that the nucleation rate exponentially increases with time. The shape of CSD (F_0 and R_S in Eq. (9.1)) is determined by the crystallization kinetics (G, J) and the crystallization time (t_x); therefore, difference in CSD is attributed to the dependence of G, J, and t_x on the cooling process (represented by cooling rate for simplicity).

9.1.2 Attaching Physical Significance to Exponential CSD

To examine relationships between three parameters (G, J, t_x) and cooling process, Cashman and Marsh (1988) analyzed boring core samples from a lava lake in Hawaii (Makaopuhi lava lake), which is an example in nature from which aspects of crystallization process, such as cooling rate, can be estimated under field conditions. At that time, based on the assumption of batch crystallization (a closed system), the crystallization time of the sample was estimated by correlating it to the depth of the sample and using a heat conduction model, i.e., the timescale of crystallization tx was estimated from the inverse of the cooling rate estimated using the heat conduction model. Moreover, the tx value was used to estimate values of nucleation rate and crystal growth rate from the intercept and slope of CSD of individual samples (Table 9.1). For plagioclase, $J = 1.6 \times 10^{-3} \sim 3.4 \times 10^{-2}$ (no./cm^{3}s) and $G = 6.0 \times 10^{-11} \sim 9.9 \times 10^{-11}$ (cm/s) were obtained; moreover, as the depths of samples are greater, J and G are larger. For ilmenite, $J = 2.0 \times 10^{-4} \sim 2.2 \times 10^{-3}$ (no./cm^{3}s) and $G = 3.4 \times 10^{-10} \sim 4.9 \times 10^{-10}$ (cm/s) were obtained. Note that as the depths of samples are greater, J and G are larger. Furthermore, assuming that the nucleation rate and crystal growth rate are determined by power functions of the effective degree of supercooling ΔT ($J \propto \Delta T^{\xi_J}$, $G \propto \Delta T^{\xi_G}$), $\xi_G = 1.59$ is determined by the crystal growth experiment by Kirkpatrick (1976) and dependence of the nucleation rate on the degree of supercooling is estimated to be $\xi_J = 7.5$ from CSD in nature. However, this approach of thinking yields a larger effective degree of supercooling for the slower cooling rate (samples at greater depths), which is contrary to intuition. This problem is related to a problem of what point in time J and G, which are estimated from CSD, are from. This has to be examined while including the significance and interpretation of the petrogenetic model of exponential CSD.

Table 9.1 Estimated nucleation rate and crystal growth rate of magma crystallized by cooling in a lava lake in Hawaii (Makaopuhi lava lake) (Cashman and Marsh 1988). The slope $1/R_S$ is the corrected value

Sample	Depth (ft)	$1/R_S$ (cm^{-1})	F_0 (cm^{-4})	J (no./cm^{3}s)	G (cm/s)	ϕ (%)
Plagioclase						
68-2-20	51.5	1290	3.42×10^8	3.4×10^{-2}	9.9×10^{-11}	38.3
68-1-19	49.5	890	6.59×10^7	5.2×10^{-3}	7.9×10^{-11}	47.0
68-1-18	48.0	672	5.08×10^7	4.0×10^{-3}	8.0×10^{-11}	66.1
68-1-17	46.1	751	3.16×10^7	1.7×10^{-3}	5.4×10^{-11}	79.7
68-1-16	44.2	544	2.68×10^7	1.6×10^{-3}	6.0×10^{-11}	82.8
Ilmenite						
68-2-20	51.5	–	–	2.2×10^{-3}	4.9×10^{-10}	38.3
68-1-19	49.5	–	–	0.8×10^{-3}	3.9×10^{-10}	47.0
68-1-16	44.2	–	–	0.2×10^{-3}	3.4×10^{-10}	82.8
Magnetite						
68-1-16	44.2	–	–	7.6×10^{-2}	2.9×10^{-10}	82.8

Table 9.2 Nucleation rate and crystal growth rate of phenocrysts and microlites of the 1980–1986 eruptions of Mount St. Helens, estimated by giving time constants (Cashman 1988)

Sample	Time constant	J(no./cm^{3}s)	G(cm/s)
Iron-titanium oxide phenocry			
Blast dacite	30 (years)	3.35×10^{-4}	1.33×10^{-12}
Blast dacite	150 (years)	6.69×10^{-5}	2.66×10^{-13}
Dome in March 1984	30 (years)	3.25×10^{-4}	9.75×10^{-13}
Dome in March 1984	150 (years)	6.50×10^{-5}	1.95×10^{-13}
Plagioclase phenocryst			
Blast dacite	30 (years)	1.569×10^{-5}	1.046×10^{-11}
Blast dacite	150 (years)	4.72×10^{-6}	3.15×10^{-12}
Dome in March 1984	30 (years)	2.114×10^{-5}	9.88×10^{-12}
Dome in March 1984	150 (years)	6.36×10^{-6}	2.97×10^{-12}
Blast dacite microlite			
BD 8 Horizontal	52 (days)	1.64×10^{4}	1.19×10^{-11}
BD 8 Horizontal	1 (hour)	2.05×10^{7}	1.48×10^{-8}

Cashman (1988) measured the CSD of microlites of Mount St. Helens, considered precursory events such as precursory earthquakes of the eruption as signals of magma ascent, provided an appropriate crystallization time from the timing of the signals, and calculated J and G. Consequently, the nucleation rate of microlites was found to be remarkably larger than that of phenocrysts (Table 9.2). From change of the microlite CSD, (Cashman 1992) revealed that the nucleation rate decreased from 1.2×10^3 (no./cm^3 s) to 2.7×10^{-2} (no./cm^3 s) over the period from June 1980 to October 1986.

Advancing the discussion further, Cashman (1993) reported the dependence of the nucleation rate J and the crystal growth rate G on the cooling rate (dT/dt) in the following form and attempted to estimate parameters from geological occurrence and situations.

$$\log G = \beta_1 + \beta_2 \log \left| \frac{dT}{dt} \right| \tag{9.8}$$

$$\log J = \beta_3 + \beta_4 \log \left| \frac{dT}{dt} \right| \tag{9.9}$$

She used a heat conduction model to estimate the cooling rate at the crystallization front as a function of the distance from the boundary y and divided data of the crystal size $R(y)$ as a function of the distance by the characteristic time of cooling $\tau = \kappa/y^2$ to estimate G from $G = R/\tau$. Moreover, the nucleation rate was estimated from J = G/R4, which was obtained from a general scaling relation $R = (G/J)^{1/4}$ (Box in Sect. 7.2). Consequently, she estimated the dependence of the nucleation rate and the growth rate on the cooling rate as shown in Table 9.3. Moreover, using these data, the cooling rate exponent ξ that expresses the dependence of the crystal number density on the cooling rate

$$N \propto \left| \frac{dT}{dt} \right|^{\xi} \tag{9.10}$$

can be estimated from a general relational equation about the crystal number density $N = (J/G)^{3/4}$. The cooling rate exponent is an extremely important parameter for interpreting rock texture and is listed in the table.

Subsequently, although Marsh (1998) calculated CSDs in various cases where the nucleation rate and the growth rate are functions of time, he did not focus on the dependence of J and G, as well as N on the cooling rate and did not quantitatively link the difference in CSDs with the difference

Table 9.3 Exponentsβ_4 and β_2, estimated by Cashman (1993), represent the dependence of the nucleation rate and the crystal growth rate on the cooling rate The indices and dependence of crystal number density N on cooling rate, $\xi = 3(\beta_4 - \beta_2)/4$, obtained from $N = (J/G)^{3/4}$ are listed

Natural sample	β_2	β_4	ξ
Salvador dikes (data by Gray)	0.87	1.37	0.37
Prehistoric Makaopuhi lava lake	0.88	1.35	0.35
Data by Lofgren et al. (pyroxene)	0.64	1.79	0.86

in cooling rate. Furthermore, because he calculated on the assumption that growth rate is constant when the growth rate $G = G_0 \exp(t/t_G)$ exponentially increases with time, he was unable to obtain an accurate mathematical expression of CSD. Resmini (2007) modeled CSD and demonstrated that when $G \propto 1/t$, CSD would be a straight line on a semilogarithmic plot using numerical calculation; however, he did not derive exponential CSD. Giving J and G, Spillar and Dolejs (2014) modeled crystallization that considers spatial occupation of particles and numerically calculated CSD.

9.1.3 Perographic Studies of CSD

There have been numerous petrographic studies of CSD. Studies applying CSD to phenocryst are listed below. Mangan (1990) analyzed the CSD of olivine phenocrysts of Hawaii, reported differences in the CSD between xenocrysts and in-situ crystals, and discussed the magma feeding system. Tomiya and Takahashi (1995) then measured the CSD of volcanic rocks of Usu volcano and concluded that a bent curve in the CSD is attributed to the mixing of the CSD of crystals originating from two different magmas and confirms that magma mixing occurred. Resmini and Marsh (1995) obtained many CSDs of plagioclase crystals in lava with a wide range from about 0.05 mm to about 2 mm, and then demonstrated change of texture as a function of the magma chamber processes or the order of eruption (the order of strata). Higgins (1996) obtained CSDs from phenocrysts to microlites and argued that the slope of the microlite part and that of the phenocryst part differed. He determined G and estimated the time constant at several years and a dozen or so years, respectively.

Studies applying CSD to groundmass texture are listed below. Wilhelm and Worner (1996) obtained CSDs of sill and lava flow as a function of space and estimated G and J as a function of space using crystallization time calculated from a heat conduction model. Hammer et al. (1999) obtained CSD up to 0.01 (mm) and demonstrated CSD takes exponential distribution. Moreover, they presented a microlite systematics relation (Sect. 10.3.4) between N_A (the crystal number density per unit area) and ϕ (the amount of crystals). Hammer et al. (2000) seems to be the first study quantitatively comparing discharge rate and texture, although they did not draw CSD plots. Higgins (2002) considered that the correlation between the intercept and slope, which is often observed in CSD, originates from the volume fraction remaining constant. Zieg and Marsh (2002) obtained many CSDs and demonstrated that there is a correlation between the intercept and slope, thus indicating a constant crystallinity. They considered that the slope of CSD depends on the cooling rate, although they did not quantitatively discuss it. Piochi et al. (2005) determined the CSD of pyroclastic products of Campi Flegrei (Italy) up to about 0.1 mm, reported a bend on the CSD at a slightly less than 0.1 mm, and argued that magma ascent and crystallization have two stages.

Morgan et al. (2007) investigated the relationship between the characteristics of Sr isotopes and CSD of recent eruptive materials of Stromboli volcano (Italy) as well as discussed the timescale of magma supply. Salisbury et al. (2008) measured CSD with a quite wide range and obtained slopes of microlites, microphenocrysts, and phenocrysts by distinguishing them. They argued that the distinction is consistent with difference of trends in a correlation diagram of An# and the size, as well as discussed the timescale in the magma chamber processes. Vinet and Higgins (2011) measured the CSDs of drill core samples from Kilauea crater in Hawaii and distinguished between the CSD of deformed olivine and that of undeformed olivine. Preece et al. (2013) measured the CSDs of samples from a dome-forming eruption of Merapi (Indonesia) and stated that CSDs can be classified into growth-dominated CSD and nucleation-dominated CSD as per the slope and intercept.

9.1.4 Problems in CSD Studies and Establishment of Calculation Methods of CSD

One of the limitations of past studies is that CSD calculation in which J or G is given as a function of time or size has not been mathematically accurate. For example, in Marsh (1998), in some places, G was assumed to be constant; however, in other places G was an exponential function of time. Thus, a mathematically incorrect CSD was obtained and discussion based on the incorrect CSD was performed.

Another limitation is that experimental studies demonstrating the type of difference arising in CSD depending on the difference of cooling processes had not been advanced. However, in recent years, such studies have been rapidly advancing (Since Burkhard 2002, Pupier 2008, Brugger and Hammer 2010, and Shea and Hammer 2013 have followed). To effectively use such experimental results, a method of calculating CSD in a mathematically correct approach using a kinetic model has to be established. This will connect the parameters appearing in a mathematical expression of the CSD to the parameters in experiments and allow us to evaluate which kinetic model actually acts by comparing using experimental CSD.

For example, for a closed-system batch crystallization, there are three kinetic models resulting in exponential CSDs. The first model is that the nucleation rate exponentially increases with time ($J = J_0 \exp(t/t_J)$) and the growth rate is constant ($G =$ const.). The second one is that the nucleation rate and the crystallization rate are constant ($J =$ const. $\dot{\phi}$ =const., the crystal growth rate is in inverse proportion to time $G = G_0(t/t_0)^{-1}$). Moreover, the third one is that the nucleation rate is constant and the growth rate is an exponential function against the size ($J =$ const., $G = G_0 \exp(R/R_G)$). Although these three models share the same slope that is the inverse of the characteristic size R_S, their physical significance is different. In the first case, the characteristic size R_S is determined as the product of time characterizing the increasing rate of nucleation t_J and the growth rate, $R_S = Gt^* = Gt_J$. In the second case, the characteristic size R_S is determined by the cube root of the ratio of the nucleation rate to the crystallization rate $\dot{\phi}/J$. Change in the slope arising from different crystallization processes (e.g., the cooling rate) has to be interpreted by the influence on Gt_J in the first case and on $\dot{\phi}/J$ in the second case. To be concrete, we have to examine the way in which parameters appearing here, i.e., t_J, J_0, G_0, t_0, and $\dot{\phi}$, depend on the cooling rate. For example, t_J, a time constant characterizing exponential increase in the nucleation rate, includes coefficients determined by interfacial energy and activation energy of diffusion and is inversely proportional to cooling rate. By substituting values into them, the examination of kinetic models and interpretation of CSDs become practically possible.

In the following, we will derive CSDs for the kinetic models of various nucleation and growth laws, including what conditions in which this exponential CSD arises. First, we will present a general derivation using an Eulerian approach for a closed system in which crystals do not enter or exit, and we will show certain concrete examples. Next, we will explain a general derivation using a Lagrangian approach and will try to practically calculate several examples. Some models derived by this method with different assumptions will be applied to CSD data obtained from laboratory experiments to evaluate each model. Then, we will make an Eulerian description for CSD of an open system and obtain an analytical solution.

9.2 Analytical Solution of CSD Based on an Eulerian Description

9.2.1 Solution by Separation of Variables: A General Solution and Examples of Formation of Exponential Distribution

We will consider homogeneous crystallization in a closed system. The temporal development of size distribution using an Eulerian approach (an equation of population balance) is described by

$$\frac{\partial F(R,t)}{\partial t}+\frac{\partial}{\partial R}(G(R,t)F(R,t))=0 \tag{9.11}$$

Here, the growth rate G is assumed to be a function only of size $G(R)$. Separating variables in $F(R,t)$ into a function only of R, $f_1(R)$, and another function only of t, $f_2(t)$, yields

$$F(R,t)=f_1(R)\cdot f_2(t) \tag{9.12}$$

Then, substituting this into Eq. (9.11) gives

$$f_1\frac{\partial f_2}{\partial t}+f_2\frac{\partial}{\partial R}(G(R)f_1(R))=0 \tag{9.13}$$

Moreover, rearranging this equation with $\varphi(R)=G(R)f_1(R)$ yields

$$\frac{1}{f_2(t)}\frac{\partial f_2}{\partial t}=-\frac{G(R)}{\varphi(R)}\frac{\partial\varphi}{\partial R}=c \tag{9.14}$$

For each side comprising only an independent variable to be equal, each side has to be a constant, c.

Because the left-hand side and the right-hand side should be equal,

$$f_2(t)=f_2(0)e^{ct} \tag{9.15}$$

is obtained, and because the middle part and the right-hand side should be equal,

$$\varphi(R)=\varphi(R_0)\exp\left(-\int_{R_0}^{R}\frac{c}{G(R)}dR\right) \tag{9.16}$$

is obtained. Here, $\varphi(R_0)=G(R_0)f_1(R_0)$ holds.

From these equations, the size distribution function $(F(R,t)=f_1f_2=\varphi f_2/G)$ is obtained as follows:

$$F(R,t)=F(R_0,0)\frac{G(R_0)}{G(R)}\exp\left[c\left(t-\int_{R_0}^{R}G(R)^{-1}dR\right)\right] \tag{9.17}$$

The Case Where the Growth Rate Is Constant Regardless of Size ($G=G_0$)

The general solution is

$$F(R,t)=F(R_0,0)\exp\left[c\left(t+\frac{R_0}{G_0}\right)-\frac{c}{G_0}R\right] \tag{9.18}$$

where R_0 is a constant determined by the boundary condition. This appears to be exponential CSD. However, depending on the boundary condition, c does not become a constant and sometimes breaks the assumption; therefore, the application range is largely limited by solving using the separation of variables.[1]

Example of Formation of Exponential Distribution

When the growth rate is constant, regardless of size ($G = G_0$), the nucleation rate exponentially increases with time, c becomes a constant and the solution can be obtained using this method. In this case,

$$J(t) = J_0 e^{t/t_J} = G_0 F(0, t) = G_0 F(R_0, 0) e^{c(t+R_0/G_0)} \tag{9.21}$$

holds, where the time constant of the nucleation rate t_J is a time constant that indicates the exponential increase in the rate of nucleation (see Box).

Time constant of the nucleation rate that exponentially increases

When the nucleation rate is given by Eq. (7.48), the time constant of the nucleation rate that exponentially increases with time, t_J, is determined by considering temperature as a function of time and differentiating the equation as follows:

$$t_J^{-1} = \left(-\frac{\tilde{q}}{\tilde{T}_1^2} + \frac{(3\tilde{T}_1 - 1)\Lambda}{\tilde{T}_1^2 (1 - \tilde{T}_1)^3} \right) \frac{|\dot{T}|}{T_0} = \tilde{A} \frac{|\dot{T}|}{T_0} = \frac{\tilde{A}}{t_{\text{th}}} \tag{9.22}$$

which is proportional to the cooling rate. Here, $\tilde{T}_1$ is a dimensionless, scaled temperature corresponding to a temperature T_1 at which nucleation becomes active. t_{th} is a characteristic cooling time and defined as $t_{\text{th}} = T_0/|\dot{T}|$. $\tilde{A}$ is the inside of the square brackets and is a constant representing the magnitude of the influence of cooling on increase in nucleation rate. The relationship between the $\tilde{A}$ value and dimensionless supercooling $\Delta\tilde{T}_1 = \Delta T_1/T_0$ is shown in the figure below. The value of $\tilde{A}$ against three different nucleation barrier $\log(\Gamma)$. It depends on $\Delta\tilde{T}_1^{-3}$ when the degree of supercooling is small. The numbers by the curves indicate the value of $\log(\Gamma)$.

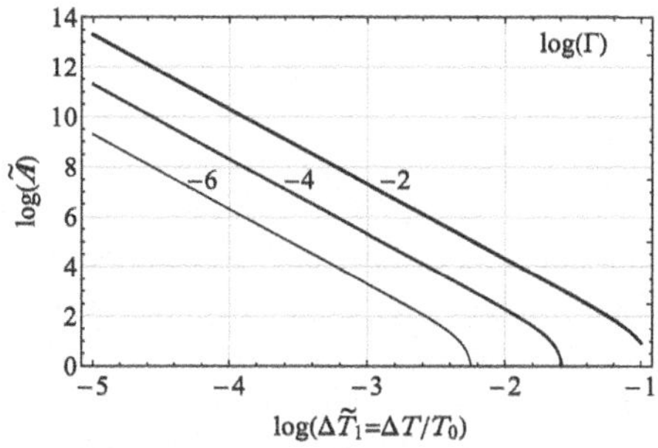

---

[1]When a constant nucleation rate ($J = J_0$ = constant) is assumed,

$$J(t) = J_0 = G_0 F(0, t) = G_0 F(R_0, 0) \exp\left[c\left(t + \frac{R_0}{G_0}\right)\right] \tag{9.19}$$

is obtained from the boundary condition. Therefore,

$$\exp\left[c\left(t + \frac{R_0}{G_0}\right)\right] = \frac{J_0}{G_0 F(R_0, 0)}, \quad c = \frac{\ln J_0/(G_0 F(R_0, 0))}{t + R_0/G_0} \tag{9.20}$$

is obtained, which indicates that c depends on time t.

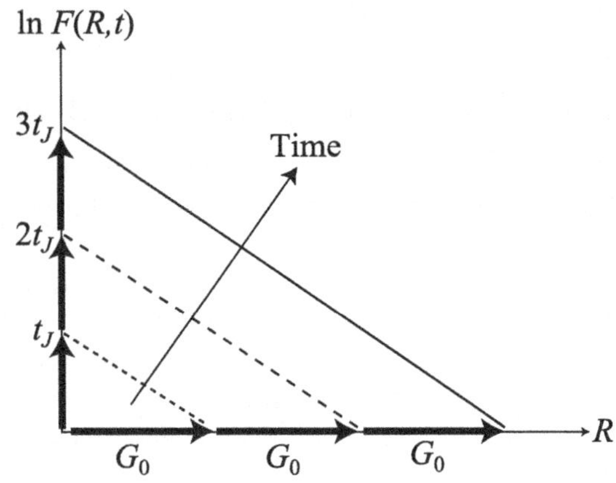

Fig. 9.2 Example of formation of exponential distribution (No. 1). Schematic of the temporal development of CSD in the case where the nucleation rate exponentially increases with time ($J(t) = J_0 \exp(t/t_J)$) and the growth rate is constant (G_0)

From this equation, the constants c and R_0 and $F(R_0, 0)$ are determined by being related to the nucleation rate as follows.

$$c = t_J^{-1} \tag{9.23}$$

$$\frac{J_0}{G_0} = F(R_0, 0)e^{R_0/G_0 t_J} \tag{9.24}$$

Finally, CSD will be

$$F(R, t) = \frac{J_0}{G_0} \exp\left[\frac{t}{t_J} - \frac{R}{G_0 t_J}\right] \tag{9.25}$$

which is an exponential CSD. The intercept and the slope of the CSD are

$$F_0 = \frac{J_0 e^{t/t_J}}{G_0} \tag{9.26}$$

$$|\text{slope}| = \frac{1}{G_0 t_J} = 1/R_S \tag{9.27}$$

where R_S $(= G_0 t_J)$ is a characteristic size and matches a mean particle size.

Let us qualitatively understand the fact that CSD becomes exponential distribution in this case. The nucleation density F_0 or $F(0, t)$ increases with the ratio of the nucleation rate to the growth rate $J_0(t)/G_0 = J_0 \exp(t/t_J)/G_0$. The increase in the rate on a natural logarithmic scale $d \ln F_0/dt$ will be constant on the logarithmic scale because $d(t/t_J)/dt = t_J^{-1}$. The increase in the rate of crystal size that formed at a size 0 is G_0, which is constant on a linear scale. Consequently, on a semilogarithmic graph, the size distribution function will move at the same rate on the horizontal axis and the vertical axis and will result in exponential distribution (Fig. 9.2).

9.2.2 A General Solution in the Case Where the Growth Rate Depends only on Time

In the case where the growth rate depends only on time, the separation of variables cannot be used. To obtain a general solution, let us use the Laplace transformation. When the growth rate is a function of only time, G in the second term on the left-hand side of Eq. (9.11) can be moved outside the derivative.

Therefore, the conservation law of size distribution will be as follows:

$$\frac{\partial F(R,t)}{\partial t} + G(t)\frac{\partial F(R,t)}{\partial R} = 0 \tag{9.28}$$

The Laplace transformation with respect to space (the size space R here) will be applied to this equation. Multiplying the equation by e^{-sR} (s is a dummy variable), integrating it, and using the definition of the Laplace transformation

$$\hat{F} = \int_0^{\infty} e^{-sR} F dR \tag{9.29}$$

yield

$$\text{The first term} = \frac{\partial \hat{F}}{\partial t} \tag{9.30}$$

$$\text{The second term} = G(t)\left[-F(0,t) + s\hat{F}\right] \tag{9.31}$$

and Eq. (9.28) can be written as

$$\frac{\partial \hat{F}}{\partial t} + sG(t)\hat{F} = G(t)F(0,t) \tag{9.32}$$

This is the first-order linear differential equation of $\hat{F}$ with respect to time and the general solution can be obtained as follows:

$$\hat{F} = \int_0^t G(t')F(0,t')\exp\left[-s\int_{t'}^t G(t'')dt''\right]dt' + c\exp\left[-s\int_0^t G(t')dt'\right] \tag{9.33}$$

where c is a constant of integration and can be written as follows:

$$c = \int_0^{\infty} e^{-sR} F_0(R)dR = \hat{F}_0(s) \tag{9.34}$$

when the initial condition is $F_0(R) = F(R,0)$.

Because F can be obtained by the inverse Laplace transform of $\hat{F}$,

$$F(R,t) = \mathcal{L}^{-1}(\hat{F}) \tag{9.35}$$

$$= \int_0^t G(t')F(0,t')\delta(R - \int_{t'}^t G(t'')dt'')dt' + \int_0^R \delta(R' - \int_0^t G(t')dt')F(R-R',0)dR' \tag{9.36}$$

are obtained. Here, $\mathcal{L}^{-1}(e^{-sX}) = \delta(R-X)$ is used. Using a property of the delta function

$$\int f(x)\delta(g(x))dx = \int f(g)\left(\frac{\partial g}{\partial x}\right)^{-1}\delta(g)dg = f(g=0)\left[\left(\frac{\partial g}{\partial x}\right)^{-1}\right]_{g=0} \tag{9.37}$$

on the first term on the right-hand side yields

$$F(R,t) = F(0, t'(R,t)) + F(R - \int_0^t G(t'')dt'', 0) \tag{9.38}$$

where $t'(R,t)$ is a function of R and t, which satisfies

$$R - \int_{t'}^t G(t'')dt'' = 0 \tag{9.39}$$

The first term on the right-hand side of Eq. (9.38) is determined by the boundary condition, i.e., time change of nucleation:

$$G(0,t')F(0,t') = J(t') \tag{9.40}$$

Finally, the general solution is expressed as follows:

$$F(R,t) = \underbrace{\frac{J(t')}{G(t')}}_{Boundarycondition} + \underbrace{F(R - \int_0^t G(t'')dt'', 0)}_{Initialcondition} \tag{9.41}$$

The first term reflects the boundary condition in the size space and represents inflow at $R = 0$, i.e., the influence of the nucleation rate, whereas the second term represents the influence of initial condition.

Example of Formation of Exponential Distribution: No. 1

The case where growth rate depends only on time includes the case where growth rate is constant ($G = G_0$). Although this case is the same as that where the separation of variables is used, described earlier, we describe it again because a base equation differs.

For simplicity, the case where crystals do not exist is considered as the initial condition. In the case, Eq. (9.41) is written as follows:

$$F(R,t) = \frac{J(t')}{G_0} \tag{9.42}$$

where the relationship between t' and R is given by Eq. (9.39), i.e.,

$$R = G_0(t - t') \tag{9.43}$$

Assuming that time dependence of the nucleation rate is an exponential function $J(t) = J_0 \exp(at)$,

$$J(t') = J(t - R/G_0) = J_0 \exp\left(t/t_J - \frac{R}{G_0 t_J}\right) \tag{9.44}$$

is obtained, which is exponential distribution. Qualitative explanation of formation of exponential distribution in this case was described in Fig. 9.2 and the corresponding body text.

Example of Formation of Exponential Distribution: No. 2

Similarly, assuming the case in which crystals do not exist as the initial condition, we consider the case in which the nucleation rate is constant J_0 and the growth rate is inversely proportional to time. Eq. (9.41) can then be written as follows:

$$F(R,t) = \frac{J_0}{G(t')} \tag{9.45}$$

As seen later, this case corresponds to that in which the crystallization and nucleation rates are constant, and the pre-exponential factor of the nucleation rate is extremely small. Substituting an equation of the growth rate

$$G(t) = \frac{R_0}{t} \tag{9.46}$$

into Eq. (9.39) and integrating it on the assumption of $G(t) = dR/dt$ yields

$$R = R_0 \ln \frac{t}{t'} \tag{9.47}$$

Substituting the nucleation time t' obtained from this equation into Eqs. (9.46) and (9.45) yields

$$F(R,t) = \frac{J_0 t}{R_0} \exp\left(-\frac{R}{R_0}\right) \tag{9.48}$$

which is an exponential distribution.

Let us qualitatively understand the fact that CSD becomes exponential distribution in this case. Because the nucleation density F_0 or $F(0,t)$ is determined by the ratio of the nucleation rate to growth rate at the nucleation time t', $J(t')/G(t')$, or $J_0 t/R_0$, it linearly increases with time. Because the growth rate is infinitely large mathematically when $t = 0$, even if crystals form at a constant nucleation rate, it immediately grows (mathematically to the infinite size) and the nucleus density will be 0 at $R = 0$. Because the increasing rate of the crystal size generated at size 0 is inversely proportional to time, the growth rate of nuclei slows down with time and the crystal comes to remain at $R = 0$. Therefore, the nucleus density is proportional to time and increases. Because, as the size is large, nucleation occurs earlier at time t' and goes through a history in which the growth rate was large, the number of crystals that was in the size bin of dR at the size 0 in the early stage will be diluted to the size bin width t/t' at the size R at the time t. This is easier to understand by looking at it through the Lagrangian approach described later. Consequently, the size distribution function linearly decreases with increase in size, and the intercept of the function on the vertical axis is constant on a semilogarithmic graph; i.e., the size distribution function shifts to exponential distribution (Fig. 9.3).

9.2.3 A General Solution in the Case Where Growth Rate Depends only on the Size

In the case where growth rate is a function only of size, the second term on the left-hand side of Eq. (9.11) becomes a function only of R and the conservation law of size distribution will be

$$\frac{\partial F(R,t)}{\partial t} + \frac{\partial}{\partial R}(G(R)F(R,t)) = 0 \tag{9.49}$$

The Laplace transformation with respect to time will then be applied to this equation. Multiplying the equation by e^{-st}, integrating it, and using the definition of the Laplace transformation

$$\hat{F} = \int_0^\infty e^{-st} F dt \tag{9.50}$$

yield

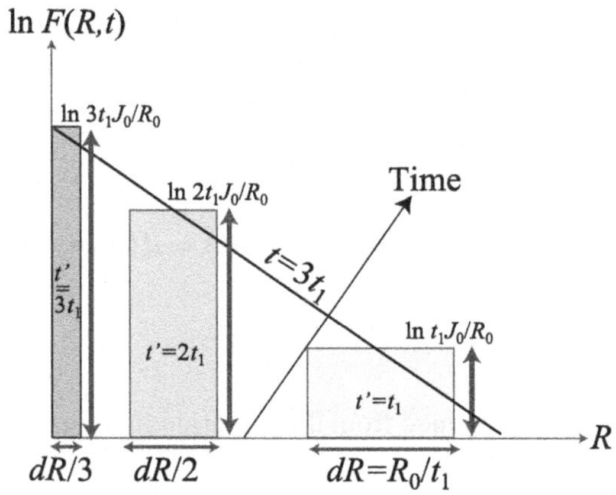

Fig. 9.3 Example of formation of exponential distribution (No. 2). Schematic of exponential CSD in cases where the nucleation rate ($J_0 \times 1$ s $= F(R) \times dR$: Area of the rectangle) is constant and growth rate decreases in inverse proportion to time ($G(t) = R_0/t$). The size bin of crystals nucleating at the time t_1 is R_0/t_1, whereas that of crystals nucleating at the time twice t_1 is a half of R_0/t_1; i.e., the bin width becomes narrow with time. The height of F increases with time and J_0 is constant

$$\text{The first term} = -F(R,0) + s\hat{F} \tag{9.51}$$

$$\text{The second term} = \frac{\partial}{\partial R}\left(G(R)\hat{F}\right) \tag{9.52}$$

and Eq. (9.49) can be written as

$$-F(R,0) + s\hat{F} + \frac{\partial}{\partial R}\left(G(R)\hat{F}\right) = 0 \tag{9.53}$$

Rearranging this equation gives

$$\frac{d\hat{F}}{dR} + \frac{s + \frac{dG}{dR}}{G}\hat{F} = \frac{F(R,0)}{G} \tag{9.54}$$

This is the first-order linear differential equation of $\hat{F}$ with respect to space (size space) and the general solution can be obtained as

$$\hat{F} = e^{-(st_A(R)+\tilde{b}(R))}\left(\int_0^R \frac{F(R',0)}{G(R')} e^{st_A(R')+\tilde{b}(R')} dR' + c\right) \tag{9.55}$$

where

$$t_A(R) = \int_0^R \frac{dR''}{G(R'')} \tag{9.56}$$

$$\tilde{b}(R) = \ln\left(\frac{G(R)}{G(0)}\right) \tag{9.57}$$

c is a constant of integration and determined from the Laplace transformation of the boundary condition $F(0,t)$

$$c = \hat{F}(0,s) = \int_0^\infty e^{-st} F(0,t)dt \tag{9.58}$$

In this case, the boundary condition is $G(0)F(0,t) = J(t)$ and the Laplace transformation of it gives c as

$$c = \frac{\hat{J}(s)}{G(0)} \tag{9.59}$$

Because F can be obtained by the inverse Laplace transformation of $\hat{F}$,

$$F(R,t) = \mathcal{L}^{-1}(\hat{F}) \tag{9.60}$$

$$= \int_0^R \frac{F(R',0)}{G(R')} e^{\tilde{b}(R,R')} \mathcal{L}^{-1}(e^{-st_{\mathrm{A}}(R,R')}) dR' + \frac{e^{-\tilde{b}(R)}}{G(0)} \mathcal{L}^{-1}(e^{-st_{\mathrm{A}}(R)} \hat{J}(s)) \tag{9.61}$$

$$t_{\mathrm{A}}(R, R') = t_{\mathrm{A}}(R) - t_{\mathrm{A}}(R') \tag{9.62}$$

$$\tilde{b}(R, R') = \tilde{b}(R') - \tilde{b}(R) \tag{9.63}$$

are obtained. Using $\mathcal{L}^{-1}(e^{-sX}) = \delta(t - X)$ for the first term and convolution for the second term yields

$$F(R,t) = \underbrace{\frac{J(t - t_{\mathrm{A}}(R))}{G(R)}}_{Boundary condition} + \underbrace{\frac{G(R')}{G(R)} F(R',0)}_{Initial condition} \tag{9.64}$$

A property of the delta function, described earlier, is used here. R' in this equation is a variable connecting R and t through

$$t = \int_{R'}^{R} \frac{dR''}{G(R'')} \tag{9.65}$$

Example of Formation of Exponential Distribution: No. 3

For simplicity, we assume the case where crystals do not exist as the initial condition and the nucleation rate is constant ($J(t) = J_0$). Equation (9.64) is then written as follows:

$$F(R,t) = \frac{J_0}{G(R)} \tag{9.66}$$

When the crystal growth rate exponentially increases with respect to the size, we assume that

$$G(R) = G_0 \exp(R/R_{\mathrm{G}}) \tag{9.67}$$

holds. Here, R_{G} is a constant characterizing the size dependence of the growth rate. In this case, Eq. (9.66) is obtained as follows:

$$F(R,t) = \frac{J_0}{G_0} \exp(-R/R_{\mathrm{G}}) \tag{9.68}$$

which is an exponential distribution.

Let us qualitatively understand the fact that CSD becomes exponential distribution in this case. Because the nucleation density F_0 or $F(0,t)$ is determined by the ratio of the nucleation rate to growth rate $J_0(t)/G(R)|_{R=0} = J_0/(G_0 \exp(R/R_{\mathrm{G}}))|_{R=0}$, it is always constant at J_0/G_0 and does not change

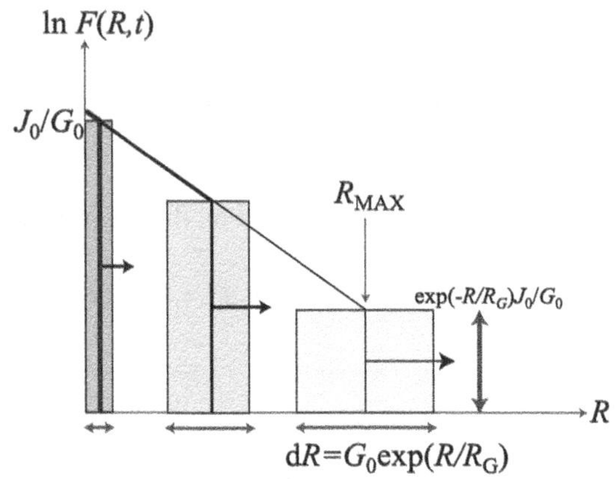

Fig. 9.4 Example of formation of exponential distribution (No. 3). Schematic of exponential CSD in the case where the nucleation rate is constant and the growth rate is given as a function of size ($G(R) = G_0 \exp(R/R_G)$). R_{MAX} is given by $R_G \ln\left(\frac{G_0}{R_G}t + 1\right)$, which progresses with time

with time. The increasing rate of crystal size forming at size 0 is $G_0 \exp(R/R_G)$, which exponentially increases with increase in size at an accelerated pace. Because, as the size is large, the increasing rate becomes large, the size bin that was dR at the size 0 in the early stage will be expanded to $\exp(-R/R_G)$ times when it grows to the size R. Moreover, because the size bin contains J_0/G_0 crystals, the size bin on the linear scale dR must contain 1/exp(R/R_G), i.e., $\exp(-R/R_G)$ times crystals. That is, the crystal number density will be diluted with increase in the size of size space. Consequently, the size distribution function linearly decreases with increase in size, and the intercept of the function on the vertical axis is constant on a semilogarithmic graph; i.e., the size distribution function becomes exponential distribution (Fig. 9.4). In this case, assuming that nucleation starts at $t = 0$, because crystals at $t < 0$ do not exist, the size corresponding to the nucleation at $t = 0$ will propagate as a size front with growth.

9.3 Analytic Solution of CSD Based on a Lagrangian Description

9.3.1 Method

The same result as obtained using the Eulerian approach can be obtained by the Lagrangian approach in the time–size space. An advantage of the Lagrangian approach is that if the nucleation rate and the growth rate are given as functions of time, a size distribution function can be relatively easily obtained in an analytical manner. The procedure is described as follows.

Equation (5.50) of the bubble size distribution explained in the section on temporal development of the vesiculation process, can be replaced with the crystal size distribution $F(R, t)$ as it is.

$$F(R(t', t), t) = \frac{\delta t' J(t')}{\delta R(t', t)} \tag{9.69}$$

Here, the nucleation rate $J(t')$ is the nucleation rate at the nucleation time t'. Note that the size R of the crystal size distribution at a certain time t, $F(R, t)$ explicitly includes the nucleation time t'. Because $\delta R(t', t)$ on the right-hand side is given by Eq. (5.51), it is written using the growth rate $G(R, t)$ as follows:

$$\delta R(t', t) = \int_{t'}^{t} G(R(t', \tau), \tau) d\tau - \int_{t'+\delta t'}^{t} G(R(t' + \delta t', \tau), \tau) d\tau \tag{9.70}$$

Although this can be calculated if an expression of the growth rate is given, the growth rate depends on the degree of supersaturation at the time. Therefore, to calculate $\delta R(t', t)$ at the time, the degree of supersaturation at all times up to that time have to be identified. Because the degree of supersaturation is a function of temperature and concentration, and the concentration depends on the amount of crystallization, F until the time has to be identified, which is not straightforward to do.

Now, aside from coupling with temperature or concentration described above, we simply assume that the growth rate G explicitly depends only on time. Then,

$$\delta R(t', t) = \int_{t'}^{t} G(\tau)d\tau - \int_{t'+\delta t'}^{t} G(\tau)d\tau = \int_{t'}^{t'+\delta t'} G(\tau)d\tau \tag{9.71}$$

$$= G(t')\delta t' \tag{9.72}$$

are obtained. If the last integral is approximated with $G(t')\delta t'$, $\delta t'$ in the denominator of Eq. (9.69) and that in the numerator are cancelled out and

$$F(R(t', t), t) = \frac{J(t')}{G(t')} \tag{9.73}$$

is obtained. This is an important equation that will serve as a basis from now onward and represents that the size distribution at the present time t is determined by the ratio of the nucleation rate to the growth rate at a past time t'.

However, note that the past nucleation time t' has to be calculated. The only clues to the past nucleation time are the current crystal size $R(t', t)$ and the growth rate. Retracing the growth will reach the birth of the crystal, i.e., the time at $R = 0$. That is, the current crystal size is calculated by

$$R(t', t) = \int_{t'}^{t} G(\tau)d\tau \tag{9.74}$$

Solving this equation with respect to t' yields t' as a function of R and t, because the right-hand side is a function of t' and t.

To close the problem of obtaining the size distribution function, the termination time of nucleation has to be determined. The termination of nucleation is approximated with the time of maximum nucleation. The time of maximum nucleation agrees with the time of the maximum value of the degree of supercooling (the degree of supersaturation). This condition is used when the number density of bubbles is semi-analytically obtained (Sects. 5.5.3 and 5.5.4). The degree of supercooling ΔT_L liquid at the temperature T and the concentration of a crystallization component C is defined by the difference between the temperature T and the melting point of the liquid (liquidus) $T_L(C)$ as follows:

$$\Delta T_L(C, T) = T_L(C) - T \tag{9.75}$$

The liquidus temperature as a function of concentration can then be simplified and written as follows:

$$T_L(C) = T_0 \left(1 + \frac{k_B}{\Delta s}\left(\frac{C}{C_0} - 1\right)\right) \tag{9.76}$$

From the conservation of mass, the concentration of the crystallization component is expressed by using the volume fra in chapter "Temporal development of vesiculation"ction of the crystal phase ϕ as follows:

$$C = \frac{C_0 - \phi/v_S}{1-\phi} \tag{9.77}$$

(a single solid phase is assumed). Because the volume fraction at the maximum nucleation is sufficiently smaller than 1, the following approximation holds.

$$C = C_0 - \frac{\phi}{v_S} \tag{9.78}$$

Substituting this equation into Eqs. (9.76) and (9.75), differentiating it with respect to time, and taking the extremum condition yield

$$\frac{d\Delta T_L(C,T)}{dt} = \underbrace{-\frac{kT_0}{\phi_0 \Delta s}\frac{d\phi}{dt}}_{\text{Decrease}} \underbrace{-\frac{dT}{dt}}_{\text{Increase}} = 0 \tag{9.79}$$

where $\phi_0 = C_0 v_S$, the initial volumetric concentration. The first term on the right-hand side corresponds to decrease in the degree of supersaturation due to crystal growth and the second term corresponds to increase in the degree of supersaturation due to cooling. The volume fraction ϕ is then determined using the size distribution function $F(R,t)$ as

$$\phi = \int_0^\infty \alpha_s R^3 F(R,t)dR \tag{9.80}$$

where α_s is the shape factor and is $4\pi/3$ for a sphere. Because $F(R,t)$ has already been obtained, this integral can be calculated. Solving the equation of the middle part and the right-hand side of Eq. (9.79) with respect to t yields the time t_n^* of the maximum nucleation ($\approx$ termination of nucleation). CSD at the termination of nucleation is obtained as $F(R,t_n^*)$. In the case where observed CSD has gone through the subsequent growth, we have to substitute it into a general solution obtained by the Eulerian description using the growth law and calculate the case where the term derived from the boundary condition is 0. To realize the above procedure, we will show certain calculation examples of the case where the nucleation or growth rate as a function of time is given.

9.3.2 Case 1: $J = J_0 \exp(t/t_J)$ and $G = G_1$ = Constant; Exponential Distribution

We consider the case where G is a constant G_1. In this case, Eq. (9.74) is simplified as follows:

$$R(t',t) = G_1(t-t') \tag{9.81}$$

and

$$t' = t - \frac{R(t',t)}{G_1} \tag{9.82}$$

is then obtained. Substituting this equation into Eq. (9.73) algebraically gives

$$F(R(t',t),t) = \frac{J(t - \frac{R(t',t)}{G_1})}{G_1} \tag{9.83}$$

We consider the case where the nucleation rate exponentially increases with time t';

$$J(t') = J_0 e^{t'/t_J} \tag{9.84}$$

where the constant t_J related to the nucleation rate is a characteristic time of increase in the degree of supersaturation because of cooling and is related to the cooling rate (Eq. 9.22).

Substituting Eq. (9.82) given as a function of R into t' in this equation gives the size distribution as follows:

$$F(R(t', t), t) = \frac{J_0 \exp\left(\frac{t}{t_J} - \frac{R(t', t))}{G_1 t_J}\right)}{G_1} \tag{9.85}$$

which is an exponential CSD. In this case, the intercept, the slope, and the characteristic size R_1 are as follows:

$$F_0 = \frac{J_0 e^{t/t_J}}{G_1} \tag{9.86}$$

$$R_1 = G_1 t_J (= 1/|\text{slope}|) \tag{9.87}$$

where the characteristic size ズ R_1 is size change corresponding to change in the population density by 1 on a natural logarithmic scale and can be written as follows:

$$R_1 = G_1 \times t_1 \tag{9.88}$$

using the time constant at that time, $t_1/t_J = 1$ ($J = J_0 \times e$ at the time).[2]

Although t increases with time and the intercept also increases, a rough shape of the size distribution is determined at the termination time of nucleation t_n^* and the number density is also determined. To see the complete picture of CSD and the parameter dependence, we examine the manner in which the termination time of nucleation t_n^* depends on parameters. Because the termination of nucleation comes immediately after the nucleation reaches the local maximum, the time of nucleation climax can be considered as the termination time of nucleation. The time of local maximum of nucleation is then determined from the extremum condition for the degree of supersaturation. From Eq. (9.79), an equation of the extremum condition, the condition for nucleation termination will be as follows:

$$\frac{1}{R_X^*} \frac{1}{\phi_0} \frac{d\phi}{dt} - \frac{1}{R_T^*} \frac{1}{T_0} \frac{dT}{dt} = 0 \tag{9.89}$$

where $\phi_0 = C_0 v_c$ is the initial volumetric concentration of the crystallization component and R_X^* and R_T^* are chemical and thermal expressions of the scale of the critical nucleus radius defined by Eqs. (7.21) and (7.18).

The amount of crystals ϕ is calculated by

$$\phi = \alpha_\text{s} \int_0^\infty R^3 F(R, t) dR \tag{9.90}$$

using the size distribution function. Here, α_s is the shape factor and, when the shape is a sphere and the size is the radius, $\alpha_\text{s} = 4\pi/3$. Substituting Eq. (9.85) into the equation above yields

[2] Note that this is different from the maximum particle size in the system, R_max (described later).

$$\phi = \frac{6\alpha_s J_0 e^{t/t_J}}{G_1}(G_1 t_J)^4 \tag{9.91}$$

This equation is substituted into the equation of the extremum condition (Eq. 9.89) and t_n^* is given as it satisfies

$$e^{t_n^*/t_J} = \frac{\phi_n^* G_1}{6\alpha_s J_0}(G_1 t_J)^{-4} \tag{9.92}$$

where ϕ_n^* is an effective crystal volume fraction that determines termination of nucleation and is

$$\phi_n^* = \frac{\phi_0}{\tilde{A}} \frac{R_X^*}{R_T^*} \tag{9.93}$$

in this case.[3] $\tilde{A}$ is given by Eq. (9.22) and is a constant representing the magnitude of the influence of cooling on the increasing rate of the nucleation rate.

Rewriting Eq. (9.85) using ϕ_n^* and the characteristic size R_1 gives

$$F(R) = F(R, t^*) = \frac{\phi_n^*}{6\alpha_s}(G_1 t_J)^{-4} e^{-\frac{R}{G_1 t_J}} = \frac{\phi_n^*}{6\alpha} R_1^{-4} e^{-R/R_1} \tag{9.94}$$

where

$$R_1 = G_1 t_J \tag{9.95}$$

Moreover, the intercept and the slope will be

$$F_0 = \frac{\phi_n^*}{6\alpha}(G_1 t_J)^{-4} \tag{9.96}$$

$$|\text{slope}| = \frac{1}{G_1 t_J} \tag{9.97}$$

This is shown in Fig. 9.5. This CSD is uniquely determined by $G_1 t_J$, the product of the exponential part of nucleation, which is an inverse estimate of the cooling rate, t_J, and the growth rate G_1. Moreover, as this ratio is large, the slope becomes steep and the value of the intercept becomes large. However, the maximum particle size[4] that can be observed increases with decrease in this ratio and corresponds to increase in the effective growth rate and the maximum particle size is $R_{\max} = G_1 t_n^*$.

In this case, the crystal number density N is

$$N = \int_0^\infty F(R)dR = \frac{\phi_n^*}{6\alpha_s}(G_1 t_J)^{-3} = \frac{\phi_n^*}{6\alpha} R_1^{-3} \tag{9.98}$$

ϕ_n^* is a constant independent of cooling rate, which comes into play through t_J, i.e., R_1 (Eq. 9.22). As found from this, the crystal number density is proportional to the third power of the cooling rate $t_J^{-1} \approx |\dot{T}|$. However, it does not agree with the dependence on the number density observed in

[3]This is a condition by which increase in the degree of supersaturation due to cooling and decrease in the degree of supersaturation due to decrease in the concentration of components in liquid with progress of crystallization are brought into balance. Although a similar result can be obtained by a simple condition ($\phi = \phi_n^*$) that nucleation terminates when the concentration of the crystallization component in liquid decreases by more than a certain critical value, whether or not a critical concentration depends on the cooling rate or other parameters at that time is unclear at this point in time. In other words, the dependence of ϕ_n^* on parameters will be clarified when the extremum condition is used (Eq. 9.93).

[4]That is, a particle size the population density of which exceeds a certain value.

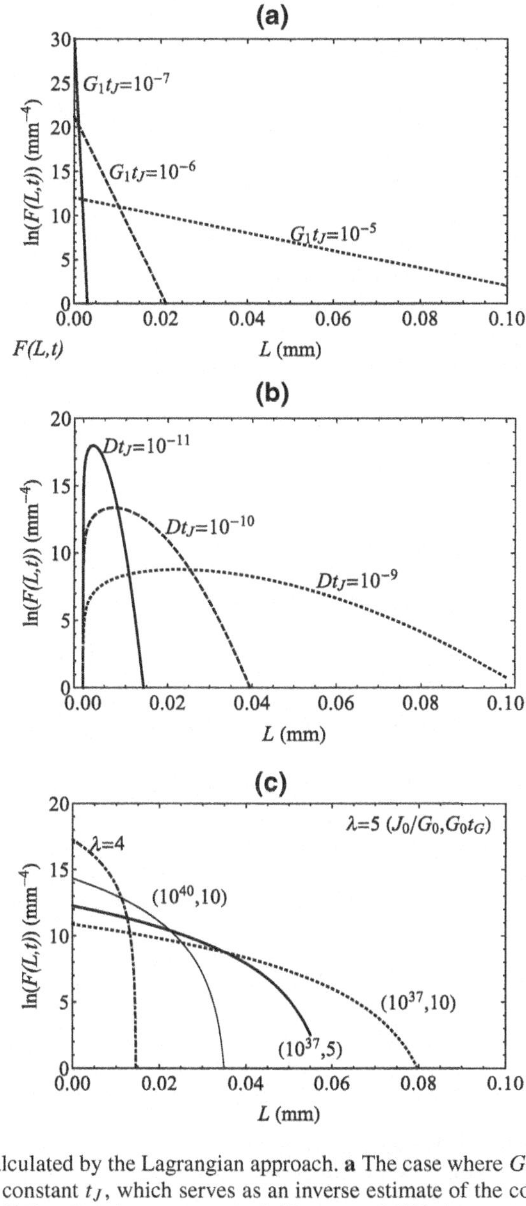

Fig. 9.5 Examples of CSD calculated by the Lagrangian approach. **a** The case where G is a constant G_1 (Eq. 9.94). As $G_1 t_J$, the product of the time constant t_J, which serves as an inverse estimate of the cooling rate, and the growth rate G_1 (= constant), becomes small, the slope becomes steep. **b** The case of diffusion growth $G = (D(t - t'))^{1/2}$. As DtJ, the product of t_J and diffusivity D, becomes small, the slope becomes large. **c** CSD calculated from Eq. (9.133). The cases of $t_G > 0$. λ, J_0/G_0, and $G_0 t_G$ are varied. The dotted line is the reference case ($\lambda = t_G/t_J = 5$, $J_0/G_0 = 10^{37}$, $G_0 t_G = 10$). The thick solid line is the case where the value of $G_0 t_G$ is half the reference. The thin solid line is the case where the value of J_0/G_0 is 10^{-3} times the reference. The chain line is the case of $\lambda = 4$. $\phi_n^* = 0.1$ is assumed

laboratory experiments or natural dikes. The case that agrees with it is the case of diffusion-limited growth or the case where the growth rate is given exponentially under a special condition, as seen below. In this case, the product of the intercept of the CSD $F_0 (= F(0))$ and the mean particle size defined by

$$R_m = \frac{\int_0^\infty RF(R)dR}{N} = G_1 t_J = R_1 \tag{9.99}$$

gives the crystal number density as follows:

$$N = R_{\mathrm{m}} F_0 = \frac{F_0}{|\mathrm{slope}|} \tag{9.100}$$

The mean particle size is the inverse of the slope.

9.3.3 Case 2: $J = J_0 \exp(t/t_J)$ and Diffusion Growth

In the case of diffusion growth,

$$R(t', t) = \sqrt{D(t - t')} \tag{9.101}$$

holds and

$$t' = t - \frac{R^2(t', t)}{D} \tag{9.102}$$

$$\delta R = \frac{R(t', t)\delta t'}{2(t - t')} = \frac{D\delta t'}{2R(t', t)} \tag{9.103}$$

hold. Therefore,

$$F(R, t) = \frac{2RJ(t')}{D} = \frac{2J_0 e^{t/t_J}}{D} R e^{-\frac{R^2}{Dt_J}} \tag{9.104}$$

is obtained. In the same manner as the previous section, we will obtain the termination time of nucleation t_{n}^*. The amount of crystals ϕ is then written as follows:

$$\phi = \frac{4\pi}{3} \int_0^\infty R^3 F(R, t) dR = \frac{6\alpha_{\mathrm{s}} J_0 e^{t/t_J}}{3D} \int_0^\infty R^4 e^{-\frac{R^2}{Dt_J}} dR \tag{9.105}$$

and the last integral can be integrated to be $3\pi^{1/2}(Dt_J)^{5/2}/8$. Similar to the previous section, using the extremum condition yields

$$e^{t_{\mathrm{n}}^*/t_J} = \frac{\phi^* D}{\pi^{3/2} J_0} (Dt_J)^{-\frac{5}{2}} \tag{9.106}$$

and ϕ_{n}^* in this case is equal to what is given by Eq. (9.93).

Therefore, the size distribution $F(R)$ is obtained as follows:

$$F(R) = \frac{2\phi_{\mathrm{n}}^*}{\pi^{3/2}} \left(\frac{a}{D}\right)^{\frac{5}{2}} R e^{-\frac{R^2}{Dt_J}} = \frac{8\phi_{\mathrm{n}}^*}{3\alpha\pi^{1/2}} R_2^{-4} \left(\frac{R}{R_2}\right) e^{-(R/R_2)^2} \tag{9.107}$$

Here, the characteristic size R_2 is defined as follows:

$$R_2 = (Dt_J)^{1/2} \tag{9.108}$$

he values of the parameter Dt_J are shown in Fig. 9.6.

This CSD is shown in Fig. 9.5. A feature of the case of diffusion growth is that even if snapshots are taken when nucleation continues on, the CSD decreases when it approaches the particle size 0. The

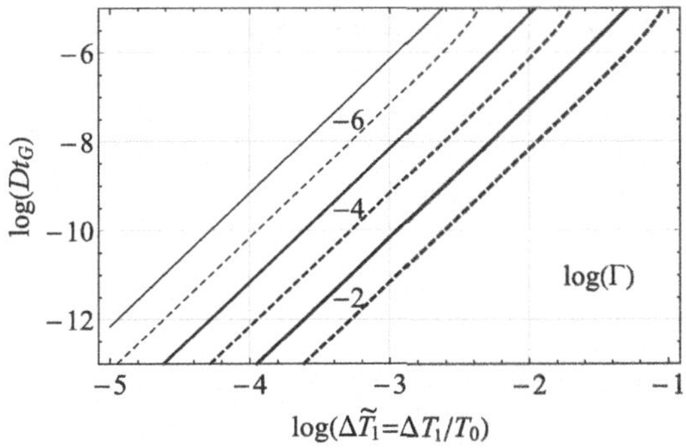

Fig. 9.6 The relationship between the parameter Dt_J and the degree of supercooling ($\Delta\tilde{T}_1$) giving nucleation. The solid lines represent diffusivity $D = 10^{-11}$ (m^2/s) and the broken lines represent $D = 10^{-12}$ (m^2/s). T_0= 1400 (K) and $|\dot{T}| = 10^{-5}$ (K/s) are used in all cases. The numbers next to the lines represent the values of $\log \Gamma$

reason for this behavior is that because the growth rate is infinite around the particle size 0 and δR in Eq. (9.69) becomes infinite, the number density will be diluted.[5]

Similarly, the crystal number density is obtained as follows:

$$N = \frac{4\phi_{\rm n}^*}{3\alpha\pi^{1/2}} (Dt_J)^{\frac{3}{2}} = \frac{4\phi_{\rm n}^*}{3\alpha\pi^{1/2}} R_2^{-3} \tag{9.109}$$

Thus, in diffusion-limited growth, the crystal number density increases in proportion to the 3/2 power of t_J^{-1} ($\propto |\dot{T}|$) (Eq. 9.22), that is, the cooling rate.

In this case, the mean particle size will be

$$R_{\rm m} = \frac{1}{2} (\pi Dt_J)^{\frac{1}{2}} \tag{9.110}$$

In addition to the mean particle size, a particle size that gives the peak of the CSD, $R_{\rm p}$, characterizes the shape of the CSD.

$$R_{\rm p} = \left(\frac{Dt_J}{2}\right)^{\frac{1}{2}} \tag{9.111}$$

Between $R_{\rm m}$ and $R_{\rm p}$ there is a relationship of $R_{\rm m} = (\pi/2)^{1/2} R_{\rm p} \approx 1.25 R_{\rm p}$. The product of the characteristic particle size scale and the peak value $F(R_{\rm p})$ gives the crystal number density. The peak CSD is as follows:

$$F(R_{\rm p}) = \frac{8\phi_{\rm n}^*}{3\alpha(2\pi e)^{1/2}} (Dt_J)^{-2} \tag{9.112}$$

and a relationship

$$N = e^{1/2} R_{\rm p} \cdot F(R_{\rm p}) \approx 1.6 R_{\rm p} \cdot F(R_{\rm p}) \approx 1.3 R_{\rm m} \cdot F(R_{\rm p}) \tag{9.113}$$

holds.

[5]Moreover, the decrease of CSD on the smaller side of particle size occurs spontaneously with progress of crystallization. Its causes include (1) decrease in the nucleation rate associated with decrease in the degree of supersaturation and (2) decrease in the effective nucleation area due to increase in the amount of crystals.

9.3.4 Case 3: $J = J_0 \exp(t/t_J)$ and $G = G_0 \exp(t/t_G)$

Let us consider another case in which the growth rate exponentially increases with time.[6]

$$G(t') = G_0 e^{t'/t_G} \tag{9.114}$$

where the parameter t_G is the dependence of the crystal growth rate on the degree of supercooling (see Box in this section).

In this case, if $\delta t'$ is sufficiently short, $\delta R(t') = R(t', t) - R(t' + \delta t', t)$ will become

$$\delta R(t') = \int_{t'}^{t'+\delta t'} G_0 e^{t''/t_G} dt'' = G_0 t_G e^{t'/t_G} \left(e^{\delta t'/t_G} - 1\right) \approx G_0 e^{t'/t_G} \delta t' \tag{9.115}$$

and the size distribution function can be obtained by calculating

$$F(R(t', t), t) = \frac{\delta t' J(t')}{\delta R(t')} = \frac{J(t')}{G_0 e^{t'/t_G}} = \frac{J(t')}{G(t')} = \frac{J_0 e^{t'/t_G}}{G_0 e^{t'/t_G}} \tag{9.116}$$

In the following, we separately calculate the cases where $t_G > 0$ and $t_G < 0$.[7]

The Case of $t_G > 0$

The size $R(t', t)$ is obtained by integrating the growth law as follows:

$$R(t', t) = G_0 t_G \left(e^{t/t_G} - e^{t'/t_G}\right) \tag{9.119}$$

Transforming this equation to obtain t'

$$t' = t + t_G \ln\left(1 - \frac{R(t', t)}{G_0 t_G e^{t/t_G}}\right) \tag{9.120}$$

and substituting this into Eq. (9.116) yields the size distribution function

$$F(R, t) = \frac{J_0}{G_0} e^{(t_J^{-1} - t_G^{-1})t} \left(1 - \frac{R}{G_0 t_G e^{t/t_G}}\right)^{\frac{t_G}{t_J} - 1} \tag{9.121}$$

---

[6]Note that to express unified time change of the degree of supercooling of the entire system, the part of et' /tG in this equation should not be $e^{(t'-t'')/t_G}$.

[7]Spohn et al. (1988) linearly approximated Eq. (7.103). In that case, $G = G_0(1 + t/t_G)$ is obtained and the size distribution is obtained as follows:

$$F(R) = \frac{J_0 \exp\left(-\frac{t_G}{t_J}\left(1 + \sqrt{1 + \frac{2}{G_0 t_G}(R_{\max} - R)}\right)\right)}{G_0 \sqrt{1 + \frac{2}{G_0 t_G}(R_{\max} - R)}} \tag{9.117}$$

$$R_{\max} = G_0 t \left(1 + \frac{1}{2}\frac{t}{t_G}\right) \tag{9.118}$$

However, the integrated result of the volume fraction is complicated and it is analytically difficult to evaluate the termination time of nucleation t^*.

A time constant of the exponentially increasing growth rate

Assuming that the crystal growth follows the crystal growth law of reaction-limited growth, Eq. (7.103), a time constant t_G is obtained by differentiating it considering temperature as a function of time,

$$t_G^{-1} = \left(-\frac{\tilde{q}}{\tilde{T}_1^2} + \frac{\varepsilon \Delta \tilde{h}}{1-\varepsilon}\right)\frac{|\dot{T}|}{T_0} = \tilde{B}\frac{|\dot{T}|}{T_0} = \frac{\tilde{B}}{t_{th}} \tag{9.122}$$

$$\Delta \tilde{h} = \frac{\Delta h}{k_B T_0} \tag{9.123}$$

$$\varepsilon = e^{\Delta \tilde{h}(1-1/\tilde{T}_1)} \tag{9.124}$$

[a].

Here, $\tilde{B}$ is the inside of the parentheses and a constant representing the influence of cooling on increase in the growth rate. The figure below is a plot of $\tilde{B}$ against the dimensionless degree of supercooling $\Delta\tilde{T}_1 = \Delta T_1/T_0$ for different latent heat $\Delta\tilde{h}$. This figure reveals that $\tilde{B}$ shows dependence on $\Delta\tilde{T}_1^{-1}$ regardless of the latent heat.(continued to the next page)

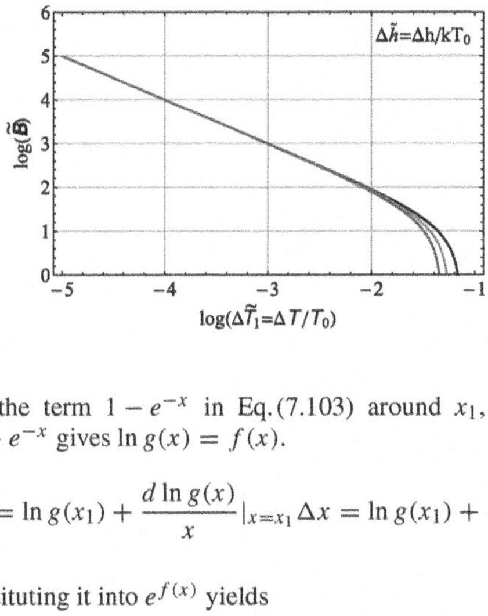

[a]We consider that we expand the term $1-e^{-x}$ in Eq. (7.103) around x_1, i.e., $x = x_1 + \Delta x$, and express $e^{f(x \approx x_1 + \Delta x)}$. Letting $g(x) = 1 - e^{-x}$ gives $\ln g(x) = f(x)$.

$$\ln g(x) = \ln g(x_1) + \frac{d \ln g(x)}{x}|_{x=x_1} \Delta x = \ln g(x_1) + \frac{e^{-x_1}\Delta x}{1-e^{-x_1}} \tag{9.125}$$

Letting this equal $f(x)$ and substituting it into $e^{f(x)}$ yields

$$e^{f(x)} = e^{\ln g(x_1) + \frac{e^{-x_1}\Delta x}{1-e^{-x_1}}} = g(x_1) e^{\frac{e^{-x_1}}{1-e^{-x_1}}\Delta x} \tag{9.126}$$

That is, b corresponds to $e^{-x_1}/(1-e^{-x_1})$.

In the same manner as the previous section and the section before it, we obtain the termination time of nucleation t_n^*. The amount of crystals as a function of time is given as follows:

$$\phi = \alpha_s \int_0^\infty R^3 F(R,t) dR = \frac{\alpha J_0 e^{(t_J^{-1} - t_G^{-1})t}}{G_0} \int_0^{R_3} R^3 \left[1 - \frac{R}{R_3}\right]^{\frac{t_G}{t_J}-1} dR \tag{9.127}$$

Note that R_3 is defined as $G_0 t_G \exp(t/t_G)$ and is a function of time here. The parameter λ is then defined as follows:

$$\lambda = \frac{t_G}{t_J} \tag{9.128}$$

The integral part can be calculated to be

$$\int_0^{R_3} R^3 \left[1 - \frac{R}{R_3}\right]^{\lambda-1} dR = \frac{6(R_3)^4\Gamma(\lambda)}{\Gamma(\lambda+4)} \tag{9.129}$$

From the extremum condition for the maximum nucleation,

$$e^{t_n^*/t_G} = \left[\left(\frac{\phi_3^*}{6\alpha}\right)\left(\frac{J_0}{G_0}\right)^{-1}(G_0 t_G)^{-4}\right]^{\frac{1}{\lambda+3}} \tag{9.130}$$

Here, ϕ_3^* is

$$\phi_3^* = \frac{\phi_0 \Delta\tilde{h}}{kT_0 B(\lambda+3)} \frac{\Gamma(\lambda+4)}{\Gamma(\lambda)} \tag{9.131}$$

This is what is obtained by multiplying the volume fraction ϕ_n^* corresponding to the maximum nucleation rate by $\Gamma(\lambda+4)/\Gamma(\lambda)$. At this point of time, the characteristic size R_3 is determined to be

$$R_3 = \left[\left(\frac{6\alpha}{\phi_3^*}\right)\left(\frac{J_0}{G_0}\right)(G_0 t_G)^{(1-\lambda)}\right]^{-\frac{1}{\lambda+3}} \tag{9.132}$$

By using this, the size distribution function is obtained as follows:

$$F(R) = \frac{\phi_3^*}{6\alpha} R_3^{-4}\left(1 - \frac{R}{R_3}\right)^{\lambda-1} \tag{9.133}$$

(Figure 9.5c).

The Case of $t_G < 0$

In this case, the growth rate decreases with time, and the size $R(t', t)$ obtained by integrating the growth law as follows:

$$R(t', t) = G_0|t_G|\left(e^{t'/t_G} - e^{t/t_G}\right) \tag{9.134}$$

In the same manner as the case of $t_G > 0$,

$$t' = t + t_G \ln\left(1 + \frac{R(t', t)}{G_0|t_G|e^{t/t_G}}\right) \tag{9.135}$$

$$F(R, t) = \frac{J_0}{G_0} e^{(t_J^{-1} - t_G^{-1})t}\left(1 + \frac{R}{G_0|t_G|e^{t/t_G}}\right)^{\frac{t_G}{t_J}-1} \tag{9.136}$$

$$\begin{aligned}\phi &= \alpha_s \int_0^\infty R^3 F(R, t) dR \\ &= \frac{\alpha_s J_0 e^{(t_J^{-1} - t_G^{-1})t}}{G_0}\left(G_0|t_G|e^{t/t_G}\right)^4 \int_0^\infty x^3(1+x)^{\lambda-1} dx\end{aligned} \tag{9.137}$$

The integral part can be calculated to be,[8]

$$\phi = \frac{\alpha_s J_0}{G_0} (G_0|b|)^4 e^{(t_J^{-1}+3t_G^{-1})t} f_\lambda \tag{9.140}$$

From the extremum condition for the maximum nucleation, $e^{t_n^*/t_G}$ is obtained as the same expression as the case of $t_G > 0$ and ϕ_3^* is obtained as follows:

$$\phi_3^* = \frac{\phi_0 \Delta \tilde{h}}{kT_0 B(\lambda + 3)} f_\lambda^{-1} \tag{9.141}$$

When this ϕ_3^* is used, the characteristic size R_3 in the case of $t_G < 0$ will be the same expression as the case of $t_G > 0$. The size distribution function is then obtained as follows:

$$F(R) = \frac{\phi_3^*}{6\alpha_s} R_3^{-4} \left(1 + \frac{R}{R_3}\right)^{\lambda-1} \tag{9.142}$$

On a semilogarithmic scale, CSD (that is, ln $F(R)$) will be convex upward for $t_G/t_J < 1$ and convex downward for $t_G/t_J > 1$.

The number density is then obtained as follows:

$$N = \frac{1}{\lambda} \left(\frac{\phi_3^*}{6\alpha_s}\right) R_3^{-3} \tag{9.143}$$

The influence of the cooling rate is exerted through t_J and t_G. In λ, the cooling rate does not have any influence on λ because the denominator and the numerator are separately proportional to the cooling rate and their influence is cancelled. Ultimately, the influence of the cooling rate on the number density is exerted only through t_G in R_3 (Eq. 9.132). Extracting the t_G dependence of the number density yields

$$N \propto t_G^{-\frac{3(\lambda-1)}{\lambda+3}} \propto |\dot{T}|^\xi \tag{9.144}$$

$$\xi = \frac{3(\lambda-1)}{\lambda+3} = \frac{3(1 - t_J/t_G)}{1 + 3t_J/t_G} \tag{9.145}$$

The exponential part of Equation (9.144) is shown in Fig. 9.7 as the cooling rate exponent ξ. The cooling rate exponent estimated from dike cooling in nature and laboratory experiments is $0 \sim 1.5$, which correspond to $1\sim 5$ for λ and $1\sim 0.2$ for t_J/t_G.[9] Moreover, for $t_J/t_G = 0$, i.e., crystallization around the temperature of the maximum growth rate, the cooling rate exponent is 3 and the distribution function corresponds to exponential CSD in which $G =$ constant. Toramaru (2001) deals with this case as reaction-limited growth and obtains $\xi = 3$.

8

$$\int_0^\infty x^3(1+x)^{\lambda-1}dx = 6f_\lambda \tag{9.138}$$

$$f_\lambda = \frac{1}{\lambda(\lambda+1)(\lambda+2)(\lambda+3)} \tag{9.139}$$

are obtained.

9 At this time, Γ_1/Γ_2 monotonously changes from 1/24 to 1/1680.

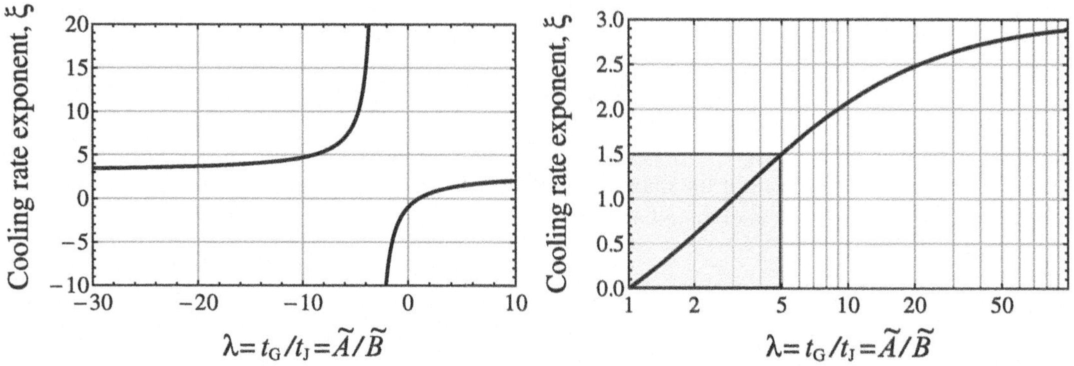

Fig. 9.7 The relationship between the cooling rate exponent calculated from Eq. (9.143) and λ. (Left) Range where λ is from -10 to 10. The cooling rate exponent diverges to $\pm\infty$ at $\lambda = -3$. (Right) The relationship between the cooling rate exponent and λ in the range where λ is from 0 to 3. A realistic range of λ is shown in the shaded square ($1 < \lambda < 5$)

Moreover, in this case, a simple relationship holds among the intercept of the CSD $F(0)$, the mean particle size R_{m}, and the crystal number density N.

$$F(0) = \left(\frac{\phi_3^*}{6\alpha_s}\right) R_3^{-4} \propto |\dot{T}|^{\frac{4}{3}\xi} \tag{9.146}$$

$$R_{\mathrm{m}} = \frac{1}{(\lambda+1)} R_3 \propto |\dot{T}|^{-\frac{1}{3}\xi} \tag{9.147}$$

$$N = \left(1 + \frac{1}{\lambda}\right) R_{\mathrm{m}} \cdot F(0) \tag{9.148}$$

Furthermore, the maximum particle size $R_{\max}$ and the mean particle size R_{m} are mutually related through $R_{\max} = (\lambda + 1)R_{\mathrm{m}}$. The slope of the CSD at the intercept $F'(0) = dF(R)/dR|_{R=0}$ is

$$F'(0) = -(\lambda - 1)R_3^{-5} \propto |\dot{T}|^{\frac{5}{3}\xi} = -(\lambda - 1)\frac{F(0)}{R_{\max}} \tag{9.149}$$

which is the quotient of the value of the intercept by the scale of the particle size.

9.3.5 Case 4: The Nucleation Rate and the Crystallization Rate Are Constant; Exponential Distribution

As an interesting case, let us consider the case where the nucleation rate and the crystallization rate are constant. The crystallization rate $\dot{\phi} = d\phi/dt$ is connected with the crystal growth rate $G = dR/dt$ through the size distribution $F(R(t', t), t)$ as follows:

$$\dot{\phi} = 3\alpha_s G(t) \int_0^{R(0,t)} R(t', t)^2 F(R(t', t), t) dR(t', t) \tag{9.150}$$

where $R(t', t)$ is the size of crystals at time t, which have nucleated at time t'. To obtain the size distribution, we have to obtain a $G(t)$ that gives a constant $\dot{\phi}$ regardless of time. However, because the size distribution function in question is included in the integral part, we have to simultaneously solve

$G(t)$ and the integral equation.[10] Here, we will not directly solve them but give constraints to $\bar{n}$ on the assumption that the growth rate has the dependence $G(t) \propto t^{\bar{n}}$.

Avoiding the integral above, we start from the moment equations for the size distribution, mentioned in the previous chapter. The moment equations not containing Ostwald ripening are given by

$$\frac{dM_0}{dt} = J_4 \tag{9.152}$$

$$\frac{dM_1}{dt} = G \cdot M_0 \tag{9.153}$$

$$\frac{dM_2}{dt} = 2G \cdot M_1 \tag{9.154}$$

$$\frac{dM_3}{dt} = 3G \cdot M_2 \tag{9.155}$$

Because the crystallization rate is constant, $\alpha_s \cdot dM_3/dt$ is constant. Furthermore, because the nucleation rate is a constant value J_1, $M_0 = J_1 t$. Assuming that the growth rate is $G(t) = ct^{\bar{n}}$ (c is a constant), substituting these into the moment equations and integrating in succession yields

$$\frac{dM_3}{dt} = \frac{6J_1c^3}{(\bar{n}+2)(2\bar{n}+3)} t^{3\bar{n}+3} \tag{9.156}$$

In order for this to not depend on time, $\bar{n}$ must be - 1. We then obtain the size distribution in the case where the growth rate is inversely proportional to time.

$$G(t) = \frac{dR}{dt} = \frac{R_4}{t} \tag{9.157}$$

where R_4 is a characteristic crystal size and will be related to the nucleation rate J_4 and the crystallization rate $\dot{\phi}$ later. Integrating Eq. (9.157) yields the following:

$$R(t', t) - R(t', t') = R_4 \ln \frac{t}{t'} \tag{9.158}$$

Assuming that $R(t', t') = 0$, the relationship between the nucleation time t' and the crystal size R at time t is as follows:

$$t' = t \cdot e^{-R/R_4} \tag{9.159}$$

From Eq. (9.73),

[10] Assuming $G(t) = dR/dt = G_0 t^{\bar{n}}$, integrating it from $t = t'$ to t, and letting $R(t') = 0$ yields $t' = (t - R/G_0)^{1/(\bar{n}+1)}$. Moreover, $F(R(t', t), t) = J_4/G(t')$ holds, and substituting this into the integral in question and calculating it gives

$$\begin{aligned} \dot{\phi} &= 3\alpha_s J_4 t^{\bar{n}} \int_0^{R(0,t)} R(0,t')^2 \left[t^{\bar{n}+1} - \frac{R(0,t')}{G_0} \right]^{-\frac{\bar{n}}{\bar{n}+1}} dR(0,t') \\ &= \frac{6\alpha_s J_4 \left[G_0(1+\bar{n}) t^{1+\bar{n}} \right]^3}{(\bar{n}+2)(2\bar{n}+3)} \end{aligned} \tag{9.151}$$

For this equation to be independent of time, the exponent of t, $3(\bar{n}+1)$, should be 0, i.e., $\bar{n} = -1$. However, at this time, the coefficient $(1+\bar{n})^3$ will be 0. Because when we integrated dR/dt to obtain the relationship among R, t, and t', we assumed $\bar{n} \neq -1$, we have to return to that point and confirm it.

$$F(R) = \frac{J_4}{G(t')} = \frac{J_4 t}{R_4} e^{-\frac{R}{R_4}} \tag{9.160}$$

is then obtained, which is an exponential distribution. Because this model assumes that nucleation occurs at a constant rate, the extremum condition of the degree of supersaturation that determines the termination time of nucleation itself is meaningless.

The volume fraction of crystals ϕ is then determined as follows:

$$\phi = \alpha_s \int_0^\infty R^3 F(R) dR = 6\alpha_s J_4 t R_4^3 \tag{9.161}$$

From this, $\dot{\phi} = 6\alpha_s J_4 R_4^3$ is obtained. Therefore,

$$R_4 = \left(\frac{\dot{\phi}}{6\alpha_s J_4} \right)^{1/3} \tag{9.162}$$

holds.

If we consider that the amount of crystals observed ϕ_{obs} and the crystal number density N correspond to the termination time of nucleation $t = t_n^*$,

$$N = J_4 t_n^* = \frac{\phi_{obs}}{6\alpha_s R_4^3} \tag{9.163}$$

is obtained. Although there seems to be a self-evident proportional relation between the crystal number density and the amount of crystals, because the nucleation rate J_4 has dependence on the cooling rate, the proportional relations are changed if the cooling conditions differ, even if the magma has the same composition.

The size distribution is then obtained as follows:

$$F(R) = \frac{\phi_{obs}}{6\alpha_s} R_4^{-4} e^{-R/R_4} \tag{9.164}$$

In this model, the intercept and slope are connected with the crystallization rate $\dot{\phi}$ and nucleation rate J_4 through the characteristic size scale R_4.

$$F_0 = \frac{\phi_{obs}}{6\alpha_s R_4^4} = \frac{\phi_{obs}}{6\alpha_s} \left(\frac{6\alpha_s J_4}{\dot{\phi}} \right)^{4/3} \tag{9.165}$$

$$|\text{slope}| = \frac{1}{R_4} = \left(\frac{6\alpha_s J_4}{\dot{\phi}} \right)^{1/3} \tag{9.166}$$

Moroever, in this case,

$$F_0 = \frac{\Phi_{obs}}{6\alpha} (|\text{slope}|)^4 \tag{9.167}$$

which is characteristic of an exponential distribution, holds.

As the crystallization rate becomes small and the nucleation rate becomes large, the characteristic size scale R_4 becomes small, the slope becomes steep, and the intercept becomes large. The crystallization rate $\dot{\phi}$ can then be related to the change in temperature when the amount of crystals is determined

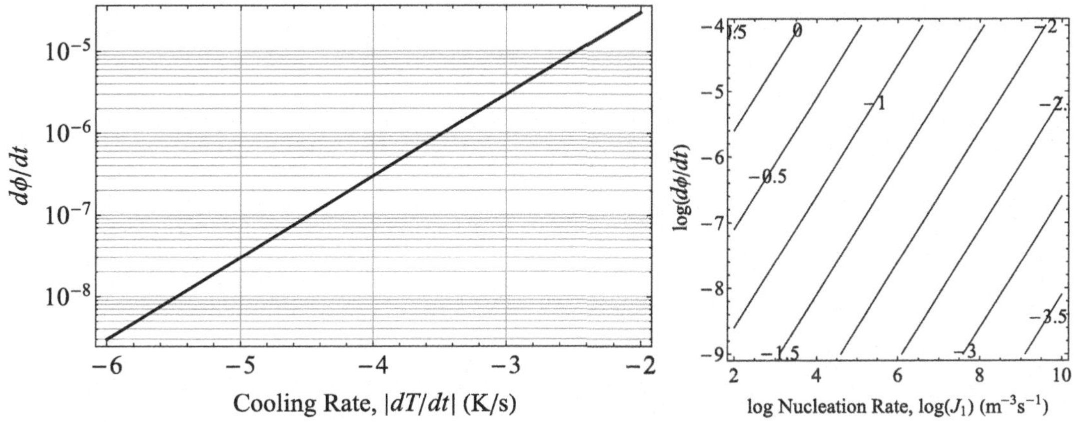

Fig. 9.8 Values of parameters in the model in which the crystallization rate and the nucleation rate are constant. (Left) The crystallization rate as a function of the cooling rate on the assumption that the system is in equilibrium. (Right) The contour plot of the logarithm of the characteristic size as a function of the nucleation rate and the crystallization rate, $\log R_4$; $\alpha_s = 4.2$, equivalent to a sphere, is used for the shape factor

by quasi-equilibrium by which the distance from the liquidus (the degree of supersaturation) is kept constant.

$$\dot{\phi} = \frac{d\phi}{dT}\frac{dT}{dt} = \phi'_T \cdot \left|\frac{dT}{dt}\right| \tag{9.168}$$

where ϕ'_T is $-d\phi/dT$, a thermodynamic factor determined by the path in the phase diagram. An approximate value of it is $\phi'_T = \Delta\Phi_{\mathrm{obs}}/(T_L - T_S) = 3 \times 10^{-4} \sim 5 \times 10^{-3}$. The relationship between the cooling rate and crystallization rate is shown in the left chart of Fig. 9.8. Moreover, the right chart shows the characteristic size R_4 as a function of realistic values of the nucleation rate and crystallization rate. The characteristic size takes on values from 1 mm to 1 μm.

Evaluation of this model should be conducted by comparing with the case where $J = J_0 \exp(t/t_J)$ and G are constant, which yields an exponential CSD.

9.4 Comparison with Experiments

9.4.1 Experimental Studies on CSD

There have been fewer experimental studies on CSD than petrographic studies on CSD. Burkhard (2002) determined the time dependence of G from CSD by re-heating basaltic glass to crystallize (anneal) it. This is based on the author's idea of explaining crystallization on the surface of natural lava from an eruption occurring in 1994 in Hawaii by re-heating. Moreover, this study determined the dependence of G on cooling rate on the assumption of heat conduction. Although this may be examined using the model for the forward problem, time change of J was estimated from CSD in this study. This is probably the first study that experimentally connects CSD and G and J under the condition of $T =$ constant.

Zieg and Lofgren (2006) observed the development of CSDs in cooling experiments. However, because the final temperatures differ because samples were cooled at a constant rate, the dependence on the cooling rate cannot be accurately extracted. Pupier (2008) conducted systematic experiments using initial temperatures, duration of the initial temperatures, and cooling rate as parameters, calculated

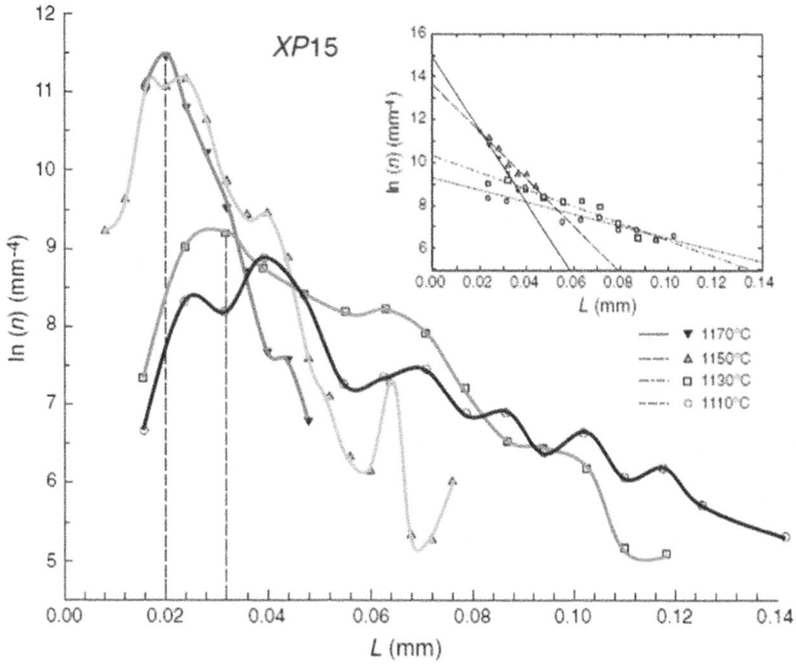

Fig. 9.9 Experimental CSDs with different quenching temperatures by Pupier (2008). The cooling rate is 1 °C/h, the annealing temperature is 1190 °C, and the annealing time is 68 h. Adapted by permission from Springer Nature [Pupier (2008)], COPYRIGHT (2008)

CSDs on the assumption that the nucleation rate exponentially increases with time, and estimated rate constants of the nucleation rate. Schiavi et al. (2009) was the first study experimentally examine the relationship between CSD and nucleation and growth by in-situ observation and claimed that change of the shape of CSDs is attributed to Ostwald ripening and coalescence. Brugger and Hammer (2010) performed decompression-induced crystallization experiments to compare the direct measurements of J and G obtained by the experiments with those obtained from CSDs. Iezzi et al. (2011) quantitatively showed difference in CSDs according to cooling rates. Ni et al. (2014) compared the growth rate of crystals and time change of the shape of CSDs by in-situ observation and concluded that the intercept and the slope of CSDs are heavily influenced by secondary modification after nucleation.

Here, we introduce the experimental results of Pupier (2008), which include conditions that are easy to compare with theory. Using basaltic melt, they conducted experiments in which (1) the cooling rate (1 °C/h), (2) the quenching temperature, (3) annealing temperature (1190 °C) higher than the liquidus (about 1175 °C) and annealing time (9 h), and (4) annealing time (0 s) at the annealing temperature were used as parameters. Values in parentheses are standard values used as constant values when other parameters were changed.

Figure 9.9 shows CSDs obtained in an experiment in which only the quenching temperature was used as a parameter and others were constant. This figure revealed that the slope of the intercept of the CSDs becomes gentle as the quenching temperature becomes low. Pupier et al. stated that this is the effect of Ostwald ripening; moreover, the value of the intercept decreases. This means decrease in the number of crystals. Based on these results, it is possible that the nucleation of crystals has already terminated at 1170 °C.

Among the intercept F_0, the nucleation rate J, and the nucleus growth rate G, there is a relationship of $J = F_0 G$ (Eq. 9.6). Moreover, if appropriate crystallization (nucleation) time Δt is used, the slope is expressed as $1/G\Delta t$. Therefore, the nucleation rate can be written as follows:

$$J = \frac{F_0}{|\text{slope}|\Delta t} \tag{9.169}$$

(see Cashman and Marsh 1988 and Sect. 9.1.2). They estimated the crystallization time from the cooling rate, and then estimated the nucleation rate using this equation. Consequently, an average nucleation rate was estimated at $4.83 \times 10^8\,\mathrm{m^{-3}\,s^{-1}}$ in this case. They calculated the mean value of the nucleation rate to be $2.9 \times 10^7\,\mathrm{m^{-3}\,s^{-1}}$ by dividing the number of crystals by crystallization time. Moreover, assuming that time change of the nucleation rate is $J = J_0 \exp(t/t_J)$, they numerically estimated unknown quantities (the coefficient J_0, the time constant t_J, and the growth rate G) from known quantities (the intercept and the slope of the CSD and the crystal fraction). In this case, the mean nucleation rate was estimated at $3.11 \times 10^8\,\mathrm{m^{-3}\,s^{-1}}$. They argued that the reason why the method estimated from the number density yields a smaller value is that small crystals were not completely detected.

Figure 9.10 shows some of their experimental results in which the cooling rate was varied. Focusing on the result of the highest quenching temperature because nucleation completes around 1170 °C, we find that as the cooling rate increases, the intercept increases, and the slope increases. This relationship is shown in Fig. 9.11. Then, let us apply the theoretical consideration to this experimental result.

9.4.2 Application of a Closed-System CSD to Experiments

Summary of CSD Model

The four examples (hereinafter referred to as models) dealt with by the Lagrangian description in Sect. 9.3 are classified in terms of parameters as follows:

Model 1 **(Equation** (9.94)) CSD takes an exponential form and only the parameter $G_1 t_J$ determines the intercept and slope, which are characteristics of CSD (Eqs. 9.96 and 9.97). t_J in the parameter is a time constant of an exponential increasing rate of nucleation and is inversely proportional to cooling rate: $t_J^{-1} = \tilde{A}|\dot{T}|/T_0$. $\tilde{A}$ is given as a constant related to the interfacial tension (barrier of nucleation) or latent heat of crystallization (Eq. 9.22). G_1 is a free parameter and can be dependent on the cooling rate. For example,

$$G_1 = c_G |\dot{T}|^{\chi_G} \tag{9.170}$$

where c_G and χ_G are constants.

Model 2 **(Equation** (9.107)) This distribution function has a convex-upward shape and only the parameter Dt_J determines the shape of the distribution function. t_J is the same as that of the Model 1 and D is the effective diffusivity of the crystallization component. In this case, the crystal number density is determined by Eq. (9.109).

Model 3 **(Equation** (9.133) **or** (9.142)) This distribution function has a convex-upward shape when $t_G > 0$ and a convex-downward shape when $t_G < 0$. The shape of this distribution function is governed by three parameters, $\lambda(= t_G/t_J)$, J_0/G_0, and $G_0 t_G$, which are determined by the cooling rate and physical properties.

Model 4 **(Equation** (9.164)) For this distribution function, the ratio of the nucleation rate to the crystallization rate, $J_4/\dot{\phi}$, is the only parameter. The crystallization rate $\dot{\phi}$ is determined by the thermodynamic factor and the cooling rate (Eq. 9.168). The nucleation rate can then depend on the cooling rate.

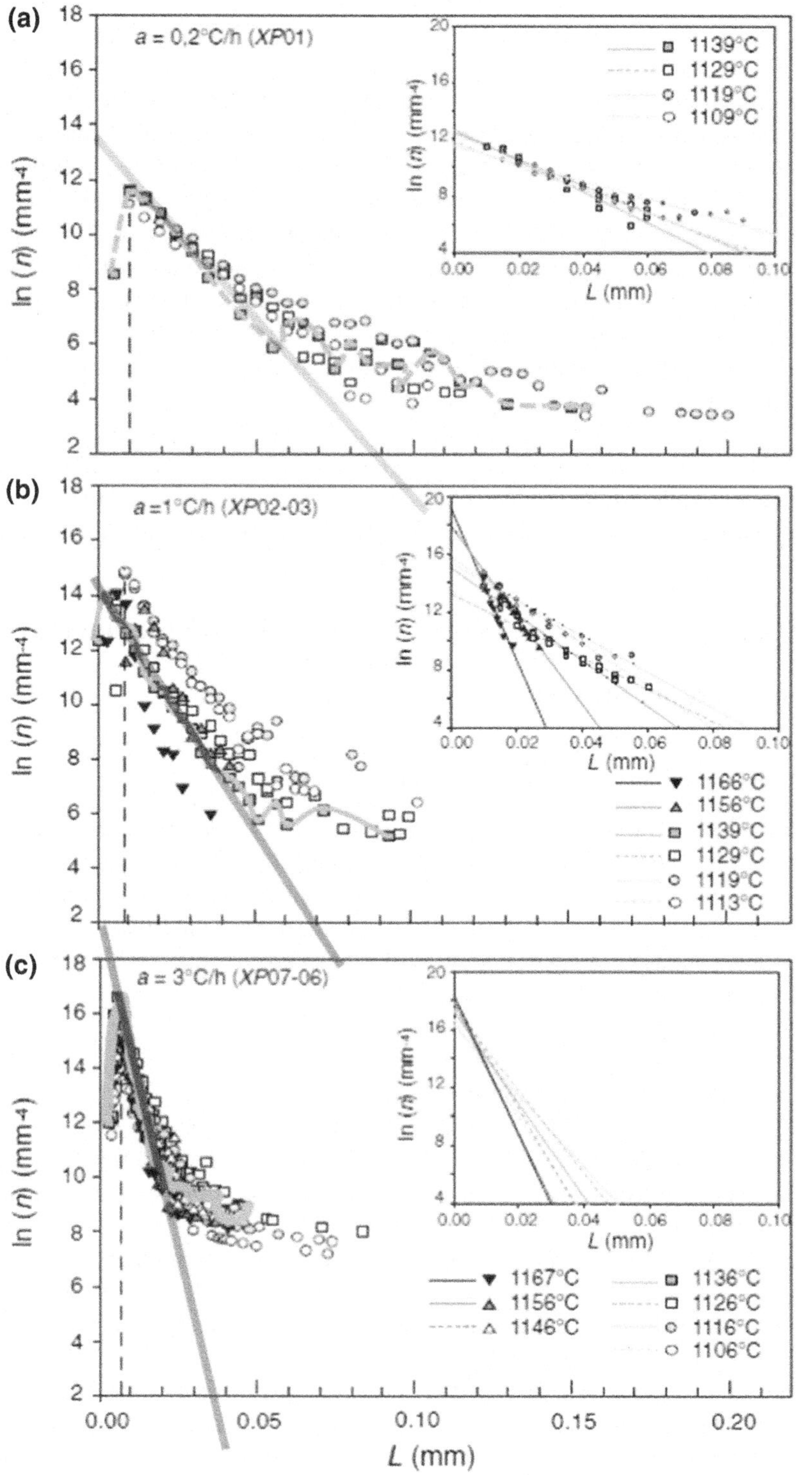

Fig. 9.10 CSDs obtained by experiments in which the cooling rate was varied, conducted by Pupier (2008). CSDs of various quenching temperatures are plotted for one cooling rate. Difference in (**a**), (**b**), and (**c**) is difference in the cooling rate. Adapted by permission from Springer Nature [Pupier (2008)], COPYRIGHT (2008)

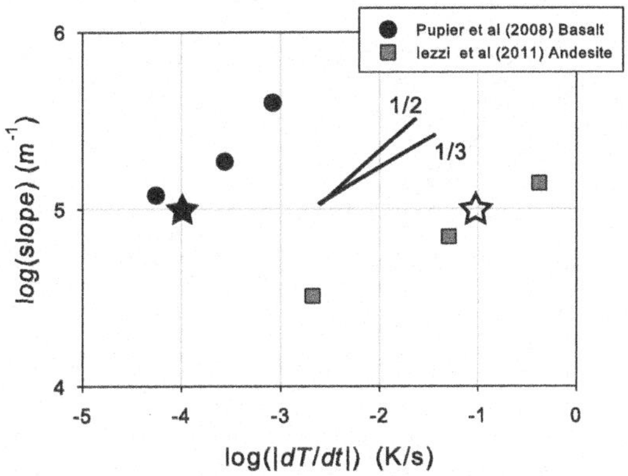

Fig. 9.11 Relationship between the cooling rate and the slope of CSDs in experiments in which the cooling rate was varied. Pupier (2008), Iezzi et al. (2011)

Model 1

In this model, the slope of an exponential CSD is governed by the parameter $G_1 t_J$ and determined as follows:

$$|\text{slope}| = \frac{1}{G_1 t_J} = \frac{A}{T_0 c_G} |\dot{T}|^{1-\chi_G} \tag{9.171}$$

considering the dependence on the cooling rate. In the experimental result of Pupier (2008), the slope increases with the square root of cooling rate. Therefore, $1 - \chi_G = 1/2$, i.e., $\chi_G = 1/2$ is expected.

Moreover, Eq. (9.98) indicates that the crystal number density has dependence of

$$N \propto \left|\dot{T}\right|^{3(1-\chi_G)} \tag{9.172}$$

on the cooling rate. Therefore, the cooling rate exponent of this experiment is expected to be 3/2. This value agrees with the cooling rate exponent on the assumption of diffusion-limited growth and homogeneous nucleation. If the dependence of the growth rate on the cooling rate is not considered, the cooling rate exponent will be 3.

Figure 9.12a shows a comparison between theoretical CSDs and experimental ones. $\alpha_s = 10$ (assuming a rectangular solid having a ratio of the minor axis:the intermediate axis:the major axis = 1:2.22:4.75, on the basis of Pupier (2008)) and $\phi = 0.1$ are assumed. This value is not the volume fraction at immediately after the termination of crystal nucleation and drift by later growth is considered. The fact that the value of the intercept apparently increase because of growth is also considered. This can be done by translating CSD at the termination of nucleation along the crystal size axis. The amount of translation is then determined by the ratio of the volume fraction at the termination of nucleation $\phi_n^* \approx 0.01$ to ϕ at rapid cooling ($= 0.1$).

Parameter $G_1 t_J$ is then determined such that it fits the experimental CSD, and the dependence on the cooling rate is $\chi_G \approx 1/2$. In Fig. 9.11, a reference point is 10^{-4} K/s against the slope 10^5 m^{-1}. $\tilde{A}/T_0 c_G$ at this time is 10^7. In the same manner, for an experiment for andesitic melt (Iezzi et al. 2011), assuming that a reference point in Fig. 9.11 is the slope 10^5 m^{-1} against the cooling rate 10^{-1} K/s, $\tilde{A}/T_0 C_1$ is obtained as $10^{5.5}$, and this model explains the experimental CSDs well.

Model 2

In this model, the parameter $D t_J$ determines everything and, because the diffusivity D is a physical property independent of the cooling rate, only t_J includes the influence of the cooling rate. t_J is then defined by $t_J^{-1} = \tilde{A}|\dot{T}|/T_0$ and $\tilde{A}$ is determined by the shape of the nucleation rate; i.e., $\tilde{A}$ is a physical

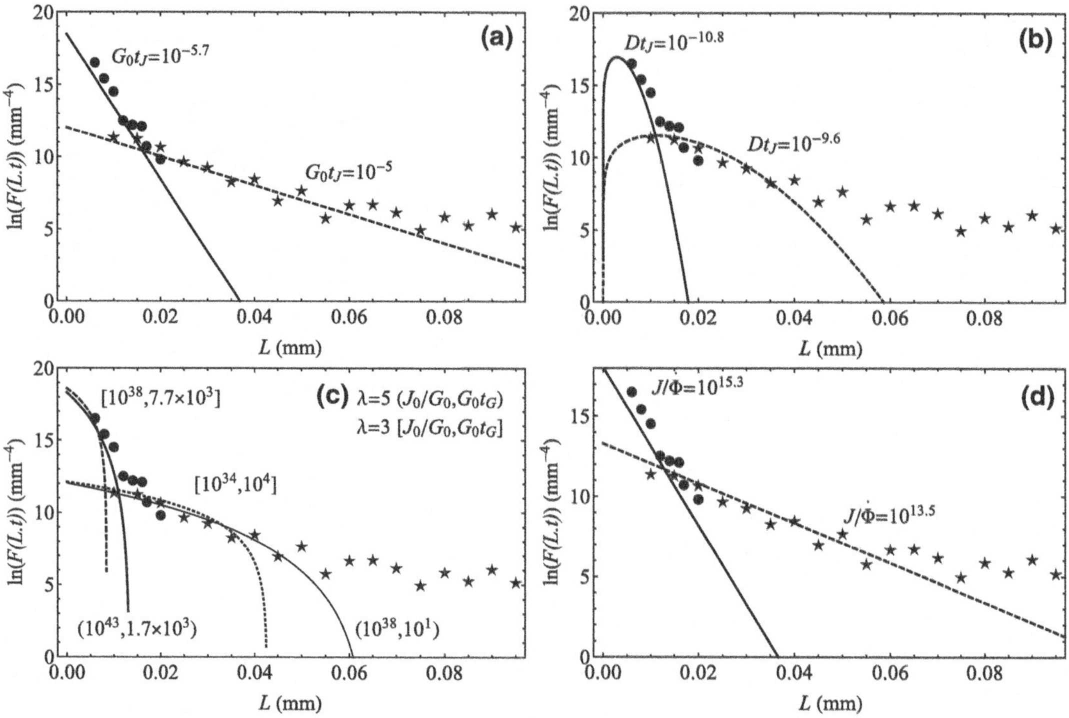

Fig. 9.12 Comparison between experimentally obtained plagioclase CSDs (Pupier 2008) and model CSDs at different cooling rates (0.2 (°C/h) = 5.6×10^{-5} (K/s) (star symbols) and 3 (°C/h) = 8.3×10^{-4} (K/s) (black circles)). $\phi^* = 0.1$ is assumed. Parameters in each model are shown in the figures and the main body. **a** Model 1: The case where the nucleation rate exponentially increases and the growth rate is independent of time (but depends on the cooling rate). **b** Model 2: The case where the nucleation rate exponentially increases and the growth is diffusion-limited growth. **c** Model 3: The case where the nucleation rate and the growth rate exponentially increase. **d** Model 4: The case where the nucleation rate and the crystallization rate are constant

property. Therefore, t_J^{-1} will linearly depend on the cooling rate. Consequently, the crystal number density is determined from Eq. (9.109) to have dependence of

$$N \propto \left|\dot{T}\right|^{3/2} \tag{9.173}$$

on the cooling rate. Although this has already seen in chapter "Cooling Crystallization of Magma" from the numerical aspect, we will understand the entire picture of it including CSD in this chapter.

Figure 9.12b shows CSDs obtained by calculation and experimental results. As in the case of Model 1, the shape factor $\alpha_s = 10$ is used. The ratio of the cooling rate of two CSDs, 8.3×10^{-4} K/s and 5.6×10^{-5} K/s, is the ratio of the parameter$(Dt_J)^{-1}$. In this case, $(Dt_G)^{-1} = 10^{9.6}$ for the CSD of 5.6×10^{-5} K/s and $(Dt_J)^{-1} = 10^{10.8}$ for the experiment with the cooling rate 8.3×10^{-4} K/s are used. By deducting the effect of cooling rate, $\tilde{A}/D$ value is uniquely determined at $\tilde{A}/D \approx 7.3 \times 10^{13}$. Assuming that the diffusivity is $10^{-12}\,\mathrm{m^2/s}$ and the dimensionless degree of supercooling $\Delta\tilde{T} = \Delta T/T_0$ is 0.01, the $\tilde{A}$ value will be about 100 and an estimate of the solid–liquid interfacial energy, Γ, will be obtained at 10^{-4} from the figure in the box in Sect. 9.2.1. From Fig. 7.5, this value is found to correspond to a value of the solid - liquid interfacial energy, $10^{-2}\,\mathrm{J/m^2}$.

Although this model can explain CSDs with an intermediate size, it disagrees with those with the smallest size and largest size. This may be because Ostwald ripening occurs after nucleation in the experiments. Moreover, while CSDs of the model have a convex-upward shape, the experimental CSDs have a convex-downward shape. This difference has to be discussed further.

Model 3

CSDs in this case are determined by three parameters, λ, J_0/G_0, and $G_0 t_G$. Because the crystal number density is determined by Eqs. (9.144) and (9.145) and the cooling rate exponent ξ corresponding to realistic dependence on the cooling rate must take values from 0 to 1.5, as described earlier, the corresponding λ should take values from 1 to 5. Because the dependence of CSDs of this experiment on the cooling rate is larger than the intermediate, we assume that $\lambda = 3$ and 5 here. As seen from Fig. 9.12 (c), CSD of $\lambda = 5$ reproduces the experimental result.

This model CSDs have a convex-upward shape and differ from the experimental CSDs. Moreover, the ranges of other parameters are wide, $10^{38} < J_0/G_0 < 10^{43}$ and $10^1 < G_0 t_G < 10^4$ (for $\lambda = 3$ and 5), and it is difficult to provide constraints on the interpretation of experimental results.

Model 4

This model corresponds to the case where the pre-exponential factor of the crystal nucleation rate is extremely small and nucleation is heterogeneous. CSDs are exponential and governed by the only parameter, the ratio between the nucleation rate and the crystallization rate, $J_1/\dot{\phi}$. This parameter is then considered to strongly depend on cooling rate. As given by Eq. (9.168), the crystallization rate is as follows:

$$\dot{\phi} = \phi'_T |\dot{T}| \tag{9.174}$$

where ϕ'_T is the thermodynamic factor and determined from an equilibrium phase diagram. Assuming that the volume fraction of the main solid phase such as plagioclase and pyroxene, $\Delta\phi$, is from 0.1 to 0.5 and that temperature differences between the liquidus and the solidus ΔT is about from 100 to 300 K, ϕ'_T is estimated at $\phi'_T = \Delta\phi/\Delta T \approx (0.1 \sim 0.5)/(100 \sim 300) \approx 3 \times 10^{-4} \sim 5 \times 10^{-3}$. Because this is a constant, the crystallization rate $\dot{\phi}$ will linearly depend on the cooling rate.

Generally, the nucleation rate strongly depends on the effective degree of supercooling. As seen in chapter "Cooling Crystallization of Magma", the effect of the cooling rate is exerted on this effective degree of supercooling through the time dependence of crystal growth, and the effective crystal growth is suppressed as the cooling rate increases. Consequently, the degree of supercooling required for the maximum nucleation becomes also large and the nucleation rate also increases. Now, we express the effect as follows:

$$J = c_J |\dot{T}|^{\chi_J}, \tag{9.175}$$

where the coefficient c_J is the kinetic factor influenced by, for example, the surface tension and diffusivity. The exponent χ_J is then a constant representing the magnitude of the dependence on the cooling rate.

From these equations, the dependence of the ratio between the nucleation rate and the crystallization rate on the cooling rate can be written as follows:

$$\frac{J}{\dot{\phi}} = \frac{c_J}{\phi'_T} |\dot{T}|^{\chi_J - 1} \tag{9.176}$$

Moreover, the slope of CSD $-(6\alpha_s J/\dot{\phi})^{1/3}$ is written as follows:

$$|\mathrm{slope}| = -\left(\frac{6\alpha_s c_J}{\phi'_T}\right)^{1/3} |\dot{T}|^{(\chi_J - 1)/3} \tag{9.177}$$

Because the exponent $(\chi_J - 1)/3$ ranges from 1/3 to 1/2 from Fig. 9.11, the value of the exponent χ_J ranges from 2 to 5/2. Using this value and fitting to the graph, we can estimate the kinetic factor for the experiment by Pupier (2008). Consequently, the ratio of the controlling parameters $J/\dot{\phi}$ will be $10^{13.5}$ when the cooling rate is 0.2 °C/h and 1015.3 when the cooling rate is 3 °C/h. The kinetic factor is then estimated for each cooling rate at $c_J = 1.5 \times 10^{17}$ and 1.86×10^{17} when $\chi_J = 2.5$ ($c_J = 1.5 \times 10^{15}$ and 4.8×10^{15} when $\chi_J = 2$). Thus, the kinetic factor of melt having the same composition will take an extremely similar value and we can find that the internal consistency of this model is being maintained.

After estimating the controlling parameter, we can estimate the nucleation rate from the slope of CSD by giving the crystallization rate from thermodynamics or experimental results. Using the crystallization rate $\dot{\phi} = 2\times10^{-3}$, the nucleation rate can be estimated at $J = 3.3 \times 10^6\,\mathrm{m^{-3}s^{-1}}$ from the slope $10^5\,\mathrm{m^{-1}}$ for 10^{-4} K/s. Moreover, when the cooling rate is 10^{-2} K/s, the slope of CSD will be $10^6\,\mathrm{m^{-1}}$ and the nucleation rate can be estimated at $J = 3.3 \times 10^{11}\,\mathrm{m^{-3}\,s^{-1}}$. These values overlap with the range of values estimated by Pupier (2008) by a different method but are slightly larger than them.

Figure 9.11 also contains CSD data (after correction for the original data) of andesitic melt by Iezzi et al. (2011). Difference due to melt composition can be considered as difference in the kinetic factor c_J (there is not much difference in the thermodynamic factor). Estimation of c_J from the data of Iezzi et al. (2011) by the same manner provides $c_J \approx 10^{12}$ $(\chi_J = 2.5) - 10^{11}(\chi_J = 2.0)$. Thus, the kinetic factor of andesitic melt is smaller than that of basaltic melt. This difference may reflect difference in the effective surface energy or difference between homogeneous nucleation and heterogeneous nucleation.

9.5 Open-System CSD

Randolph and Larson (1988) and Marsh (1988) demonstrated that, considering the extraction of crystals in a chemical engineering plant, the CSD of crystals remaining in the chemical reactor becomes exponential distribution. Under the constant volume condition, temporal development of the crystal size distribution considering injection and extraction of crystals can be written as follows:

$$\frac{\partial F(R)}{\partial t} + \frac{\partial}{\partial R}(G(R)F(R)) = I_{\mathrm{in}} - I_{\mathrm{out}} \tag{9.178}$$

9.5.1 Exponential Distribution as a Steady-State Solution: In the Case Where the Extraction Rate Is Constant

Now, we assume the case with no injection ($I_{\mathrm{in}} = 0$) and a constant extraction rate ($\mathrm{s^{-1}}$) over all crystals; $I_{\mathrm{out}} = F(R)/|t_{\mathrm{ex}}|$, where $|t_{\mathrm{ex}}|$ is a time constant that defines the extraction rate of crystals $1/|t_{\mathrm{ex}}|$. Moreover, assuming that the crystal growth rate G is independent of the crystal particle size and time and is constant (G_0) and that the system is in a steady state, Equation (9.178) is rewritten as follows:

$$G_0 \frac{\partial F(R)}{\partial R} = -F(R)/|t_{\mathrm{ex}}| \tag{9.179}$$

This differential equation can be immediately integrated to give a solution under the boundary conditions $F = F_0$ when $R = 0$ and $F(\infty) = 0$ when $R = \infty$ as follows:

$$F(R) = F_0 e^{-\frac{R}{G|t_{\mathrm{ex}}|}} \tag{9.180}$$

This is an exponential distribution. Here, the intercept F_0 is connected with the nucleation rate J_0 determined by

$$J_0 = \int_0^{\infty} I_{\rm out} dR = \text{const.}, \tag{9.181}$$

which is obtained by applying $G(0)F(0) = J_0$ and $G(\infty)F(\infty) = 0$ to the integral of the equation for a steady state (Eq. 9.179) from 0 to ∞ of R. Ultimately, the intercept and the slope of CSD is connected with the nucleation rate, the growth rate, and the residence time by

$$F_0 = \frac{J_0}{G_0} \tag{9.182}$$

$$L = G_0 t_{\rm ex} \quad (= 1/|\text{slope}|) \tag{9.183}$$

When the extraction rate of crystals depends on the crystal particle size ($I_{\rm out} = F(R)/|t_{\rm ex}(R)|$),

$$F(R) = F_0 e^{-\int_0^R \frac{1}{G_0 |t_{\rm ex}(R)|} dR} \tag{9.184}$$

is obtained.

In the case where the extraction rate is constant, the solution will be exponential distribution as follows:

$$F(R) = F_0 e^{-\frac{R}{G_0 t_{\rm ex}}} \tag{9.185}$$

Using this exponential distribution as an example, we will explain a mechanism of an open-system CSD (Fig. 9.13).

When the extraction rate is constant, the number of particles extracted from the system is proportional to size distribution or the population density $F(R)$, and can be written as $F(R)/|t_{\rm ex}|$. In the growth process of a group of crystals of size of interest (i.e., with the progress of time), the number of the crystals decreases by $1/|t_{\rm ex}|$. Consequently, the number of crystals of groups of larger-sized crystals, which are influenced by extraction for longer time, will decrease more. In the case where the extraction rate increases with the crystal particle size, such as crystal settling, CSD of phenocryst will be a convex-upward curve and $F(R)$ of larger-sized crystals becomes scarcer (Marsh 1988).

9.5.2 Non-steady State Solution

We assume a general case where difference between the supply rate and the extraction rate is a function only of size: $I_{\rm in} - I_{\rm out} = F(R, t)/t_{\rm ex}$. Hereafter, we assume both the case where $t_{\rm ex}$ is positive and the case where $t_{\rm ex}$ is negative and conduct calculation without the absolute value symbol. Moreover, when the growth rate is a function only of size, the conservation law of size distribution is obtained as follows:

$$\frac{\partial F(R,t)}{\partial t} + \frac{\partial}{\partial R}(G(R)F(R,t)) = \frac{F(R,t)}{t_{\rm ex}(R)} \tag{9.186}$$

The Laplace transformation with respect to time will then be applied to this equation. Because only the right-hand side differs from Sect. 9.2.3,

$$\text{R.H.S.} = t_{\rm ex}(R)^{-1}\mathcal{L}(F) = t_{\rm ex}(R)^{-1}\hat{F} \tag{9.187}$$

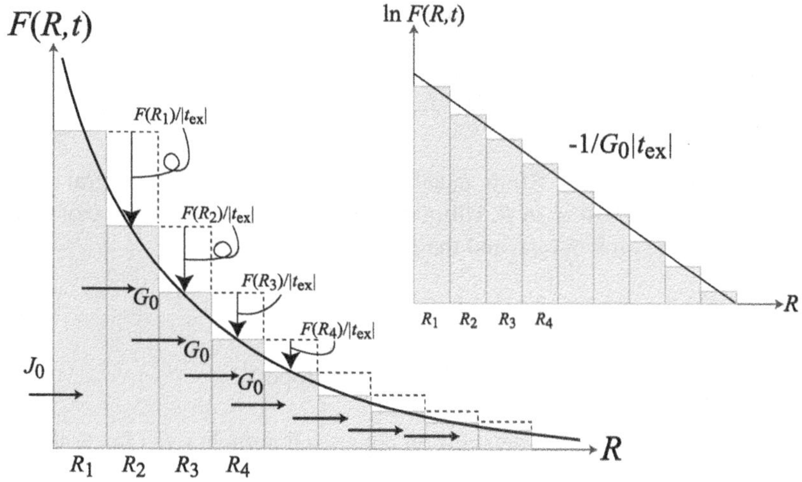

Fig. 9.13 Explanation of why an open-system CSD becomes exponential distribution. In the case where the extraction rate $1/|t_{ex}|$, the nucleation rate J0, and the growth rate G0 are constant. The size distribution $F(R)$ on a linear scale (larger graph) exponentially decreases with the size R because larger-sized crystals nucleated earlier, spent longer crystallization time, and extracted more. The CSD becomes linear when it is plotted on a semilogarithmic graph (smaller graph at the upper right)

is obtained, and Eq. (9.186) is written as follows:

$$-F(R,0) + s\hat{F} + \frac{\partial}{\partial R}\left(G(R)\hat{F}\right) = t_{ex}(R)^{-1}\hat{F} \tag{9.188}$$

Here, rearranging this equation using

$$\varphi(R,s) = G(R)\hat{F} \tag{9.189}$$

gives

$$\frac{d\varphi}{dR} + \left(\frac{s}{G} - \frac{1}{Gt_{ex}}\right)\varphi = F(R,0) \tag{9.190}$$

As in the case of Sect. 9.2.3, the general solution is obtained as follows:

$$\varphi = e^{-st_A(R,0)+\tilde{b}(R,0)}\left(\int_0^R F(R',0)e^{st_A(R',0)-\tilde{b}(R',0)}dR' + c\right) \tag{9.191}$$

where

$$t_A(R,0) = \int_0^R \frac{dR''}{G(R'')} \tag{9.192}$$

$$\tilde{b}(R,0) = \int_0^R \frac{dR'}{G(R')t_{ex}(R')} \tag{9.193}$$

c is a constant of integration and determined from the Laplace transformation of the boundary condition $F(0,t)$

$$\varphi(0,s) = c = \int_0^\infty e^{-st}G(0)F(0,t)dt \tag{9.194}$$

In this case, the boundary condition is $G(0)F(0,t) = J(t)$ and the Laplace transformation of it yields c as follows:

$$c = \hat{J}(s) \tag{9.195}$$

Because F can be obtained by the inverse Laplace transformation of $\varphi/G(R)$,

$$\begin{aligned} F(R,t) = &= \frac{1}{G(R)}\left[\int_0^R F(R',0)e^{\tilde{b}(R,R')}\mathcal{L}^{-1}(e^{-s\tilde{b}(R,R')})dR' \right. \\ &\left. + e^{\tilde{b}(R,0)}\mathcal{L}^{-1}(e^{-st_A(R,0)}\hat{J}(s))\right] \end{aligned} \tag{9.196}$$

$$t_A(R,R') = \int_{R'}^R \frac{dR''}{G(R'')} \tag{9.197}$$

$$\tilde{b}(R,R') = \int_{R'}^R \frac{dR''}{G(R'')t_{ex}(R'')} \tag{9.198}$$

are obtained. Using $\mathcal{L}^{-1}(e^{-sX}) = \delta(R-X)$ for the first term and convolution theorem for the second term yields

$$F(R,t) = \underbrace{\frac{G(R')}{G(R)}F(R',0)e^{\int_{R'}^R \frac{dR''}{G(R'')t_{ex}(R'')}}}_{Initialcondition} + \underbrace{\frac{J(t-\int_0^R \frac{dR''}{G(R'')})}{G(R)}e^{\int_0^R \frac{dR''}{G(R'')t_{ex}(R'')}}}_{Boundarycondition} \tag{9.199}$$

A property of the delta function, described earlier, is used here. R' in this equation is a variable $R'(R,t)$ connecting R and t via

$$t - \int_{R'}^R \frac{dR''}{G(R'')} = 0 \tag{9.200}$$

Let us examine certain simple cases in the following.

(1) The Case Where $G = G_0 =$ Constant and $t_{ex}^{-1} = 0$

From Eq. (9.200),

$$t - \int_{R'}^R \frac{dR''}{G_0} = t - \frac{R-R'}{G_0} = 0 \tag{9.201}$$

is obtained and then

$$R' = R - G_0 t \tag{9.202}$$

is obtained. Because the term of $e^{\int}$ equals 1,

$$F(R,t) = F(R - G_0 t, 0) + \frac{J(t-\frac{R}{G_0})}{G_0} \tag{9.203}$$

is obtained.

(2) The Case Where $G = G_0 =$ Constant and t_{ex}= Constant

R' and the relationship between t and R are the same as above (Eq. 9.202). Because the integral terms of $e^{\int}$ are

$$\int_{R'}^{R} \frac{dR''}{G_0 t_{ex}} = \frac{R - R'}{G_0 t_{ex}} = t/t_{ex} \tag{9.204}$$

$$\int_{0}^{R} \frac{1}{G_0 t_{ex}} dR'' = \frac{R}{G_0} R \tag{9.205}$$

Therefore,

$$F(R, t) = F(R - G_0 t, 0)e^{t/t_{ex}} + \frac{J(t - \frac{R}{G_0})}{G_0} e^{\frac{R}{G_0 t_{ex}}} \tag{9.206}$$

is obtained.

(3) The Case Where J$J = J_0 \exp(t/t_J)$, $G = G_0 =$ Constant, and t_{ex}= Constant

Because R', the relationship between t and R, and the integral terms of $e^{\int}$ are the same as above,

$$F(R, t) = F(R - G_0 t, 0)e^{t/t_{ex}} + \frac{J_0 e^{t/t_J}}{G_0} e^{-\frac{t_J^{-1} - t_{ex}^{-1}}{G_0} R} \tag{9.207}$$

is obtained.

(4) A General Case Where $J = J_0 \exp(t/t_{ex})$ and G and t_{ex} are a Function of R

$$\begin{aligned} F(R, t) = {} & \frac{G(R')}{G(R)} F(R', 0) e^{\int_{R'}^{R} \frac{dR''}{G(R'')t_{ex}(R'')}} \\ & + \frac{J_0 e^{t/t_J}}{G(R)} e^{-\int_0^R \left(\frac{1}{G(R')t_J} - \frac{1}{G(R')t_{ex}(R')}\right) dR'} \end{aligned} \tag{9.208}$$

holds. Here, R' is determined from Eq. (9.200).

9.6 Avrami Model and Size Distribution

Avrami (1939, 1940, 1941) discussed the relationship between time change in the volume of a phase newly generated by phase change and the nucleation/growth rate. He evaluated the influence of the newly generated phase on nucleation and growth. We consider a rate of phase change at the point of time where the volume of the new phase becomes V in the initial volume before the phase change, V_0. The phase change rate dV/dt is different from an increasing rate of the volume under a state where there is no geometrical effect such as impingement of new phase, dV_{ext}/dt. This V_{ext} is called an extended volume and a normal volume fraction V_{ext}/V_0 is evaluated as follows:

$$\frac{V_{ext}}{V_0} = \frac{4\pi}{3} \int_0^t J(t') \left[\int_{t'}^{t} G(t'')dt'' \right]^3 dt' \equiv \phi_{ext} \tag{9.209}$$

Space where the new phase can nucleate or grow is decreased by the volume of the already-generated new phase to $V_0 - V$. Therefore, the influence of it can be evaluated by giving a weight of the ratio of the available volume to the initial volume, $(V_0 - V)/V_0$, to the increasing rate of the volume without the new phase. That is,

$$\frac{dV}{dt} = \frac{V_0 - V}{V_0}\frac{dV_{\text{ext}}}{dt} \tag{9.210}$$

This is a differential equation regarding V as a function of V_{ext} and can be analytically solved by the separation of variables under the condition where $V = 0$ when $V_{\text{ext}} = 0$ and $V = V_0$ when $V_{\text{ext}} = \infty$ as follows:

$$V = V_0\left(1 - \exp(-\frac{V_{\text{ext}}}{V_0})\right) \tag{9.211}$$

Because of $\phi = V/V_0$ and substituting Eq. (9.209) into V_{ext}/V_0 yield

$$\phi = 1 - \exp\left(-\frac{4\pi}{3}\int_0^t J(t')\left[\int_{t'}^t G(t'')\right]^3 dt''dt'\right) \tag{9.212}$$

Let us consider a special case when calculating the extended volume. The nucleation rate is proportional to the power of time: $J = \bar{J}_0 t^{\bar{m}}$. The growth rate is proportional to the power of time $(t - t')$ elapsed from nucleation (at time t'): $G = \bar{G}_0(t - t')^{\bar{n}}$.[11] Here, $\bar{J}_0$ and $\bar{G}_0$ are coefficients in the Avrami model and their dimensions depend on the exponents $\bar{m}$ and $\bar{n}$. Using the nucleation and growth rates, the extended volume can be written as follows:

$$\phi_{\text{ext}} = \frac{4\pi B_0 \bar{J}_0 \bar{G}_0^3}{3(1+\bar{n})} t^{\bar{m}+3(\bar{n}+1)+1} \tag{9.213}$$

$$B_0 = B(\bar{m}+1, 3(\bar{n}+1)+1) \tag{9.214}$$

where $B(a, b)$ is the Beta function.[12] Here, we focused on the power of time, $\bar{n}+1$ represents the time dependence of the change in the particle size and $\bar{m}+3(\bar{n}+1)+1$ represents the time dependence of crystallization due to nucleation and growth. Then, we define them as follows:

$$\bar{n}_a = \bar{m} + 3(\bar{n}+1) + 1 = \bar{m} + 3\bar{n}_r + 1 \tag{9.217}$$

$$\bar{n}_r = \bar{n} + 1 \tag{9.218}$$

Using them yields time change in the volume fraction due to phase change as follows:

$$\phi = 1 - \exp\left(-\frac{4\pi B_0 \bar{J}_0 \bar{G}_0^3}{3\bar{n}_r} t^{\bar{n}_a}\right) \tag{9.219}$$

---

[11] Note that we do not assume $G = G_0 t^{\bar{n}}$. In the case where $G = G_0 t^{\bar{n}}$, all crystals have uniform growth rate in the unified time flow regardless of nucleation time.

[12] The Beta function is then defined by

$$B(a, b) = \int_0^1 x^{a-1}(1-x)^{b-1}dx \tag{9.215}$$

and has the following relationship with the Gamma function.

$$B(a, b) = \frac{\Gamma(a)\Gamma(b)}{\Gamma(a+b)} \tag{9.216}$$

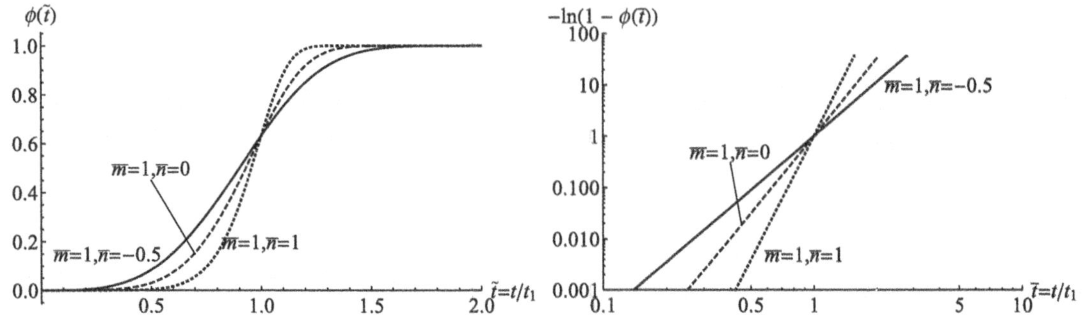

Fig. 9.14 (Left) Avrami curve. (Right) Avrami plot

The power of time in this equation, $\bar{n}_a$, is normally called the Avrami exponent and represents the time dependence of nucleation and growth. When evaluating phase change experiments, a logarithmic plot of $-\ln(1-\phi)$ against t is called the Avrami plot and the slope of the plot will be the Avrami exponent (Fig. 9.14).

Now, what type of size distribution will be obtained in this case? Let us obtain a size distribution function with parameters $\bar{m}$ and $\bar{n}$ or $\bar{n}_a$ and $\bar{n}_r$, by the same method as the previous section. The size distribution is written as follows:

$$F_{\text{ext}}(R(t',t),t) = \frac{J(t')}{G(t')} = \frac{\bar{J}_0(t')^{\bar{m}}}{\bar{G}_0(t-t')^{\bar{n}}} \tag{9.220}$$

by using the nucleation rate and the growth rate. Note that because this size distribution ignores the influence of collisions of crystals, and the size distribution is that corresponding to the extended volume. Therefore, we use F_{ext} as a symbol of the size distribution.

Because the particle size is $R(t',t) = \bar{G}_0(t-t')^{\bar{n}_r}$,

$$t' = t - \left(\frac{\bar{n}_r}{\bar{G}_0}R\right)^{1/\bar{n}_r} \tag{9.221}$$

is obtained, and substituting this equation into Eq. (9.220) and rearranging it yields

$$F_{\text{ext}}(R,t;\bar{m},\bar{n}) = \frac{\bar{J}_0}{\bar{G}_0}\left(\frac{\bar{n}_r}{\bar{G}_0}R\right)^{-\frac{\bar{n}}{\bar{n}_r}}\left[t-\left(\frac{\bar{n}_r}{\bar{G}_0}R\right)^{\frac{1}{\bar{n}_r}}\right]^{\bar{m}} \tag{9.222}$$

Here, we assume that, in the Avrami model, the time when the extended volume becomes 1 (the absolute value of the inside of the exponential part of Eq. (9.219) becomes (1) is t_{Av}, then scale the time by t_{Av}, and scale the crystal size by R_{Av}, i.e. the size of crystals at t_{Av}, which have nucleated at $t=0$. t_{Av} and R_{Av} are then obtained as follows:

$$t_{\text{Av}} = \left(\frac{4\pi B_0\bar{J}_0\bar{G}_0^3}{3\bar{n}_r}\right)^{-1/\bar{n}_a} \tag{9.223}$$

$$R_{\text{Av}} = \frac{\bar{G}_0 t_{\text{Av}}^{\bar{n}_r}}{\bar{n}_r} = \frac{\bar{G}_0}{\bar{n}_r}\left(\frac{4\pi B_0\bar{J}_0\bar{G}_0^3}{3\bar{n}_r}\right)^{-\bar{n}_r/\bar{n}_a} \tag{9.224}$$

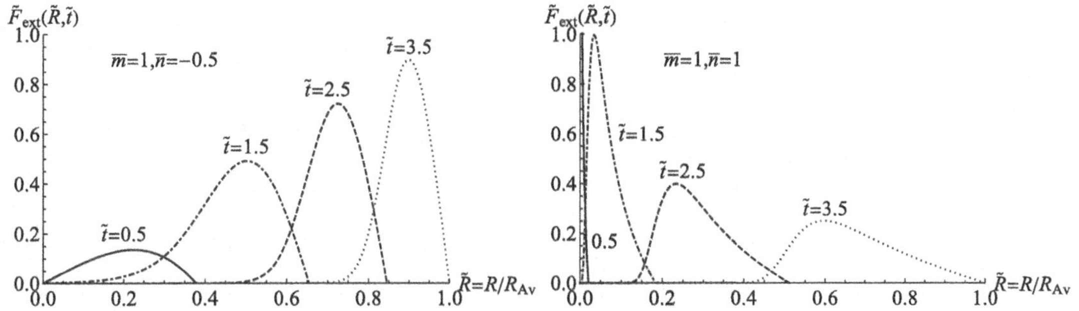

Fig. 9.15 (Left) Change of CSD when$\bar{m} = 1$ and $\bar{n} = -0.5$. (Right) Change of CSD when $\bar{m} = 1$ and $\bar{n} = 1$. The size is scaled by the maximum particle size at $\tilde{t} = 3.5$

By using dimensionless time and size using them, $\tilde{t} = t/t_{Av}$ and $\tilde{R} = R/R_{Av}$, the size distribution function is written as follows:

$$F_{ext}(\tilde{R}, \tilde{t}; \bar{m}, \bar{n}) = F_{ext0}(\bar{m}, \bar{n}) \cdot \tilde{F}_{ext}(\tilde{R}, \tilde{t}; \bar{m}, \bar{n}) \tag{9.225}$$

$$F_{ext0}(\bar{m}, \bar{n}) = \frac{\bar{J}_0}{\bar{G}_0} \left(\frac{4\pi B_0(\bar{m}, \bar{n}) \bar{J}_0 \bar{G}_0^3}{3\bar{n}_r} \right)^{-(\bar{m}-\bar{n})/(\bar{n}_a)} \tag{9.226}$$

$$\tilde{F}_{ext}(\tilde{R}, \tilde{t}; \bar{m}, \bar{n}) = \tilde{t}^{\bar{m}-\bar{n}} \tilde{R}^{-\bar{n}/\bar{n}_a} \left[1 - \tilde{R}^{1/\bar{n}_a} \right]^{\bar{m}} \tag{9.227}$$

If the effect of change in the volume fraction on the nucleation rate is considered,

$$F_{ext}(R(t', t), t; \bar{m}, \bar{n}) = \frac{(1 - \phi(t'))J(t')}{G(t')} \tag{9.228}$$

is obtained and similarly using the scaled time $\tilde{t}$ yields

$$1 - \phi(t') = \exp\left[-\tilde{t}^{\bar{n}_a} \left(1 - \tilde{R}^{1/\bar{n}_r} \right)^{\bar{n}_a} \right] \equiv \tilde{F}_A(\tilde{R}, \tilde{t}; \bar{m}, \bar{n}) \tag{9.229}$$

Using this $\tilde{F}_A$, the size distribution function is obtained as follows:

$$F_{ext}(\tilde{R}, \tilde{t}; \bar{m}, \bar{n}) = F_{ext0}(\bar{m}, \bar{n}) \cdot \tilde{F}_A(\tilde{R}, \tilde{t}; \bar{m}, \bar{n}) \cdot F_{ext}(\tilde{R}, \tilde{t}; \bar{m}, \bar{n}) \tag{9.230}$$

(Figure 9.15). Note that the size scale R_{Av} does not depend on time.

References

Avrami M (1940) Kinetics of phase change. II transformation–time relations for random distribution of nuclei. J Chem Phys 8, 212

Avrami M (1939) Kinetics of phase change. I general theory. J Chem Phys 7:1103

Avrami M (1941) Granulation, phase change, and microstructure kinetics of phase change. III. J Chem Phys 9:177

Brugger CR, Hammer JE (2010) Crystallization kinetics in continuous decompression experiments: implications for interpreting natural magma ascent processes. J Petrol 51:1941–1965

Burkhard DJ (2002) Kinetics of crystallization: example of micro-crystallization in basalt lava. Contrib Mineral Petrol 142(6):724–737
Cashman KV (1988) Crystallization of Mount St. Helens 1980–1986 dacite: A quantitative textural approach. Bull Volcanol 50:194–209
Cashman KV (1992) Groundmass crystallization of Mount St. Helens dacite, 1980–1986: a tool for interpreting shallow magmatic processes. Contrib Mineral Petrol 109:431–449
Cashman KV (1993) Relationship between plagioclase crystallization and cooling rate in basaltic melts. Contrib Mineral Petrol 113:126–142
Cashman KV, Marsh BD (1988) Crystal size distribution (CSD) in rocks and the kinetics and dynamics of crystallization II, Makaopuhi lava lake. Contrib Mineral Petrol 99:292–305
Hammer JE, Cashman KV, Hoblitt RP, Newman S (1999) Degassing and microlite crystallization during pre-climactic events of the 1991 eruption of Mt. Pinatubo, Philippines. Bull Volcanol 60:355–380
Hammer JE, Cashman KV, Voight B (2000) Magmatic processes revealed by textural and compositional trends in Merapi dome lavas. J Volcanol Geotherm Res 100:165–192
Higgins MD (1996) Magma dynamics beneath Kameni volcano, Thera, Greece, as revealed by crystal size and shape measurements. J Volcanol Geotherm Res 70:37–48
Higgins MD (2002) A crystal size-distribution study of the Kiglapait layered mafic intrusion, Labrador, Canada: evidence for textural coarsening. Contrib Mineral Petrol 144:314–330
Iezzi G, Mollo S, Torresi G, Ventura G, Cavallo A, Scarlato P (2011) Experimental solidification of an andesitic melt by cooling. Chem Geol 283(3):261–273
Kirkpatrick RJ (1976) Towards a kinetic model for the crystallization of magma bodies. J Geophys Res 81(14):2565–2571
Mangan MT (1990) Crystal size distribution systematics and the determination of magma storage times: The 1959 eruption of Kilauea volcano. Hawaii J Volcanol Geotherm Res 44:295–302
Marsh BD (1988) Crystal size distribution (CSD) in rocks and the kinetics and dynamics of crystallization I. Theory Contrib Mineral Petrol 99:277–291
Marsh BD (1998) On the interpretation of crystal size distributions in magmatic systems. J Petrol 39:553–599
Morgan DJ, Jerram DA, Chertkoff DG, Davidson JP, Pearson DG, Kronz A, Nowell GM (2007) Combining CSD and isotopic microanalysis: magma supply and mixing processes at Stromboli Volcano, Aeolian Islands, Italy. Earth Planet Sci Lett 260(3):419–431
Ni H, Keppler H, Walte N, Schiavi F, Chen Y, Masotta M, Li Z (2014) In situ observation of crystal growth in a basalt melt and the development of crystal size distribution in igneous rocks. Contrib Mineral Petrol 167:1003. https://doi.org/10.1007/s00410-014-1003-9
Piochi M, Mastrolorenzo G, Pappalardo L (2005) Magma ascent and eruptive processes from textural and compositional features of Monte Nuovo pyroclastic products, Campi Flegrei, Italy. Bull Volcanol 67(7):663–678
Preece K, Barclay J, Gertisser R, Herd RA (2013) Textural and micro-petrological variations in the eruptive products of the 2006 dome-forming eruption of Merapi volcano, Indonesia: Implications for sub-surface processes. J Volcanol Geotherm Res 261:98–120
Pupier E, Duchene S, Toplis MJ (2008) Experimental quantification of plagioclase crystal size distribution during cooling of a basaltic liquid. Contrib Mineral Petrol 155(5):555–570
Randolph AD, Larson MA (1988) Theory of particulate processes, 2nd edn. Academic Press, New York, pp 369
Resmini RG (2007) Modeling of crystal size distributions (CSDs) in sills. J Volcanol Geotherm Res 161(1):118–130
Resmini RG, Marsh BD (1995) Steady-state volcanism, paleoeffusion rates, and magma system volume inferred from plagioclase crystal size distributions in mafic lavas: Dome Mountain. Nevada J Volcanol Geotherm Res 68(4):273–296
Salisbury MJ, Bohrson WA, Clynne MA, Ramos FC, Hoskin P (2008) Multiple plagioclase crystal populations identified by crystal size distribution and in situ chemical data: implications for timescales of magma chamber processes associated with the 1915 eruption of Lassen Peak, CA. J Petrol 49(10):1755–1780
Schiavi F, Walte N, Keppler H (2009) First in situ observation of crystallization processes in a basaltic-andesitic melt with the moissanite cell. Geology 37(11):963–966
Shea T, Hammer JE (2013) Kinetics of cooling- and decompression-induced crystallization in hydrous mafic-intermediate magmas. J Volcanol Geotherm Res 260:127–145
Spillar V, Dolejs D (2014) Kinetic model of nucleation and growth in silicate melts: Implications for igneous textures and their quantitative description. Geochim Cosmochim Acta 131:164–183
Spohn T, Hort M, Fischer H (1988) Numerical simulation of the crystallization of multicomponent melts in thin dikes or sills. 1: the liquidus phase. J Geophys Res 93:4880–4894
Tomiya A, Takahashi E (1995) Reconstruction of an evolving magma chamber beneath Usu Volcano since the 1663 eruption. J Petrol 36(3):617–636
Toramaru A (2001) A numerical experiment of crystallization for a binary eutectic system with application to igneous textures. J Geophys Res 106:4037–4060
Vinet N, Higgins MD (2011) What can crystal size distributions and olivine compositions tell us about magma solidification processes inside Kilauea Iki lava lake, Hawaii? J Volcanol Geotherm Res 208:136–162

Wilhelm S, Worner G (1996) Crystal size distribution in Jurassic Ferrar flows and sills (Victoria Land, Antarctica): evidence for processes of cooling, nucleation, and crystallisation. Contrib Mineral Petrol 125(1):1–15

Zieg MJ, Lofgren GE (2006) An experimental investigation of texture evolution during continuous cooling. J Volcanol Geotherm Res 154(1):74–88

Zieg MJ, Marsh BD (2002) Crystal size distributions and scaling laws in the quantification of igneous textures. J Petrol 43(1):85–101

Part IV
Application

10 Exploring Eruptive Phenomena from Vesiculation and Crystallization

Abstract

The previous chapters explained the mechanism of each individual process (elementary process) involved in the vesiculation and crystallization of magma; the overall time evolution of the vesiculation process and the crystallization process comprising the elementary processes; the roles of physical properties in the processes; and shared the accuracy and limits of theoretical understanding compared with experiments. This chapter introduces the examples of interpreting the texture of volcanic products and readings of quantitative information from them by applying the understanding achieved to date. Keeping correspondence with eruption styles in mind, this chapter deals with actual data obtained from the texture analysis of eruptive products and applies the understanding achieved to date to the actual data to quantitatively understand eruptive phenomena in the order of explosive eruptions and nonexplosive eruptions. The last section describes the application to intrusions. Moreover, the bubble texture and crystalline texture are separately explained.

10.1 Mode of Occurrence of Eruptive Products and Rock Texture

10.1.1 Geological Occurrence

Geological occurrence is related to the present state of rock in nature. The modes of occurrences of igneous rock include pyroclastic deposits (deposit composed of fragmented magmatic particles or pyroclasts), lava flow and lava dome (rock that appeared at the ground surface or sea floor as a continuum and solidified there), and intrusions (rock that cooled and solidified under the ground without appearing at the ground surface).

Pyroclasts are eruptive products obtained from the so-called explosive eruptions. They are roughly classified by the relations to types of eruptions which produced them, as follows: (1) Products contained in Plinian eruptions and large-scale pyroclastic flow deposits (ignimbrite)[1] with relatively high-viscosity magma (pumice and volcanic ash), (2) Products formed by Strombolian eruptions and

[1] Pyroclastic flow deposits that are welded like lava after deposition are called welded tuff.

A. Toramaru, *Vesiculation and Crystallization of Magma*, Advances in Volcanology,
https://doi.org/10.1007/978-981-16-4209-8_10

lava fountain eruptions with relatively low-viscosity magma (such as scoria, spatter, Pele's tears, and Pele's hair), (3) Products formed by single Vulcanian eruptions involving shock waves in the air (volcanic ash), (4) Products contained in pyroclastic flow deposits caused by the collapse of lava domes (volcanic ash and rock fragments), and (5) Products formed by phreatomagmatic eruptions (volcanic ash and volcanic particles).

Lava flows and lava domes are eruptive materials produced from effusive eruptions or nonexplosive eruptions. Although eruptive deposits among the pyroclasts described in (4) are fragmented in terms of modes of occurrence, the eruption style related to formation of the rock texture should be considered as an effusive one.

The intrusions are formed when magma cools down and solidifies at depth without appearing at the ground surface; they are primarily a product of cooling crystallization. Igneous rock intruding at shallow depths often shows systematic textural change from the boundary with the country rock (host rock)[2] toward the interior of the intrusions, which are broadly classified into sills (a nearly-horizontal sheet of igneous rock) and dikes (dykes; a sheet of igneous rock having an orientation that is obviously different from a horizontal orientation) based on the consistency between country rock and the intrusion and geometric relation between a horizontal line and the shape of the igneous rock.[3] If a country rock is a sedimentary rock, the layered structure of the country rock often becomes horizontal similar to that at the time of sedimentation. There are some cases where a sheet of intrusive rock concordant with the layer structure of the country rock is called a sill and discordant one is called a dike.

Such modes of occurrence of the igneous rock are very helpful to understand texture in rock and consider the physical process of eruptions. Because the modes of occurrence are often linked directly with eruption styles, the internal texture of contained materials will be an important key for identifying factors controlling the eruption styles.

10.1.2 Convenient Classification of Bubbles and Crystals

Bubbles and crystals in igneous rock in various modes of occurrence can be classified as follows based on the differences in the relative size:

Matrix-bubble Small bubbles that are found in highly vesicular volcanic rock and volcanic ash described in (1) and (2), which are invisible to the naked eye. From the eruption style that produces matrix bubbles or texture analysis mentioned later, matrix bubbles are considered to be formed in the volcanic conduit when magma ascends to the ground surface or in some decompressing processes such as collapse of volcanic edifices, i.e., decompression-induced vesiculation process.

Pheno-bubble Bubbles that are distinct from matrix bubbles and visible to the naked eye. Pheno-bubbles are considered to be formed in magma chambers before the formation of matrix bubbles, i.e., before ascending in the volcanic conduit (second boiling; Sect. 1.4.2) or by the coalescence of bubbles regardless of where they form.

Matrix crystal (microlite) Groundmass crystal. Microlites are groundmass crystals that are smaller than about 100 μm. They are considered to be formed under relatively higher cooling rate and by cooling-induced crystallization at around the ground surface or decompression-induced crystallization when magma ascends in the volcanic conduit.

---

[2] Surrounding rock that existed before magma intrudes.

[3] Stock that has a horizontal lower boundary and a spherical upper boundary is also a type of intrusion.

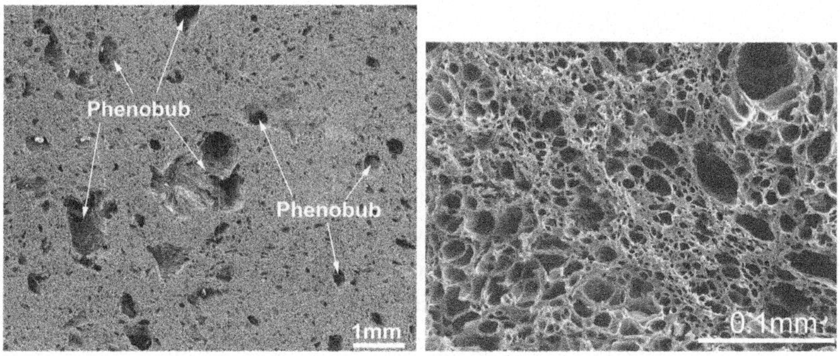

Fig. 10.1 SEM images of a cross section of pumice of Osumi pumice fall, which is a precursory Plinian eruption of the Aira caldera-forming eruption (Ito pyroclastic flow) that occurred 29,000 years ago. (left) Texture on a scale of about 1 cm width. (right) Enlarged view of a matrix part. (Toramaru 2014)

Phenocryst Crystals that are visible to the naked eye. Phenocrysts are considered to be formed under relatively lower cooling rate and by cooling-induced crystallization in magma chambers deep underground.[4]

Although matrix bubbles and pheno-bubbles correspond in terms of relative size to microlites and phenocrysts among crystals, their classification is for convenience. There seem to be distinct background differences in the causes of their formation.[5] In intrusions such as dikes, there are bubbles (filled with secondary minerals; amygdale) formed by second boiling, which are classified into matrix-bubbles and pheno-bubbles.

10.2 Pumice and Scoria (Highly Vesicular Volcanic Rock) Erupted from Plinian Eruptions

10.2.1 Characteristics of Bubble Texture

Highly vesicular volcanic rock is produced from explosive eruptions, particularly Plinian eruptions and attendant pyroclastic flows. The erupted mass of Plinian eruptions is often 0.1 km^3 or more (DRE volume).[6] The volcanic explosivity index (VEI) (Newhall and Self 1982) of such eruptions is often 4 or higher. The total erupted mass of caldera-forming eruptions is more than several tens of cubic kilometers, and most of them are pyroclastic flow eruptions. However, they are frequently accompanied with precursory Plinian eruptions and shift to pyroclastic flows due to collapse of eruption columns of the Plinian eruptions (Sparks and Wilson 1976). The eruption duration is generally more than several hours and is sometimes several days, although it is intermittent.

[4] Because formation of glass inclusions (melt inclusions) in phenocrysts requires hopper and dendrite crystals, described in Chap. 8, some researchers consider that the glass inclusions are not necessarily formed at low cooling rates but formed when magma ascends in volcanic conduits (Waters et al. 2015).

[5] Phenocrysts and groundmass are based on an obvious difference in crystal size (CSD is bimodal); sometimes they are not based on absolute sizes. The same seems to hold true for pheno-bubbles and matrix-bubbles. If crystal size distribution does not show obvious bimodal but continuous distribution, the texture is called seriate. The corresponding bubble texture may have a power law distribution that is continuous from matrix-bubbles to pheno-bubbles.

[6] An abbreviation of dense rock equivalent. Because highly vesicular volcanic rock contains much void spaces, the volume is converted to a DRE volume by excluding void spaces. A volume we can directly obtain from a deposit is an apparent volume with void spaces.

For example, Fig. 10.1 shows the SEM images of a cross-section of pumice of Osumi pumice fall, which is a precursory Plinian eruption of the Aira caldera formation at 29 ka. This is high silica rhyolite, containing about 78 wt% of SiO_2. The images show that within the pumice, large bubbles of about 1 mm are interspersed in a matrix densely filled with small bubbles smaller than 10 μm. The former, large bubbles, are pheno-bubbles and the latter, small bubbles, are matrix-bubbles.

There are about 10–20 pheno-bubbles in this field of view, and there are many relatively small bubbles measuring several hundred micrometers, which can be rather classified as matrix-bubbles. In practice, these small bubbles cannot be distinguished by the naked eye. In crystalline texture, there is the same problem in distinguishing between phenocrysts and groundmass crystals. In terms of the shape of pheno-bubbles, some pheno-bubbles are nearly spherical and some are deformed. Thus, it is suggested that, because the surrounding matrix-bubbles are far smaller, the irregularly shaped pheno-bubbles are not formed by the deformation by interaction with surrounding individual bubbles.

The size of matrix-bubbles ranges from several micrometers to 100 μm, and smaller ones are predominant. Although larger ones are found in large numbers because they are conspicuous (probably by the psychological visual effect), conversion to the number per unit volume of melt before vesiculation confirms that they are fewer in number than expected. Furthermore, matrix-bubbles seem to monotonously and exponentially increase in number from larger ones to smaller ones. The shape of matrix-bubbles is largely spherical, oval, or close to these shapes, although some are polyhedral and some are crushed. The fact that the length of a deformed part of the crushed bubbles is nearly equal to the size of an adjacent bubble suggests that the crushed bubbles are deformed by an interaction with adjacent bubbles.

Sometimes, somewhat elongated bubbles localize, although it is uncertain whether they are pheno-bubbles or matrix-bubbles. Moreover, the elongated bubbles localize in a pumice grain, and in some cases they occupy the whole body of a pumice grain. Although the abundance of pumice grains containing the elongated bubbles varies in the Osumi fallout deposits, it is obviously less than the abundance of pumice grains containing the elongated bubbles in Ito pyroclastic flows deposits (ignimbrite).

In pumice or scoria containing less SiO_2, although the boundary between pheno-bubbles and matrix-bubbles is unclear, matrix-bubbles occupy a certain area. In scoriae having basaltic to basaltic andesite composition, irregularly shaped bubbles become conspicuous because of the influence of coalescence and complex deformation process.

Generally, microlites are not contained in high silica pumice, and the crystallinity of microlites in less silica (than dacite) pumice and scoria is diverse. Crystallinity often shows correlations with crystal number density and vesicularity (microlite systematics; described later).

In pumices contained in large-scale pyroclastic flows as accompanied with caldera formation, bubbles often elongate in one direction. A feature of these bubbles is that the size of the bubbles (the volume of the single bubble) is much larger than bubbles that do not elongated.

10.2.2 Matrix-Bubble

Application of BND Decompression Rate Meter

The analyses of vesicular texture of pumice or scoria produced from Subplinian or Plinian explosive eruptions revealed that bubble number density correlates with the eruption column height, i.e., the volcanic intensity (left graph of Fig. 1.9) (Toramaru 2006). This relationship in various eruptions has been recently investigated (Alfano et al. 2012). However, a doubt remained that the relationship may only reflect magma composition through viscosity or diffusion coefficient of water (right graph of Fig. 1.9). Now, we have a means of resolving the doubt.

As described in Sect. 5.5 bubble number density N is a quantitative indicator of decompression rate and their relationship is represented by Eq. (5.71). Using a more accurate coefficient, Eq. (5.71) can be written as follows:

$$N = 34 \cdot C \cdot \alpha_1^{-2} \cdot \alpha_2^{-1/4} \cdot \alpha_3^{-3/2} \tag{10.1}$$

where C is the number of water molecules (concentration defined by the number of molecules per unit volume). Parameters α_1, α_2, and α_3 are given by Eqs. (5.56), (5.58), and (5.59), respectively, and α_3 is inversely proportional to decompression rate. Rewriting Eq. (10.1) to an equation giving decompression rate gives

$$\left|\frac{dP}{dt}\right|_{\mathrm{BN}} = 3.5 \times 10^{14} \cdot D \cdot \gamma^2 \cdot P_{\mathrm{W}}^{1/3} \cdot T^{1/2} \cdot N^{2/3} \tag{10.2}$$

The value of the coefficient in this equation is determined by the Boltzmann constant and the partial molecular volume of water. If the number density N is given by an observed value, the decompression rate at bubble nucleation $|dP/dt|_{\mathrm{BN}}$ can be obtained.

To calculate the decompression rate using Eq. (10.2), D, γ, P_{W}, and T should be given (see Sect. 11.6 in Appendix (Chap. 11), regarding physical properties of magma). The saturation pressure of water, P_{W}, and temperature of magma, T, can be estimated from mineral assemblages and mineral compositions using phase equilibrium relation. The diffusion coefficient of water, D, (Fig. 11.4) can be calculated from Zhang and Behrens (2000) and Beherns et al. (2004) as a function of water content, SiO_2 content, and temperature. The saturation concentration of water can be obtained from P_{W} using the simplified Burnham model (Eq. (2.26)). Moreover, the surface tension γ can be calculated from the experiment using rhyolitic melt (Bagdassarov et al. 2000) as a function of water saturation (pressure and water content) and temperature (Fig. 11.2). Because SiO_2 content correlates with temperature and the correlation can be assumed to hold, the decompression rate is a function of SiO_2 content and bubble number density with the initial water content as a parameter. The left graph of Fig. 10.2 is a contour plot representing the relationship. The approximate decompression rate can be estimated by plotting observed N and SiO_2 content in this graph.

To apply this method to an Osumi pumice fall, the bubble number density N should be estimated first. Let us assume that average size (radius) of matrix-bubbles in Fig. 10.1 is approximately 10 μm.[7] Among bubble number density N_ϕ, bubble radius R, and volume fraction of bubbles ϕ, a relation $\phi = 4\pi R^3 N_\phi/3$ holds. In this relation N_ϕ is number density per unit volume of highly vesicular volcanic rocks containing bubbles. As number density per unit volume of melt before vesiculation N is required, N_ϕ is converted to N using the relation $N = N_\phi/(1-\phi)$. Letting the volume fraction of bubbles (vesicularity) be 70%, $N \approx 5 \times 10^{15}\ \mathrm{m}^{-3}$. Because SiO_2 content is 78 wt%, the decompression rate of Osumi pumice fall will be located at a star mark in Fig. 10.2 and is found to be on the order of about 10^8 Pa/s. This graph contains other data on various Plinian eruptions and is applicable to decompression rate between 10^6 and 10^8 Pa/s.

As described in Chap. 1, the eruption column height of Plinian eruptions can be estimated using the dependence of the size of eruptive materials, found by geological surveys, on the distance from a source of supply. The right graph of Fig. 10.2 shows correlation between the eruption column height and the decompression rate at bubble nucleation. This graph shows that the decompression rate is higher as the intensity of eruptions is higher and that the decompression rate is proportional to the eruption column height H_{T} to the power of four.

---

[7] Although bubble number density should be calculated by the method to estimate three-dimensional morphological information from a two-dimensional image, as described in Sect. 11.9 in the Appendix (Chap. 11), a simplified method was used here.

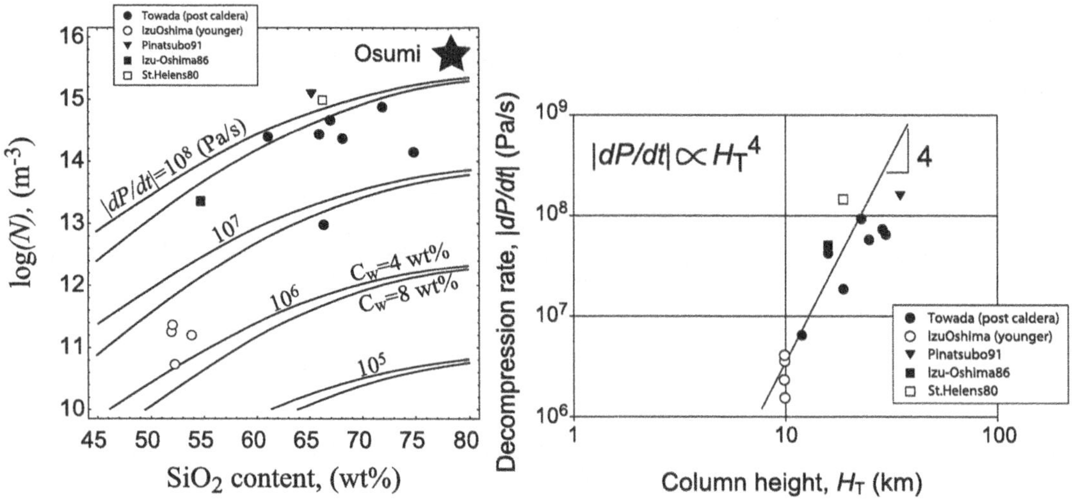

Fig. 10.2 (left) A contour plot of the decompression rate as a function of bubble number density and SiO_2 content. The cases of 4 wt% and 8 wt% of the initial water content are plotted. Marks in the figure represent data of pumice and scoria produced from eruptions specified in the legend (right). Relationship between the decompression rate at bubble nucleation depth, which was calculated using an observable quantity of bubble number density N and physical properties of saturation pressure, assumed temperature, and the eruption column height. Adapted from Toramaru (2006), copyright 2006, with permission from Elsevier

$$\left|\frac{dP}{dt}\right|_{\mathrm{BN}} = c_{\mathrm{H}} H_{\mathrm{T}}^{4} \tag{10.3}$$

where c_{H} is a constant of proportionality and is about $10^{-9.5}(\mathrm{Pa\,s^{-1}\,m^{-4}})$.

A scaling law of eruption intensity

When plumes are continuously supplied, it is empirically considered that the eruption column height is proportional to the discharge rate $\dot{M}$ (kg/s) to the power of 1/4 and that the discharge rate is proportional to the discharge velocity U_{VE} at the vent; thus, the following equation holds:

$$H_{\mathrm{T}} = c_{\dot{M}} \dot{M}^{1/4} = c_{\mathrm{U}} U_{\mathrm{VE}}^{1/4} \tag{10.4}$$

where $c_{\dot{M}}$ and c_{U} are coefficients of proportionality (See Koyaguchi 2008; Sparks et al. 1997, for more detail).

Interestingly, the decompression rate $|dP/dt|_{\mathrm{BN}}$ at the bubble nucleation depth and the discharge velocity U_{VE} at the vent or the discharge rate $\dot{M}$ are in proportion to each other. This is a type of scaling law connecting the discharge velocity, which is a surface phenomenon, with dynamics in volcanic conduits, which cannot be directly observed. That is,

$$\left|\frac{dP}{dt}\right|_{\mathrm{BN}} = c_{\dot{M}} c_{\mathrm{U}}^{1/4} U_{\mathrm{VE}} \tag{10.5}$$

As described in Sect. 5.5, in the case of steady flow in volcanic conduits, the decompression rate in a volcanic conduit is proportional to the discharge velocity from the vent and is simply given by Eq. (5.67). In this case, the coefficient of Eq. (10.5), $c_{\mathrm{H}} c_{\mathrm{U}}^{1/4}$, is equal to ρg. Assuming that the density ρ

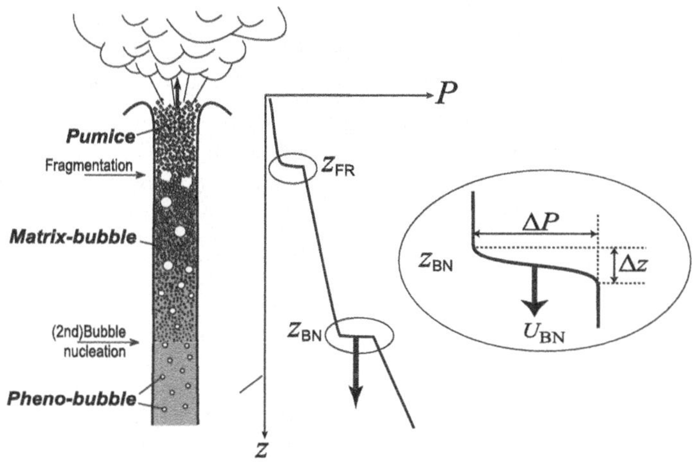

Fig. 10.3 Two rapid decompression processes that can explain high bubble number density observed in explosive eruptions. One is rapid decompression accompanied by fragmentation of magma at a depth of Z_{FR}. Another one is rapid decompression caused by propagation of rarefaction shock wave related to bubble nucleation at a depth of Z_{BN}

is 1000 kg/m^3 and the velocity at the vent is 100 m/s, the same as the velocity of sound[8] of the mixture of pyroclasts and gases as rough estimates, the decompression rate will be 10^6 Pa/s. Although a simplified estimation, this is not much different from a more accurate estimation based on the conduit flow model (Wilson et al. 1980). However, the decompression rate estimated from the bubble number density is 10^6–10^8 Pa/s, which is much larger than the estimate obtained based on the steady flow in volcanic conduits. This difference indicates that the decompression process controlling bubble nucleation differs from the decompression process in the steady flow in volcanic conduits. Let us consider a mechanism of how a scaling law holds, which may explain the difference.

A mechanism of a scaling law

One possibility is a decompression process right above the fragmentation surface. Right above the fragmentation surface, pressure drastically changes and the decompression rate increases. However, the fragmentation of magma is considered to occur when vesicularity (described later) becomes remarkably large after vesiculation considerably proceeds. For this purpose, phenobubbles have to exist there and grow exceedingly. This does not suit the formation of matrix-bubbles on the fragmentation surface caused by second nucleation because of the decrease of water content in melt. However, if second nucleation and fragmentation of magma correlate to each other and simultaneously occur, this can be a mechanism that produces high bubble number density. This may be related to the third possibility or the role of pheno-bubbles in fragmentation of magma (See the section of pheno-bubbles below).

The second possibility is that gas-liquid surface tension is smaller than that obtained from an experiment by Bagdassarov et al. (2000). From Eq. (10.2), if surface tension decreases by one order of magnitude, decompression rate estimated from the bubble number density observed will decrease by two orders of magnitude and decompression under steady flow in volcanic conduits is acceptable (Campagnola et al 2016).

The third possibility is that a type of shock wave (rarefaction shock wave) that causes rapid decompression triggers bubble nucleation (Toramaru 2006) (Fig. 10.3). Shock waves are produced in a normal fluid only when a substance changes from low pressure to high pressure, i.e., a substance is compressed. This simultaneously generates entropy and satisfies the second law of thermodynamics (see Box in Chap. 2). However, if a substance moves from high pressure to low pressure (as is the case of magma ascending in a volcanic conduit), a normal substance simply expands adiabatically and does

[8] A choked flow is assumed. A choked flow is a condition where a flux becomes limited by the velocity of sound when compressed fluid passes through a pipe.

not produce entropy.[9] The reason why shock waves are not generated when a normal fluid moves from high pressure to low pressure is that if shock waves existed, entropy would decrease; this phenomenon goes against the second law of thermodynamics (Zeldovich and Raizer 1967).[10] However, if a specific heat is extremely large or a nonequilibrium process such as phase changes occurs with decompression, entropy is generated and shock waves can be formed (Borisov et al. 1983; Thompson et al. 1986). In this case, the decompression rate is the product of the propagation velocity of a bubble nucleation wave U_{BN} and the pressure gradient $\Delta P/\Delta Z$. If we assume that the pressure gradient is 10 MPa/m, the propagation velocity is about 10 m/s.

10.2.3 Pheno-Bubble

Pheno-bubbles have two important meanings: (1) they bear overpressure of magma chambers for triggering eruptions, and (2) they bear buoyancy, which is a driving force for the magma ascent from an underground magma chamber to the ground surface. To examine this, let us look at the pumice of Osumi pumice fall, which is a precursory Plinian eruption of the Aira caldera formation (Fig. 10.1). Moreover, let us discuss whether second nucleation occurs or not, using Fig. 5.21.

The state of pheno-bubbles in magma chambers
In Fig. 10.1, 10–100 pheno-bubbles measuring about 1 mm exist in 1 cm^2 of a cross-section of the sample. Using this figure and assuming that the vesicularity is 70%, $10 \sim 100/(0.01^2\,(m^2))/0.001\,(m)/(1-0.7) \approx 3 \times 10^8 \sim 3 \times 10^9\ m^{-3}$ of pheno-bubbles exist in a unit volume of melt under atmospheric pressure (assumed pressure).

To convert the size of bubbles into that in the magma chamber, the depth and the pressure of the magma chamber are assumed to be 10 km (250 MPa). As the magma ascent can be considered as an isothermal process, based on an equation of state of an ideal gas, the bubble size R_{PE} in the magma chamber can be estimated at $(P_f/P_0)^{1/3} \times 10^{-3}$ m, where P_f is the pressure at which pumice solidifies and P_0 is 250 MPa, the pressure in the magma chamber. If the solidification pressure of pumice is atmospheric pressure (0.1 MPa), R_{PE} is 0.074 mm. As will be discussed in Sect. 10.2.4 about vesicularity, the solidification pressure of pumice is probably higher than 1 atm. Thus, if the solidification pressure is assumed to be 25 MPa and vesicularity is assumed to be 70%, R_{PE} is 0.46 mm. Let us consider the case where magma containing about $10^9/m^3$ of the pheno-bubbles of that size ascends just before eruption. Whether water exsolution from melt proceeds by the growth of these existing pheno-bubbles or by the generation of matrix-bubbles triggered by new nucleation can be determined from Fig. 5.21.

Figure 5.21 shows the state of pheno-bubbles in Osumi pumice estimated in this manner. The volume fraction of the pheno-bubbles is almost equivalent to $10^{-3\pm1}$, which is included in the transition regime from the low-volume fraction regime to the high-volume fraction regime. In the case where the magma in this state (gas-liquid surface tension $\gamma = 0.1$ N/m) is decompressed, second nucleation occurs and then matrix-bubbles form if the decompression rate is 10^4 Pa/s or higher. If gas-liquid surface tension is one order of magnitude smaller, the limit decompression rate becomes 10^2 Pa/s, i.e., two orders of magnitude smaller. As seen earlier, the decompression rate estimated from matrix-bubbles (10^8 Pa/s) is sufficiently larger than this figure. This is not in conflict with the idea that matrix-bubbles are formed by second nucleation.

---

[9] In the case where viscous dissipation is not involved.

[10] It is related to the fact that the specific volume—pressure curve of ordinary fluid is always convex downward.

Occurrence of second nucleation and the ascent velocity from magma chambers

Let us estimate the initial ascent velocity of magma containing such pheno-bubbles. If the ascent of magma from a magma chamber is determined by balance between buoyancy and viscous drag, the initial velocity $U(0)$ can be estimated from $L_D^2 \Delta\rho g/\eta$, where L_D is the open width of a fissure and η is the viscosity of magma. Assuming that $L_D = 1$ m and $\eta = 10^5$ Pas, the initial velocity is 2.5×10^{-4} m/s.[11] The decompression rate of magma at this time can be evaluated by $\rho g U(0)$, resulting in $2.5^2 \approx 6.25$ Pa/s when $\rho = 2500$ kg/m^3. This is sufficiently smaller than the limit decompression rate at which second nucleation occurs; therefore, the magma starts to ascend without occurrence of second nucleation of bubbles. Subsequently, the magma increases the ascent velocity and the decompression rate with expansion and growth of bubbles,. At a certain point in time, the magma exceeds the limit because of some causes and starts second nucleation. Because the decompression rate at that time is between 10 and 100 MPa/s, some causes cannot be explained by a simple steady-state conduit flow model, as described earlier.

Overpressure of the magma chamber just before an eruption

Although there are no guarantees that the magma chamber is in mechanical equilibrium before formation of pheno-bubbles, if the magma chamber tries to expand by the volume fraction of pheno-bubble from the mechanical equilibrium state against the surrounding rock, overpressure ΔP is approximately the product of the shear modulus of surrounding rock and the increasing ratio of volume, i.e., 0.001. Assuming that the shear modulus of surrounding rock is 10 GPa and overpressure is 10 MPa, i.e., about 100 atm, there is a possibility that this sort of overpressure causes a fissure in surrounding rock and triggers an eruption.

If this overpressure is the driving force of the ascent velocity of magma, the initial velocity can be estimated by $U(0) \approx \Delta P L_D^2/(L_E \eta)$, where L_E is the length scale of extension of the initial fissure, which is larger than the width L_D. Assuming that $L_D \approx L_E$ is the upper limit of $U(0)$ estimation, the initial velocity is in the order of 100 m/s. The corresponding decompression rate, $\Delta P \times U(0)/L_E$, is about 10^3 MPa/s, which is considerably large and can probably trigger second nucleation. However, because this estimation does not consider some factors such as an unsteady process at the initial phase of fissure formation and the shape of an open fissure (Koyaguchi 2008) and contains considerable uncertainty, further examination is required.

The role of pheno-bubbles in the fragmentation process of magma

Recently, studies revealed that pheno-bubbles may have a further important meaning in the mechanism of magma fragmentation. Although the magma fragmentation process is an important elemental process that separates explosive eruptions and nonexplosive eruptions, much has remained unknown to date (Gonnermann 2015). Using bubble-bearing syrup having rheological properties similar to those of magma as an analog material and using a device in Spring 8, Kameda et al. (2017) conducted the in situ observation of fragmentation process caused by rapid decompression and explained that the inhomogeneous distribution of the bubble size triggers brittle fracture. They found, based on both theory and experiment, the wall of large bubbles (which can be considered as pheno-bubbles) surrounded by small-sized bubbles (which can be regarded as matrix-bubbles) is broken and that the break triggers the fragmentation of the whole specimen. This result indicates that there is a close relationship among magma fragmentation process, which controls an explosive eruption, matrix-bubble formation process, and the presence of pheno-bubbles. Moreover, if pheno-bubbles are ubiquitous, this mechanism seems to generally work, and it gives hints on attaching a meaning to the size distribution of tephra (a general term for pyroclasts such as highly vesicular volcanic rocks and volcanic ash), which is produced by fragmentation.

---

[11] In a steady-state conduit flow model, the initial velocity is one of the important parameters influencing degassing efficiency as well as gas permeability and controls whether the eruption becomes explosive or nonexplosive.

10.2.4 Vesicularity

Vesicularity of highly vesicular volcanic rocks (pumice and scoria) is an observable that is most difficult to give a physical meaning. For example, assuming that the initial water content is 3–6 wt%, which is a common estimate, vesiculation to 1 atm in a closed system results in 99.88% of the volume fraction of bubbles. Note that there are no such eruptive materials other than reticulite, which are produced from lava fountain eruption of basaltic magma. This indicates that if pumice or scoria solidify at 1 atm, the mass conservation does not hold in terms of water (as a representative of volatiles) between pre-eruptive magma and pumice or scoria (eruptive materials). This suggests that the vesiculation process in explosive eruptions that produce pumice or scoria is not a closed system but an open system, i.e., water has gone out of magma in some form during the vesiculation process. Although this is a big paradox in vesiculation in a closed system, a discussion with the assumption that the system is closed can lead to the following suggestions.

If assuming that pumice and scoria solidify at a pressure higher than 1 atm, the mass conservation seems to hold in terms of water even in a closed system. Figure 10.4 shows a graph of vesicularity as a function of solidification pressure at various saturation pressure. While the vesicularity is higher than 99% when solidification pressure is 1 atm (0.1 MPa), the vesicularity ranges from 70% to 90% when solidification pressure is 100 atm. Houghton and Wilson (1989) believed that the variety of vesicularity of pumice or scoria was affected by a pressure at which they solidified. For Plinian eruptions, if solidification pressure is within 100–300 atm for 250 MPa of saturation pressure, for example, the vesicularity will be within the range between about 90 and 70%, which are realistic values. The vesicularity of eruptive materials from subaqueous eruptions and phreatomagmatic eruptions considerably varies, and it indicates that the range of solidification pressure in these eruptions is considerable. In phreatomagmatic eruptions, such cases can occur because magma undergoes a complicated temperature—pressure history because of interaction between water and magma.

Gardner et al. (1996) estimated viscosity from magma composition and obtained data, indicating that vesicularity approaches a certain value when viscosity becomes low. Moreover, they reported that vesicularity of highly viscous magma is low because bubbles cannot expand to the equilibrium size because of viscous resistance. Furthermore, Kaminski and Jaupar (1997) assumed a fluid-dynamic behavior from a vent of a Plinian eruption and calculated expansion of pumice by giving a thermal history and pressure change of gas in a volcanic plume with consideration given to expansion of the gas and viscous resistance and cooling process of pumice. Consequently, vesicularity in pumice and scoria

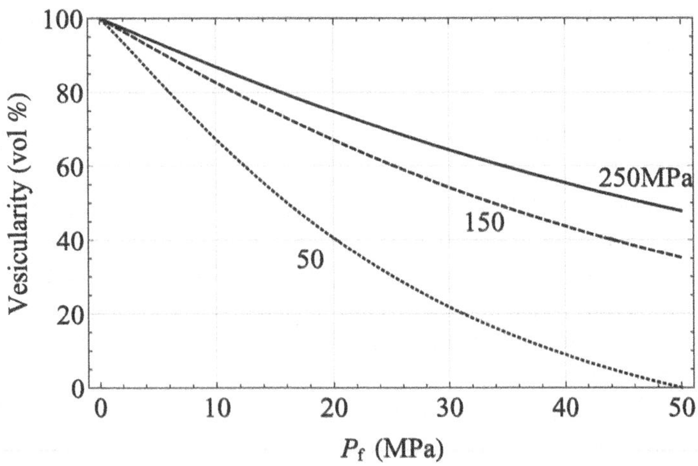

Fig. 10.4 Relationship between the vesicularity and the solidification pressure P_f in closed equilibrium vesiculation process. Numbers next to curves represent the initial saturation pressure (MPa)

or within a volcano was found to become considerably inhomogeneous. The results of these studies suggest that the viscosity of magma is a factor controlling vesicularity of highly vesicular volcanic rocks.

10.3 Microlite Crystalline Texture of Plinian or Subplinian Eruptions

As described in the first section of this chapter, the mode of occurrence of eruptive products and rock texture, among pumice and scoria produced from Subplinian or Plinian eruptions, silica-rich ones such as high silica rhyolite often contain no microlites. On the other hand, andesite and basaltic andesite scoriae often contain microlites. Silica-poor (sub-)Plinian eruptions show a tendency toward relatively lower eruption column height. A rough correlation among subplinian eruptions, low-silica content, and microlite-containing appears to hold.

10.3.1 Characteristics of Crystalline Texture

Microlites, groundmass crystals in pumice and scoriae, are formed as a result of water exsolution from melt associated with decompression. Therefore, the characteristics of the shape and texture of microlites can be an indicator of decompression process. In particular, the crystal number density will be a function of the effective cooling rate if nuclei of crystals are formed under near homogeneous nucleation conditions, as can be seen in Chap. 8. The effective cooling rate is equivalent to the water exsolution rate and can be converted from the decompression rate in the equilibrium vesiculation regime. Because the decompression rate can be translated to the magma ascent velocity, the ascent velocity can be estimated.

Let us use the 1986 subplinian eruption at Izu-Oshima, which is basaltic andesite in the chemical compositions, as an example. As introduced in Chap. 1, changes in this eruption, including the change in the eruption column height over time, have already been known in detail. To use this advantage, the scoria deposit layer was divided into several sub-layers with several centimeters to a dozen or so centimeters for sampling, from which samples were collected. Then, particles with sizes of about 1 cm were selected from the whole samples in each sub-layer for analysis. By taking similar size of particles for analysis from sub-layers, we can relate the position (stratigraphic height) in the deposit from lower to upper layers to the order of time when the particles were released from the vent. Although in general proportionality between the time released from the vent and the stratigraphic position of a particle in the deposit are not guaranteed, by taking similar size of particles, we can minimize the influence of size sorting during the settling in the atmosphere.[12]

Scoriae from this eruption contain plagioclase, pyroxene, and a small amount of iron-titanium oxides as microlites. Figure 10.5 shows the results of textural analysis (relationship between the crystallinity and the crystal number density). The crystallinity and the number density extensively vary from microlite-poor to microlite-rich. This variety is characteristically found in basaltic to andesitic Plinian (Subplinian) eruptions. The microlite-poor type is called Type A and microlite-rich one is called Type B. Note that the transition between the two types is continuous. A positive correlation between the crystallinity and the number density is found in pyroxene but is not found in plagioclase.

[12] If the settling flux on the ground (the uppermost depositional surface) is constant, the increasing rate of thickness of the deposit is constant. Therefore, a stratigraphic positions in the deposit is exactly proportional to time. However, if the particle size distribution or the eruption column height at a source of supply change with time, the settling flux will change with time. In this case, to determine the scale of time on vertical positions in the deposit, we have to calculate it by the method of Iriyama et al. (2018).

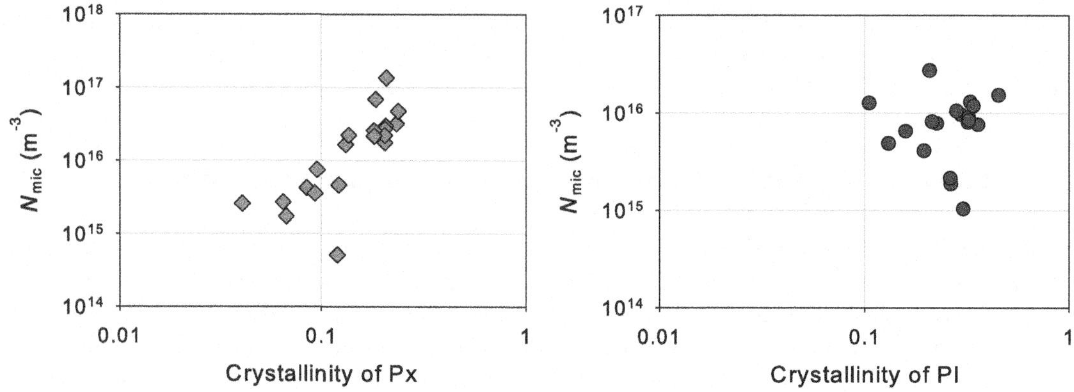

Fig. 10.5 Relationship between the crystallinity and the crystal number density in the 1986B Subplinian eruption at Izu-Oshima. (Left) pyroxene, (Right) plagioclase

10.3.2 Crystal Number Density, Water Exsolution Rate, Decompression Rate, and Ascent Velocity: Application of MND Water Exsolution Rate Meter

Because nucleation of pyroxene occurs under conditions closer to homogeneous nucleation conditions than nucleation of plagioclase, as seen in Chaps. 7 and 8, let us advance the discussion with the focus on pyroxene. The number density of Type-A pyroxene microlites is approximately $10^{15}\,m^{-3}$ and that of Type-B is approximately$10^{17}\,m^{-3}$. Moreover, the amount of Type-A pyroxene crystals is several percent, whereas that of Type-B pyroxene crystals is 20%.

The microlite number density (MND) N_{mic} is connected with the water exsolution rate $|\dot{C}_W|$ by Eq. (8.12). From this equation, the water exsolution rate is given by Eq. (8.14) as a function of the observable N_{mic}:

$$|\dot{C}_W| = \frac{1}{T_L' a^{2/3}} N_{mic}^{2/3} \tag{10.6}$$

where a coefficient a is a kinetic factor given by Eq. (7.151) and a thermodynamic factor T_L' is given by Eq. (8.7). The substitution of these factors into the coefficient $(T_L' a^{2/3})^{-1}$ of Eq. (10.6) leads to

$$(T_L' a^{2/3})^{-1} = 5.9 \times 10^{-12-0.23\Delta C_{Si}+0.43C_W} \tag{10.7}$$

Assuming that the water content at which nucleation of pyroxene occurs is 1 wt% and the difference with 50 wt%, the standard silica content, ΔC_{Si}, is 8. Note that substituting these values into Eq. (10.7) yields

$$(T_L' a^{2/3})_{Py}^{-1} = 2.3 \times 10^{-13} \tag{10.8}$$

Using this value, the water exsolution rate of Type A is calculated at about 0.0023 wt%/s and that of Type B is calculated at 0.05 wt%/s. If they are in the equilibrium vesiculation regime, the decompression rate is calculated by Eq. (8.16) at 2.6×10^4 Pa/s for Type A and 5.6×10^5 Pa/s for Type B. Their conversion into the magma ascent velocity $U = |\dot{P}|/(\rho g)$ (the density is converted into an order of 10^3 kg/m^3) yields the order of 2 m/s for Type A and the order of 60 m/s for Type B.

10.3.3 Relationship with Shift of Eruption Style

The variety of crystalline texture correlates with shift in eruptions. Figure 10.6 shows changes in the eruption column height with time, which were obtained via observation, and the ratio of types of crystalline texture in each stratigraphic sub-layer. This figure informs two important points. The first point is that microlite-poor Type A is predominant in the initial and end phases of the eruption and microlite-rich Type B is predominant in the peak phase. Another point is that Type A and Type B always coexist. Although there is a clear contrast of the ratio of types between the initial (Units 1 and 2) and end (Units 5 and 6) phases and the peak phase (Units 3 and 4), the crystallinity of microlites varies in any horizon (unit). These facts impose a constraint on dynamics in volcanic conduits.

Under the two assumptions that crystallization occurs in the equilibrium vesiculation regime and that the nucleation of microlite is close to homogeneous nucleation, the variety of the crystal number density results in the variety of the magma ascent velocity at microlite nucleation. Thus, in this eruption, magma that ascended the volcanic conduit at various velocities was simultaneously discharged from the vent. Considering that in such a state of the volcanic conduit, magma ascends with a turbulent state where magma moves up and down and magma undergoes various water exsolution rates. Is it really true that such a conclusion is unique? In the next section, we will have a deeper discussion on it in the process of examining correlation between the microlite number density and the crystallinity.

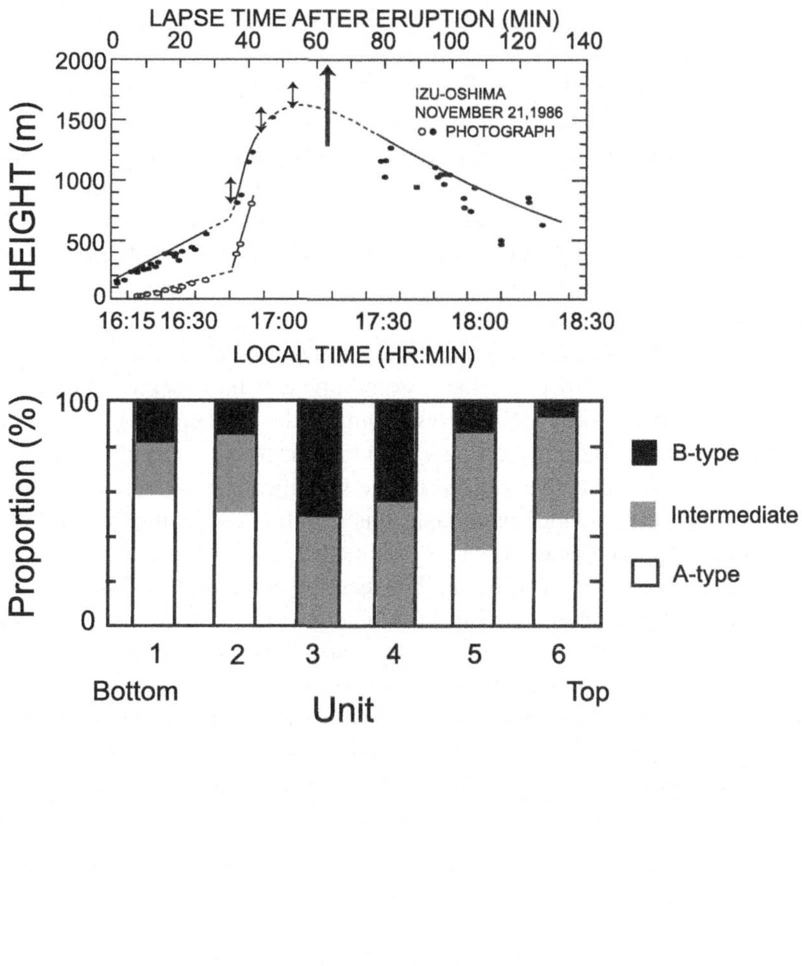

Fig. 10.6 (Upper graph) Change in the eruption column height with time of the 1986B Subplinian eruption at Izu-Oshima. (Lower graph) The ratio of microlite texture in each deposit layer. At a point 2 km east of the vent. Thickness: about 30 cm. (adapted from Toramaru et al. 2008)

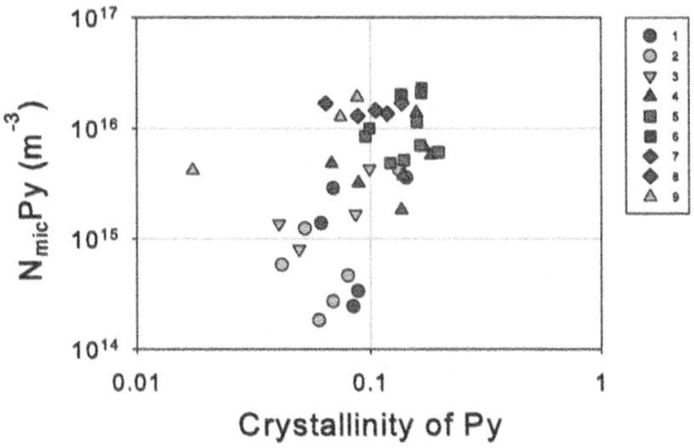

Fig. 10.7 Relationship between the crystallinity and the number density of pyroxene microlites emitted from the Hoei eruption of Mt. Fuji

10.3.4 Relationship Between the Crystal Number Density and the Crystallinity

Positive correlation between the microlite number density and the crystallinity of pyroxene microlites found in the 1986 Subplinian eruption at Izu-Oshima, which is basaltic andesite in the chemical compositions, is found in scoriae emitted from the Hoei eruption of Mt. Fuji (A.D.1707), which shares similar features in the chemical composition and the eruption style (Fig. 10.7). Because this positive correlation between the crystal number density and the crystallinity is observed in several other eruptions (Hammer et al. 1999; Miwa et al. 2009; Martel 2012; Pichavant et al. 2013) and seems to be a universal relationship, let us call this correlation microlite systematics. To understand this correlation, it should be based on a decompression-induced crystallization process in which both the vesiculation process and the crystallization process explained to date are involved.

When nucleation of crystals occurs in the equilibrium vesiculation regime under kinetic conditions close to homogeneous nucleation conditions, the amount of crystals is determined by the water content of magma at the time when the magma effectively solidifies. As the water content is uniquely determined by pressure, when the solidification pressure is almost the same, the amount of crystals would be the same. This is observed in the single-step decompression experiment (SDE) (Fig. 8.8) that the amount of crystals has a considerable range, which indicates that solidification pressure varies. A smaller amount of crystals means higher solidification pressure. As explained in the section on vesicularity, solidification pressure can take values in a wide range.

If the vesicularity is an index of the solidification pressure in a closed system, a positive correlation between the vesicularity and the crystallinity will be observed. Figure 10.8 shows the correlation between the vesicularity and the crystallinity of the Hoei eruption of Mt. Fuji. The correlation between the vesicularity and the crystallinity found in this figure is negative rather than positive. Although the crystallinity should be determined by the solidification pressure if it follows the equilibrium theory described in the previous paragraph, this negative correlation suggests that the crystallinity is not determined by solidification pressure. Note that a closed system is assumed here. Klug and Cashman (1994) focused on the relationship between the amount of crystals and the viscosity and considered that the greater the amount of crystals, the higher the viscosity and the lower the vesicularity.

Let us consider that the vesicularity is controlled by another factor such as the out-gassing process.[13] The positive correlation between the crystal number density and the amount of crystals indicates that the lower the solidification pressure, the higher the decompression rate. Assuming that the amount of decompression that corresponds to the solidification pressure is ΔP_f, the decompression rate $|d\Delta P_f/dt|$ will have a positive correlation with ΔP_f. If the correlation is a proportional relation,

[13] In other words, negative correlation between the vesicularity and the crystallinity shall be excluded from consideration here.

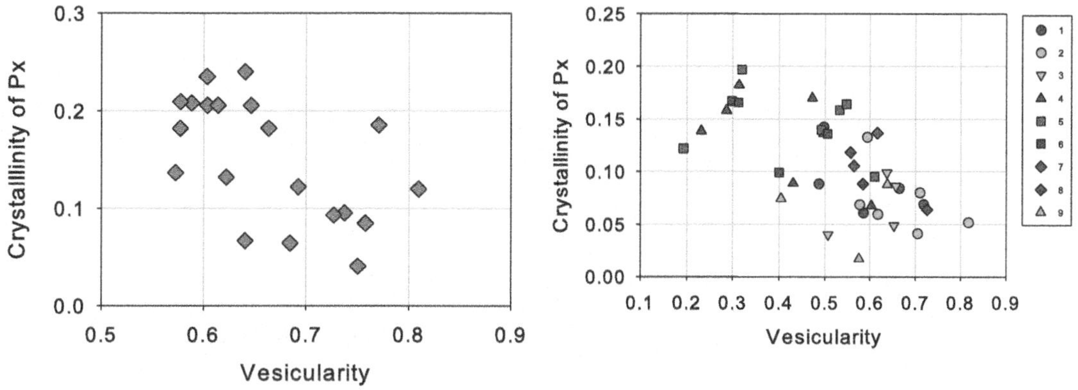

Fig. 10.8 Relationship between the crystallinity and the vesicularity of pyroxene microlites found in Subplinian eruptions involving basaltic andesite. (Left chart) The 1986B eruption at Izu-Oshima. (Right chart) The Hoei eruption of Mt. Fuji. Negative correlations were found in both charts

$$\left|\frac{d\Delta P_{\mathrm{f}}}{dt}\right| \propto \Delta P_{\mathrm{f}} \tag{10.9}$$

holds. This apparent relation means that the ascent rate is proportional to the amount of decompression.

The decompression rate has a relationship of $N_{\mathrm{mic}} \approx |d\Delta P_{\mathrm{f}}/dt|^{3/2}$ with the crystal number density. Moreover, because the amount of decompression is almost proportional to the amount of crystals ϕ_{obs}, $\Delta P_{\mathrm{f}} \propto \phi_{\mathrm{obs}}$ holds. From these, the following relationship is obtained.

$$N_{\mathrm{mic}} \propto \phi_{\mathrm{obs}}^{3/2} \tag{10.10}$$

Because the relationships found in the 1986 eruption at Izu-Oshima and the Hoei eruption of Mt. Fuji are a little steeper than the power of 3/2, it is suggested that the decompression rate is somewhat strongly dependent on the amount of decompression compared with a proportional relation as in Eq. (10.9).

Note that for homogeneous nucleation, because the nucleation rate that determines the crystal number density exponentially depends on the amount of decompression, the crystal number density changes almost ten orders of magnitude when the amount of decompression changes only several times. The change of the crystal number density that would be found in observation and experiments appears when heterogeneous nucleation occurs (Fig. 8.6).

In fact, when a pre-exponential factor of the nucleation rate in heterogeneous nucleation is sufficiently small, a positive correlation between the crystal number density and the amount of crystals appears. As developed in Sect. 9.3.5 in Chap. 9 on CSD, this relationship will be proportional, as expressed in Eq. (9.163). In this case, the nucleation rate is almost constant and the crystal growth rate is inversely proportional to time. The actual correlation is steeper than proportional relation (the crystal number density is proportional to the amount of crystals to the power of two to three) and this model cannot be applied to it as it is; however, there is room for further examination.

Furthermore, uneven temperature in magma just before an eruption is a possible candidate. Lower temperature gives higher effective supercooling in decompression-induced crystallization and higher crystallinity. Moreover, because higher crystallinity gives higher number density, it is expected that the ascent rate of the lower-temperature part of the magma chamber just before an eruption is high. Frankly, we believe that this is the most likely possibility.

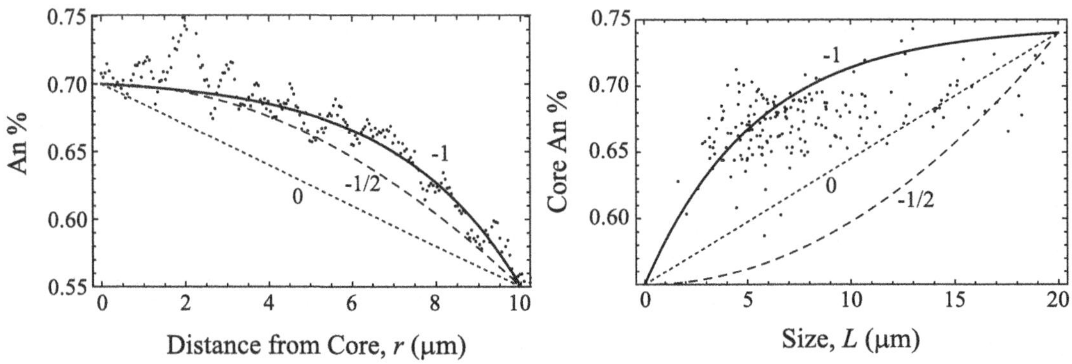

Fig. 10.9 Chemical composition and zonal structure of microlites found in scoriae emitted from the 2011 Subplinian eruption of Shinmoedake in the Kirishima volcano group. (Left chart) Zoning profile. (Right chart) Correlation between the core composition and the size. Figures next to curves represent exponents of time of growth laws: 0 (constant growth rate), −1/2 (diffusion-controlled growth), −1 (CSD Model 4)

Although several possibilities explaining the positive correlation between the crystal number density and the amount of crystals are mentioned above, most of them have not yet been sufficiently examined. One thing we should remember when discussing the magma ascent rate is that the interpretation of the microlite number density may change based on the results of future studies.

10.3.5 Chemical Composition of Microlite Emitted from the 2011 Subplinian Eruption of Shinmoedake

The 2011 eruption of Shinmoedake in the Kirishima volcanoes started with a phreatomagmatic eruption, followed by three Subplinian eruptions in 24 h and then intermittent Vulcanian eruptions and simultaneous lava dome formation at the vent (Kozono et al. 2013; Nakada et al. 2013). Whole-rock chemical composition of most of scoriae produced from the Subplinian eruptions is andesite with a SiO_2 content of 58 wt%. A semilogarithmic graph of the crystal size distribution (CSD) of microlites in the scoriae is gently convex downward (Suzuki et al. 2018). Many microlites are tabular to needle-like in shape.

Generally, microlites formed under a relatively higher cooling rate have a variety of compositional zoning because of their complex crystal growth (Sect. 7.6), and the same holds true for the products of the 2011 eruption of Shinmoedake. Kichise (2015) examined the composition of the microlites and found interesting facts. Among them, some microlites have marked normal zoning (Fig. 10.9; left chart), and there is a positive correlation between the core composition and the size (Fig. 10.9; right chart). As described in Sect. 7.7.2, when a zonal structure is considered as growth zoning, the composition profile of the zonal structure is related to the growth law (see Fig. 7.29 and relevant equations). Moreover, the relationship between the core composition and the size is influenced by the growth law (Fig. 7.30 and relevant equations). Figure 10.9 shows the relationship compared with the following three theoretical predictions: a constant growth rate (reaction-controlled growth), a growth law that is inversely proportional to time (CSD Model 4), and a growth law that is proportional to time to the power of −1/2 (diffusion-controlled growth). While it is difficult to distinguish between diffusion-controlled growth and a growth that is inversely proportional to time in the chart of composition profile (Fig. 10.9; left chart), the relationship between the core composition and the size can explain well the case where the growth rate is inversely proportional to time (Fig. 10.9; right chart). These results suggest that decompression-induced crystallization during magma ascent in the 2011 Subplinian eruption of Shinmoedake occurred under the condition of CSD Model 4 (Sect. 9.3.5), i.e., the condition in which the crystallization rate and the nucleation rate are constant (Toramaru and Kichise, manuscript in preparation).

10.4 Volcanic Ash Emitted from Vulcanian Eruptions

10.4.1 Characteristics of Vulcanian Eruptions

A Vulcanian eruption emits volcanic plumes to a height of several kilometers or lower with shock waves and volcanic ash. Before the eruption, ground deformation indicating the increase in pressure immediately underneath the vent, followed by contraction just before the eruption occurs (Iguchi et al. 2008). The eruption occurs suddenly and often lasts less than several dozen seconds. We consider the mechanism of how this type of eruption occurs is that a cap formed around the vent comes off and gas and magma compressed in a conduit burst forth into the air. Note that the shock-tube model often applies to this mechanism (Ishihara 1985; Turcotte et al. 1990).

Volcanic ash emitted from Vulcanian eruptions differs from that emitted from caldera-forming eruptions in constituent materials. While most of volcanic ash emitted from caldera-forming eruptions is bubble-wall types of glass and micropumice, as seen in Chap. 1 (Fig. 1.11), volcanic ash emitted from Vulcanian eruptions contains many rock fragments. Let us consider Vulcanian eruptions that occurred at Sakurajima volcano as an example and think about what is found from volcanic ash formed by this type of eruption.

10.4.2 Vulcanian Eruptions at Sakurajima Volcano and Characteristics of Its Volcanic Ash

From the 1970s onward, the volcanic ash of Sakurajima volcano has been archived by Sakurajima Volcano Research Center, Disaster Prevention Research Institute, Kyoto University. Sakurajima volcano has an important advantage that geophysical data, such as seismograms (on smoked papers) of volcanic earthquakes and data obtained by tiltmeters and linear strainmeters, has been archived in the Center since then. Therefore, Sakurajima volcano, which frequently erupts is an ideal location for comparison between materials scientific data and geophysical data.[14]

A difficult point in analyzing volcanic ash from Vulcanian eruptions is that a variety of volcanic ash types having different histories are simultaneously emitted. Vulcanian eruptions occur frequently and repeatedly from the identical open conduit. Therefore, volcanic ash includes not only one derived from magma that directly contributes to the driving force of the eruption of interest but also one derived from old magma. Volcanic ash should be classified first according to the freshness by careful observation to select volcanic ash that seems to directly contribute to the eruption event of interest. Miwa et al. (2009) classified volcanic ash emitted by Vulcanian eruptions from Sakurajima volcano from the 1970s to 1980s using such a method and compared textural data of the volcanic ash and geophysical data using the advantage of Sakurajima volcano.

[14] Currently, scientific monitoring of materials has been continuously conducted (Arzilli and Carroll 2013).

10.4.3 Relationship Between Crystal Texture and Surface Phenomena

Figure 10.10 shows the correlation between the amplitude of explosion earthquakes and the fraction of fresh particles contained in volcanic ash (left chart). The chart shows that, although several trends are observed, the higher the fraction of fresh particles, the larger the amplitude of explosion earthquakes. This is the first instance that actually shows that the more the gas supplied from fresh magma, the higher the explosion intensity (the difference between internal pressure in a conduit and the atmospheric pressure). The trends appear to be related to BL-type volcanic earthquake swarms, which is a characteristic volcanic earthquake.

BL-type volcanic earthquakes are B-type earthquakes (that do not exhibit clear signals, such as P-wave and S-wave, that characterize fracture of rock) exhibiting long-period earthquakes (dominant frequency: 1–3 Hz) (Nishimura and Iguchi 2006). Sometimes, BL-type volcanic earthquakes occur in swarms before eruption and seem to be continuous tremors. A close comparison of characteristics found in both such volcanic earthquakes and volcanic ash allows us to identify physical processes the materials involved in the eruption underwent.

Moreover, Miwa and Toramar (2013) focused only on volcanic ash emitted from eruption events preceded by BL-type volcanic earthquake swarms and analyzed ash texture and water content in groundmass glass of 29 ~ 81 particles of the volcanic ash from each event. Consequently, they reported an obvious positive correlation between the amplitude of air pressure waves and the bubble number density (Fig. 10.10; right chart). Although details on the state of gas in the shallow conduit of Vulcanian eruptions have remained unknown, the outgassing of vesicular magma should continuously occur. The gas removed from vesicular magma becomes fumarolic gas or enters the ground and does not contribute to eruption. The number of bubbles in volcanic ash can be used as an indicator of outgassing; thus, it is considered that the greater the number of bubbles is, the less outgassing occurs. The positive correlation seen in this figure suggests that less-outgassed magma results in larger amplitude of air pressure waves when the volcano explodes.

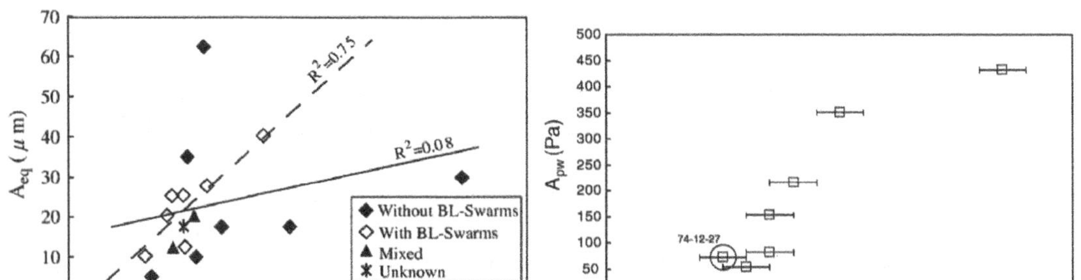

Fig. 10.10 Result of analysis of volcanic ash emitted from Vulcanian eruptions (1974–1987) at Sakurajima. (Left chart) Relationship between the fraction of volcanic ash particles having fresh surfaces (S-fraction) and the amplitude of explosion earthquakes (A_{eq}). Reprinted from Miwa et al. (2009), copyright 2009, with permission from Elsevier. (Right chart) Relationship between the representative value (VNDD) of the bubble number density (VND) of fresh volcanic ash particles containing many bubbles, emitted from events accompanied by BL-type volcanic earthquake swarms (BL-Swarm), and the amplitude of air pressure waves (A_{pw}). Reprinted by permission from Springer Nature (Miwa and Toramar 2013), copyright (2013)

10.5 Microlite Texture in Lava Domes

10.5.1 A Lava Dome of the Heisei Eruption of Unzen

Relationship between plagioclase microlites and the discharge rate

As already described in Sect. 1.3.2, particles in a deposit of a dome-collapse pyroclastic flow (block-and-ash flow) that occurred in the Heisei eruption of Unzen contained very few bubbles and a lot of microlites. The characteristics of growth of a lava dome of the Heisei eruption of Unzen is that the magma discharge rate shows two cycles, as shown in Fig. 1.16. Noguchi et al. (2008) conducted a texture analysis of eruptive materials on the time series of eruptions and reported correlations between plagioclase microlite texture, including the crystallinity, the size, and the number density, and variation of the discharge rate (Fig. 10.11).

Using data of the number density of plagioclase microlite, the water exsolution rate can be obtained from Eq. (10.6) of the microlite number density (MND) water exsolution rate meter. Assuming equilibrium vesiculation, the decompression rate can be estimated from Eq. (8.15). Moreover, assuming that pressure is determined by magma-static pressure ($P = \rho g z$, where z is the depth of the magma), the decompression rate can be related to the ascent velocity $U = dz/dt$ (Eq. (5.67)). Therefore, the ascent velocity can be estimated. For calculations, the values of chemical composition were obtained by direct analysis and water content at microlite nucleation was estimated by an experimental result of Anorthite content in plagioclase microlite (Couch et al. 2003a; Holtz et al. 2004). The estimation revealed that at

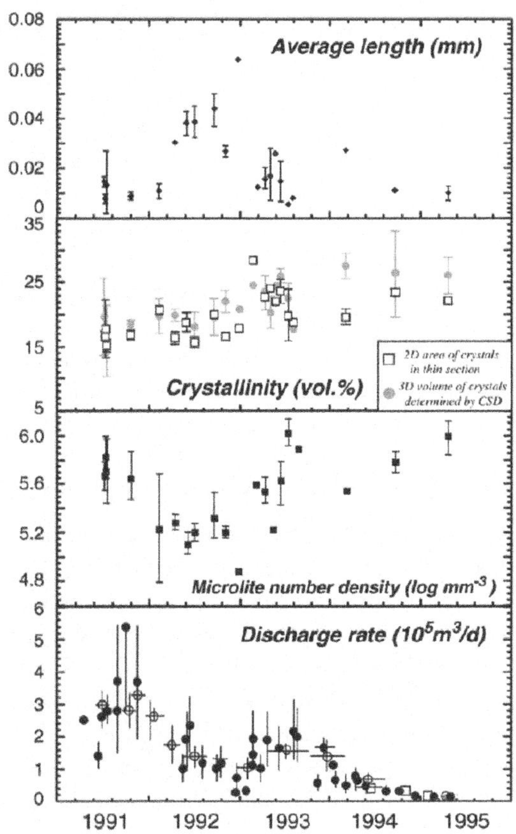

Fig. 10.11 Change of texture of plagioclase microlites produced from the Heisei eruptions of Fugendake, Unzen volcano, with respect to time (Noguchi et al. 2008). Size (average length), crystallinity, crystal number density, and discharge rate, in the order from top to bottom. Reprinted from Noguchi et al. (2008), copyright 2008, with permission from Elsevier

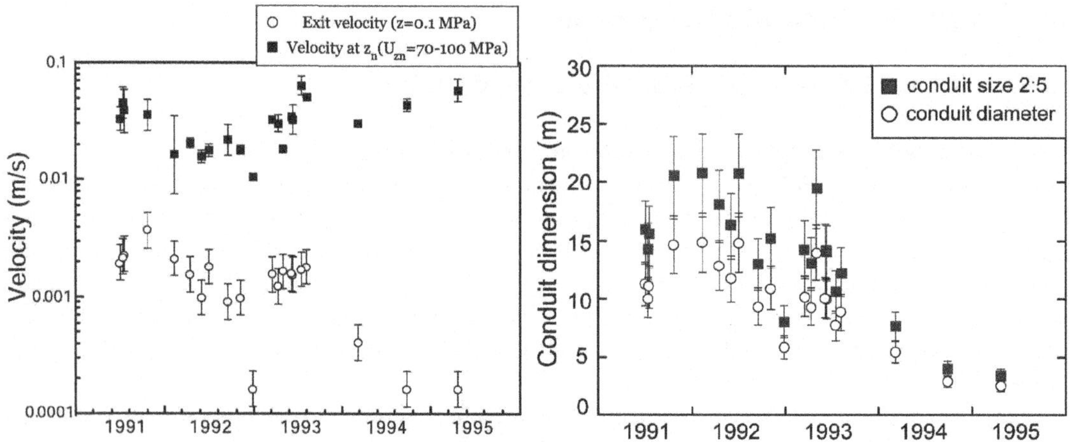

Fig. 10.12 (Left chart) Change of the ascent rate with respect to time (black squares) at 150 MPa (microlite nucleation pressure) estimated from the number density of plagioclase microlites produced from the Heisei eruptions of Fugendake, Unzen volcano, and change of the ascent velocity at the surface with respect to time (white circles) estimated from the discharge rate. The cross-sectional area at the surface is assumed to be constant at 1000–2000 m^2 (Nakada et al. 1999) (corresponding to an error bar of the rate at the surface). (Right chart) Change in the size of the conduit crosssection under the ground, determined based on both the mass discharge rate at the surface and the ascent velocity under the ground, to satisfy the conservation of mass (Eq. (10.11)). The shape is assumed to be an oval or a circle. (Noguchi et al. 2008)

a depth of 70–100 MPa (3–4 km depth), i.e., the nucleation depth of plagioclase microlites, the ascent velocity changes in the order of 0.01–0.1 m/s (Fig. 10.12).

The left chart of Fig. 10.12 informs us the following interesting points:

1. In terms of absolute values of the ascent velocity at the surface and under the ground, the ascent velocity under the ground is about one order of magnitude higher than that at the surface.
2. Up to 1993, the exit velocity at the surface and the ascent velocity under the ground changed with the same tendency.
3. Change of the ascent velocity under the ground occurred several months earlier than that of the exit velocity at the surface.
4. From 1994 onward, the tendencies of the ascent velocity at the surface and that under the ground were opposite. In 1995, when the eruption ended, while the ascent velocity at the surface naturally goes toward 0, that under the ground tended to increase.

Next, let us consider reasons for these points.

Estimation of dynamics in volcanic conduits under the ground

In a magma ascending process in conduits, the law of conservation of mass holds for parts of crystals and silicate melt except a gas phase. The conservation of mass means that as seen from a coordinate z that moves with magma (Lagrangian description), the mass of magma passing through a cross section of the conduit (cross-sectional area: $\mathcal{A}$), $\rho U \mathcal{A}$, is constant at any position along the conduit. Moreover, for nonexplosive eruptions such as lava domes, because outgassing from the system seems to be effective, the remaining part comprising silicate melt and crystals contains almost no gas and can be approximated as incompressible; moreover, the density is assumed to be constant. Consequently, the conservation of mass can be written as follows:

$$U(z_n, t')\mathcal{A}(z_n, t') = \text{const.} = U(0, t)\mathcal{A}(0, t) \tag{10.11}$$

where z_n is the microlite nucleation depth and $z = 0$ means the surface. Time t' is the time when magma that appeared at the surface at the time t was at the depth of z_n, and is related by

$$t = t' + \int_{z_n}^{0} \frac{dz}{U(z, t(z))} \tag{10.12}$$

For simplicity, we assume that the velocity is not non-uniform in the cross section of the conduit. In fact, magma is successively extruded from the conduit, such as Japanese agar jelly called tokoroten. For magma, the cross-sectional area can vary spatiotemporally, and the driving force of extrusion includes pressure from the surrounding, its own buoyancy, and overpressure. On the basis of Eq. (10.11), let us consider the results obtained from analysis of microlites.

Inserting data of the discharge rate at the surface $U(0, t)\mathcal{A}(0, t)$ and the ascent velocity under the ground $U(z_n, t')$ into the equation to estimate the change of the conduit size with respect to time gives the right chart of Fig. 10.12. Phase lag, whereby the variation at the surface lags several months behind the variation under the ground, would be the difference between t and t', i.e., the time required for the change of velocity to propagate from depth to the surface. The propagation time of the change of velocity would result from compressibility of magma, viscous drag of magma, and mechanical interaction between magma and the surrounding rocks, which are neglected in the abovementioned equation. The reason why the ascent rate at the surface $U(0, t)$ is lower than the ascent rate under the ground $U(z_n, t')$ in the whole time domain is that the cross-sectional area of the conduit at the surface is larger than that under the ground. Because the inverse of the velocity is almost equal to the ratio of the cross-sectional area ($\mathcal{A}(0, t)/\mathcal{A}(z_n, t') = U(z_n, t')/U(0, t)$), the cross-sectional area of the conduit under the ground is determined to be almost one-tenth of that at the surface.

The fact that the ascent velocity at depth and that at the surface change with the same tendency means that the cross-sectional area under the ground does not change largely. In the first half, however, because change of the cross-sectional area under the ground is linked with the ascent velocity under the ground, increase in the ascent velocity seems to act on expansion of the conduit. The reason for increase in the ascent velocity under the ground in the latter half is that the cross-sectional area of the conduit shrinks more rapidly than the rate of decrease in inflow of magma. That suggests that the response of the cross-sectional area of the conduit at that location to change in the ascent rate under the ground is highly nonlinear. From these results, it is revealed that the cross-sectional area of the conduit at depth varies in association with the intensity of the discharge rate (Fig. 10.13).

10.5.2 Controlling Factors Dividing Explosive Eruptions and Nonexplosive Eruptions

A driving force for volcanic explosions is the expansion of gas that exsolves from magma by the vesiculation of volatiles dissolving in magma at depths (Sect. 1.4.3). As gas, which is a driving force for volcanic explosions, comes out of magma, magma loses the potential for explosion. Thus, a degassing efficiency is a factor determining whether an eruption is explosive or nonexplosive.

Since Jaupart and Allègre (1991), many studies have focused on identifying factors controlling the outgassing efficiency and theoretically predicted that two factors, the ascent velocity of magma and permeability of magma, are important. With regards to permeability of magma, the roles of processes involving bubbles, such as bubble coalescence and bubble deformation, have been experimentally and

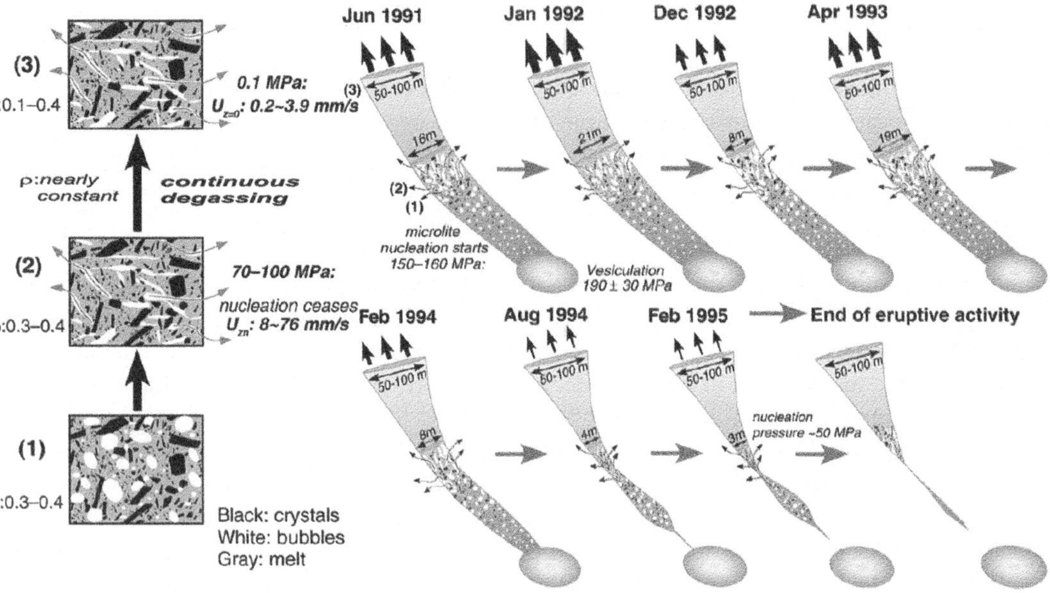

Fig. 10.13 Change of the volcanic conduit system of the Heisei eruptions of Fugendake, Unzen volcano, with respect to time, estimated from microlite analysis (Noguchi et al. 2008)

theoretically studied, as introduced in chapter "Other Bubble-Related Processes". Furthermore, the beginning of this problem is the measurement of permeability of natural samples by Eichelberger et al. (1986) (Sect. 10.7).

Constraints on the ascent velocity of magma are given by the bubble number density (BND) and the microlite number density (MND) handled in the previous chapters. Figure 10.14 shows estimates for the magma ascent rate in various eruption styles and various chemical composition, summarized by Cassidy et al. (2018). These estimates were obtained not only by the methods based on BND and MND but also other methods such as a method using characteristic texture (breakdown rim) in the rim of crystals formed because amphibole becomes unstable at low pressure (Rutherford and Hill 1993) and a method using concentration gradient of volatiles in glass inclusions (Liu et al. 2007).[15]

These data show that there is an obvious correlation between the rates at depths (the decompression rate and the ascent rate) and explosivity of eruptions at the surface (whether explosive or nonexplosive). This evidently supports theoretical prediction, and many researchers have actively been working on to elucidate this mechanism.

10.6 Bubbles and Crystalline Texture in Rhyolite Lava

10.6.1 Rhyolite Lava

Rhyolite lava has diverse structures each of which has its vesicularity and vesicle texture. Generally, rhyolite lava flow is a single flow unit, which often comprises the following four-layered structure: bubble-poor glassy obsidian; bubble- and crystal-rich rhyolite; bubble-poor glassy obsidian; and

[15] Among them is a method using crystal fracturing, although not included here (Miwa and Geshi 2012).

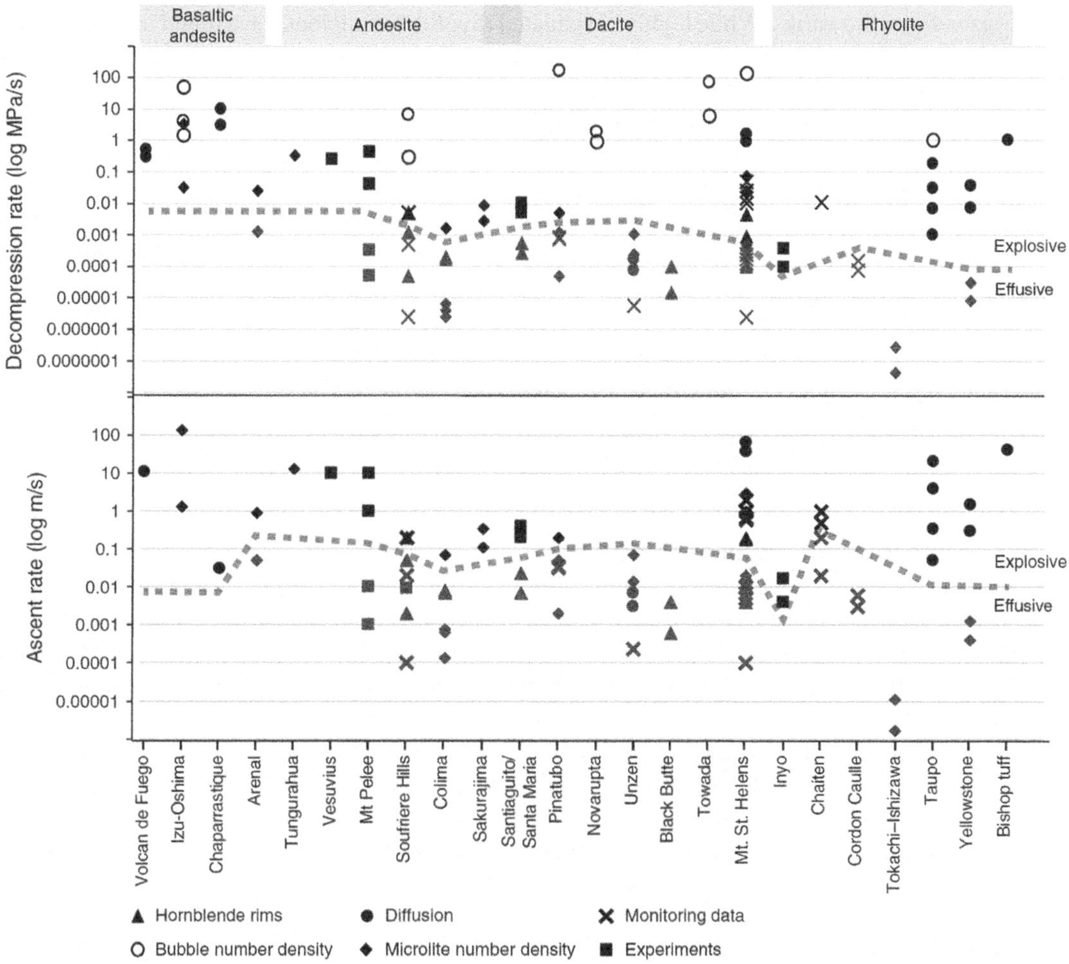

Fig. 10.14 Relationship among the magma decompression rate and the magma ascent rate, eruption styles, and chemical composition, summarized by Cassidy et al. (2018)

pumiceous well-vesiculated lava layer, from the bottom to the top (Manley and Fink 1987; Stevenson et al. 1994). In some cases, however, only parts of them are present or observable (Sano et al. 2015; Furukawa et al. 2019).

There are many mysteries in rhyolite, which has features such as low water content, low temperature, high SiO_2 content, and high viscosity. For example, it is still unknown what timescale rhyolite magma ascends in the conduit and flows on the surface. Rhyolite typically shows banded structure on various scales. The contrasts of bands are attributed to the differences in the amount of crystals and the amount of vesicles. Where and how such differences in the amount of crystals and the amount of vesicles are generated is the largest question of rhyolite structure and seems to have a key to solve the mystery of rhyolite emission.

The vesicularity and the crystallinity (the volume fraction of crystals) of the bands are also diverse. For example, obsidian found in Shirataki, Hokkaido, shows nearly 0% of the vesicularity, and Newberry lava in Oregon, USA, shows a complex layered structure in which highly vesiculated lava layers of up to 90% vesicularity and obsidian layers of almost 0% vesicularity alternate on various scales. However, they form a relatively simple lava lobe as a single flow unit in the whole.

The size of vesicles varies. While highly vesiculated rhyolite (pumicious) lava found in Himeshima, Oita, is fine, uniform one measuring several hundred micrometers, Newberry lava measures several millimeters to one centimeter. Vesicles typically somewhat elongate their spherical shape, although to a lesser extent compared to pumice in ignimbrite.

The microlite crystallinity (the volume fraction of microlites) also varies. Vesiculated volcanic rocks of rhyolite contain no microlites or extremely little, if any, although rhyolite lava contains them somewhat. The crystallinity of obsidian with extremely low vesicularity ranges from about 1% or less microlite (obsidian found in Shirataki, Hokkaido) to more than 50% (obsidian found in Kannonzaki, Himeshima, Oita).

It is a mystery as to how highly viscous rhyolite lava with low water content ascends to the surface and flows and why the slowly ascended lava contains parts that did not crystallize but solidified in a glassy state. It can be simply considered that crystals are not formed because the lava slowly ascended and cooled because time for crystal nucleation and growth in silicate melt with high viscosity and low molecular diffusivity is extremely long; however, it would require further consideration.

10.6.2 Deformation Index of Bubbles

In rhyolite lava, shear-deformed bubbles are typically observed. Rust et al. (2003) observed glassy obsidian containing bubbles moderately, found in Newberry, and applied a result of a bubble deformation experiment to the observed result. Observables are the geometry of bubbles (three axial length). From the mode of occurrence (the state where samples are deformed) in the field, a planar fabric can often be determined based on the arrangements of bubbles and crystals and contrast in the volume fraction. Samples are cut as an observation plane of a thin section is perpendicular to the planar fabric and is parallel to the elongation direction. In this plane, a major axis and a minor axis observed by transmitted light is L_1 and L_3. An intermediate axis L_2 cannot be determined by observation only from this plane. They reported correlation between L_2 and L_3 by observing a plane cut perpendicular to the extension direction, estimated the radius of an undeformed, spherical bubble, R, from the correlation and the relationship $R = (L_1 L_2 L_3)^{1/3}$, and then used it for further analysis. As a result of their analysis, the deformation index $DF = (L_1 - L_3)/(L_1 + L_3)$ (Eq. (6.9)) has different dependence on the Capillary number (Ca), which is calculated by

$$\mathrm{Ca} = \frac{\eta \dot{\epsilon} R}{\gamma} \tag{10.13}$$

whether deformation mode is pure shear or simple shear. Therefore, the distinction of the deformation mode (pure shear or simple shear) was found to be possible.

When the type of deformation is determined, shear rate can be estimated. For less-deformed simple shear, $DF = \mathrm{Ca}$ holds; therefore, shear rate can be written as follows:

$$\dot{\epsilon} = \frac{\gamma DF}{\eta R} \tag{10.14}$$

When surface tension γ and viscosity η are known, the shear rate can be calculated from this equation with observables DF and R. Moreover, if γ and η are the same in the same obsidian, the difference in slopes of plots of DF against R will mean difference in shear rate (Fig. 10.15). Assuming that interfacial tension is constant, Rust and Cashman (2007) measured volatile contents in obsidian found in Newberry volcano, estimated the viscosity when obsidian is deformed, and estimated the shear rate at 10^{-2}–10^{-4}. The lower limit of deformation time can be estimated by dividing deformation obtained

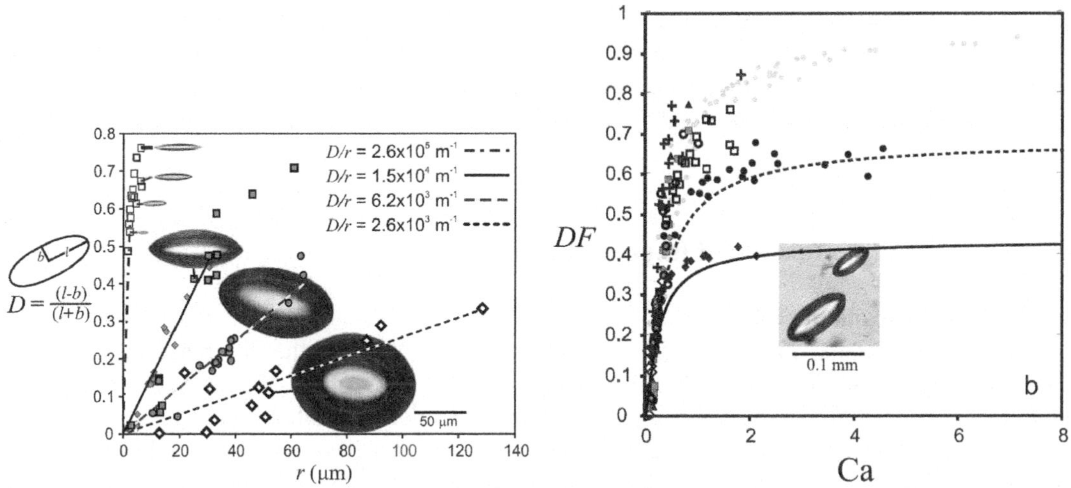

Fig. 10.15 (Left chart) Linear relationship between DF (denoted as D in the chart) and the radius of an undeformed bubble, R (denoted as r in the chart). The steady model is applicable. (Right chart) Relationship between DF and the Ca number. Because DF remains in the range between 0.4 and 0.8 even though Ca exceeds 1, the steady model based on linear relationship is not applicable to the region (adapted from (Rust and Cashman 2007)). The curves represent the unsteady deformation model (the MJT model) (Ohashi et al. 2018), which is applicable to the region outside linear relationship. The reason why the unsteady model curves and data points based on the steady model do not agree is that the shear rate used for calculating the Ca number from the geometry of bubbles differs between the steady model (Rust and Cashman 2007) and the unsteady model (Ohashi et al. 2018). Reprinted by permission from Springer Nature (Rust and Cashman 2007), copyright (2007)

from the geometry by the shear rate, and the estimates are within 0.2–8 h. Because the system is in a steady state, the geometry does not change even if deformation time elapses longer than the lower limit of time. Therefore, constraints on the actual deformation time are limited.

The abovementioned analysis is limited to the case where the geometry of deformed bubbles can be assumed to be in a steady state, i.e., plots of DF against R are linear. This shows that when the radius of bubbles becomes large, the effect of interfacial tension becomes weak in a balance between interfacial tension and deformation force; therefore, the bubbles get longer well. In many cases, DF actually is within 0.4–0.8 and does not largely change even if R becomes large (the Ca number becomes large) (Fig. 10.15; right chart). This means that the deformation of bubbles does not reach a steady state in which interfacial tension and deformation force is in balance. Moreover, because stress at the bubble interface further deforms the bubbles, the bubble shapes do not agree with strain ellipsoids placed on the deformation field of liquid. Therefore, the estimation of the time elapsed by using strain measured by deviation from sphere of bubbles has errors in this term.

As introduced in Chap. 6, Ohashi et al. (2018) developed a method of simultaneously estimating shear rate and the amount of deformation (deformation time) by applying a model involving an unsteady state (the MJT model). This model uses the geometry of bubbles in an unsteady state and handles the relationship between the bubble geometry and strain of liquid appropriately, which was an unsatisfactory point in the steady state model. Procedures in the model are as follows:

First, give sense of strain, viscosity, and interfacial tension. Next, read a value to which DF converges from a scatter plot in which DF and R, obtained from observation, are plotted. Compare the value to which DF converges with a theoretical curve of unsteady deformation to estimate an unknown parameter ϵ. Moreover, adjust and determine another unknown parameter, shear rate $\dot{\epsilon}$, such that the theoretical curve fits the observed data. Consequently, the shear rate was estimated at about 10^{-2}, which is equal to that obtained from the steady-state model, and the strain time was estimated at 2.7–4 min, which is well short. In practice, because of difference of the bubble size distribution among samples, estimates for the shear rate and the amount of strain greatly vary according to samples. To estimate a strain history that the whole lava underwent when it flowed, such spatial inhomogeneity has to be handled appropriately.

10.6.3 Crystalline Texture of Rhyolite Lava

Banded structure and microlites

Contrast of bands containing almost no vesicles is often attributed to the difference in the crystal number density of mafic minerals such as iron-titanium oxides. Some of crystalline textures included in rhyolite show characteristic spherical material (Breitkreuz 2013). One is called lithophysa, in which pre-existing bubbles or cavities newly formed in lava are secondarily filled with silica minerals. The interior structure is generally a concentric banded structure and the core often remains hollow. Another is called spherulite, in which liquid crystallized under a glass transition point (devitrification) and formed a radial structure with silica minerals and feldspar. This was reproduced experimentally (Lofgren 1971).

Because rhyolite is considered to have slow ascent rate, it is hard to determine whether decompression-induced crystallization or cooling crystallization. This makes it difficult to understand the origin of the banded structure of lava. Actually, in examples of measuring the crystal number density of microlites in rhyolite, no distinguished changes are observed in the data for different distances from the cooling surface determined from the occurrence and for different lithofacies (Befus et al. 2015; Sano et al. 2015). Such non-correlation with the occurrence at the surface suggests that microlites crystallized at depths.

Projection

Focusing on projections, which are finer texture than microlites in rhyolite lava from Tokachi-Ishizawa, shirataki, Hokkaido, Sano and Toramaru (2017) reported an interesting correlation between natural layered structure and texture (Fig. 10.16; the inset image in the left chart). Projections, that is a needle-like texture which develop on the crystal surface of microlites, often grow up from the edges of the crystal face (Lofgren 1974b). Careful measurement revealed that there is an obvious positive correlation between the length L_{p} and the spacing d_{p} of projections and that the correlation increases as the texture is closer to banded rhyolite layer, from obsidian. The correlation is expressed roughly by the following equation:

$$L_{\mathrm{p}} \propto d_{\mathrm{p}}^{1.5\pm0.5} \tag{10.15}$$

(left chart of Fig. 10.16).

The length of projection is determined by the product of growth rate G and crystallization time t_{x}.

$$L_{\mathrm{p}} = G t_{\mathrm{x}} \tag{10.16}$$

The spacing between projections is proportional to the diffusion coefficient D of crystallizing components and is inversely proportional to the growth rate (Keith and Padden 1963; Kurz and Fishe 1986):

$$d_{\mathrm{p}} = \frac{D}{G} \tag{10.17}$$

Furthermore, because the development of unstable interfaces such as projections is known to have dependence on the cooling rate ($|\dot{T}|$), dp shall be written as follows using a constant of proportionality

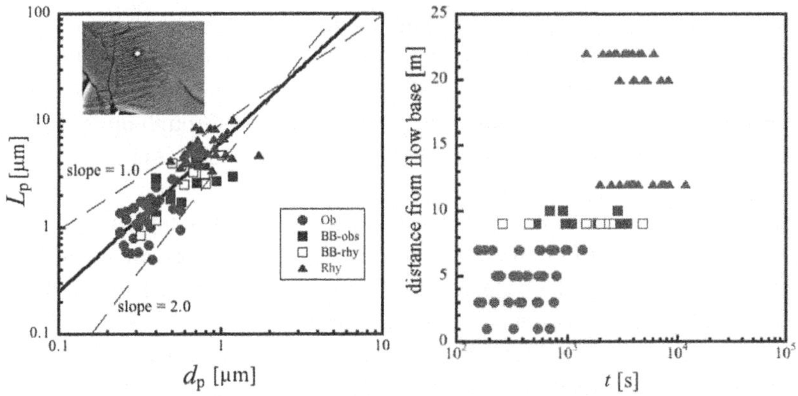

Fig. 10.16 (Left chart) Relationship between the length of projection L_p (μm) and the spacing d_p (μm). The inset image (about 20 micrometers in width) is a BSE image of typical projections, which was taken by an electron microscope. (Right chart) Relationship between the crystallization time estimated by Eq. (10.21), assuming the observed values in the left chart and the diffusion coefficient of 10^{-15} m^2/s, and the location in lava flows (adapted from Sano and Toramaru 2017). Reprinted from Sano and Toramaru 2017, copyright 2017, with permission from Elsevier

c_p:

$$d_p = c_p|\dot{T}|^{-\chi_p} \tag{10.18}$$

In this equation χ_p is positive, and the equation indicates that the spacing becomes short when the cooling rate increases. Assuming that in the cooling process, temperature range ΔT_x in which projections can grow is determined from a phase equilibrium diagram. Note that the cooling rate has a relationship of $|\dot{T}| = \Delta T_x/t_x$ with crystallization time t_x. From Eqs. (10.16)–(10.18),

$$L_p = D\Delta T_x c_p^{-1/\chi_p} d_p^{-1+1/\chi_p} \tag{10.19}$$

is obtained. Comparison of this equation with the actual observation result (Eq. (10.15)) yields $-1 + 1/\chi_p = 1.5 \pm 0.5$, i.e.,

$$\chi_p \approx \frac{1}{3} \sim \frac{1}{2} \tag{10.20}$$

is inferred. This exponent in the cooling rate is within the range found in instabilities of dendrite arm spacing of crystals (Horwath and Mondolfo 1962; Suzuki et al. 1968), and it is a probable formation process of projections.

Estimation of crystallization time

The likeliness concluded above is important in application, i.e., if the length L_p and the spacing d_p is given as observables and a plausible diffusion coefficient D (not so difficult although depending on precision) is given, the growth rate G can be estimated from Eq. (10.17). Moreover, using the growth rate, crystallization time t_x can be estimated from Eq. (10.16), i.e.,

$$t_x = \frac{L}{G} = \frac{L_p d_p}{D} \tag{10.21}$$

holds. In other words, t_x is obtained by dividing the product of observables L_p and d_p by the diffusion coefficient D. Relationship between the crystallization time obtained in this manner and the location in lava is shown in the right chart of Fig. 10.16. The crystallization time ranges from 100 s to 10^4

seconds in rhyolite lava, whereas the crystallization time in obsidian is one order of magnitude shorter. However, although there is a correlation between the cooling time (the inverse of the cooling rate) and the location, the correlation is not a monotone function, e.g., time varies with the square of the distance; therefore, it is difficult to explain it by cooling by heat conduction through an uniform medium after the emplacement of lava (Appendix 11.7). Instead, different lithofacies seem to have different cooling time.

This means that the condition is that respective lithofacies, i.e., vesiculated rhyolite layers and obsidian layers, have an obviously different thermal history when projections are formed. One possibility is that parts near the conduit wall, which extremely quickly quenched on the way of magma ascent and solidify under the ground, did not appear at the surface; in other words, only relatively inner parts may be extruded to the surface. Extreme differences by lithofacies may be originated from heat transport mechanisms other than heat conduction such as gas phase convection because vesiculated rhyolite is a porous medium. Complex rhyolite layered structure reflects the state in which vesiculated parts and unvesiculated parts ascend being mixed together.

10.7 Phenocryst Texture of Lava Flow

10.7.1 Porphyritic Texture: Phenocrysts and Groundmass

Rock texture of lava flow often shows porphyritic texture comprising phenocrysts and groundmass (or matrix). It is generally considered that, in porphyritic texture, phenocrysts are crystallized at depths and groundmass is crystallized by rapid cooling near the surface. As seen in Chap. 8, groundmass crystals in pumice and scoria, which are the products of explosive eruptions, are produced by decompression-induced crystallization in which effective supercooling results from rise of liquidus caused by water exsolution from melt. There is a high possibility that the nuclei of groundmass in lava flow are formed by decompression-induced crystallization.

Magma erupts as lava flows when the magma cannot obtain a sufficient ascent velocity because of the low bubble content or when fragmentation does not occur in the magma. This means that water exsolution from the magma occurred somewhere at depths or water content of the magma has been low from the first.[16] In any case, as magma crosses the liquidus curve, shown in Fig. 8.2, on the way to the surface, the magma goes into the effective supercooling state and decompression-induced crystallization occurs.[17] The nucleation of groundmass crystals only by cooling crystallization after eruptions occurs only when the initial temperature is higher than the liquidus of dry magma at a pressure of 1 atm (Fig. 8.2).

Because phenocrysts and groundmass are generally formed from the same minerals, phenocrysts may grow when magma ascends or is supercooled near the surface. In fact, however, whether by decompression-induced crystallization or cooling crystallization, nucleation and growth of groundmass crystals newly occur to form groundmass. This problem is essentially the same as the problem of second nucleation of bubbles, addressed in Chap. 5. Replacing the word bubbles described in a summary of second nucleation of bubbles (Sect. 5.9) by the word crystals gives the following descriptions:

[16] Because magma with low water content cannot obtain overpressure that triggers ascent and buoyancy that is a driving force of ascent by itself, the magma may appear at the surface together with magma emitted by explosive eruptions in the same event. In practice, typical stratovolcanoes in island arcs are composed of pyroclasts such as vesiculated rocks (pumice and scoria) and lava flows, both of which constitute one eruption event (Moriya 1983).

[17] If the initial water content is low, pressure corresponding to T_{BN}, at which the liquidus curve turns upward in Fig. 8.2, shifts to lower on the curve, that is, temperature shifts to higher.

1. When the volume fraction of pre-existing crystals (phenocrysts) is small, the critical effective cooling rate at which second nucleation occurs is determined only by the number density of the pre-existing crystals. In this case, the critical effective supercooling rate must be about one order of magnitude higher than the cooling rate needed to form the number density of the pre-existing crystals.
2. When the volume fraction of pre-existing crystals (phenocrysts) is large, the critical cooling rate at which second nucleation occurs is proportional to the number density and the size of the pre-existing crystals.
3. If second nucleation of crystals occurs, the number density of secondary crystals (groundmass crystals) becomes one order of magnitude or more higher than the number density of primary crystals (phenocrysts).
4. To the number density of secondary crystals (groundmass crystals), the MND cooling rate meter can be adapted.

As groundmass crystals (microlites) are addressed in other sections (Sects. 10.3, 10.5, and 10.8), let us look at phenocrysts in lava at Sakurajima as an example to discuss in detail below.

10.7.2 Lava of Sakurajima Volcano in Historical Times

At Sakurajima volcano, four large-scale eruptions (larger than 0.1 km^3 DRE) occurred during the past 550 years. The four eruptions are the Bunmei (A.D. 1471–1476), An-ei (A.D. 1779–1782), Taisho (1914–1915), and Showa (A.D. 1946) eruptions. Among them, eruptions other than the Showa eruption began as Plinian eruptions, followed by lava flows, and the Showa eruption primarily discharged only lava flows. The volume of these eruptions is about 1–2 km^3 DRE except the Showa eruption, indicated as follows, and the volume of vesiculated rocks and that of lava are almost the same order of magnitude: the Bunmei 1.2 (0.5) km^3, An-ei 2.1 (0.4) km^3, Taisho 2.0 (0.5) km^3, and Showa 0.1 (0.02) km^3. Figures in parentheses are DRE volume of pyroclasts or vesiculated rocks (Ishihara et al. 1981; Kobayashi and Tameike 2002; Kobayashi et al. 2013). From analysis of the whole-rock chemical composition, SiO_2 content is 66–68 wt% for Bunmei, 64–66 wt% for An-ei, 59–64 wt% for Taisho, and 61–62 wt% for Showa (Nakagawa et al. 2011). All lava contains plagioclase, orthopyroxene, clinopyroxene, and magnetite as phenocrysts, and lava of Taisho and after that contain olivine. The fraction of phenocrysts in lava is 15% for Bunmei, 20% for An-ei, 25–28% for Taisho, and 29% for Showa (Yanagi et al. 1991). Groundmass comprises plagioclase, orthopyroxene, clinopyroxene, titanomagnetite, and glass.

10.7.3 CSD of Phenocrysts

Chemical composition of phenocrysts show that lava from Sakurajima volcano is a mixture of at least two different magma (Yanagi et al. 1991). For example, there are two kinds of plagioclase: one originating from low-temperature magma around An#58 and another originating from high-temperature magma around An#85. Therefore, when studying CSD of phenocrysts, individual phenocrysts have to be measured after being distinguished. Yamashita and Toramaru (2021) conducted detailed crystalline texture observations of each lava and measured CSD after classifying phenocrysts into ones originating from low-temperature magma and ones originating from high-temperature magma. They reported that while intercept and slope of CSD of low-temperature components (Type A) have not changed for the past 550 years, the intercept and slope of CSD of high-temperature components (Type B) have increased with time (Fig. 10.17).

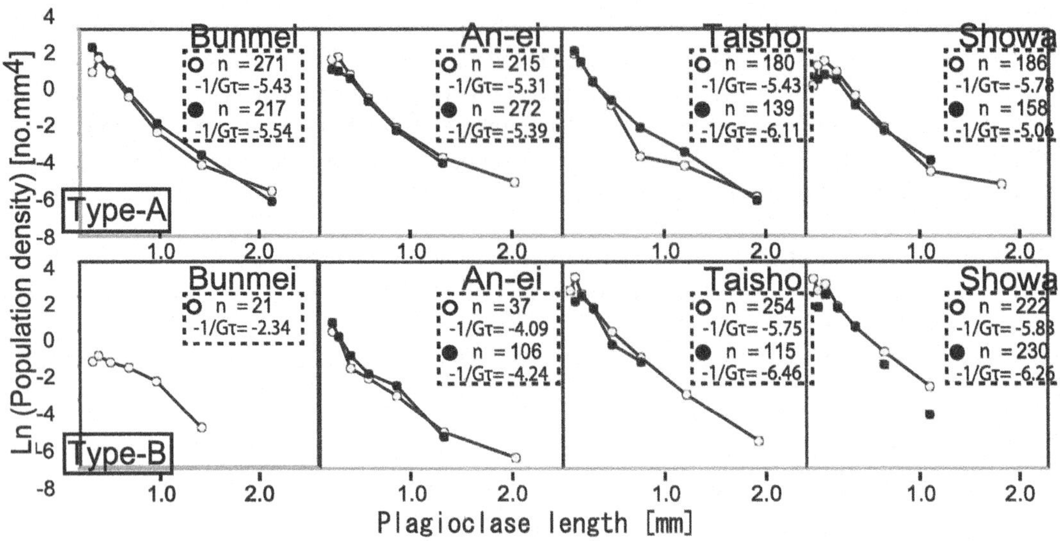

Fig. 10.17 CSD of plagioclase phenocrysts in lava of eruptions of Sakurajima volcano in historical times (Yamashita and Toramaru 2021). The upper charts are of plagioclase phenocrysts originating from low-temperature magma (Type-A). The lower charts are of plagioclase phenocrysts originating from high-temperature magma (Type-B). The charts are arranged from left to right in the order of Bunmei, An-ei, Taisho, and Showa eruptions. White circles and black circles represent different samples collected from different lava units. From Yamashita and Toramaru (2021) copyright 2021 by John Wiley Sons, Inc. Reprinted by permission of John Wiley & Sons, Inc.

10.7.4 Interpretation of Phenocryst CSD

As a premise for interpreting CSD, we have two systems for crystallization: a closed system and an open system. In the case where crystals are small in size and time between the start of crystallization and the end of solidification is relatively short, such as microlites, in-situ crystallization occurs and CSD is determined by nucleation and growth process in a closed system. In the case where crystals have obviously experienced a variety of things, such as crystal settling in a magma chamber, convection, and mixing with magma having different temperature or composition, such as phenocrysts, it should be treated as an open system. In this case, the system is within a scale we observe, i.e., a thin section or a sample size.

For the open system a general solution is given by Eq. (9.199) in Chap. 9 on CSD. In the case with the initial condition where there are no phenocrysts, no particles are supplied from outside the system, and nucleation rate (J_0), growth rate (G_0), and extraction rate (note: $t_{ex} < 0$) are constant, CSD shows a certain exponential distribution after sufficient time elapsed.[18]

$$F(R) = \frac{J_0}{G_0} \exp\left(-\frac{R}{G_0|t_{ex}|}\right) \tag{10.22}$$

[18] If the nucleation rate, as a function of time, becomes J_0 like a step function when $t = 0$, the maximum particle size R_{max} present at time t is $G_0 t$ and the value of its CSD is written as $J_0 \exp(-R_{max}/G_0|t_{ex}|)/G_0 = J_0 \exp(-t/|t_{ex}|)/G_0$, which decreases exponentially with time. Because CSD is generally measured from about several to ten times the characteristic size defined by slope, CSD can be regarded as being in a steady state, from the observational point of view, after the elapse of several to ten times $|t_{ex}|$. Although the value of $|t_{ex}|$ is unknown, the discussion in the main body supposes that intervals of large-scale eruptions satisfy the condition for a steady state. Strict examination will be conducted using unsteady solutions for parameters that are a function of time and for plausible values.

From this equation, the intercept and slope of CSD are determined by J_0/G_0, the ratio between nucleation rate and growth rate, and $G_0|t_{ex}|$, the product of timescale defining extraction rate $|t_{ex}|$ and growth rate G_0. In other words, the fact that CSD is constant indicates that these parameters are constant and the fact that the slope increases indicates that $G_0|t_{ex}|$ decreases.

Because the CSD of plagioclase phenocrysts originating from low-temperature magma of Sakurajima volcano has been constant for 550 years, the parameters described above have been constant. For the CSD of plagioclase phenocrysts originating from high-temperature magma, the intercept and the slope have increased. That means that in the case of a steady-state model[19] in which the nucleation rate and the growth rate are constant, J_0/G_0 and $1/(G_0|t_{ex}|)$ have increased. This examination leads to the following two possibilities:

Possibility 1: The case where G_0 decreases with time. In this case, increase in the slope of CSD corresponds to decrease in characteristic size caused by decrease in the growth rate. When looking at size space, it corresponds to the fact that although nucleation proceeds at a constant rate and then smallest-sized crystals form at a constant rate, the crystals cannot grow and therefore smaller size bins are accumulated. Note that the extraction of crystals proceeds at a constant rate.

Possibility 2: The case where $|t_{ex}|$ decreases with time. The case where residence time that characterizes extraction rate $|t_{ex}|$ becomes shorter is assumed. Although extraction seems to be attributed to crystal settling, the density of plagioclase, which influences crystal settling, is not likely to change materially because the composition is almost constant. Moreover, decrease in $|t_{ex}|$ alone cannot explain increase in the intercept.

As described above, the causes of change in CSD of plagioclase phenocrysts originating from high-temperature magma seem to arise from decrease in effective growth rate over several hundred years.

10.7.5 Growth Rate and Magma Supply Rate in Open-System CSD

Growth rate in the formulation of open-system CSD is an effective measure to represent how efficiently crystals can grow (Marsh 1988, 1998). Let us assume that high-temperature magma experiences a series of processes forming CSD, such as crystallization by cooling, convection, and crystal settling, in a high-temperature magma chamber. It is considered that high-temperature magma generates in the mantle, is injected into a high-temperature magma chamber, and stays there for a certain time. Subsequently, the high-temperature magma moves to a low-temperature magma chamber, is mixed with low-temperature magma, and finally erupts. If the volume of the high-temperature magma chamber is considered not to change greatly, the amount of magma supplied from the mantle and the amount of magma moving to the low-temperature magma chamber should be almost equal.

In such a situation, if the high-temperature magma chamber is, for example, cooled efficiently, effective growth rate will increase. Moreover, if residence time ($t_{residence}$) in the high-temperature magma chamber becomes short, effective growth rate will decrease. Residence time is influenced by magma supply rate from the mantle. If the magma supply rate increases, residence time in the high-temperature magma chamber will decrease, i.e.,

$$G_0 \propto t_{residence} \propto I_{Mantle}^{-1} \tag{10.23}$$

[19] This means "a steady-state in a timescale shorter than an interval of eruptions" Different steady-state CSD values can be given to different eruptions.

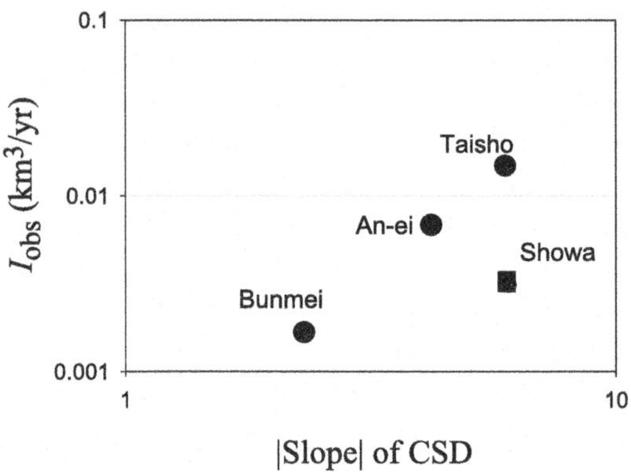

Fig. 10.18 Relationship between the geologically determined discharge rate at the surface I_{obs} and the slope of CSD of plagioclase originating from high-temperature magma. From Yamashita and Toramaru (2021) copyright 2021 by John Wiley Sons, Inc. Reprinted by permission of John Wiley & Sons, Inc.

where I_{Mantle} is the magma supply rate from the mantle. Thus, for the intercept and the slope of CSD,

$$\text{intercept} \propto I_{Mantle} \tag{10.24}$$

$$|\text{slope}| \propto I_{Mantle} \tag{10.25}$$

hold, and both are expected to be proportional to the magma supply rate from the mantle.

Although the magma supplied from the mantle is somewhat influenced by, for example, modification of chemical composition, losses such as crystal settling and change in the geometry of the magma plumbing system over the years, variation in the supply rate should be reflected in variation in the emitted mass at the surface. Figure 10.18 shows the correlation between the discharge rate at the surface $I_{obs} = M/\Delta t$, obtained from the actual amount of eruptive materials M in a single eruption and the dormant period Δt from the preceding large-scale eruption, and the slope of CSD. Except the Showa lava, the amount of eruptive materials of which is an order or more of magnitude less than the others, obvious positive correlation between the discharge rate of large-scale eruptions and the slope of CSD is found. This relationship can be roughly expressed as follows:

$$I_{obs} \propto |\text{slope}|^2 \tag{10.26}$$

This equation indicates that between the supply rate from the mantle estimated from CSD and the actual discharge rate is the following relationship:

$$I_{obs} \propto I_{Mantle}^2 \tag{10.27}$$

This equation indicates that the discharge rate at the surface is apparently amplified compared to the supply rate from the mantle estimated from CSD, and it is very interesting that there is a possibility that supply from the mantle controls a surface phenomenon. The causes of amplification have to be examined from various aspects, such as assumption of the steady state in which nucleation and growth rate are constant, and of interaction with low-temperature magma.

10.7.6 Possibility of Long-Term Prediction Using Phenocryst CSD

For current volcanic eruption prediction, the prediction of time, i.e., when a volcano will erupt, is difficult. Although it is difficult, there are two possibilities. One is short-term prediction through

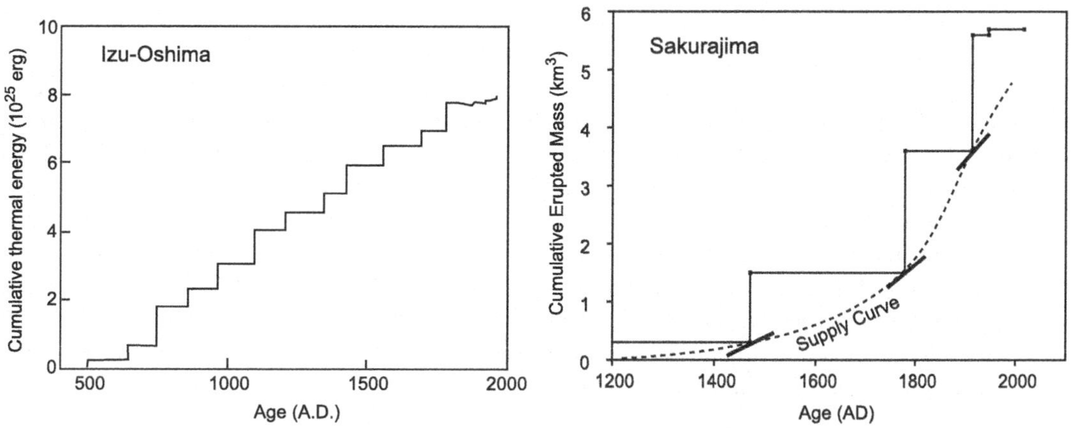

Fig. 10.19 (Left chart) Step diagram of Izu-Oshima volcano by Nakamura (1965). Thermal energy is used instead of erupted mass. (Right chart) Step diagram of Sakurajima volcano. Lines at each eruption point are the supply rates estimated from the slope of CSD of plagioclase phenocrysts originating from high-temperature magma. Because absolute values of the supply rates are not found from CSD, they are determined as agreeing with a line connecting the Bunmei eruption and the An-ei eruption. The broken line is the hypothetical supply rate curve

monitoring by various observations and modeling that explains the observation results. Another is long-term prediction through grasp of eruption history by geological survey of past eruptions. The grasp of eruption history is summarized in a plot of cumulative erupted mass against time, a so-called step-diagram. Nakamura (1965) investigated in detail the distribution of eruptive materials from large-scale eruptions of Izu-Oshima volcano that occurred during the past 2000 years and whose eruption age is known, estimated the erupted mass, and plotted values obtained by converting the cumulative erupted mass into the thermal energy against the eruption age (Left chart of Fig. 10.19).[20] The step diagrams reflect the conservation of mass and dynamic conditions of volcano systems in the long term and are extremely significant (Characteristics of step-diagrams are described in detail in Koyama and Yoshida (1994)). For example, the slopes of a step diagram provide us a rough estimate of the average supply rate of magma to the volcano. From the step diagram, the approximate magma supply rate peculiar to the volcano can be grasped, and the imminence of the next eruption can be known by the approximate magma supply rate and the dormant period from the preceding eruption.

There are some difficulties in this long-term prediction. One is that to estimate erupted mass is difficult. Even though small-scale eruptions are negligible because of their minor contribution to the step diagram, the accurate estimation of the erupted mass of large-scale eruptions (more than 0.1 km^3 DRE) is indispensable. Eruptive materials usually comprise tephra (pyroclasts) and lava flow. Between them, the erupted mass of tephra can be relatively accurately estimated from geological surveys owing to characteristics of Plinian eruptions and areal distribution of eruptive materials (Pyle 1989). It is, however, difficult to accurately estimate the erupted mass of lava flow because of local distribution when old lava is covered with new lava. Moreover, there are many imponderables such as deposits in the vent. For the eruptions of Sakurajima volcano in historical times, because lava of an eruption flowed as it avoided the past eruptions, the mass can be relatively accurately estimated. As for this point, Sakurajima volcano has a considerable advantage.

Another difficulty is that the slope of a step-diagram obviously changes with time. Let us consider the case where the supply rate is constant. If the volume of a magma chamber is constant and the volcano erupts when the same amount of magma as the amount discharged is newly supplied (i.e.,

[20] This is based on the thought of Kazuaki Nakamura that thermal energy is the primordial energy of volcanic eruptions.

the system in which conservation of mass holds), the lower points of the step diagram will be placed on a line corresponding to the supply rate. In this case, the time of the next eruption will be given by the point where this line intersects with a line drawn horizontally from the point showing the latest eruption, i.e., this is a time-predictable type.[21] In this case, however, the volume of discharge cannot be predicted. On the other hand, if the upper points of the step-diagram are placed on a line of the supply rate, the volume of magma that will be discharged when the volcano erupts can be predicted, although the time when the volcano erupts cannot be predicted, i.e., this is a mass (volume)-predictable type. Volcanos in nature, in any case, are usually not so simple (Koyama and Yoshida 1994). One of the reasons is probably change with time in the supply rate from the mantle. If change with time in the supply rate from the mantle can be estimated by the method using CSD, a supply rate curve will be drawn in the step-diagram (the broken line in the right chart of Fig. 10.19) and the time when the upcoming eruption will occur can be predicted.

10.8 Rock Texture of Intrusions

10.8.1 Difference in Characteristics of Structure and Texture Based on the Size of Intrusive Body

Dikes and sills intruding into the shallow depth often show characteristic change in texture with a distance from the cooling surface. When the size (width or thickness) of intrusive body is small, the whole intrusive body shows a porphyritic texture and spatial variation is primarily observed only in groundmass texture. In intrusions measuring smaller than dozens of meters, the particle diameters of groundmass increase in the intrusive body and the groundmass texture comes to define the rock texture. Thus, for relatively small-sized intrusions, their structure and texture can be explained only by in situ crystallization, in which macroscopic motion of liquid and crystals in the process of cooling and solidifying can be ignored.[22]

Moreover, for slightly larger intrusions measuring more than 100 m, segregation structure and fluid motion occurring in the cooling process produce spatially complex rock texture. In further larger intrusions measuring more than several kilometers, the ascent of liquid by buoyancy and crystal settling become more prominent; the temperature decrease and crystallization by cooling from country rocks causes convection. Thus, layered structures form in the large body. In relatively large intrusions like this, noticeable differences in structure and texture are observed between the outer part and the inner part (McBirney and Noyes 1979). These fossil magma chambers are called layered intrusions and they have been actively studied (Charlier et al. 2015). The following section describes relatively small intrusions in which complex motion occurring in large magma chambers does not occur and in situ crystallization is predominant.

[21] For the Old Faithful Geyser in Yellowstone National Park in United States the time-predictable type holds. Geysers and volcanic eruptions share precursory earthquakes and decompression-induced vesiculation processes (Kieffer 1984).

[22] There are some exceptions. Settling of olivine phenocrysts is obviously observed in the picrite sill found at Ogi in Sado Island, Niigata (Toramaru et al. 1996, 1997) and the sheet-like intrusion at Nosappumisaki, Hokkaido (Simura and Ozawa 2011). Others are the accumulation of phenocrysts into the central part of dike by Bagnold effect (Bagnold 1954; Bhattacharji and Smith 1964).

10.8.2 Spatial Change in the Crystal Number Density of Groundmass in Narrow Dikes

In Sect. 1.4.1 spatial variation of groundmass texture in dikes was introduced. In Sect. 7.6.6 kinetics of crystallization was discussed on the basis of data of distance-dependent change in the number density in dikes, obtained by Gray (1970, 1978), as experiment results in nature. At the time, a solution in which the width is assumed to be infinite, not finite as in the case of dikes, was referred. Using the dike at Hirado as an example, let us think about the case with a finite width.

A solution in the case of dikes with finite width is given by Eq. (11.130) in the Appendix. The left chart of Fig. 10.20 shows a relationship, based on this solution, between the cooling rate and the distance from the cooling surface. In the greater part of the center region of the intrusive body, the cooling rate becomes constant. The number density theoretically inferred from Eq. (7.152) using this cooling rate is plotted in the right chart. The actual data of the number density obtained from the dike at Hirado are shown in the chart. This chart seems to explain well a trend in which the number density decreases with the power of −3 of the distance from the contact (the cooling surface) and a trend in which the number density does not change around the center. These trends are seen in spatial variation of the crystal size, shown in Fig. 1.21.

Narrow dikes and a decreasing trend in the number density near the boundary surface are influenced by where the actual cooling boundary is set. In this chart, the actual cooling surface is set at 6 mm outside the material boundary. Attention is required in applying the crystal number density cooling rate meter (Eq. (7.152)) derived in Chap. 7 to the neighborhood (several centimeters) of the boundary surface of the actual dikes.

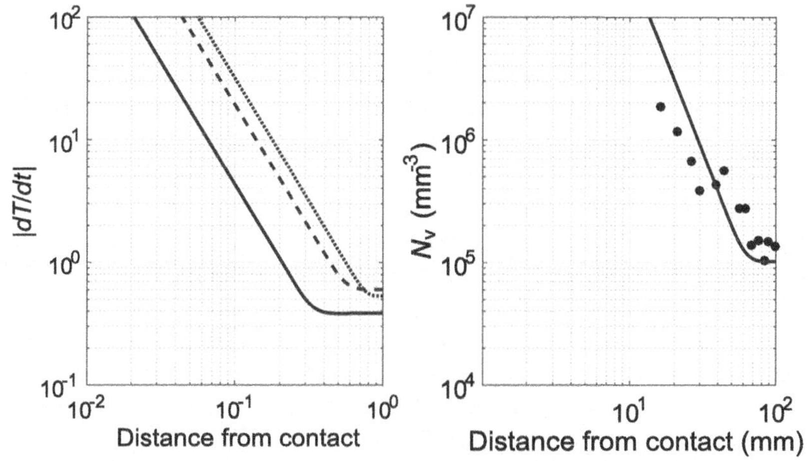

Fig. 10.20 (Left chart) The cooling rate as a function of the distance from the contact in dikes. Both the horizontal and vertical axes are dimensionless. The dotted line represents $T_m/T_{m0} = 0.98$, the broken line 0.9, and the solid line 0.8. (Right chart) Change in the crystal number density expected on the basis of the left chart in the case of diffusion-controlled growth. Actual data of the basaltic dike at Hirado are shown in black circles. $T_m/T_{m0} = 0.93$. $1 - T_{s0}/T_{m0} = 0.7$ is assumed in common. The boundary is set at 6 mm to the side of the country rock from the actual material boundary. T_{m0} and T_{s0} are the initial temperatures of magma and country rock. T_m is a temperature at which crystals nucleate

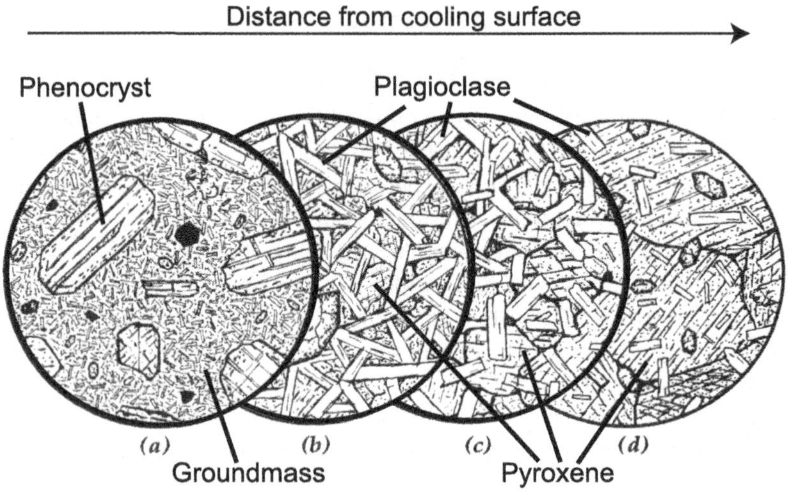

Fig. 10.21 Rock texture variation peculiar to basaltic hypabyssal rocks. With the distance from the cooling surface, textures change from **a** porphyritic to **b** intergranular to **c** ophitic to **d** poikilitic. It should be noted that scales of respective sketches differ (adapted from McBirney 1993). In Prof. McBirney's textbooks, drawings of an older man are always added to sketches of thin sections. McBirney, A. Igneous Petrology (3rd ed 151). Burlington, MA: Jones & Bartlett Learning, 2007

10.8.3 Spatial Variation of Groundmass Textures, Focusing on the Mutual Relationship Between Plagioclase and Pyroxene

Figure 10.21 is an example of typical texture variation of basaltic intrusions (McBirney 1993). With the distance from the cooling surface, the textures often change from porphyritic to intergranular to ophitic to poikilitic. Plagioclase and pyroxene are representative constituent minerals of basaltic intrusions, and a mutual relationship between them produces such characteristic textures.

In a porphyritic texture, phenocrysts and groundmass are easily distinguishable in size. This is because, as in the case of a texture observed in eruptive rocks, groundmass is formed by rapid cooling from the cooling surface. In an intergranular texture, groundmass plagioclase and pyroxene of a similar size and number density exist in spaces between the others. In an ophitic texture, it is often observed that a part of plagioclase single crystals intrude into larger pyroxene single crystals. In a poikilitic texture, it is often observed that plagioclase single crystals are enclosed in further larger pyroxene single crystals.[23] Such a systematic change as a function of the distance from the cooling surface is not always observed. For example, in sills at the Isle of Skye, Scotland, ophitic textures are observed even in the vicinity of the cooling surface.

An interesting finding in an ophitic texture is that while plagioclase poikilitically enclosed in the innermost region of pyroxene is small in size, plagioclase which partially intrudes into pyroxene, or which is present in spaces between pyroxene crystals, are large in size. This situation can be interpreted that plagioclase has already existed in magma before pyroxene nucleates. As pyroxene grows, plagioclase enclosed in pyroxene in the early stage of growth cannot grow further and retain its size, while plagioclase enclosed in pyroxene in the later stage of growth increase its size until it is enclosed. You can notice that if you carefully look at the sketch drawn by Prof. McBirney.

[23] Part of textures called ophitic sometimes include plagioclase poikilitically and completely enclosed in pyroxene. The same holds true for textures called poikilitic.

Such textures characteristically observed in hypabyssal rocks are considered to be evidence that in situ crystallization occurred. Among these textures, intergranular textures and poikilitic textures can be formed in crystal settling or secondary growth under high temperature. However, ophitic textures in which plagioclase is naturally enclosed in pyroxene are only formed by in situ crystallization. In other words, the presence of ophitic textures proves that convection and crystal settling did not occur when magma crystallized. Therefore, how ophitic textures peculiar to intrusions form has been discussed from long in the past (Wager 1961).

One idea is the classical understanding using master curves of the nucleation rate and the growth rate, introduced in Chap. 7 (Sect. 7.2), i.e., nucleation is dominant earlier in plagioclase; on the contrary growth is dominant in pyroxene. Leaving aside the question of why this is so, this understanding appears to be qualitatively much to the point, i.e., plagioclase forms a large number of small crystals (i.e., the number density becomes large) and pyroxene forms a small number of large crystals. In this crystallization process, some plagioclase crystals are poikilitically and completely enclosed in pyroxene but some are not wholly enclosed, resulting in the formation of ophitic texture. However, the reason why the texture systematically changes with the distance cannot be understood from this, i.e., the dependence on the cooling rate is not considered.

The main factor of texture variation shown in Fig. 10.21 is the contrast in the number density between plagioclase and pyroxene.[24] Thus,

$$\frac{N_{\text{pl}}}{N_{\text{px}}} \tag{10.28}$$

can be used as an index. In this equation, N_{pl} and N_{px} are the crystal number density of plagioclase and pyroxene, respectively. When the index is close to 1, the texture is intergranular, and as the index increases, the texture changes from intergranular to ophitic to poikilitic. As discussed in Sect. 7.6, observation in nature tells us that plagioclase and pyroxene have different dependence of the number density on the cooling rate. Moreover, this texture variation is one of expressions of the fact.

Let us review the discussion in Sect. 7.6. In the case where nucleation occurs under a condition close to a homogeneous nucleation condition (i.e., the pre-exponential factor of the nucleation rate and solid-liquid interfacial energy are sufficiently large) and crystal growth at the nucleation stage is diffusion-controlled growth, the crystal number density depends on the power of 3/2 of the cooling rate. In this case, the fact that the crystal number density depends on the power of −3 of the distance can explain the crystal number density of pyroxene and iron-titanium oxides obtained from observation. In the case where nucleation occurs under a condition close to a heterogeneous nucleation (i.e., the pre-exponential factor of the nucleation rate and solid-liquid interfacial energy are sufficiently small) (regardless of growth law), the dependence of the crystal number density on the cooling rate becomes smaller than the power of 3/2. In extreme cases (the nucleation rate and the crystallization rate are constant regardless of time: CSD Model 4), the dependence becomes negative. Plagioclase fits this case. Although why plagioclase has such dependence is not understood, the possibility that the initial condition is under the liquidus (there are many molecular clusters) or crystal structure of plagioclase (network silicate) seem to account for that. It is considered from the above that difference in dependence of the crystal number density on the cooling rate between plagioclase and pyroxene is the primary factor in systematic texture change from the cooling surface.

[24] The order of crystallization is considered to be another factor.

References

Alfano F, Bonadonna C, Gurioli L (2012) Insights into eruption dynamics from textural analysis: the case of the May, 2008. Chaitén eruption. Bull Volcanol 74(9):2095–2108

Arzilli F, Carroll MR (2013) Crystallization kinetics of alkali feldspars in cooling and decompression-induced crystallization experiments in trachytic melt. Contrib Mineral Petrol 166(4):1011–1027

Bagdassarov N, Dorfman A, Dingwell D (2000) Effect of alkalis, phosphorus, andwater on the surface tension of haplogranite melt. Am Mineral 85:33–40

Bagnold RA (1954) Experiments on a gravity-free dispersion of large solid spheres in a Newtonian fluid under shear. Proc Royal Soc London. Series A 225(1160):49–63

Befus KS, Manga M, Gardner JE, Williams M (2015) Ascent and emplacement dynamics of obsidian lavas inferred from microlite textures. Bull Volcanol 77–88. https://doi.org/10.1007/s00445-015-0971-6

Beherns H, Zhang Y, Xu Z (2004) H_2O diffusion in dacitic and andesitic melts. Geochim Cosmochim Acta 68:5139–5150

Bhattacharji S, Smith CH (1964) Flowage differentiation. Science 145(3628):150–153

Borisov AA, Borisov Al A, Kutateladze SS, Nakoryakov VE (1983) Rarefaction shock wave near the critical liquid-vapor point. J Fluid Mech 126:59–73

Breitkreuz C (2013) Spherulites and lithophysae: 200 years of investigation on high-temperature crystallization domains in silica-rich volcanic rocks. Bull Volcanol 75(4):705

Campagnola S, Romano C, Mastin LG, Vona A (2016) Confort 15 model of conduit dynamics: applications to Pantelleria Green Tuff and Etna 122 BC eruptions. Contrib Mineral Petrol 171(6):60

Cassidy M, Manga M, Cashman K, Bachmann O (2018) Controls on explosive-effusive volcanic eruption styles. Nature comm 9(1):2839

Charlier B, Namur O, Latypov R, Tegner C (eds) (2015) Layered intrusions. Springer, Berlin

Couch S, Harford CL, Sparks RSJ, Carroll MR (2003) Experimental constraints on the conditions of formation of highly calcic plagioclase microlites at the Soufriere Hills Volcano. Montserrat J Petrol 44:1455–1475

Eichelberger JC, Carrigan CR, Westrich HR, Price RH (1986) Non-explosive silicic volcanism. Nature 323(6089):598

Furukawa K, Uno K, Kanamaru T, Nakai K (2019) Structural variation and the development of thick rhyolite lava: a case study of the Sanukayama rhyolite lava on Kozushima Island. Jpn J Volcanol Geotherm Res 369:1–20

Gardner JE, Thomas RME, Jaupart C, Tait S (1996) Fragmentation of magma during Plinian volcanic eruptions. Bull Volcanol 58:144–162

Gonnermann HM (2015) Magma fragmentation. Ann Rev Earth Planet Sci 43:431–458

Gray NH (1970) Crystal growth and nucleation in two large diabase dikes. Can J Earth Sci 7:366–375

Gray NH (1978) Crystal growth and nucleation in flash-injected diabase dikes. Can J Earth Sci 15:1904–1923

Hammer JE, Cashman KV, Hoblitt RP, Newman S (1999) Degassing and microlite crystallization during pre-climactic events of the 1991 eruption of Mt. Pinatubo. Philippines Bull Volcanol 60:355–380

Holtz F, Sato H, Lewis J, Behrens H, Nakada S (2004) Experimental petrology of the 1991–1995 Unzen dacite, Japan. Part I: phase relations, phase composition and pre-eruptive conditions. J Petrol 46(2):319–337

Horwath JA, Mondolfo LF (1962) Dendritic growth. Acta Metall 10(11):1037–1042

Houghton BF, Wilson CJN (1989) A vesicularity index for pyroclastic deposits. Bull Volcanol 51(6):451–462

Iguchi M, Yakiwara H, Tameguri T, Hendrasto M, Hirabayashi JI (2008) Mechanism of explosive eruption revealed by geophysical observations at the Sakurajima, Suwanosejima and Semeru volcanoes. J Volcanol Geotherm Res 178(1):1–9

Iriyama Y, Toramaru A, Yamamoto T (2018) Theory for deducing volcanic activity from size distributions in plinian pyroclastic fall deposits. J Geophys Res Solid Earth 123(3):2199–2213

Ishihara K (1985) Dynamical analysis of volcanic explosion. J Geodyn 3(3–4):327–349

Ishihara K, Takayama T, Tanaka Y, Hirabayashi J (1981) Lava flows at Sakurajima volcano (1)—volume of the historical lava flows. Ann DPRI, Kyoto Univ 24(B-1):1–10

Jaupart C, Allègre CJ (1991) Gas content, eruption rate and instabilities of eruption regime in silicic volcanoes. Earth Planet Sci Lett 102:413–429

Kameda M, Ichihara M, Maruyama S, Kurokawa N, Aoki Y, Okumura S, Uesugi K (2017) Advancement of magma fragmentation by inhomogeneous bubble distribution. Sci Rep 7(1):16755

Kaminski E, Jaupart C (1997) Expansion and quenching of vesicular magma fragments in Plinian eruptions. J Geophys Res 102:12187–12203

Keith HD, Padden FJ (1963) A phenomenological theory of spherulitic crystallization. J Appl Phys 34:2009–2421

Kichise, T. (2015) Microlite textural and chemical evolution during magma ascent, applications of a new crystallization model to pyroclasts of Shinmoe-dake 2011 eruption, Kyushu University Doctoral Dissertation

Kieffer SW (1984) Seismicity at old faithful geyser: an isolated source of geothermal noise and possible analogue of volcanic seismicity. J Volcanol Geotherm Res 22(1–2):59–95
Klug C, Cashman KV (1994) Vesiculation of May 18, 1980, Mount St. Helens magma. Geology 22(5):468–472
Kobayashi T, Tameike T (2002) History of eruptions and volcanic damage from Sakurajima Volcano. Southern Kyushu. Jpn Quat Res 41(4):269–278
Kobayashi T, Miki D, Sasaki H, Iguchi M, Yamamoto T, Uto K (2013) Geological map of Sakurajima volcano (2nd edition) 1:25000. Geological Survey of Japan
Koyaguchi T (2008) Modeling of volcanic phenomena. University of Tokyo Press, pp 637. (in Japanese)
Koyama M, Yoshida Y (1994) Relationship between eruptive history and crustal stress field Based on cumulative changes in magma discharge from volcanoes. Kazan 39(4):177–190 (in Japanese with English abstract)
Kozono T, Ueda H, Ozawa T, Koyaguchi T, Fujita E, Tomiya A, Suzuki YJ (2013) Magma discharge variations during the 2011 eruptions of Shinmoe-dake volcano, Japan, revealed by geodetic and satellite observations. Bull Volcanol 75(3):695
Kurz W, Fisher DJ (1986) Fundamentals of solidification. Trans Tech Publications, Switzerland
Liu Y, Anderson AT, Wilson CJN (2007) Melt pockets in phenocrysts and decompression rates of silicic magmas before fragmentation. J Geophys Res Solid Earth 112:B06204
Lofgren G (1971) Spherulitic textures in glassy and crystalline rocks. J Geophys Res 76(23):5635–5648
Lofgren G (1974) An experimental study of plagioclase crystal morphology: isothermal crystallization. Am J Sci 274(3):243–273
Manley CR, Fink JH (1987) Internal textures of rhyolite flows as revealed by research drilling. Geology 15(6):549–552
Marsh BD (1998) On the interpretation of crystal size distributions in magmatic systems. J Petrol 39:553–599
Marsh BD (1988) Crystal size distribution (CSD) in rocks and the kinetics and dynamics of crystallization I. Theory Contrib Mineral Petrol 99:277–291
Martel, C. (2012) Eruption dynamics inferred from microlite crystallization experiments: application to Plinian and dome-forming eruptions of Mt. Pelée (Martinique, Lesser Antilles). J Petrol 53(4):699–725
McBirney A (1993) Igneous petrology, 2nd edn. Jones and Barlett Publishers, Boston
McBirney AR, Noyes RM (1979) Crystallization and layering of the Skaergaard intrusion. J Petrol 20(3):487–554
Miwa T, Geshi N (2012) Decompression rate of magma at fragmentation: Inference from broken crystals in pumice of vulcanian eruption. J Volcanol Geotherm Res 227:76–84
Miwa T, Toramaru A (2013) Conduit process in vulcanian eruptions at Sakurajima volcano, Japan: inference from comparison of volcanic ash with pressure wave and seismic data. Bull Volcanol 75(1):685
Miwa T, Toramaru A, Iguchi M (2009) Correlations of volcanic ash texture with explosion earthquakes from vulcanian eruptions at Sakurajima volcano. Jpn J Volcanol Geotherm Res 184(3–4):473–486
Moriya I (1983) Volcano morphology in Japan. University of Tokyo Press, p 135. (in Japanese)
Nakada S, Shimizu H, Ohta K (1999) Overview of the 1990–1995 eruption at Unzen Volcano. J Volcanol Geotherm Res 89:1–22
Nakada S, Nagai M, Kaneko T, Suzuki Y, Maeno F (2013) The outline of the 2011 eruption at Shinmoe-dake (Kirishima). Jpn Earth Planets Space 65(6):1
Nakagawa M, Matsumoto A, Amma-Miyashita M, Togashi Y, Iguchi M (2011) Change of mode of eruptive activity and the magma plumbing system of Sakurajima Volcano since the 20th century. In: Study on preparation process of volcanic eruption based on intergrated volcano observation 2010. Sakurajima Volcano Research Center, pp 85–94
Nakamura K (1965) Volcano-stratigraphic study of Oshima volcano, Izu. Bull Earthq Res Inst Univ Tokyo 42(4):649–728
Newhall CG, Self S (1982) The volcanic explosivity index (VEI) an estimate of explosive magnitude for historical volcanism. J Geophys Res Oceans 87(C2):1231–1238
Nishimura T, Iguchi M (2006) Nippon no kazansei jishin to bido (Volcanic earthquake and tremor in Japan). University of Kyoto Academic Press, p 242. (in Japanese)
Noguchi S, Toramaru A, Nakada S (2008) The relation between microlite textures and discharge rate for the 1991–1995 eruptions at Unzen. Jpn J Volcanol Geotherm Res 175:141–155
Ohashi M, Ichihara M, Toramaru A (2018) Bubble deformation in magma under transient flow conditions. J Volcanol Geotherm Res 364:59–75
Pichavant M, Di Carlo I, Rotolo SG, Scaillet B, Burgisser A, Le Gall N, Martel C (2013) Generation of CO_2-rich melts during basalt magma ascent and degassing. Contrib Mineral Petrol 166(2):545–561
Pyle DM (1989) The thickness, volume and grainsize of tephra fall deposits. Bull Volcanol 51(1):1–15
Rust AC, Cashman KV (2007) Multiple origins of obsidian pyroclasts and implications for changes in the dynamics of the 1300 BP eruption of Newberry Volcano. USA Bull Volcanol 69(8):825–845
Rust AC, Manga M, Cashman KV (2003) Determining flow type, shear rate and shear stress in magmas from bubble shapes and orientations. J Volcanol Geotherm Res 122:111–132
Rutherford MJ, Hill PM (1993) Magma ascent rates from amphibole breakdown: an experimental study applied to the 1980–1986 Mount St. Helens eruptions. J Geophys Res 98:19667–19685

Sano K, Toramaru A (2017) Cooling and crystallization of rhyolite-obsidian lava: insights from micron-scale projections on plagioclase microlites. J Volcanol Geotherm Res 341:158–171
Sano K, Wada K, Sato E (2015) Rates of water exsolution and magma ascent inferred from microstructures and chemical analyses of the Tokachi-Ishizawa obsidian lava, Shirataki, northern Hokkaido. Jpn J Volcanol Geotherm Res 292:29–40
Simura R, Ozawa K (2011) Magmatic fractionation by compositional convection in a sheet-like magma body: constraints from the Nosappumisaki Intrusion, northern Japan. J Petrol 52(10):1887–1925
Sparks RSJ, Bursik MI, Carey SN, Gilbert J, Glaze LS, Sigurdsson H, Woods AW (1997) Volcanic plumes. Wiley
Sparks RSJ, Wilson L (1976) A model for the formation of ignimbrite by gravitational column collapse. J Geol Soc 132(4):441–451
Stevenson RJ, Briggs RM, Hodder APW (1994) Physical volcanology and emplacement history of the Ben Lomond rhyolite lava flow, Taupo Volcanic Centre, New Zealand. N Z J Geol Geophys 37(3):345–358
Suzuki A, Suzuki T, Nagaoka Y, Iwata Y (1968) On secondary dendrite arm spacing in commercial carbon steel with different carbon content (in Japanese with English abstract). J Jpn Inst Metals 12:1301–1305
Suzuki Y, Maeno F, Nagai M, Shibutani H, Shimizu S, Nakada S (2018) Conduit processes during the climactic phase of the Shinmoe-dake 2011 eruption (Japan): insights into intermittent explosive activity and transition in eruption style of andesitic magma. J Volcanol Geotherm Res 358:87–104
Thompson PA, Carofano GC, Kim Y-G (1986) Shock waves and phase changes in a large-heat-capacity fluid emerging from a tube. J Fluid Mech 166:57–92
Toramaru A (2006) BND (bubble number density) decompression rate meter for explosive volcanic eruptions. J Volcanol Geotherm Res 154:303–316
Toramaru A (2014) On the second nucleation of bubbles in magmas under sudden decompression. Earth Planet Sci Lett 404:190–199
Toramaru A, Ishiwatari A, Matsuzawa M, Nakamura M, Arai S (1996) Vesicle layering in solidified intrusive magma bodies: a newly recognized type of igneous structure. Bull Volcanol 58(5):393–400
Toramaru A, Ishiwatari A, Matsuzawa M, Nakamura M, Arai S (1997) Corrections to vesicle layering in solidified intrusive magma bodies: a newly recognized type of igneous structure. Bull Volcanol 58(8):655–656
Toramaru A, Noguchi S, Oyoshihara S, Tsune A (2008) MND (microlite number density) water exsolution rate meter. J Volcanol Geotherm Res 175:156–167
Turcotte DL, Ockendon H, Ockendon JR, Cowley SJ (1990) A mathematical model of vulcanian eruptions. Geophys J Intern 103(1):211–217
Wager LR (1961) A note on the origin of ophitic texture in the chilled olivine gabbro of the Skaergaard intrusion. Geol Mag 1008:353–369
Waters LE, Andrews BJ, Lange RA (2015) Rapid crystallization of plagioclase phenocrysts in silicic melts during fluid-saturated ascent: phase equilibrium and decompression experiments. J Petrol 56(5):981–1006
Wilson L, Sparks RSJ, Walker GPL (1980) Explosive volcanic eruptions–IV the control of magma properties and conduit geometory on eruption column behavior. Geophys J Roy Astr Soc 63:117–148
Yamashita S, Toramaru A (2021) Control of magma plumbing systems on long term eruptive behavior of Sakurajima volcano: insights from CSD (Crystal Size Distribution) analysis. In: Vetere F (ed) Dynamic magma evolution, geophysical monograph 254, 1st edn. Wiley, American Geophysical Union, pp 89–98
Yanagi T, Ichimaru Y, Hirahara S (1991) Petrochemical evidence for coupled m agma chambers beneath the Sakurajima volcano, Kyushu. Jpn Geochem J 25:17–30
Zeldovich YB, Raizer YP (1967) Physics of shock waves and high-temperature hydrodynamic phenomena. Academic press, New York
Zhang Y, Behrens H (2000) H_2O diffusion in rhyolitic melts and glasses. Chem Geol 169:243–262

Appendix 11

Abstract
In this chapter as appendix, some fundamental knowledge and concepts, derivation of equations, material properties related to magmas, and practical methods of textural observation are briefly summarized for reader's convenience. For more comprehensive understanding, readers will find the helpful illustration in cited articles.

11.1 Thermodynamic Potential

The concept of thermodynamic potentials or thermodynamic functions was developed during the process of establishing thermodynamics. Thermodynamic potentials are functions that take the minimum when the system is in a stable equilibrium; moreover, they are potential energy extended to thermodynamic consideration. The functions include intensive variables T, P, and μ,[1] and extensive variables S, V, and N, as independent variables; for example, the internal energy $\mathcal{E}$ of a single-component system, which is written as follows:

$$d\mathcal{E} = TdS - PdV + \mu dN \tag{11.1}$$

Because this thermodynamic potential $\mathcal{E}$ follows the total differential form of the independent variables S, V, and N, the above equation is developed to be equivalent to the following:

$$d\mathcal{E} = \left(\frac{\partial \mathcal{E}}{\partial S}\right)_{V,N} dS + \left(\frac{\partial \mathcal{E}}{\partial V}\right)_{N,S} dV + \left(\frac{\partial \mathcal{E}}{\partial N}\right)_{S,V} dN \tag{11.2}$$

The common thermodynamic potentials are internal energy $\mathcal{E}(S, V)$, Helmholtz free energy $\mathcal{F}(V, T)$, Gibbs free energy $\mathcal{G}(T, P)$, and enthalpy $\mathcal{H}(P, S)$. The letters in parentheses represent independent variables and thermodynamic potentials are selected depending on which independent variables are handled. Among these thermodynamic potentials, there is a relationship $\mathcal{G} = \mathcal{F} + PV = \mathcal{H} - TS = \mathcal{E} + PV - TS$.

This is the Legendre transformation, which, in terms of mathematics, transforms independent variables. In addition to this equation, if you recall the total differential form (e.g., Eq. (11.1)) that explicitly

[1] For chemical potential μ, see Box in Sect. 2.5.1.

A. Toramaru, *Vesiculation and Crystallization of Magma*, Advances in Volcanology,
https://doi.org/10.1007/978-981-16-4209-8_11

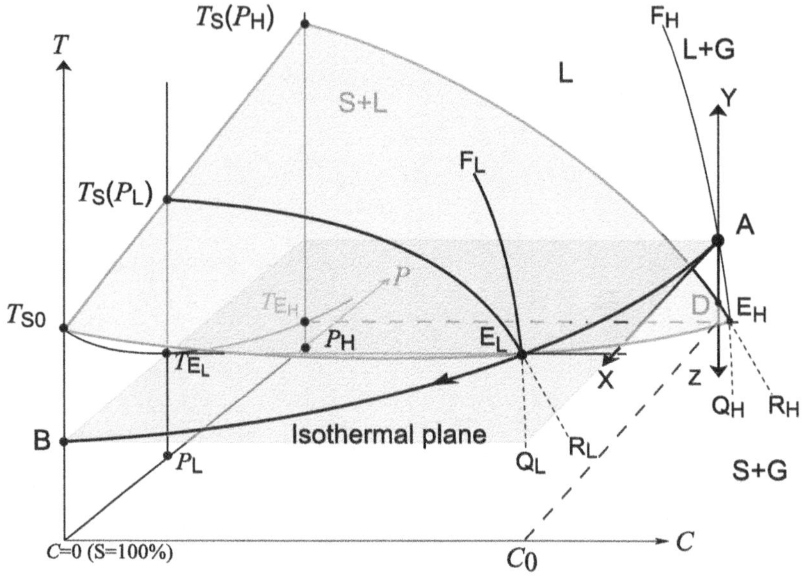

Fig. 11.1 Schematic equilibrium phase diagram of the albite—H_2O system in a temperature T-pressure P-water content C space

contains independent variables, you can calculate thermodynamic potentials by integrating it. For this purpose, you have to memorize the relationships between sets of independent variables and potentials.

A method of easily memorizing the combination between the thermodynamic potentials and corresponding independent variables is as follows: First, say "Lucky seven" and write the opening "S $\mathcal{E}$ V" in "seven" (Guggenheim 1950, modified by Wada 1981). Next, write the intensive variables T and P, which make conjugate pairs[2] with the extensive variables S and V, respectively, as located at the corresponding opposite apex of the square. Then, in a clockwise direction, fill the four sides of the square with letters "$\mathcal{EFGH}$" in an alphabetic order from the top.

$$\begin{array}{ccc} S & \mathcal{E} & \textcircled{V} \\ \mathcal{H} & & \mathcal{F} \\ P & \mathcal{G} & \textcircled{T} \end{array}$$

Now this is completed. Both sides of each thermodynamic potential are its independent variables. You then have to insert a minus sign before the circled variables, e.g.,

$$d\mathcal{G} = -SdT + VdP + \mu dN \tag{11.3}$$

If S, V, and μ have been determined as functions of temperature, pressure, and composition, this equation can be immediately integrated. The basic knowledge used here is just that S and T, as well as V and P, are a conjugate pair.

[2] A special combination between an intensive variable and an extensive variable whose product (the intensive variable) × (the extensive variable) makes one with the unit of energy. In particular, $T \times S$, $P \times V$, and $\mu \times N$. Memorizing these combinations helps you manipulate thermodynamic equations.

11.2 Summary of an Equilibrium Phase Diagram of a Water-Bearing System

To date, we observed a vesiculation-related phase change processes in a 2D equilibrium phase diagram; let us now see them in a 3D, temperature—pressure -water content space to summarize them (Fig. 11.1). There is an isothermal plane that passes through the state A (pressure P_H; temperature T_{E_L}; and H_2O content in melt, C_0). The saturation curve (solubility curve) of H_2O at a given temperature is A–E_L–B, which is used to explain a system that changes with pressure in an isothermal process. The curve F_H–A–E_H–Q_H is the L/L+G equilibrium curve (saturation curve) on the T–C plane at a pressure P_H. Here, L is the liquid phase and G is the gas phase. The curve $T_S(P_H)$–D–E_H–R_H is the L/S+L equilibrium curve on the T–C plane at a pressure P_H. Here, S is the solid phase, and these two curves are then used for examining phase change in isobaric processes. Both albite and H_2O form a eutectic system, and the eutectic point (the point of intersection of the two curves, L/L+G and L/S+L) is denoted using E_H. When the pressure changes to P_L, the L/L+G equilibrium curve and the L/S+L equilibrium curve shift to F_L–E_L–Q_L and $T_S(P_L)$–E_L–R_L, respectively, and the eutectic point shifts to E_L. The trajectory of the eutectic point associated with the change in pressure will then be E_H–E_L–T_{S0}. Moreover, the projection on the T–P plane will be T_{H_E}–T_{E_L}–T_{S0}, and the melting point increases as the pressure decreases. The melting point of albite in a dry system ($C = 0$) increases as T_{S0}–$T_S(P_L)$–$T_S(P_H)$ with increase in pressure. Moreover, the curved surface passing through F_H–Q_H–F_L–Q_L–T_{S0}–B (not shaded in the figure) will then be the L/L+G equilibrium surface in the T–P–C space, whereas the curved surface passing via $T_S(P_H)$–R_H–$T_S(P_L)$–R_L–T_{S0} will be the L/S+L equilibrium surface.

11.3 Derivation of the Fokker-Planck Equation

11.3.1 Derivation of an Equation in Terms of the Number of Molecules as Size, l

We will now derive Eq. (3.29). When we write the transition probability from the size l' to the size l as $W_o(l' \to l)$, change in the number of bubbles $N(l)$ (number/m^3) can be written as follows (corresponding to Eq. (3.27)):

$$\frac{\partial N(l,t)}{\partial t} = \sum_{l'} \left(W_o(l' \to l) \cdot N(l',t) - W_o(l \to l') \cdot N(l,t) \right) \tag{11.4}$$

This equation is then called a master equation.

We then write such transition probability of coagulation/fragmentation as $\Psi_C(l, l')$ and $\Psi_F(l, l')$, respectively, and rewrite W_o representing gain in the number and loss in number.

$$\text{gain} : W_o(l' \to l) = \underbrace{\Psi_C(l - l', l') \cdot N(l - l', t)}_{\text{Coagulation}} + \underbrace{\Psi_F(l', l)}_{\text{Fragmentation}} \tag{11.5}$$

$$\text{loss} : W_o(l \to l') = \underbrace{\Psi_C(l' - l, l) \cdot N(l' - l, t)}_{\text{Coagulation}} + \underbrace{\Psi_F(l, l')}_{\text{Fragmentation}} \tag{11.6}$$

Substituting these equations in Eq. (11.4) yields the following:

$$\frac{\partial N(l,t)}{\partial t} = \sum_{l'} \left(\underbrace{\left(\Psi_{\mathrm{C}}(l-l',l') \cdot N(l-l',t) + \Psi_{\mathrm{F}}(l',l)\right) \cdot N(l',t)}_{\text{gain}} \right.$$
$$\left. - \underbrace{\left(\Psi_{\mathrm{C}}(l'-l,l) \cdot N(l'-l,t) + \Psi_{\mathrm{F}}(l,l')\right) \cdot N(l,t)}_{\text{loss}} \right) \quad (11.7)$$

Here, we impose a requirement that equilibrium distribution $N_{\mathrm{E}}(l)$ should be a steady-state solution of the above equation, $dN_{\mathrm{E}}(t)/dt = 0$. Moreover, between a certain size l and another size l' at the steady-state (detail balance), we assume that the ratio between coagulation and fragmentation is balanced:

$$\Psi_{\mathrm{C}}(l-l',l') \cdot N_{\mathrm{E}}(l-l') \cdot N_{\mathrm{E}}(l') = \Psi_{\mathrm{F}}(l,l') \cdot N_{\mathrm{E}}(l) \equiv W(l':l) \quad (11.8)$$

This assumption has a significance that connects disquilibrium processes (coagulation and fragmentation) and disequilibrium processes(equilibrium distribution as a property of an equilibrium state). Using $W(l':l)$ yields the following:

$$\frac{\partial N(l,t)}{\partial t} = \underbrace{\sum_{l'>l} W(l:l') \frac{N(l',t)}{N_{\mathrm{E}}(l')}}_{\text{Fragmentation1}} - \underbrace{\frac{1}{2}\sum_{l'<l} W(l':l) \frac{N(l,t)}{N_{\mathrm{E}}(l)}}_{\text{Fragmentation2}}$$
$$+ \underbrace{\frac{1}{2}\sum_{l'<l} W(l':l) \frac{N(l',t)}{N_{\mathrm{E}}(l')} \frac{N(l-l',t)}{N_{\mathrm{E}}(l-l')}}_{\text{Coagulation1}}$$
$$- \underbrace{\sum_{l'>l} W(l:l') \frac{N(l'-l,t)}{N_{\mathrm{E}}(l'-l)} \frac{N(l,t)}{N_{\mathrm{E}}(l)}}_{\text{Coagulation2}} \quad (11.9)$$

Here, we consider both the inflow and outflow of one molecule as growth and vaporization and distinguish them from coagulation and fragmentation, which is attributed to interactions among assemblages of molecules. For this purpose, we introduce change in size corresponding to change of a single molecule, Δl. Ignoring coagulation/fragmentation processes ($|l-l'| > 1$), we transform Eq. (11.9) into the following:

$$\frac{\partial N(l,t)}{\partial t} = \underbrace{W(l:l+\Delta l) \frac{N(l+\Delta l,t)}{N_{\mathrm{E}}(l+\Delta l)}}_{\text{Fragmentation1}} - \underbrace{W(l-\Delta l:l) \frac{N(l,t)}{N_{\mathrm{E}}(l)}}_{\text{Fragmentation2}}$$
$$+ \underbrace{W(l-\Delta l:l) \frac{N(l-\Delta l,t)}{N_{\mathrm{E}}(l-\Delta l)} \frac{N(\Delta l,t)}{N_{\mathrm{E}}(\Delta l)}}_{\text{Coagulation1}}$$
$$- \underbrace{W(l:l+\Delta l) \frac{N(\Delta l,t)}{N_{\mathrm{E}}(\Delta l)} \frac{N(l,t)}{N_{\mathrm{E}}(l)}}_{\text{Coagulation2}} \quad (11.10)$$

Here, Fragmentation 1 and Coagulation 2 are equivalent to the transfer of the bubble number between the size l and the size $l + \Delta l$; we will write it as $Q(l+)$. Moreover, Fragmentation 2 and Coagulation 1 are equivalent to transfer of the bubble number between the size $l - \Delta l$ and the size l; we will write it as $Q(l-)$. If we set the direction in which l increases so as for Q to be positive, the above equation will then be written as follows:

$$\frac{\partial N(l,t)}{\partial t} = Q(l-) - Q(l+) \tag{11.11}$$

where $Q(l+)$ is the total outflow from l determined from exchange between l and $l + \Delta l$ and given as follows:

$$\begin{aligned} Q(l+) &= W(l : l+\Delta l)\frac{N(\Delta l, t)}{N_{\mathrm{E}}(\Delta l)}\frac{N(l,t)}{N_{\mathrm{E}}(l)} \\ &\quad - W(l : l+\Delta l)\frac{N(l+\Delta l, t)}{N_{\mathrm{E}}(l+\Delta l)} \\ &= -W(l : l+\Delta l)\cdot \Delta l \cdot \frac{\partial}{\partial l}\left(\frac{N(l,t)}{N_{\mathrm{E}}(l)}\right) \end{aligned} \tag{11.12}$$

Here, we used the Taylor expansion of $\frac{N(l+\Delta l,t)}{N_{\mathrm{E}}(l+\Delta l)}$

$$\frac{N(l+\Delta l, t)}{N_{\mathrm{E}}(l+\Delta l)} = \frac{N(l,t)}{N_{\mathrm{E}}(l)} + \Delta l \frac{\partial}{\partial l}\left(\frac{N(l,t)}{N_{\mathrm{E}}(l)}\right) \tag{11.13}$$

Moreover, we assume that the distribution of small bubbles is close to an equilibrium state and $\frac{N(\Delta l,t)}{N_{\mathrm{E}}(\Delta l)} = 1$ holds. $Q(l-)$ can then be rewritten in the same manner as follows:

$$Q(l-) = -W(l-\Delta l : l)\cdot \Delta l \cdot \frac{\partial}{\partial l}\left(\frac{N(l,t)}{N_{\mathrm{E}}(l)}\right) \tag{11.14}$$

Using these equations to rearrange Eq. (11.11) yields

$$\frac{\partial N(l,t)}{\partial t} = (W(l : l+\Delta l) - W(l-\Delta l : l))\cdot \Delta l \cdot \frac{\partial}{\partial l}\left(\frac{N(l,t)}{N_{\mathrm{E}}(l)}\right) \tag{11.15}$$

Here, using Taylor expansion of $W(l - \Delta l : l)$ around l:

$$W(l-\Delta l : l) = W(l : l+\Delta l) - \Delta l \frac{\partial W(l : l+\Delta l)}{\partial l} \tag{11.16}$$

we can rewrite Eq. (11.15) as follows:

$$\frac{\partial N(l,t)}{\partial t} = \frac{\partial}{\partial l}\left((\Delta l)^2 \cdot W(l) \cdot \frac{\partial}{\partial l}\left(\frac{N(l,t)}{N_{\mathrm{E}}(l)}\right)\right) \tag{11.17}$$

which is a Fokker-Planck equation. Here, we rewrite $W(l) \equiv W(l : l + \Delta l)$, and Eq. (3.29) is what was obtained by rearranging this equation with respect to the term of $N(l,t)$ and the term of $\partial N(l,t)/\partial l$ in the parentheses.

11.3.2 Transformation from $N(l, t)$ to $F(R, t)$

$N(l, t)$ and $F(R, t)$ are then connected via the conservation of number, i.e.,

$$N(l, t)dl = F(R, t)dR \tag{11.18}$$

Rearranging it yields

$$F(R, t) = N(l, t)\frac{dl}{dR} \tag{11.19}$$

The number of molecules in a cluster, l, is then linked with the cluster radius R through

$$l v_{\mathrm{G}} = \frac{4\pi}{3}R^3 \tag{11.20}$$

where v_{G} is the molecular volume of gas. Therefore,

$$\frac{dl}{dR} = \frac{4\pi R^2}{v_{\mathrm{G}}} \tag{11.21}$$

holds, although this is not used here.

We then transform Eq. (11.17) into an equation with respect to R. Applying

$$\frac{\partial}{\partial l} = \frac{\partial}{\partial R}\frac{\partial R}{\partial l} \tag{11.22}$$

to Eq. (11.17) yields

$$\frac{\partial N(l, t)}{\partial t} = \frac{\partial}{\partial R}\left((\Delta l)^2 \cdot W(l) \cdot \frac{\partial}{\partial R}\left(\frac{N(l, t)}{N_{\mathrm{E}}(l)}\right)\frac{\partial R}{\partial l}\right)\frac{\partial R}{\partial l} \tag{11.23}$$

Then, transposing $\partial R/\partial l$ outside the parentheses on the right-hand side to the left-hand side and placing it into $\partial/\partial t$ yields

$$\frac{\partial}{\partial t}\left(N(l, t)\frac{\partial l}{\partial R}\right) = \frac{\partial}{\partial R}\left((\Delta l)^2 \cdot W(l)\frac{\partial R}{\partial l} \cdot \frac{\partial}{\partial R}\left(\frac{N(l, t)\frac{\partial l}{\partial R}}{N_{\mathrm{E}}(l)\frac{\partial l}{\partial R}}\right)\right) \tag{11.24}$$

where the denominator and the numerator of $N(l, t)/N_{\mathrm{E}}(l)$ is multiplied by $\partial l/\partial R$, respectively.

From Eq. (11.19),

$$\frac{\partial F(R, t)}{\partial t} = \frac{\partial}{\partial R}\left((\Delta R)^2 \cdot W(R)\frac{\partial R}{\partial l} \cdot \frac{\partial}{\partial R}\left(\frac{F(R, t)}{F_{\mathrm{E}}(R)}\right)\right) \tag{11.25}$$

is obtained, which is a Fokker-Planck equation. Here we write ΔR and $W(R)$ rather than Δl and $W(l)$, respectively. The Fokker-Planck equation for the cluster radius R is different from the case in which exchange between l and R is not considered by $\partial R/\partial l$ in the growth term. Because we collectively define it as $A(R)$, the definition of $A(R)$ correspondingly differs. However, it does not influence our subsequent discussion.

11.4 F_E^0 of Equilibrium Size Distribution

The proportional coefficient F_E^0 is the population density of bubbles with a radius $R \to 0$, i.e., infinitesimal radius, and has a unit (number of molecules)/m^4. In this case, the molecules of gas components belong to any one of clusters of bubbles with respective sizes and the number of the molecules dynamically changes. Therefore, the number cannot be described only via macroscopic investigation; however, because it is reasonable to consider that F_E^0 is related to the number of molecules per unit volume, C, we will relate it to $C = F_E^0 \times \delta R$. Here, δR is the effective partition size that determines the existence width of the critical size in the size space. δ R can then be considered as the increment ΔR_C of the size space that increases as molecules in the critical bubbles increase in number by 1. Here, we will consider what is obtained by dividing the critical size R_C by the number of molecules contained in it, $N_C \approx (R_C)^3/v_G$, as the increment qualitatively. That is, we assume $\delta R \approx (R_C)/N_C \approx (R_C)/((R_C)^3/v_G) = v_G/(R_C)^2$; consequently, $F_E^0 = C/\delta R$ is rewritten as follows:

$$F_E^0 \approx \frac{C{R_C}^2}{v_G} \tag{11.26}$$

Let us discuss this somewhat more precisely using equilibrium bubble number density distribution $N_E(l)$ (1/m^3). Although this will be a repetition of the discussion from Eqs. (11.18) to (11.19), the number density distribution

$$N_E(l)dl \tag{11.27}$$

is defined by the number of bubbles containing the number of molecules l in a unit volume of liquid and has a relationship

$$F_E(R)dR = N_E(l)dl \tag{11.28}$$

with the size distribution function $F_E(R)$, i.e.,

$$F_E(R) = N_E(l)\frac{dl}{dR} \tag{11.29}$$

is obtained.

In the same manner as $F_E(R)$, $N_E(l)$ is given as a function of change in the Helmholtz energy, described by l, as follows:

$$N_E(l) = N_1 \exp\left(-\frac{\Delta\mathcal{F}(l)}{k_B T}\right) \tag{11.30}$$

Here, as a very close approximation, we can consider the coefficient $N_E(1)$, i.e., the number of bubbles containing one molecule to be equivalent to the concentration defined by the number of molecules per unit volume, C, as follows[3]:

$$N_1 = C \tag{11.31}$$

Moreover, the number of molecules in bubbles l can then be calculated by

[3]Specifically, the total sum of the number of molecules contained in bubbles of all size is equal to the concentration; however, in the case where the degree of supersaturation is not so large, the approximation described in the text can be used.

$$l = \frac{4\pi R^3}{3v_G} \tag{11.32}$$

Assuming that, regardless of the size R, the molecular volume v_G is a constant, the value of the size distribution function of bubbles of zero size, i.e., bubbles containing one molecule, F_E^0, is obtained as follows:

$$F_E^0 = 4\pi \frac{C{R_C}^2}{v_G} \tag{11.33}$$

This equation then differs from Eq. (11.26), obtained from scaling, by 4π.

11.5 Integration to Obtain Steady-State Nucleation Rate

The integral of the denominator of Eq. (3.69), In, is as follows:

$$In = \int_0^\infty \frac{dR}{B(R)\cdot F_E(R)} \tag{11.34}$$

$$= \frac{2}{v_G DC^2 R_C} \exp\left(\frac{4\pi\gamma R_C^2}{k_B T}\right) \cdot \int_0^\infty R^2 \exp\left(\frac{\frac{1}{2}\frac{d^2\mathcal{W}}{dR^2}(R-R_C)^2}{k_B T}\right) dR \tag{11.35}$$

The integral In is then determined by the behavior of the integrand, and the dependence of the inside of the exponential part on R greatly influences the integrated value. In such a case, we use the saddle-point method in which a function in the exponential part is expanded around R_C, which gives the local maximum value. Equation (11.35) has already taken such a form,[4, 5]

[4] Furthermore, the integrand other than the exponential part, R^2, can be moved outside the integral by representing the value at R_C, which gives the local maximum value. The integral In can then be written as follows:

$$In = \frac{8\pi R_C \exp\left(\frac{4\pi\gamma (R_C)^2}{3k_B T}\right)}{v_G DC^2} \int_0^\infty \exp\left(\frac{\frac{1}{2}\frac{d^2\Delta\mathcal{F}}{dR^2}(R-R_C)^2}{k_B T}\right) dR \tag{11.36}$$

Next, using the change of variables $R_C x = R - R_C$ yields $R: 0 \to \infty$, $x: -1 \to \infty$. Moreover, because $dR = R_C dx$ holds, the integral part of Eq. (11.36) becomes

$$R_C \int_{-1}^\infty \exp\left(-ax^2\right) dx \approx R_C \int_{-\infty}^\infty \exp\left(-ax^2\right) dx \tag{11.37}$$

where

$$a = \frac{1}{2}\left|\frac{d^2\mathcal{W}}{dR^2}\right| \frac{(R_C)^2}{k_B T} = \frac{4\pi\gamma (R_C)^2}{k_B T} = \pi (Z R_C)^2 \tag{11.38}$$

The integral (11.37) then becomes $R_C\sqrt{\pi/a} = \Gamma(1/2) R_C a^{-1/2} = Z^{-1}$ using the Gaussian integral. Here, Γ is the Gamma function.

[5] The integral whose integrand is the Gaussian function e^{-x^2} is known as the Gaussian integral, and the integral from $-\infty$ to ∞ becomes $\sqrt{\pi}$. That is,

$$\int_{-\infty}^\infty \exp\left(-x^2\right) dx = \sqrt{\pi} \tag{11.39}$$

Ultimately, the integral In will be obtained as follows:

$$In = \frac{8\pi R_C}{Z v_G D C^2} \exp\left(\frac{4\pi\gamma(R_C)^2}{3k_B T}\right) = \frac{1}{Z \cdot B_C \cdot F_E^*} \tag{11.40}$$

11.6 Physical Properties of Silicate Melt

This section describes the outline of surface tension, diffusivity, and viscosity, all of which are important concepts in this book and are physical properties that change by an order of magnitude depending on the composition of silicate melt or temperature—pressure condition.

11.6.1 Surface Tension

Although surface (interface) tension and surface (interface) energy are, strictly speaking, different (Ono 1980; Defay et al. 1966), they are used in the same sense in this book. Murase and McBirney (1973) conducted a pioneering study of measuring the surface tension between gas and silicate melt. They conducted the experiment using the sessile drop method with samples of a wide range of composition from basalt to rhyolite to soda glass over a range of temperatures from 1000 to 1500 °C. The values of surface tension obtained by their experiment were 0.3–0.4 N/m in an argon atmosphere at 1 atm. Moreover, Walker and Mullins (1981) measured the surface tension of several types of melt (from basaltic to rhyolitic) using the Czochralski method at 1 atm; moreover, Taniguchi (1988) measured the surface tension of Di-An melt using the same method. Note that the values obtained by them were 0.3–0.4 N/m. As seen in the text, homogeneous nucleation does not occur at such a large surface tension but heterogeneous nucleation occurs in nature. Moreover, it is highly unlikely that homogeneous nucleation will occur in experiments. In their study, the applicability of their data was limited because gas was not water vapor and pressure was 1 atm.

Improving experimental apparatus to solve these two problems, Bagdassarov et al. (2000) measured surface tension between synthetic haplogranitic melt and water vapor using the sessile drop method and obtained the pressure dependence of surface tension (Fig. 11.2). Their experimental result demonstrated that under water-saturated conditions at a high pressure, the surface tension decreases to as less as 0.1 N/m at 200 MPa. They then expressed the pressure dependence and temperature dependence (approximated from a figure in their paper) as follows:

$$\gamma = 0.2366 \exp\left(-0.35 \times 10^{-8} \cdot P_W - 11 \times 10^3 \left(\frac{1}{T} - \frac{1}{1273}\right) / \mathcal{R}\right) \tag{11.41}$$

As shown in Chap. 5, the initial saturation pressure at 200 MPa or higher is required to cause homogeneous nucleation in decompression-induced vesiculation experiments.

There are studies that in the case where volatiles other than water are contained, they influence the surface tension and influence nucleation pressure (Yamada et al. 2005; Fiege 2014; Gardner et al. 2018). Moreover, liquid composition, i.e., difference in concentration of compositions other than the volatiles, may have an influence (Mangan and Sisson 2005). However, the surface tension under saturated vapor pressure has not yet been directly measured and is only an estimate from nucleation pressure.

The integral (11.37) then becomes $\sqrt{\pi/a}$ by change of variables $y = a^{1/2}x$.

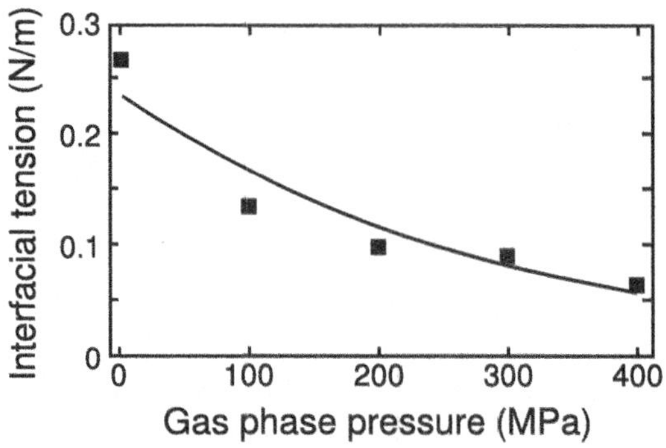

Fig. 11.2 The surface tension between synthetic haplogranitic melt and water vapor by Bagdassarov et al. (2000) with permission from the Mineralogical Society of America. Reprinted from Toramaru (2006), Copyright 2006, with permission from Elsevier

For temperature dependence of gas—liquid surface tension in a one-component system, the following van der Waals-Guggenheim equation holds:

$$\gamma = \gamma_0 \left(1 - \frac{T}{T_C}\right)^{11/9} \tag{11.42}$$

where T_C is temperature at the critical point and γ_0 is a constant temperature value distant from T_C. The critical point is then defined in terms of thermodynamics as the point where the density of gas and that of liquid become equal with increase in temperature and pressure along the saturation curve. The gas—liquid surface tension under the saturated condition is defined within the range from a low-temperature, low-pressure point to the critical point on the saturation curve ($T < T_C$) and becomes 0 at the critical point.

Moreover, the surface tension is influenced by the curvature of the interface. This is considered to arise in the case where the radius of curvature of the interface is not negligible compared with the length of interactions between molecules. This is called the Tolman correction and determined using Eq. (3.111).

11.6.2 Diffusivity of Water in Silicate Melts

Diffusivity of H_2O in silicate melt was first measured by Shaw (1974). He then placed silicate melt in a pressure vessel to bring it into contact with water, took out it from the vessel after a certain time had elapsed, cut samples from it as they correspond to the distance from the boundary, and performed chemical analysis of the samples. Then, he compared the water content obtained as a function of the distance from the boundary with solutions of a diffusion equation and obtained values of diffusivity. Subsequently, secondary-ion mass spectrometry (SIMS), an analytical technique, was developed in the 1980s and Karsten et al. (1982) and Delaney and Karsten (1981) measured Shaw's samples using SIMS at a higher spatial resolution. In the 1990s, Fourier transform infrared (FTIR) spectroscopy, an additional advanced analytical technique that allows us to measure samples with high spatial resolution under a microscope, was developed and Zhang and Stolper determined the diffusivity of water in both rhyolitic and basaltic melts using FTIR spectroscopy (Zhang and Stolper 1991). Then, using this technique, Behrens and Zhang determined the dependence of diffusivity on composition and temperature.

Zhang and Behrens (2000) dealt with the temperature dependence of water diffusivity in felsic melt and expressed diffusivity as a function of it: $D_{ZB2000}(C_W, T, P)$. Moreover, Beherns et al. (2004)

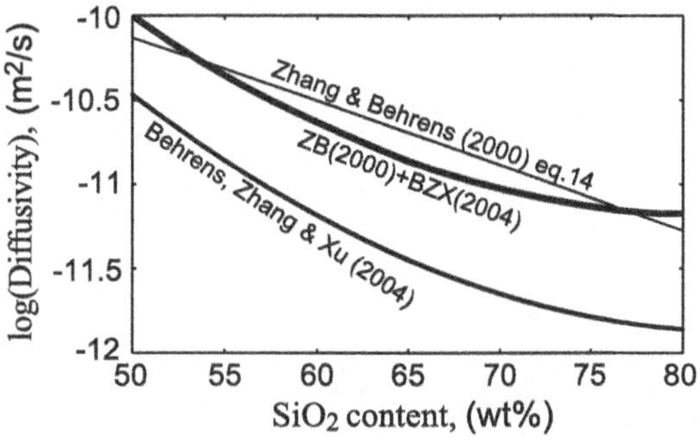

Fig. 11.3 Diffusivity of water in silicate melt by Zhang and Behrens (2000) and Beherns et al. (2004). ZB(2000)+BZX(2004) represents diffusivity taking the effect of melt composition and the effect of temperature into consideration. Water saturation at 100 MPa is assumed. Reprinted from [Toramaru 2006], Copyright 2006, with permission from Elsevier

conducted an experiment with various SiO_2 content C_{Si}(wt%) and obtained an expression for diffusivity after considering the dependence on chemical composition: $D_{BZX2004}(P, T, C_{Si})$. Integrating these two yields

$$D(P(C_W), T, C_{Si}) = D_{BZX2004}(P(C_W), T, C_{Si}) \cdot \frac{D_{ZB2000}(C_W, T, P(C_W))}{D_{ZB2000}(1\ \mathrm{wt\%}, T, P(C_W))} \quad (11.43)$$

(Figure 11.3).

Liu et al. (2004) demonstrated that the diffusivity of andesitic glass is lesser than that of rhyolitic glass at a lower temperature because of activation energy. Okumura and Nakashima (2005) conducted hydration experiments using the melt of basalt, andesite, dacite, and rhyolite; they then determined the water diffusivity and activation energy of each melt. Studies on CO_2 diffusion were initiated by Watson (1991).

The diffusion of other elements in silicate melt has been measured for many years and reviewed by Hofmann (1980) for the early stage. There is a set relationship between a pre-exponential factor of diffusivity and activation energy, which is known as a compensation law. Baker (1991, 1992) discussed this in detail and Liang et al. (1996a, b, 1997) cited it in the chapter on crystallization (Sect. 7.6.6).

11.6.3 Newtonian Viscosity

Newtonian viscosity is defined as a coefficient of a proportional relation between stress and shear rates. The viscosity of melt varies depending on temperature, pressure, chemical composition of melt, and water content and generally exhibits a non-Newtonian viscosity. Note that non-Newtonian viscosity has a feature that it changes depending on the timescale of both stress and shear rates. Qualitatively, while elasticity is predominant when the timescale is reduced (in sudden change), Newtonian viscosity is predominant when the timescale is extended (Sect. 11.7). Knowledge about the Newtonian viscosity (η) of melt not containing crystals and bubbles has been accumulated from both theoretical and experimental aspects (Kingery et al. 1976). For example, the following equation by Shaw (1972) is a convenient method described even in Chronological Scientific Tables:

$$\ln \eta = \frac{\sum Y_i (c_i \cdot Y_{SiO_2})}{1 - Y_{SiO_2}} \cdot \left(\frac{10^4}{T} - 1.5 \right) - 8.7 \quad (11.44)$$

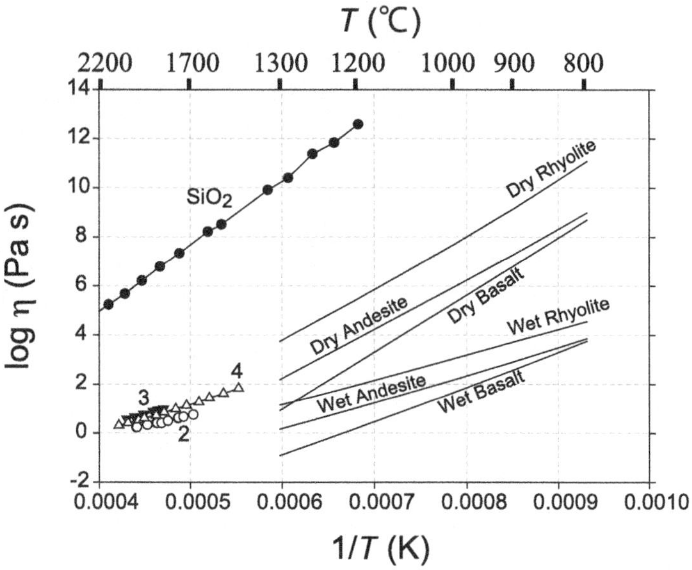

Fig. 11.4 Viscosity of silicate melt at high temperature (Urbain et al. 1982). The numbers 2, 3, and 4 correspond to the composition (2), (3), and (4) in the text. Basalt, andesite, and rhyolite correspond to the chemical composition of oceanic island basalt, calc-alkaline andesite, and calc-alkaline rhyolite described in the Chronological Scientific Tables. Moreover, dry melt does not contain water while wet melt contains 5 wt% water

where i represents an oxide component and Y_i represents mol% of the oxide (for Al_2O_3 and Fe_2O_3, the molar ratio is doubled and then converted into 100%). c_i is $SiO_2 = 0$, $Al_2O_3 = 6.7$, $TiO_2 = CaO = 4.5$, $Fe_2O_3 = FeO = MnO = MgO = 3.4$, and $Na_2O = K_2O = 2.8$, $H_2O = 2.0$.

Viscosity is then related to the structure of melt, which is almost entirely determined by the packing of oxygen ions whose ionic radius is the largest. Of these oxygen ions, the ratio of oxygen ions comprising SiO_4^{4-} tetrahedrons (T) and contributing to the whole network is important. Viscosity can then be explained by the fraction of non-bridging oxygen (NBO) that does not contribute to the network, NBO/T, to oxygen in the tetrahedrons (Mysen 1988; Taniguchi 2001). T = 100% corresponds to fused quartz. In this case, viscosity almost follows the Arrhenius equation (Urbain et al. 1982). That is,

$$\eta(SiO_2 = 100) = \exp\left(-13.4 + \frac{\mathcal{E}_\eta}{\mathcal{R}T}\right) \tag{11.45}$$

where $\mathcal{E}_\eta$ is the activation energy and is about 5.1×10^2 kJ/mol. When concentrations of major chemical components change, the value of activation energy changes. In addition to the viscosity of silica glass of SiO_2 100 wt% (1), Fig. 11.4 shows the viscosity of SiO_2 50 wt%+Al_2O_3 50 wt% (2), SiO_2 50 wt% + MgO 50 wt% (3), and SiO_2 50 wt% + Al_2O_3 25 wt% + MgO 25 wt% (4). This figure demonstrates that both Al and Mg have the effect of decrease in viscosity.

Giordano (2008) compiled experimental data on the viscosity of silicate melt in a wide range of temperatures and compositions and organized them using the following the Vogel-Fulcher theory (VFT) equation:

$$\eta = \exp\left(A_\eta + \frac{B_\eta(X_i)}{T - C_\eta(X_i)}\right) \tag{11.46}$$

Regardless of composition, A_η corresponds to a virtual viscosity at an infinite temperature and is determined as a constant value $A_\eta = -4.55$. Moreover, B_η and C_η are determined as functions of composition.

The effect of multiple components is then expressed by the fraction of the number of non-bridging oxygen (NBO) atoms to the number of bridging tetrahedrons (T). However, because elements other than SiO_2 and P_2O_5 modify (cut) the network and focus only on concentration of SiO2 and water, we can obtain the following simple empirical equation (Toramaru 1995), although it has errors to a certain extent.

$$\eta = \exp\left[0.46(C_{\text{silica}} - 72) - 1.25C_{\text{W}} + \frac{20903.6}{T}\right] \tag{11.47}$$

Figure 11.4 shows the Arrhenius plots of the viscosity of basaltic, andesitic, and rhyolitic melts (dry melt and melt containing 5 wt% of water), calculated by a convenient method by Hui and Zhang (2007). The figure shows that the plots nearly follow the Arrhenius equation; moreover, data on the viscosity of silicate melt are detailed in Takeuchi (2016).

Although the viscosity of melt containing crystals or bubbles has been investigated, a unified equation that can be applied to the case where the volume fraction of crystals and bubbles is large and that can contain the shape of crystals and the effect of size distribution has not yet been established. Because melt containing crystals or bubbles exhibits non-Newtonian behavior, it is challenging to examine the physical process of rheology and establish the relational equation for it.

11.7 A Constitutive Equation of a Viscoelastic Body

Because of its complex molecular structure, silicate melt not containing crystals or bubbles behaves as a non-Newtonian fluid or a viscoelastic property depending on the conditions. Here, we categorize and summarize the properties of an elastic body and viscous fluid.

1. A linear elastic body: The stress is linearly proportional to the displacement (strain) at the moment.
2. Newtonian fluid: The stress is linearly proportional to the rate of the displacement (the shear rate) at the moment.

A viscoelastic body can then be considered to deviate from the properties of the end members and have properties of both the end members. Major characteristics of the viscoelastic body include the following:

1. The maximum of stress does not take place at the time of the maximum displacement.
2. Response of the stress varies according to the shear rate of deformation or frequency.

Characteristic 1 is related to the fact that, the maximum of the stress of viscous fluid agrees with the maximum of the shear rate, although it has the coincidence with the maximum displacement an elastic body. Characteristic 2 shows that the predominance of viscosity and elasticity depends on the rate of deformation; in other words, a microscopic mechanism responsible for deformation.

To examine the general characteristics of viscoelastic fluid, let us set the following complex modulus of elasticity $M^*(\omega)$ and complex viscosity coefficient $\eta^*(\omega)$ for the relationship between stress $\sigma(\omega)$ and shear rate $\dot{\epsilon}(\omega)$ which vibrate at an angular frequency ω.[6]

[6]Strictly, the modulus of elasticity is distinguished between the bulk (volumetric) modulus M_{v}^* and the shear modulus M_{s}^* and the viscosity coefficient is then distinguished between the volumetric viscosity η_{v}^* and the shear viscosity η_{s}^*, which are expressed as follows:

$$M^* = M_{\text{v}}^* + \frac{4}{3}M_{\text{s}}^* \tag{11.48}$$

$$\eta^* = \eta_{\text{v}}^* + \frac{4}{3}\eta_{\text{s}}^* \tag{11.49}$$

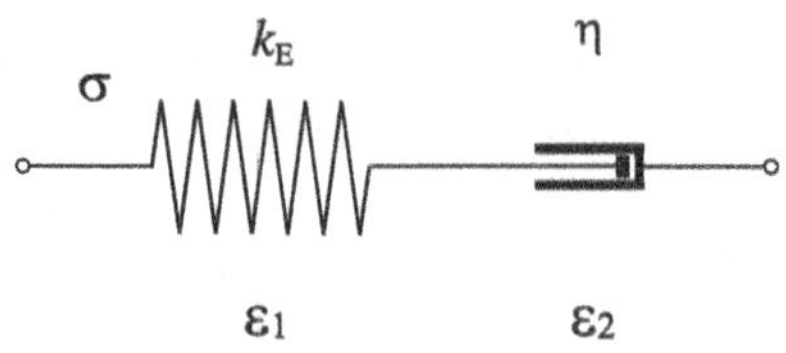

Fig. 11.5 A Maxwell model comprising a spring with an elastic constant k_E and an element dissipating viscosity (viscosity coefficient η) in series

$$\sigma(\omega) = M^*(\omega) \cdot \epsilon(\omega) \tag{11.50}$$
$$\sigma(\omega) = \eta^*(\omega) \cdot \dot{\epsilon}(\omega) \tag{11.51}$$

Here, the complex modulus of elasticity and the complex viscosity coefficient are written as follows

$$M^*(\omega) = M_R(\omega) + iM_I(\omega) \tag{11.52}$$
$$\eta^*(\omega) = \eta_R(\omega) + i\eta_I(\omega) \tag{11.53}$$

Moreover, between the real term and the imaginary term of each equation, there is a relationship $M^* = i\omega\eta^*$, i.e.,

$$M_R = -\omega\eta_I \tag{11.54}$$
$$M_I = \omega\eta_R \tag{11.55}$$

Next, to examine the manner by which the complex modulus of elasticity and the complex viscosity coefficient are concretely linked to a practical modulus of elasticity and a practical viscosity coefficient and to examine their dependence on frequency, we analyze a simple viscoelastic model. We will then consider a Maxwell viscoelastic body comprising a elastic spring and a dashpot (viscous element), as shown in Fig. 11.5. This model briefly represents the characteristics of viscoelasticity of a magma well. The relationship between the stress and strain in individual elements of this system is $\sigma = k_E\epsilon_1 = \eta\dot{\epsilon}_2$ and because the sum of the strain of the individual elements is equal to the strain of the entire system ϵ, $\dot{\epsilon} = \dot{\epsilon}_1 + \dot{\epsilon}_2$ holds.[7] From these relations, the following differential equation that connects the stress and strain of this system is obtained:

$$\frac{\dot{\sigma}}{k_E} + \frac{\sigma}{\eta} = \dot{\epsilon} \tag{11.56}$$

Substituting $\sigma(\omega) = \sigma_0 \exp(i\omega t)$ into Eq. (11.56) and rearranging it yields the real parts and imaginary parts of the complex modulus of elasticity and the complex viscosity coefficient of Eqs. (11.50)–(11.55) as follows:

However, they are not distinguished here.

[7] Although this expression is used often, the direction of stress is positive in the compression direction. If this expression is incorporated in the equation of motion, the expression should follow the direction of the common coordinate system and should be $\sigma = -k_E\epsilon_1 = -\eta\dot{\epsilon}_2$.

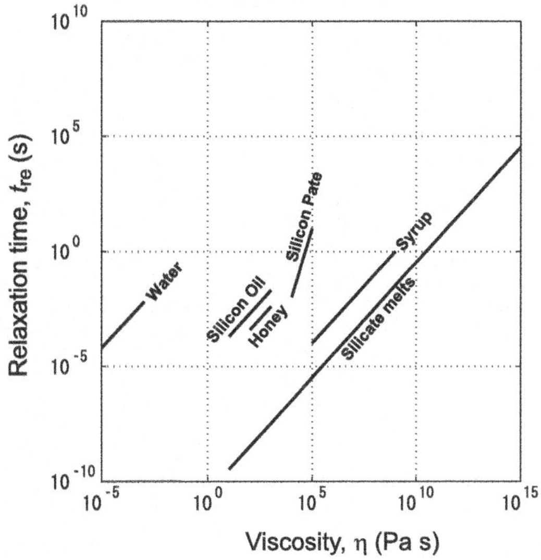

Fig. 11.6 The relationship between the relaxation time of viscoelasticity and viscosity (from data of Joseph et al. 1986)

$$M_R(\omega) = \frac{k_E \omega^2 t_{re}^2}{1 + \omega^2 t_{re}^2} \tag{11.57}$$

$$M_I(\omega) = \frac{k_E \omega t_{re}}{1 + \omega^2 t_{re}^2} \tag{11.58}$$

$$\eta_R(\omega) = \frac{\eta}{1 + \omega^2 t_{re}^2} \tag{11.59}$$

$$\eta_I(\omega) = -\frac{\eta \omega t_{re}}{1 + \omega^2 t_{re}^2} \tag{11.60}$$

where t_{re} is relaxation time of viscoelasticity defined by the following equation and ω_{re} is the corresponding angular frequency.

$$t_{re} = \frac{\eta}{k_E} = \frac{1}{\omega_{re}} \tag{11.61}$$

The relaxation time is an important index that distinguishes the behavior of a substance. By comparing the stress and pressure of the system or characteristic time of deformation and relaxation time, we can examine whether the substance shows elastic behavior or behavior of a viscous fluid.

Characteristic time of change	>	Relaxation time	→	Viscous fluid-like behavior
Characteristic time of change	<	Relaxation time	→	Elastic behavior

As evident from the equation, the relaxation time is greatly related to viscosity, which is largely influenced by temperature and compositions such as water content. However, the shear modulus (or rigidity) is not considerably dependent on temperature and quantities of trace components. Figure 11.6 shows the relaxation time of various substances.

The complex modulus of elasticity and complex viscosity coefficient can then be written as follows:

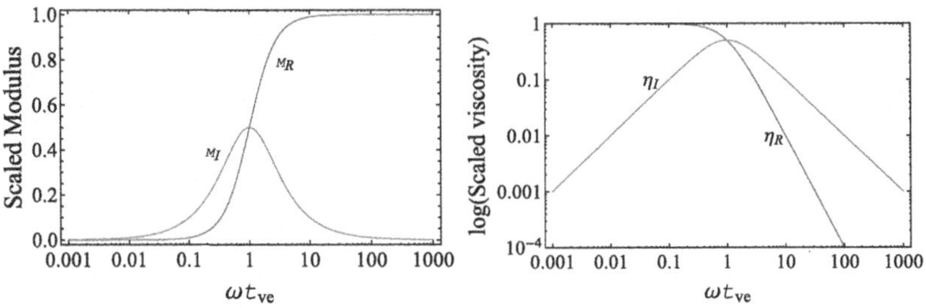

Fig. 11.7 Frequency dependence of the real part and imaginary part of the complex modulus of elasticity (left) and the complex viscosity coefficient (right). Scaled by the relaxation time

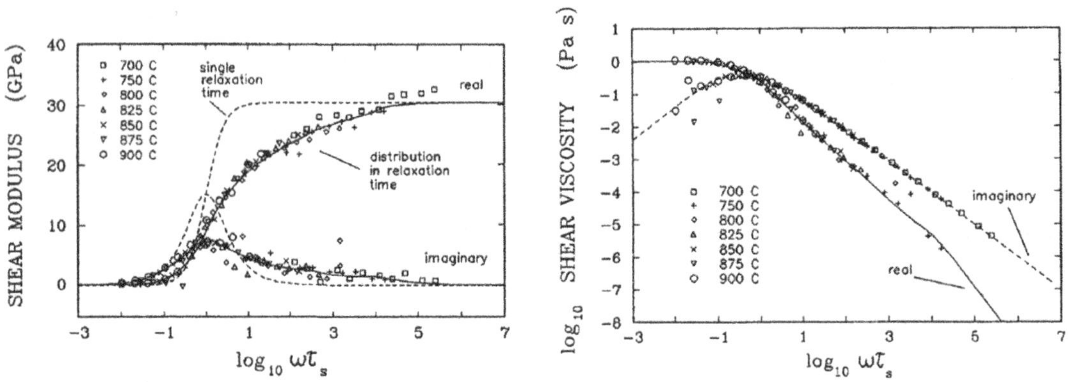

Fig. 11.8 The result of a torque test of rhyolitic melt (Webb 1992). (Left) The real part and the imaginary part of the complex shear modulus as a function of frequency of torque. (Right) The real part and the imaginary part of the complex viscosity coefficient as a function of frequency. Viscosity is scaled by the value of Newtonian viscosity measured by the micro-penetration method. Reprinted by permission from Springer Nature Webb (1992), COPYRIGHT (1992)

$$M^*(\omega) = k_E \left(\frac{\omega t_{re} \exp(-i\varphi_M)}{\sqrt{1+\omega^2 t_{re}^2}} \right) \tag{11.62}$$

$$\eta^*(\omega) = \mu \left(\frac{\exp(-i\varphi_\eta)}{\sqrt{1+\omega^2 t_{re}^2}} \right) \tag{11.63}$$

Here, the inside of the parentheses is dimensionless. Between the phase lags of φ_M and φ_η, there is a relationship $\tan\varphi_\eta = \omega t_{re} = \omega/\omega_{re} = (\tan\varphi_M)^{-1}$ $(\varphi_\eta = \varphi_M - \pi/2)$. In the Newtonian viscous fluid, $k_E \to \infty$, $t_{re} \to 0$, $\varphi_\eta \to 0$, and $\varphi_M \to \pi/2$. On the other hand, in the elastic body, $\eta \to \infty$, $t_{re} \to \infty$, $\varphi_M \to 0$, and $\varphi_\eta \to \pi/2$.

The frequency dependence of the real part of complex viscosity coefficient $\eta^*(\omega)$ is used to explain the frequency dependence of the viscosity of measured value. Newtonian viscosity is then defined as $\omega \to 0$. Plotting the angular frequency on the horizontal axis and the response of stress on the vertical axis yields a characteristic shape (Fig. 11.7). Figure 11.8 shows a result of an experiment Webb (1992) conducted by cyclically applying torqueHe interpreted that the frequency dependence cannot be explained by a single relaxation process based on the Maxwell model and that the time constant of relaxation has a distribution.

Equation (11.56) can then be written using viscoelastic relaxation time t_{re} as follows:

$$\dot{\sigma} + \frac{1}{t_{\text{re}}}\sigma = k_{\text{E}}\dot{\epsilon} \tag{11.64}$$

and if the strain rate $\dot{\epsilon}$ is considered to be a function of a general time, Eq. (11.64) can be considered as a first-order ordinary differential equation for σ. The general solution for the initial condition in which the stress is 0 is given using the following equation:

$$\sigma = \int_0^t \underbrace{\left[k_{\text{E}} \exp\left(-\frac{t-t'}{t_{\text{re}}}\right)\right]}_{\text{memory function}} \dot{\epsilon} dt' \tag{11.65}$$

The part before $\dot{\epsilon}$ in the integral is known as a memory function. When $k_{\text{E}} = 0$ and $t_{\text{re}} = \infty$, integrating it as it is and taking the limit at $t_{\text{vs}} \to \infty$ yields $\sigma = \eta\dot{\epsilon}$.

In a study on the expansion process of bubbles in a viscoelastic body, the shear rate in the radial direction $\partial u_r/\partial r$ is used instead of the shear rate $\dot{\epsilon}$ to calculate $\sigma_{rr} = -2\sigma$; the obtained σ_{rr} is then substituted into Eq. (11.74) to calculate change in the bubble radius.

11.8 Cooling of Dikes, Sills, and Lava

When we use the words "dikes" and "sills," we assume cooling by heat conduction. Because the thickness of dikes and sills is sufficiently lesser than the length in the longitudinal direction, 1D heat conduction in the thickness direction can be assumed in many cases. Although some cases consider crystal settling, it depends on the subject of research.

Here, not considering the kinetics of crystallization, we calculate temperature change as a function of time and space for cooling by 1D heat conduction under various boundary conditions to understand the influence of boundary conditions and latent heat. For this purpose, we will obtain analytical solutions for the following cases under the initial conditions given by Fig. 11.9.

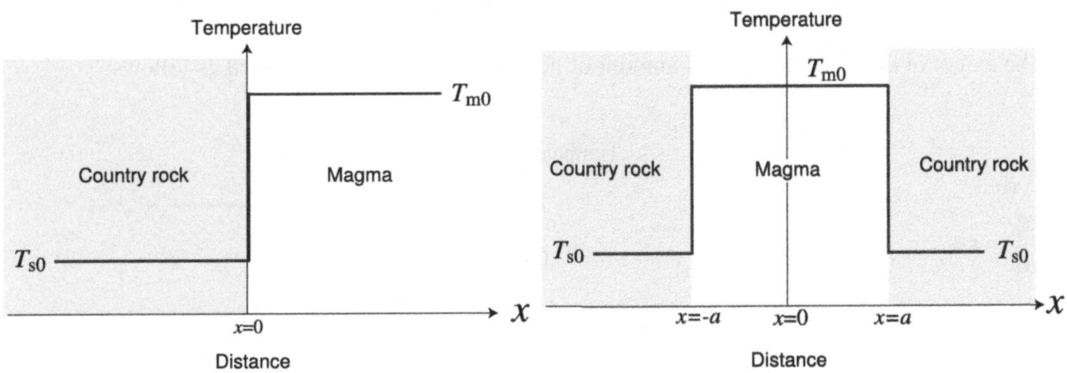

Fig. 11.9 The initial conditions of a one-dimensional heat conduction problem. (Left) The case where magma is present in semi-infinite space. (Right) The case where magma is present in space of $2a$ in width

11.8.1 The Case Where Latent Heat of Crystallization is Not Taken into Consideration

(1) The Case Where Magma Is Present in a Semi-infinite Space (Left Figure of Fig. 11.9)

The initial conditions in this case include

$$T(x,0) = T_{s0} \quad \text{at} \quad x<0, \quad t=0 \tag{11.66}$$

$$T(x,0) = T_{m0} \quad \text{at} \quad x>0, \quad t=0 \tag{11.67}$$

(1-1) The case where the country rock temperature does not change (semi-infinite space heat conduction)

We assume that the temperature at the boundary in this case is constant at the initial temperature. This is because when hydrothermal circulation occurs in the country rock, the temperature at the contact can be effectively approximated to be constant. In that case, the boundary condition will be as follows:

$$T(0,t) = T_{s0} \quad \text{at} \quad x=0 \tag{11.68}$$

The analytical solution in this case can then be solved using similarity transformation (Turcotte and Schubert 1982) or Laplace transform (Carslaw and Jaeger 1959) (see Box in Chap. 7) and obtained as follows:

$$T(x,t) = T_{s0} + (T_{m0} - T_{s0})\mathrm{Erf}\left(\frac{x}{2\sqrt{\kappa_m t}}\right) \tag{11.69}$$

(Figure 11.10). Here, κ_m is the thermal diffusivity of magma, and $\mathrm{Erf}(y)$ is the error function defined by

$$\mathrm{Erf}(y) = \frac{2}{\sqrt{\pi}}\int_0^y e^{-y'^2}dy' \tag{11.70}$$

and $\mathrm{Erfc}(y)$ is known as the complementary error function, which has a relationship $\mathrm{Erfc}(y) = 1 - \mathrm{Erf}(y)$.

The extent of decrease in the total amount of heat of magma is then obtained as follows:

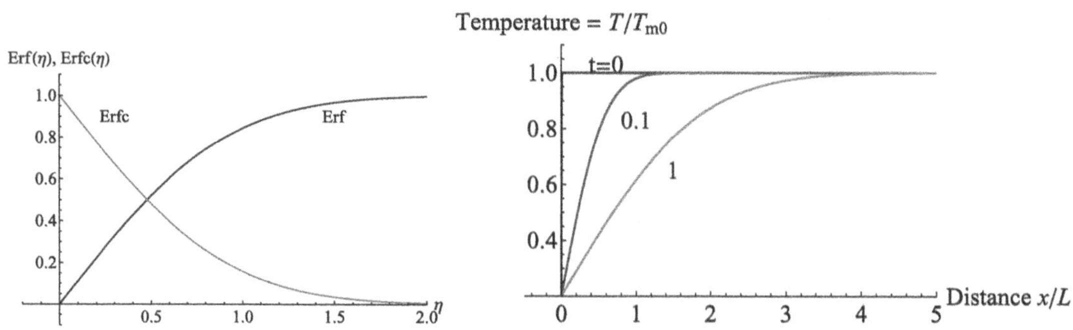

Fig. 11.10 (Left) Error functions. (Right) Solutions (Eq. (11.69)) of a one-dimensional heat conduction problem in the case where the country rock temperature does not change (Left figure of Fig. 11.9). $T_{m0} - T_{s0} = 0.8$ is assumed. Space is scaled by an arbitrary space scale l and time is scaled by l^2/κ_m

$$H_Q = \rho C_P \int_0^\infty [T_{m0} - T(x,t)]\,dx \tag{11.71}$$

$$= \rho C_P (T_{m0} - T_{s0}) \int_0^\infty \mathrm{Erfc}\left(\frac{x}{2\sqrt{\kappa_m t}}\right) dx = \frac{2\rho C_P}{\pi}(T_{m0} - T_{s0})\sqrt{\kappa t} \tag{11.72}$$

where ρ and C_P is the density of magma and the specific heat at constant pressure of magma. Moreover, the rate of decrease in the amount of heat is

$$\frac{dH_Q}{dt} = \frac{\rho C_P \sqrt{\kappa}}{\pi}(T_{m0} - T_{s0})t^{-1/2} \tag{11.73}$$

which decreases with the square root of time. Indeed, this agrees with the loss because of the heat flux at the boundary

$$k_T \left.\frac{dT}{dx}\right|_{x=0} = -k_T(T_{m0} - T_{s0})t^{-1/2} \tag{11.74}$$

where k_T is the thermal conductivity and has a relationship $\kappa = k_T/(\rho C_P)$ with the thermal diffusivity κ.

(1-2) The case where the country rock temperature changes by heat conduction
Although the temperature at the boundary T_b is constant in this case, it is determined by balance between heat flux from magma at the contact and heat flux in the country rock as follows:

$$T_b = T_{m0} \frac{1 + \frac{T_{s0}}{T_{m0}} (\kappa_m/\kappa_s)^{1/2}}{1 + (\kappa_m/\kappa_s)^{1/2}} \tag{11.75}$$

where κ_s is the thermal diffusivity of the country rock. As is evident, the steady-state temperature at the boundary is determined only by the ratio of the thermal diffusivity of country rock and that of magma, κ_m/κ_s. If heat conduction in the country rock is low, the thermal boundary layer on the country rock side has developed lesser than that on the magma side. Therefore, for heat flux to be equal via the boundary, the temperature difference between the inside of the country rock and the boundary should be small. Note that constant temperature at the boundary surface is ideally realized when both the country rock and the magma occupy a semi-infinite space. The temperature of the country rock (T_{out}) and that of magma (T_{in}) are then determined as follows:

$$T_{out}(x,t) = T_{s0} + (T_b - T_{s0})\mathrm{Erf}\left(\frac{x}{2\sqrt{\kappa_s t}}\right) \tag{11.76}$$

$$T_{in}(x,t) = T_{m0} + (T_b - T_{m0})\mathrm{Erf}\left(\frac{x}{2\sqrt{\kappa_m t}}\right) \tag{11.77}$$

where T_b is given by Eq. (11.75).

When the thermal diffusivity of country rock and that of magma is equal ($\kappa_m = \kappa_s = \kappa$), the temperature at the boundary T_b becomes the mean temperature of the two temperature, $T_b = (T_{m0} + T_{s0})/2$. Then, the temperature profile of the country rock and that in the magma are expressed by the following simple common equation (Fig. 11.11):

$$T(x,t) = \frac{T_{m0} + T_{s0}}{2} + \frac{T_{m0} - T_{s0}}{2}\mathrm{Erf}\left(\frac{x}{2\sqrt{\kappa t}}\right) \tag{11.78}$$

Ultimately, this is equivalent to an equation obtained by substituting $T_{s0} = T_b$ into Eq. (11.69).

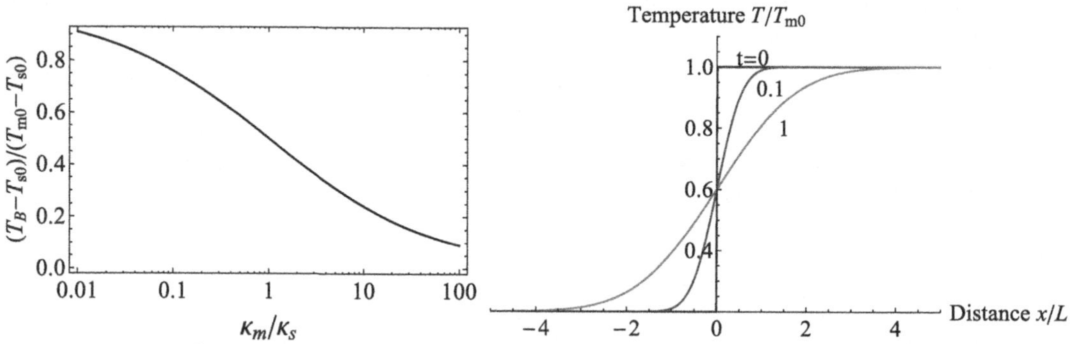

Fig. 11.11 (Left) Relationship between the temperature at the boundary T_b and the ratio between the thermal diffusivity of the country rock and that of the magma, κ_m/κ_s. The temperature at the boundary is represented by the difference with the initial temperature of the country rock ($T_b - T_{s0}$) scaled by the difference between the initial temperature of the country rock and that of the magma ($T_{m0} - T_{s0}$). Using this scaling, when the temperature at the boundary is equal to the initial temperature of the magma, the scaled value is 1, and when the temperature at the boundary is equal to the initial temperature of the country rock, the scaled value is 0. When the diffusivity is equal ($\kappa_m/\kappa_s = 1$), the temperature at the boundary becomes the mean temperature of the temperature of the country rock and that of the magma $(T_b - T_{s0})/(T_{m0} - T_{s0}) = 0.5$. $T_{m0} - T_{s0} = 0.8$ is assumed. (Right) Solution (Equation) for a problem of one-dimensional heat conduction in the case where magma is present in a semi-infinite space (Left figure of Fig. 11.9) and the country rock temperature changes. $T_{m0} - T_{s0} = 0.8$ is then assumed

(2) The Case Where Magma Is Present in Space of $2b$ in Width (Right Figure of Fig. 11.9)

The initial conditions in this case are

$$T(x,0) = T_{s0} \quad \text{at} \quad |x| > b, \quad t = 0 \tag{11.79}$$

$$T(x,0) = T_{m0} \quad \text{at} \quad |x| < b, \quad t = 0 \tag{11.80}$$

(2-1) The case where the country rock temperature does not change

The boundary conditions are that the temperature at the boundary with the country rock always remains constant:

$$T(\pm b, t) = T_{s0} \quad \text{at} \quad x = b \tag{11.81}$$

and that the heat flux (thermal diffusivity) at the boundary on the country rock side and that on the magma side are equal:

$$\kappa_s \left.\frac{dT}{dx}\right|_{x\to\pm b(|x|>b)} = \kappa_m \left.\frac{dT}{dx}\right|_{x\to\pm b(|x|<b)} \tag{11.82}$$

As in the cases considered to date, solving by the Laplace transform yields the following.

$$T(x,t) = T_{m0} - (T_{m0} - T_{s0}) \sum_0^{\infty} (-1)^n \left(\mathrm{Erfc}\left(\frac{(2n+1)b + x}{2\sqrt{\kappa t}} \right) + \mathrm{Erfc}\left(\frac{(2n+1)b - x}{2\sqrt{\kappa t}} \right) \right) \tag{11.83}$$

Because the series in this equation is immediately attenuated with increase in n, it would be sufficient to consider only the first several terms. Figure 11.12 shows the result. The cooling rate when passing

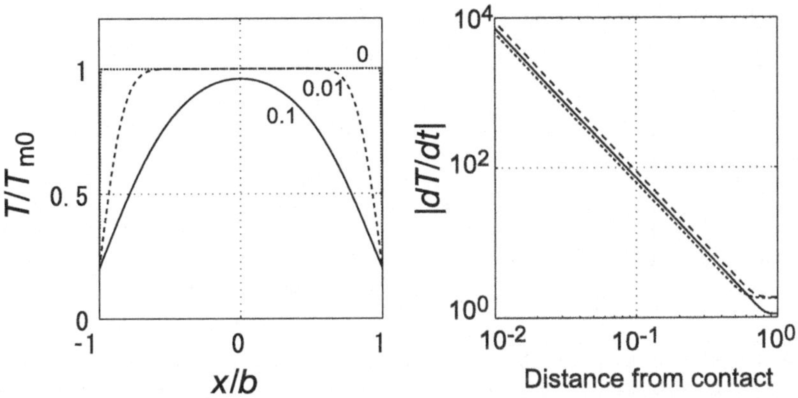

Fig. 11.12 (Left) Change in temperature in a dike cooled by heat conduction in the case where the temperature at the boundary is constant (Eq. (11.83)). The numbers next to the lines represent the values of dimensionless time $\kappa t/b^2$. (Right) Dependence of the cooling rate on the distance when passing through a certain temperature T_1. The distance is scaled by b of the width of the dike $2b$. The right end of the horizontal scale corresponds to the center of the dike. The cooling rate is scaled by $T_{m0}\kappa/b^2$. The solid line represents $T_1 = 0.98$, the broken line $T_1 = 0.9$, and the dotted line $T_1 = 0.8$. $T_{m0} - T_{s0} = 0.8$ is assumed

via a certain temperature decreases with the power of -2 of the distance and the distance dependence almost disappears at the center of dikes.

(2-2) The case where the country rock temperature changes

The boundary conditions are that the temperature of country rock at infinity is equal to the initial temperature of country rock:

$$T(\infty, t) = T_{s0} \quad \text{at} \quad x = \infty \tag{11.84}$$

The temperature of magma and that of the country rock are equal at the boundary:

$$T(\pm b, t)|_{x \to \pm b(|x|>b)} = T(\pm b, t)|_{x \to \pm b(|x|<b)} \tag{11.85}$$

and the heat flux (thermal diffusivity) at the boundary at the country rock side and that at the magma side are equal:

$$\kappa_s \left.\frac{dT}{dx}\right|_{x \to \pm b(|x|>b)} = \kappa_m \left.\frac{dT}{dx}\right|_{x \to \pm b(|x|<b)} \tag{11.86}$$

When the thermal diffusivity of magma and that of country rock are equal, a solution under initial conditions and boundary conditions described above is obtained as follows (Fig. 11.13):

$$T(x, t) = T_{s0} + \frac{T_{m0} - T_{s0}}{2}\left[\mathrm{Erf}\left(\frac{x+b}{2\sqrt{\kappa t}}\right) - \mathrm{Erf}\left(\frac{x-b}{2\sqrt{\kappa t}}\right)\right] \tag{11.87}$$

In this case, the temperature at the boundary T_b is

$$T(b, t) = T_{s0} + \frac{T_{m0} - T_{s0}}{2}\mathrm{Erf}\left(\frac{b}{\sqrt{\kappa t}}\right) \tag{11.88}$$

Moreover, temperature at the center of dikes changes with time as follows:

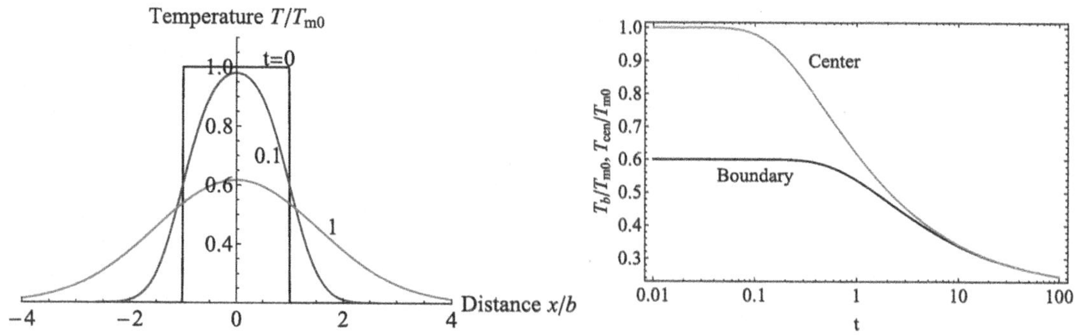

Fig. 11.13 (Left) Temperature change in a dike cooled by heat conduction. It should be noted that temperature at the boundary is not constant, unlike the case of a semi-infinite medium. (Right) Change in temperature with time at the boundary and the center of a dike. $T_{m0} - T_{s0} = 0.8$ is assumed

$$T(0, t) = T_{s0} + (T_{m0} - T_{s0})\mathrm{Erf}\left(\frac{b}{2\sqrt{\kappa t}}\right) \tag{11.89}$$

11.8.2 The Case Where Latent Heat of Crystallization is Considered (Stefan Problem)

In case where crystallization occurs on an isothermal plane (an isothermal point in a one-dimensional case) corresponding to a certain temperature (melting point), the problem of obtaining the moving velocity of the crystallization front is known as the Stefan problem.

Assuming that crystallization is completed at the melting point T_m, latent heat is generated only at the temperature, i.e., only at the position corresponding to the temperature. The released latent heat is then promptly transferred to the surrounding via conduction. The temperature profile should then be determined by diffusion equations at both sides of the position where the latent heat is released. At the position where the latent heat is released, the difference between the heat flux flowing into the position (I_+) from the higher temperature side and the heat flux flowing out of the position (I_-) to the lower temperature side (i.e., difference in the temperature gradient) is balanced with the amount of the latent heat released. The problem of obtaining the temperature profile at this time and the moving velocity of the crystallization front is known as the Stefan problemThe reader is referred to Turcotte and Schubert (1982) for details and the following summarizes only the result.

(1) The case where the country rock temperature does not change (semi-infinite heat conduction)
The analytical solution when the temperature at the boundary is constant (the boundary condition 11.68) is given by Eq. (11.69). We assume that the crystallization temperature is T_m and the position corresponding to T_m is x_m. Moreover, we assume that on the lower temperature side and the higher temperature side of x_m, different analytical solutions should hold. However, note that T_{s0} and T_{m0}, which correspond to the boundary conditions at $x = 0$ and $x = \infty$ (one boundary condition is a virtual condition), are dummy parameters ultimately determined via a consistent solution in each region (the lower temperature side and the higher temperature side).

Assuming that the temperature at the lower temperature side is $T_{out}(x, t)$ and that at the higher temperature side is $T_{in}(x, t)$, the heat flux to the lower temperature side is determined as follows:

$$I_- = k_s \frac{dT_{out}}{dx}\bigg|_{x=x_m} = k_s \frac{(T_m - T_{s0})e^{-\lambda^2}}{\sqrt{\pi \kappa_s t}\mathrm{Erf}\,(\lambda)} \tag{11.90}$$

and the heat flux from the higher temperature side is

$$I_{+} = k_{\mathrm{m}} \left. \frac{dT_{\mathrm{in}}}{dx} \right|_{x=x_{\mathrm{m}}} = k_{\mathrm{m}} \frac{(T_{\mathrm{m0}} - T_{\mathrm{m}}) e^{-\lambda^2 (\kappa_{\mathrm{s}}/\kappa_{\mathrm{m}})}}{\sqrt{\pi \kappa_{\mathrm{m}} t} \left(1 - \mathrm{Erf}\left(\lambda \sqrt{\kappa_{\mathrm{s}}/\kappa_{\mathrm{m}}}\right)\right)} \tag{11.91}$$

where λ is

$$\lambda = \frac{x_{\mathrm{m}}}{2\sqrt{\kappa_{\mathrm{s}} t}} \tag{11.92}$$

which is a constant that provides the relationship between the position of the crystallization front x_{m} and time t and is determined using the final consistent solution. The position of crystallization front and moving velocity is then determined by the constant as follows:

$$x_{\mathrm{m}} = 2\lambda \sqrt{\kappa_{\mathrm{s}} t} \tag{11.93}$$

$$\frac{dx_{\mathrm{m}}}{dt} = \lambda \sqrt{\frac{\kappa_{\mathrm{s}}}{t}} \tag{11.94}$$

The rate of release of latent heat at the crystallization front is then expressed as follows:

$$\rho \Delta H \frac{dx_{\mathrm{m}}}{dt} = \rho \Delta H \lambda \sqrt{\frac{\kappa_{\mathrm{s}}}{t}} \tag{11.95}$$

where ΔH is the latent heat of crystallization (J/kg).

Heat balance at the crystallization front is then written as follows:

$$I_{-} - I_{+} = \rho \Delta H \frac{dx_{\mathrm{m}}}{dt} \tag{11.96}$$

and substituting the above equations into this equation gives the constraint condition for the constant λ as follows:

$$\frac{(T_{\mathrm{m}} - T_{\mathrm{s0}}) e^{-\lambda^2}}{\lambda \mathrm{Erf}(\lambda)} - \left(\frac{k_{\mathrm{m}}}{k_{\mathrm{s}}}\right) \left(\frac{\kappa_{\mathrm{m}}}{\kappa_{\mathrm{s}}}\right) \frac{(T_{\mathrm{m0}} - T_{\mathrm{m}}) e^{-\lambda^2 (\kappa_{\mathrm{s}}/\kappa_{\mathrm{m}})}}{\sqrt{\frac{\kappa_{\mathrm{s}}}{\kappa_{\mathrm{m}}}} \lambda (1 - \mathrm{Erf}(\lambda))} = \frac{\sqrt{\pi} \Delta H}{C_{\mathrm{ps}}} \tag{11.97}$$

When the solid and liquid phases are equal in thermal diffusivity and thermal conductivity ($k_{\mathrm{s}} = k_{\mathrm{m}}$, $\kappa_{\mathrm{s}} = \kappa_{\mathrm{m}}$), the above equation can be rewritten as follows:

$$\frac{e^{-\lambda^2}}{\lambda} \left[\frac{T_{\mathrm{m}} - T_{\mathrm{s0}}}{\mathrm{Erf}(\lambda)} - \frac{T_{\mathrm{m0}} - T_{\mathrm{m}}}{1 - \mathrm{Erf}(\lambda)} \right] = \frac{\sqrt{\pi} \Delta H}{C_{\mathrm{P}}} \tag{11.98}$$

Moreover, if the initial temperature of the liquid is equal to the melting point ($T_{\mathrm{m0}} = T_{\mathrm{m}}$), the equation can be written as follows:

$$\frac{e^{-\lambda^2}}{\lambda \mathrm{Erf}(\lambda)} = \frac{\sqrt{\pi} \Delta H}{C_{\mathrm{P}} (T_{\mathrm{m}} - T_{\mathrm{s0}})} \equiv \frac{\sqrt{\pi} \mathrm{St}}{1 - T_{\mathrm{s0}}/T_{\mathrm{m}}} \tag{11.99}$$

where $\mathrm{St} = \Delta H / T_{\mathrm{m}} C_{\mathrm{P}}$ is a dimensionless number called Stefan number, that measures the influence of the latent heat and takes values from 0.1 to 0.3. In this case, λ for realistic values of St number and the initial temperature of country rock takes on values of $0.5 < \lambda < 1$.

(2) The case where the country rock temperature changes by heat conduction
Furthermore, in cases where the temperature at the boundary ($T_b = (T_{m0} - T_{s0})/2$ is not constant and heat conduction to the country rock is considered, an equation corresponding to Eq. (11.97) is determined as follows:

$$\frac{(T_m - T_{s0})e^{-\lambda^2}}{\lambda\,(1+\mathrm{Erf}\,(\lambda))} - \left(\frac{k_m}{k_s}\right)\left(\frac{\kappa_m}{\kappa_s}\right)\frac{(T_{m0} - T_m)e^{-\lambda^2(\kappa_s/\kappa_m)}}{\sqrt{\frac{\kappa_s}{\kappa_m}}\lambda\,(1-\mathrm{Erf}\,(\lambda))} = \frac{\sqrt{\pi}\Delta H}{C_{ps}} \tag{11.100}$$

and, when $k_s = k_m$ and $\kappa_s = \kappa_m$,

$$\frac{e^{-\lambda^2}}{\lambda}\left[\frac{T_m - T_{s0}}{1+\mathrm{Erf}\,(\lambda)} - \frac{T_{m0} - T_m}{1-\mathrm{Erf}\,(\lambda)}\right] = \frac{\sqrt{\pi}\Delta H}{C_P} \tag{11.101}$$

is obtained. Moreover, when $T_{m0} = T_m$,

$$\frac{e^{-\lambda^2}}{\lambda\,(1+\mathrm{Erf}\,(\lambda))} = \frac{\sqrt{\pi}\mathrm{St}}{1 - T_{s0}/T_m} \tag{11.102}$$

is obtained. λ in this case is $0.1 < \lambda < 1.2$, which is slightly lesser than the case where the temperature at the boundary (country rock), is constant. For realistic values of the initial temperature of the country rock and St number, $\lambda \approx 0.5$ holds and it is found from Eq. (11.93) that the progress of the crystallization front can be approximated well by

$$x_m \approx \sqrt{\kappa t} \tag{11.103}$$

11.8.3 Comparison of the Relationship Between the Cooling Rate and the Distance

As we obtained the temperature in magma in which crystallization occurs as a function of time and space in various cases, we can find the dependence of the cooling rate on space by differentiating the function with respect to time.

(1) The Case Where the Latent Heat Is Not Taken into Consideration

(1-1) The case where the country rock temperature does not change in semi-infinite space
Assuming that crystallization starts and completes at the melting point T_m, the temperature should satisfy the following relationship:

$$T_m = T_{s0} + (T_{m0} - T_{s0})\mathrm{Erf}\,(y) \tag{11.104}$$

from Eq. (11.69). Here, y is a constant restricting the relationship between the time t and the spatial position x as follows:

$$y = \frac{x}{2\sqrt{\kappa t}} \tag{11.105}$$

Although it is assumed that the solid and liquid phases have the same physical properties, if physical properties differ, λ obtained by assuming $\Delta H = 0$ in Eq. (11.97) will be y.

The cooling rate when passing via the melting point is then obtained by differentiating Eq. (11.69) with respect to time as follows:

$$\frac{dT}{dt} = -\frac{4(T_{m0} - T_{s0})\kappa y^3 e^{-y^2}}{\sqrt{\pi} x^2} \tag{11.106}$$

The equation shows that the cooling rate at the temperature at which crystallization occurs reduces in inverse proportion to the square of the distance (Fig. 11.14).

(1-2) The case where the country rock temperature changes in semi-infinite space
In this case, because temperature changes following Eq. (11.78), the temperature change can be obtained by the same procedure as that of case (1–1):

$$\frac{dT}{dt} = -\frac{2(T_{m0} - T_{s0})\kappa y^3 e^{-y^2}}{\sqrt{\pi} x^2} \tag{11.107}$$

The result is the double of case (1–1) in expression. Note that y in this equation satisfies

$$T_m = \frac{T_{m0} + T_{s0}}{2} - \frac{T_{m0} - T_{s0}}{2}\text{Erf}\,(y) \tag{11.108}$$

and it reduces by several dozen percent compared to case (1–1).

(1-3) The case of a dike with a finite width
In this case, temperature change is given as a function of distance from the center of a dike, x, and the time t by Eq. (11.87). In this case, $y = x/2\sqrt{\kappa t}$, which corresponds to the melting point T_m, does not take a unique value. In particular, deviation from a constant value increases around the center of the dike.

The cooling rate of the inside of the dike is then obtained by differentiating Eq. (11.87) with respect to time as follows:

$$\frac{dT}{dt} = -\frac{T_{m0} - T_{s0}}{4\sqrt{\pi} t(\kappa t)^{1/2}}\left((x+b)e^{-\frac{(x+b)^2}{4\kappa t}} - (x-b)e^{-\frac{(x-b)^2}{4\kappa t}}\right) \tag{11.109}$$

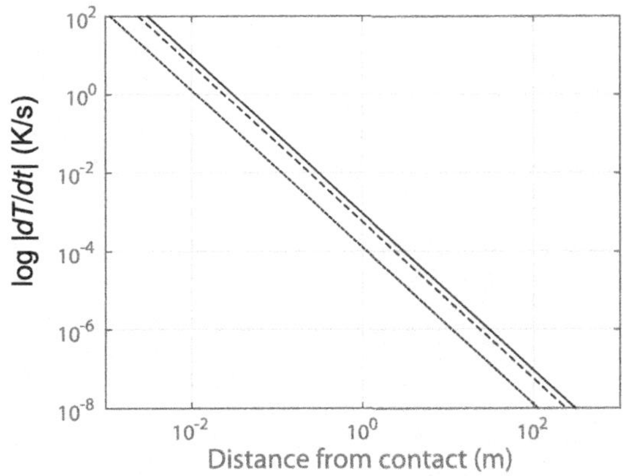

Fig. 11.14 Relationship between the distance from the boundary and the cooling rate when passing through a certain temperature, in the case where magma is present in a one-dimensional semi-infinite space. The case where difference in the initial temperature between the country rock and the magma is $0.7T_{m0}$ and the temperature of the country rock does not change (Eq. (11.106)). When the temperature of the country rock also changes by heat conduction and thermal diffusivity is equal, the cooling rate halves. The case where the initial temperature of magma is assumed to be 1200 °C. Difference in line types represents difference in passing temperature. The solid line: $0.98T_{m0}$, the broken line: $0.9T_{m0}$, and the dotted line: $0.8T_{m0}$

Substituting t as a function of x obtained assuming $T(x,t) = T_m$ in Eq. (11.87) into this equation yields the dependence of cooling rate on distance x. The dependence in some cases is shown in Fig. 10.20. Around the center of the dike, there is a region in which the cooling rate does not depend on distance. This is because, around the center of the dike, the crystallization front rapidly propagates and the difference of crystallization time disappears (Fig. 11.15). The length of the region increases as the degree of supercooling to crystallization is large.

At the center of the dike $x = 0$,

$$\left.\frac{dT}{dt}\right|_{x=0} = -\frac{8(T_{m0} - T_{s0})\kappa y_b^3 e^{-y_b^2}}{\sqrt{\pi} b^2} \tag{11.110}$$

where y_b satisfies $t = b^2/4\kappa y_b^2$ and determines the condition in which $T = T_m$ holds at $x = 0$.

11.9 The Basis of a 2D Textural Analysis

11.9.1 Introduction

The crystal number density we want to obtain is that per unit volume in a 3D space and differs from the number density on a 2D observational plane. Stereology is the field handling the problem of obtaining the quantity in a 3D space from data of a 2D cross section. In the science of rock texture, Cashman and Marsh (1988) examined the influence of crystal morphology. Higgins (2010) described observation results of CSDs and qualitative discussion about their origin. In actual observation, an application (CSDCorrections) by Higgins, available through the Internet, has been extensively used (http://www.uqac.ca/mhiggins/csdcorrections.html). In recent years, a 3D morphological information can be directly quantified using X-ray CT, although the technique has not yet become a common analytical method. Here, we explain the basis of textural analysis on the basis of 2D images, which can be more easily obtained using optical microscopy and electron microscopy. The detailed explanation on quantitative stereology is found in DeHoff and Rhines (1968) and Underwood (1972).

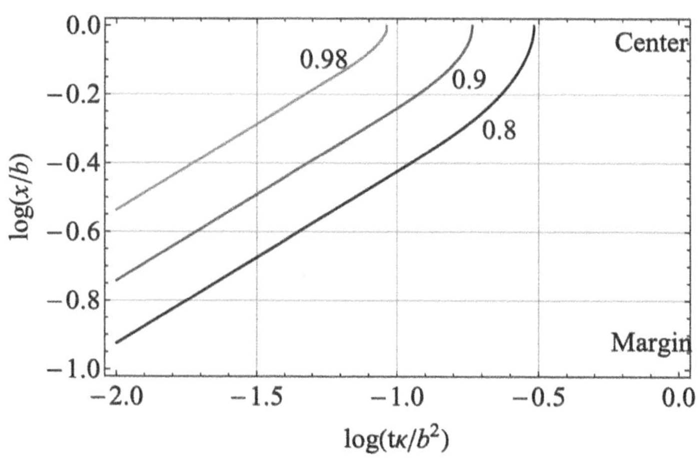

Fig. 11.15 Relationship between the position of the crystallization front in a dike, x/b, and time $t\kappa/b^2$. The numbers next to the lines indicate the difference in the value of $(T_m - T_{s0})/(T_{m0}/T_{s0})$. In straight sections of the lines, the propagation velocity of the crystallization front is proportional to the square root of time

11.9.2 Basic Measured Quantities

Normally, morphological analysis using image processing equipment is performed on a 2D cross section (hereinafter referred to as an observational plane) obtained by cutting the samples of interest. The basic measured quantities that will be important in quantitative analysis include (Fig. 11.16):

L Length of a line segment a crystal particle cuts from a line (a test line) randomly drawn on the observational plane.

L_L The ratio of length of line segments crystal particles of the phase of interest cut from lines randomly drawn on the observational plane.

N_L The number of times, per unit length, the lines randomly drawn on the observational plane intersect with crystal particles of the phase of interest. This observable is important in that $2N_L$ is equivalent to the surface area S_v, per unit volume, of the phase of interest.

A The cross-sectional area of a crystal particle of the phase of interest.

A_A Area, per unit area, occupied by crystal particles of the phase of interest. This gives a volume fraction (mode) of the phase of interest and L_L. The estimation of the mode by L_L and that by A_A should be almost nearly in agreement.

N_A The number of crystal particles of the phase of interest per unit area. This is equal to the product of the population density of crystal particles N_v and the mean particle diameter $\langle D_v \rangle$. This relationship is used to estimate the crystal number density in the following.

The letters used as symbols have the following meanings: L is the length, A is the area, and N is the number per unit volume. The subscripts indicate that the quantity is expressed by that quantity per unit of measurement, e.g., L_L is the length per unit length, i.e., a dimensionless value. Among these measured quantities, L, L_L, and N_L are obtained by measuring 1D lines and are those that can be obtained most easily when manually measured. The remaining A, A_A, and N_A are obtained by measuring 2D planes and the manual measurement requires some effort. However, they can be easily obtained using a software such as Image J (https://imagej.nih.gov/ij/).

11.9.3 Determination of the Bubble/Crystal Number Density

The bubble/crystal number density is the most fundamental quantity that characterizes rock texture. The number density is defined by the number of bubbles/crystals per unit volume (number/m^3). Although we can directly measure the population density N_A on a 2D observational plane, we cannot directly measure the population density N_v in a 3D space. Therefore, an equation that connects N_A with N_v is required; this relationship is generally written as follows:

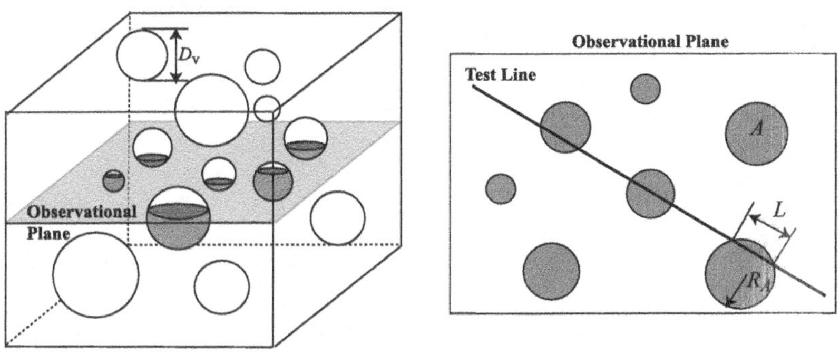

Fig. 11.16 Relationship between a two-dimensional observational plane and a three-dimensional form

$$N_{\rm v} = \frac{N_{\rm A}}{\langle D_{\rm v}\rangle} \tag{11.111}$$

where $\langle D_{\rm v}\rangle$ is the mean diameter of crystals vertically projected on the observational plane (the symbols $\langle\ \rangle$ represent that the quantity is the mean value). The problem of stereology arises when $\langle D_{\rm v}\rangle$ is to be obtained. Thus, the relationship between $\langle D_{\rm v}\rangle$ and a measured quantity on the observational plane is required. $\langle D_{\rm v}\rangle$ is determined depending on the shape of the crystals. When the crystal is spherical in shape, $\langle D_{\rm v}\rangle$ is connected with the mean radius $\langle R_{\rm A}\rangle$ of circles on the observational plane by the following equation:

$$\langle D_{\rm v}\rangle = 2\langle R_{\rm A}\rangle \tag{11.112}$$

Ultimately, $N_{\rm v}$ can be calculated from the equation

$$N_{\rm v} = \frac{N_{\rm A}}{2\langle R_{\rm A}\rangle} \tag{11.113}$$

using $\langle R_{\rm A}\rangle$ and $N_{\rm A}$, which can be obtained from measurement on the observational plane.

When $N_{\rm A}$ or $\langle R_{\rm A}\rangle$ is not measurable, the following semi-empirical equation connecting $N_{\rm A}$ to $N_{\rm v}$ is used.

$$N_{\rm v} = N_{\rm A}^{3/2} \tag{11.114}$$

11.9.4 Determination of $\langle R_{\rm A}\rangle$

The methods of determining the mean radius $\langle R_{\rm A}\rangle$ of particles (bubbles or crystals) on an observational plane using image processing equipment include:

1. Measure the total number of particles of the phase of interest and the width of individual particles (FX_i, FY_i $(i = 1, 2, \cdots, n)$ and calculate $\langle R_{\rm A}\rangle$ using the following equation:

$$\langle R_{\rm A}\rangle = \frac{1}{n}\Sigma_{i=1}^{n} FX_i \tag{11.115}$$

2. Measure the area of particles, calculate the equivalent radius HD_i by considering the shape of cross sections as circles, and calculate $\langle R_{\rm A}\rangle$ using the following equation:

$$\langle R_{\rm A}\rangle = \frac{1}{n}\Sigma_{i=1}^{n} HD_i \tag{11.116}$$

3. Assume that particles are spheres and calculate $\langle R_{\rm A}\rangle$ using $N_{\rm L}$ and $N_{\rm A}$ in the following equation:

$$\langle R_{\rm A}\rangle = \frac{4\,N_{\rm L}}{\pi N_{\rm A}} \tag{11.117}$$

References

Bagdassarov N, Dorfman A, Dingwell D (2000) Effect of alkalis, phosphorus, andwater on the surface tension of haplogranite melt. Am Mineral 85:33–40

Baker DR (1991) Interdiffusion of hydrous dacitic and rhyolitic melts and the efficacy of rhyolite contamination of dacitic enclaves. Contrib Mineral Petrol 106:462–473

Baker DR (1992) Tracer diffusion of network formers and multicomponent diffusion in dacitic and rhyolitic melts. Geochim Cosmochim Acta 56:617–631

Beherns H, Zhang Y, Xu Z (2004) H_2O diffusion in dacitic and andesitic melts. Geochim Cosmochim Acta 68:5139–5150

Carslaw HS, Jaeger JC (1959) Conduction of heat in solids. 2nd edn, Oxford University Publications

Cashman KV, Marsh BD (1988) Crystal size distribution (CSD) in rocks and the kinetics and dynamics of crystallization II, Makaopuhi lava lake. Contrib Mineral Petrol 99:292–305

DeHoff RT, Rhines FN (eds) (1968) Quantitative microscopy. McGraw-Hill, New York

Delaney JR, Karsten JL (1981) Ion microprobe studies of water in silicate melts: concentration-dependent water diffusion in obsidian. Earth Planet Sci Lett 52:191–202

Dingwell DB, Knoche R, Webb SL, Pichavant M (1992) The effect of B_2O_3 on the viscosity of haplogranitic liquids. Am Mineral 77:457–461

Fiege A, Holtz F, Cichy SB (2014) Bubble formation during decompression of andesitic melts. Am Mineral 99(5–6):1052–1062

Gardner JE, Hajimirza S, Webster JD, Gonnermann HM (2018) The impact of dissolved fluorine on bubble nucleation in hydrous rhyolite melts. Geochim Cosmochim Acta 226:174–181

Giordano D, Russell JK, Dingwell D (2008) Viscosity of magmatic liquids: a model. Earth Planet Sci Lett 271:123–134

Guggenheim EA (1950) Thermodynamics. North Holland Publ. Co., Amsterdam, p 36

Higgins M (2010) Quantitative textural measurements in igneous and metamorphic petrology. Cambridge University Press, p 276

Hofmann AW (1980) Diffusion in natural silicate melts: a critical review. In: Hargraves RB (ed) Physics of magmatic processes. Princeton University Press, Princeton, NJ, pp 385–418

Hui H, Zhang Y (2007) Toward a general viscosity equation for natural anhydrous and hydrous silicate melts. Geochim Cosmochim Acta 71(2):403–416

Joseph DD, Riccius O, Arney M (1986) Shear-wave speeds and elastic moduli for different liquids. Part 2. Exp J Fluid Mech 171:309–338

Karsten JL, Holloway JR, Delaney JR (1982) Ion microprobe studies of water in silicate melts: temperature-dependent water diffusion in obsidian. Earth Planet Sci Lett 59:420–428

Kingery WD, Bowen HK, Uhlmann DR (1976) Introduction to ceramics. Wiley, New York

Liang Y, Richter FM, Davis AM, Watson EB (1996) Diffusion in silicate melts: I. Self diffusion in Ca-Al_2O_3-SiO_2 at 1500 °C and 1 GPa. Geochim Cosmochim Acta 60:4353–4367

Liang Y, Richter FM, Watson EB (1996) Diffusion in silicate melts: II. Multicomponent diffusion in Ca-Al_2O_3-SiO_2 at 1500°C and 1 GPa. Geochim Cosmochim Acta 60:5021–5035

Liang Y, Richter FM, Chmberlin L (1997) Diffusion in silicate melts: III. Emprical models for multicomponent diffusion. Geochim Cosmochim Acta 61:5295–5312

Liu Y, Zhang Y, Behrens H (2004) H_2O diffusion in dacitic melts. Chem Geol 209:327–340

Mangan MT, Sisson TW (2005) Evolution of melt-vapor surface tension in silicic volcanic systems: experiments with hydrous melts. J Geophys Res 110:B01202. https://doi.org/10.1029/2004JB00321523-36

Murase T, McBirney A (1973) Properties of some common igneous rocks and their melts at high temperature. Geol Soc Am Bull 84:3563–3592

Mysen BO (1988) Structure and properties of silicate melts (Developments in Geochemistry 4), Elsevier Science, p 354

Okumura S, Nakashima S (2005) Water diffusion in basaltic to dacitic glasses. Chem Geol 227:70–82

Ono S (1980) Hyomenchyoryoku (Surface tension), Physics One Point series 9, Kyoritsu Press. (in Japanese)

Shaw HR (1972) Viscosities of magmatic silicate liquids; an empirical method of predition. Am J Sci 272:870–893

Shaw H (1974) Diffusion of H_2O in granitic liquids: part I. Experimental data; part II. Mass transfer in magma chambers, in geochemical transport and kinetics. Carnegie I Washington 634:139–170

Takeuchi S (2016) Pre-eruptive magma viscosity and a simplified estimating method. Bull Earthq Res Inst Univ Tokyo 91(3):55–63 (in Japanese with English abstract)

Taniguchi M (2001) Maguma kagakuheno shotai (Introduction to magma science) Shokabo, p 179. (in Japanese)

Taniguchi H (1988) Surface tension of melts in the system CaMg-Si_2O_6-$CaAl_2Si_2O_8$ and its structural significance. Contrib Mineral Petrol 100:484–489

Toramaru A (1995) Numerical study of nucleation and growth of bubbles in viscous magmas. J Geophys Res 100:1913–1931

Toramaru A (2006) BND (bubble number density) decompression rate meter for explosive volcanic eruptions. J Volcanol Geotherm Res 154:303–316

Turcotte DL, Schubert G (1982) Geodynamics: applications of continuum physics to geological problems. Wiley, New York

Underwood EE (1972) The mathematical foundations of quantitative stereology. In: Stereology and quantitative metallography, ASTM Int

Urbain G, Bottinga Y, Richet P (1982) Viscosity of liquid silica, silicates and alumino-silicates. Geochim Cosmochim Acta 46:1061–1072

Wada Y (1981) Makuronakei to tokeihousoku (Stastical laws and macroscopic system), Sangyo Press, p 176. (in Japanese)

Walker D, Mullins O Jr (1981) Surface tension of natural silicate melts from 1200°C - 1500°C and implication for melt structure. Contrib Mineral Petrol 76:455–462

Watson EB (1991) Diffusion of dissolved CO_2 and Cl in hydrous silicic to intermediate magmas. Geochim Cosmochim Acta 55(7):1897–1902

Webb SL (1992) Low-frequency shear and structure relaxation in rhyolitic melt. Phys Chem Minerals 119:240–245

Yamada K, Tanaka H, Nakazawa K, Emori H (2005) A new theory of bubble formation in magma. J Geophys Res 110:B02203. https://doi.org/10.1029/2004JB003113

Zhang Y, Behrens H (2000) H_2O diffusion in rhyolitic melts and glasses. Chem Geol 169:243–262

Zhang Y, Stolper EM (1991) Water diffusion in a basaltic melt. Nature 351:306–309

Index

A. Toramaru, *Vesiculation and Crystallization of Magma*, Advances in Volcanology,
https://doi.org/10.1007/978-981-16-4209-8

The manufacturer's authorised representative in the EU is Springer Nature Customer Service Centre GmbH, Europaplatz 3, 69115 Heidelberg, Germany. If you have any concerns regarding our products, please contact ProductSafety@springernature.com

Printed and bound by CPI Group (UK) Ltd, Croydon, CR0 4YY
30/09/2026
02225254-0020